Maximum and minimum principles

Saddle function

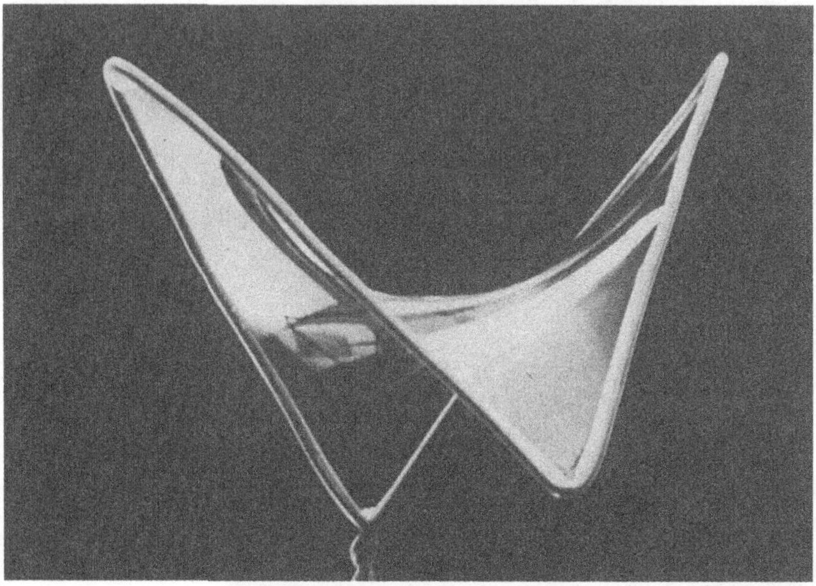

Maximum and minimum principles

A unified approach, with applications

M. J. SEWELL
Professor of Applied Mathematics, University of Reading

CAMBRIDGE UNIVERSITY PRESS
Cambridge
New York Port Chester
Melbourne Sydney

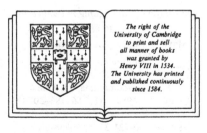

CAMBRIDGE UNIVERSITY PRESS
Cambridge, New York, Melbourne, Madrid, Cape Town, Singapore, São Paulo

Cambridge University Press
The Edinburgh Building, Cambridge CB2 8RU, UK

Published in the United States of America by Cambridge University Press, New York

www.cambridge.org
Information on this title: www.cambridge.org/9780521332446

First published 1987
Reprinted 1990
Re-issued in this digitally printed version 2007

A catalogue record for this publication is available from the British Library

Library of Congress Cataloguing in Publication data

Sewell, M.J.
Maximum and minimum principles.

Bibliography: p.
Includes index.
1. Maxima and minima. I. Title.
QA316.S35 1987 511′.66 86–20791

ISBN 978-0-521-33244-6 hardback
ISBN 978-0-521-34876-8 paperback

0

Contents

0

Preface

The purpose of this book is to describe an approach to maximum and minimum principles which is both straightforward and unified. A particular aim is to identify and illustrate the structure of the theory, and to show how that structure leads to general formulae for the construction of upper and lower bounds associated with those principles. Such bounds are important in applications.

I have found that fresh and fruitful insights have arisen repeatedly during the working out of the ideas developed in this approach. It is offered to the reader in the hope that, as he absorbs the viewpoint, he will benefit from similar experiences.

The treatment is designed to be accessible in the first three chapters to final year undergraduates in mathematics and science, and throughout to contain material which will interest postgraduates and research workers in those subjects. Some elementary prior knowledge of the calculus of variations will be helpful, but otherwise the book is self-contained. Anyone knowing more than this minimum should be able to read Chapter 3 first, with occasional references back to Chapter 1. I have given a central role to a pair of inner product spaces and, by emphasizing this simple idea, I have been able to avoid the need for any sophisticated functional analysis. The reader who does possess more technical knowledge may use it to add to what is here; he will recognise topics which I have omitted. The reader without such extra knowledge will be at no disadvantage in reading the book.

Teachers requiring a text may base an undergraduate course on material selected from Chapters 1 and 3, with a little of 2, as I have done myself. Sets of exercises are given at frequent intervals. Some exercises contain significant additions to the theory.

The book is built upon the notion of a saddle functional, say $L[x,u]$, concave in one variable x and convex in another variable u. Each variable belongs to a space having an inner product, which we denote by (x,y) and $\langle u,v\rangle$, respectively, because in general the inner products may be different. When faced with a problem *ab initio* the first step is to express its governing conditions in terms of the gradients of $L[x,u]$. These conditions may contain either equations alone, or a combination of inequalities and equations. In each case, two different subsets of the system of governing equations present themselves; for example, in mechanical contexts there will be a subset of kinematical conditions, and another subset of conditions on forces. A major feature of practical importance is that it will be easier to solve either subset considered alone, than to find a solution of the whole system. If the latter gives $L[x,u]$ a value L_0, an upper bound to L_0 can be calculated from a solution of one subset, and a lower bound from a solution of the other subset. In some problems L_0 may represent an energy expenditure which it is desirable to know; in other problems more importance may attach to solutions of the governing equations themselves, and the bounding theorems provide methods of finding approximations to these solutions.

An evolving parameter may be implicit in $L[x,u]$, ultimately weakening the saddle hypothesis in a way that only allows either an upper or a lower bound to be proved, but not both simultaneously. After further evolution a bifurcation value of the parameter may be passed, when global bounds can no longer be expected at all but, for example, local stationary values of $L[x,u]$ may still occur at solutions of the given system.

In conventional language the book may be regarded as a contribution to optimization theory, including parts both of nonlinear programming and of the calculus of variations. However, the contents show that I have taken an independent view, not following what has become, in some respects, a stereotyped format. For example, to promote understanding of the underlying structure I have given, in Chapter 2, a much fuller account of the Legendre transformation and its singularities than it seems possible to find elsewhere. I do not suggest, of course, that all optimization problems can be recast into the framework offered here; only that a worthwhile number of them can be so represented. Nor have I made a systematic attempt to repeat all the most classical problems from this viewpoint; there is scope for the reader to explore these questions.

I have tended to avoid the adjective 'variational', because its meaning seems to have been stretched in recent years to the point where it has ceased to be useful without close questioning of the context. For example, I use 'stationary principle' to describe what used to be called a 'variational principle' before variational inequalities and pseudo-variational principles came in.

The plan of the book is as follows. Chapter 1 deals with the finite dimensional case. It introduces the saddle function in three dimensions.

Gradients are used to define not only a stationary value problem, which determines a saddle point, but also problems specified by inequalities having a nonlinear 'complementarity' pattern. The main structure of the theory of upper and lower bounds is developed, and then extended to many finite dimensions in a way designed to facilitate immediate generalization to the infinite dimensional case in Chapter 3. Advantage is also taken of the simplicity inherent in finite dimensions to prove minimum principles *per se*: that is, a scalar function has a minimum where, and *only* where, an appropriate system of governing conditions (perhaps including inequalities) is satisfied. Detailed examples are given.

Chapter 2 contains the exposition of the Legendre transformation and its singularities, also in finite dimensions, with examples. Information from applied singularity theory, commonly called catastrophe theory, is integrated into this account. The place of the Legendre transformation in a generalized Lagrangian and Hamiltonian structure is explained, in particular in relation to the solution of nonlinear constraints and the expression of Euler equations, and the connection with bifurcation theory is mentioned. The chapter concludes with an outline of some aspects of network theory.

Chapter 3 is the core of the book. It gives the infinite dimensional theory of upper and lower bounds. Boundary value problems expressed by integral equations, and by ordinary and partial differential equations, and inequalities, are all treated. Cases in which an unknown 'free' boundary is to be found are included. An important intermediate role is played by a pair of operators T and T^* which are adjoint to each other in a true and not merely formal sense. For differential equations I define these operators in an original way whose simplicity has not been appreciated in the general literature. Examples are included. Emphasis is again on the structure of the theory. Detailed numerical evaluation of bounds is not my purpose; simple bounds are sometimes calculated to focus a discussion, and the reader will be able to see how they may be improved.

Chapters 4 and 5 are independent of each other, and can be interchanged in order if the reader wishes. They are each sequels to Chapter 3, but in different ways. Chapter 5 selects some general problems in the mechanics of fluids and solids. In each area the governing conditions are stated at the outset, and the methods of Chapter 3 are then applied to recast these conditions into appropriate operator form. An underlying saddle functional is then identified, whose gradients generate both the governing conditions and the theorems giving upper and lower bounds. The purpose of Chapter 5 is thus to give confidence by examples that the procedures of

Chapter 3 work in problems expressed directly in physical variables, and can organize and digest the extra algebraic complication.

Chapter 4 is at the simpler algebraic level of Chapter 3, but tries out some fresh topics for exploration: pointwise bounds, initial value problems and comparison methods. Ultimately I conclude that, before a balanced judgement is made about these topics, more detailed research is required.

I am glad to acknowledge the contribution of two particular friends in assisting me to write this book: Professor Ben Noble of the University of Wisconsin Mathematics Research Center and Dr David Porter of the University of Reading Mathematics Department. My understanding of parts of the subject matter has much increased as a result of many energetic conversations and exchanges of notes with Ben Noble over the years since 1969, when we discovered our common interest in and attitude to extremum principles. His breadth of knowledge and his pedagogic experience were often placed at my disposal when this project was starting and some early drafts were being prepared. These discussions were facilitated by my several visits to the Mathematics Research Center in Madison, and I record my appreciation of the support of its successive Directors (Professors Rosser, Noble and Nohel), and of the United States Army Research Office which funded it.

I have been fortunate that David Porter has generously given his time to read the whole book as it took shape, with many drafts of some passages. I am grateful for the numerous accurate comments and constructive suggestions of substance which he has offered for my consideration. It has been very helpful to have regular indications of how my remarks might appear to others.

Such defects as remain are entirely my responsibility.

My interest in extremum principles and Legendre transformations was first aroused in 1957 by the undergraduate lectures and a review article (1956*a*) of Professor Rodney Hill, F.R.S., who was subsequently my research supervisor at the University of Nottingham. I am happy that his deep influence has helped to shape this work.

I am indebted to my secretary, Mrs Jill Boxall, for the calm and steady way in which she assembled a well presented typescript. I am pleased to thank my son Peter, who produced all the computed curves as a basis for many of the figures. The photographs were taken at my behest by Mr Simon Johnson, of the Reading University Photographic Service.

Reading
February 1986

Michael Sewell

1

Saddle function problems

1.1. The basic idea

(i) A simple saddle function

We begin with a very straightforward illustration, using only three dimensions, of the intimate connection between a saddle function and a pair of extremum principles. This connection is the basic idea which decides the starting point of our theory and provides a central thread running through this book. We shall build an expanding theme upon it.

Consider the hyperbolic paraboloid

$$L = -x^2 + u^2 \tag{1.1}$$

referred to rectangular cartesian coordinates (see Fig. 1.1). This form is sometimes favoured by architects for the building of roof structures, such as those in the photographs collected by Faber (1963) and Joedicke (1963) and for children's climbing frames, for example at Rye in Sussex. It is

Fig. 1.1. Hyperbolic paraboloid $L = -x^2 + u^2$.

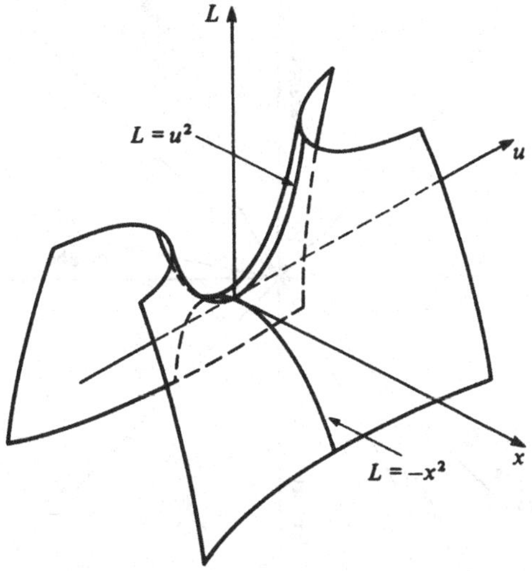

manifestly saddle-shaped in the everyday sense of the term, which is conveyed also by the photograph of the soap film in our frontispiece.

A vertical plane section $u = \lambda x$ defines, for each given λ, a path on the surface which is either a parabola or (if $\lambda^2 = 1$) one of the horizontal straight lines where $x \pm u = 0$. The latter separate those parabolae which we will call convex (if $\lambda^2 > 1$) from those which we will call concave (if $\lambda^2 < 1$). The lines $x \pm u = 0 = L$ are, in plan view (Fig. 1.2), asymptotes for all other horizontal sections $L = constant$ of (1.1). The bisectors of the asymptotes define the steepest convex and concave parabolae, and it is natural to call the intersection of these two parabolae *the saddle point* of the hyperbolic paraboloid. The saddle point can also be regarded as the intersection of *any* convex parabola with *any* concave parabola. Any other point on any convex parabola is therefore higher than the saddle point, which itself is higher than any other point on any concave parabola. In this way the height of the saddle point can be imagined to be bounded from above and below. Taking another viewpoint, the fact that minimization of all the upper bounds on a convex parabola yields the saddle point illustrates what is called a *minimum principle*; and the fact that maximization of all the lower bounds on a concave parabola also yields the saddle point is called a *maximum principle*. These together give our first example of a pair of *simultaneous extremum principles*. An extremum will always mean either a maximum or a minimum. 'Simultaneous' means simultaneously valid; alternatively one often speaks of a pair of 'dual' or of 'complementary'

Fig. 1.2. Plan view of some vertical plane sections of $L = -x^2 + u^2$.

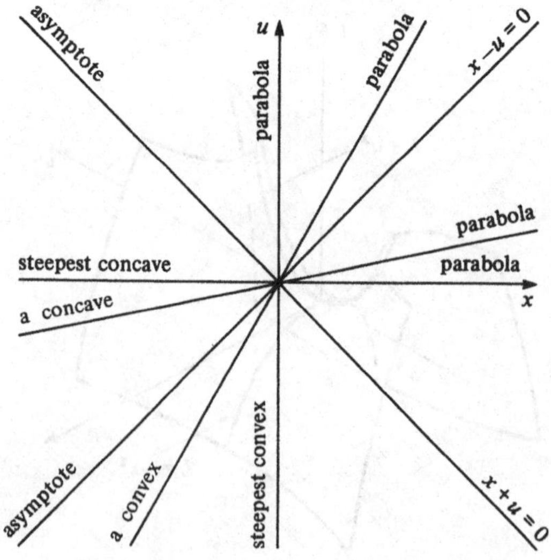

extremum principles, but these adjectives need to be justified by specific definitions such as we give later (e.g. §§2.7(i) and 3.7(xv) respectively).

More generally, consider the surface $L = L[x, u]$ defined by any given single valued C^1 function $L[x, u]$. A C^1 *function* is, by definition, continuous and all its first partial derivatives are also continuous. We shall give special attention to two particular paths on the surface. One is defined by

$$\frac{\partial L}{\partial x} = 0, \tag{1.2α}$$

and the other is defined by

$$\frac{\partial L}{\partial u} = 0. \tag{1.2β}$$

Let x_α, u_α denote any solution of (1.2α) but not necessarily (1.2β), and let x_β, u_β be any solution of (1.2β) but not necessarily (1.2α).

In the case (1.1), although not in general, these two paths become the steepest convex and concave parabolae respectively. We find that $x_\alpha = 0$, u_α is arbitrary, and $u_\beta = 0$, x_β is arbitrary. Since

$$L[0, u_\alpha] = u_\alpha{}^2, \quad L[0, 0] = 0, \quad L[x_\beta, 0] = -x_\beta{}^2,$$

the saddle point of the hyperbolic paraboloid is characterized by

$$L[0, u_\alpha] \geqslant L[0, 0] \geqslant L[x_\beta, 0]. \tag{1.3}$$

This continued inequality illustrates our definition above of a *pair of upper and lower bounds.*

Points which satisfy both of (1.2) simultaneously will be called *stationary points* of $L[x, u]$. For (1.1) the saddle point is a stationary point, and it is also the minimum of the upper bounds and the maximum of the lower bounds. In §1.2 we shall introduce simultaneous extremum principles whose extrema need *not* be stationary points, but are associated with problems governed by inequalities instead of equations.

This basic idea of bounding from above and below extends to any other surface $L = L[x, u]$ on which we can find interesecting paths, of which one is ascending and one is descending from the intersection point. The height of the interesection point can then be bounded by choosing any other point on each path. The paths need not be parabolae, nor be convex (although they often will be), nor lie in a *plane* vertical section.

The Old Man of Coniston is a mountain in the English Lake District whose ascent may be started from either the Duddon Valley or the Coniston Village end of a path which is called the Walna Scar Road, of Roman origin. This track reaches a saddle point where it tops a pass over an ascending path joining White Pike to Dow Crag. The climb may be continued via Dow Crag. Readers familiar with other mountain areas will have their own remembered saddle points, which are often major objectives because of the new perspectives which they reveal.

(ii) A convex function of one variable

Consider a single valued continuous function $F[u]$ of a single scalar variable u. We define $F[u]$ to be *strictly convex* on a single connected interval of the real line if

$$\lambda F[u_+] + (1-\lambda)F[u_-] > F[\lambda u_+ + (1-\lambda)u_-]$$

for every pair of distinct points u_+ and u_- within the interval, and for all real λ in $0 < \lambda < 1$. Geometrically this means that the straight chord joining any two points on the function always lies above the function between those points. This is illustrated in Fig. 1.3(*a*) for a *piecewise C^1 function*, i.e. a

Fig. 1.3. Convexity properties of $F[u]$. (*a*) Basic definition of strict convexity (piecewise C^1 example). (*b*) Strict convexity $F'' > 0$ ($F'' = 0$ possible at isolated points only). (*c*) Weak convexity $F'' \geqslant 0$ ($F'' = 0$ at nonisolated points). (*d*) Strict concavity.

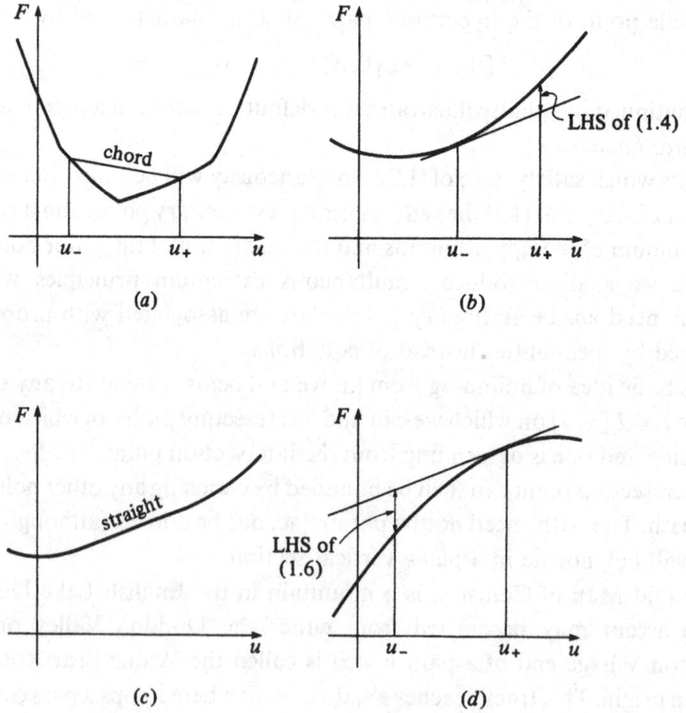

continuous function whose slope has a finite number of isolated finite discontinuities. This basic analytical definition of convexity is available regardless of whether any derivative of the function exists or is continuous.

Next suppose that $F[u]$ is a C^1 function. Then it follows by a straightforward exercise in calculus (Exercise 1.1.1) that $F[u]$ is strictly convex if and only if

$$F[u_+]-F[u_-]-(u_+-u_-)F'[u_-]>0 \tag{1.4}$$

for every pair of distinct points u_+ and u_- within the interval. A prime denotes differentiation. The geometrical interpretation (Fig. 1.3(*b*)) is that the function rises above its tangent plane at every point. By interchanging u_+ and u_- in (1.4) and adding the two results we get

$$(u_+-u_-)(F'[u_+]-F'[u_-])>0, \tag{1.5}$$

i.e. the slope of a strictly convex function is monotonically increasing.

Weakly convex is defined by writing $\geqslant$ in these inequalities so that equality can hold for *some* distinct pairs of points, thus permitting straight line segments (Fig. 1.3(*c*)). Convexity without an adjective often means weak, but can mean strict, depending on the context.

A *concave* function is a convex function turned upside down, so that the foregoing inequalities are reversed. The C^1 function never rises above its tangent plane at any point. In the format of (1.4) we can define $F[u]$ to be *strictly concave* if

$$F[u_+]-F[u_-]-(u_+-u_-)F'[u_+]>0 \tag{1.6}$$

for every pair of distinct points, the derivative now being evaluated at the opposite end of the interval compared to (1.4) (see Fig. 1.3(*d*)).

A *C^2 function* is continuous with continuous first and second derivatives. A *piecewise C^2 function* is a C^2 function except that finite jumps in the second (but not first) derivative may occur at a finite number of isolated locations. A curve drawn freehand will often be a piecewise C^2 function.

If $F[u]$ is a piecewise C^2 function, integration by parts shows that

$$\int_{u_-}^{u_+}(u_+-u)F''[u]\,du=F[u_+]-F[u_-]-(u_+-u_-)F'[u_-]. \tag{1.7}$$

Any isolated finite jumps which there may be between the values of the second derivative $F''[u]$ evaluated from the left and from the right will contribute zero to the integral in (1.7). From (1.4) and (1.7) we see that a sufficient condition for $F[u]$ to be strictly convex is that $F''[u]>0$ everywhere, except possibly at *isolated* points where $F''=0$ (e.g. the origin in the examples $F[u]=u^4$ and, for $u^2<\frac{2}{3}$, the 'flat function' e^{-1/u^2}, all of whose derivatives are zero at the origin). A sufficient condition for weak convexity is $F''[u] \geqslant 0$ at every point, where the zero values need no longer be at isolated points.

We shall need to be aware that a *nonlinear* change $u = u(v)$ of the independent variable can destroy a convexity property (or, on the other hand, create it). This is because a nonuniform compression and/or extension of the plane parallel to the abscissa can introduce (or smooth out) wrinkles in a curve $F = F[u]$ drawn on that plane. For by the chain rule applied to the new function $E[v] = F[u(v)]$ we have $\mathrm{d}E/\mathrm{d}v = F' \,\mathrm{d}u/\mathrm{d}v$ and

$$\frac{\mathrm{d}^2 E}{\mathrm{d}v^2} = \frac{\mathrm{d}^2 u}{\mathrm{d}v^2} F' + \left(\frac{\mathrm{d}u}{\mathrm{d}v}\right)^2 F''. \tag{1.8}$$

The first term on the right can cause the sign of the curvature of $E[v]$ to change even if F'' has constant sign.

(iii) A general saddle function

A C^2 function $L = L[x, u]$ of rectangular cartesian coordinates will be called *saddle-shaped at a point* if the Hessian

$$\frac{\partial^2 L}{\partial x^2}\frac{\partial^2 L}{\partial u^2} - \left(\frac{\partial^2 L}{\partial x \partial u}\right)^2 < 0 \tag{1.9}$$

at that point, irrespective of whether it is a stationary point or not. This is equivalent to requiring that the principal curvatures of the surface have opposite signs at the point, because their product is

$$\left[\frac{\partial^2 L}{\partial x^2}\frac{\partial^2 L}{\partial u^2} - \left(\frac{\partial^2 L}{\partial x \partial u}\right)^2\right]\left[1 + \left(\frac{\partial L}{\partial x}\right)^2 + \left(\frac{\partial L}{\partial u}\right)^2\right]^{-2}.$$

This result is proved in differential geometry texts (for example, see §33 of Weatherburn, 1927).

Regardless of the particular choice of cartesian coordinates (see Exercise 1.1.4), one can use the term *saddle function* to describe any function which is saddle-shaped in the sense that (1.9) holds everywhere over a simply connected domain, except that isolated points are also admitted where the Hessian in (1.9) is zero, such as the origin in the example $L = -x^4 + u^4$.

A *saddle point* is defined as any point of a saddle function where both equations $\partial L/\partial x = \partial L/\partial u = 0$ are satisfied, so that it is a stationary point.

In higher dimensional situations later we shall concentrate our attention solely upon generalizations of a particular type of saddle function $L[x, u]$, namely one for which the coordinates have already been chosen so that $L[x, u]$ is concave with respect to x at each fixed u, and convex with respect to u at each fixed x. A continuous function could merit this description, without any hypotheses about the existence of its derivatives (just as we began the account of convexity above). However, it is easier to begin our exposition with a C^2 function such that

$$\frac{\partial^2 L}{\partial x^2} < 0 \quad \text{and} \quad \frac{\partial^2 L}{\partial u^2} > 0 \quad \text{everywhere} \tag{1.10}$$

over a rectangular domain whose sides are parallel to the axes. Conditions (1.10) are sufficient not only for the function to be saddle-shaped at every point in the sense of (1.9), but *also* (recall (1.7)) for the function to have the particular strictly concave–convex property with respect to x and u, respectively, which we just indicated (as we shall prove in Theorem 1.2).

Condition (1.9) can still hold when $\partial^2 L/\partial x^2$ and $\partial^2 L/\partial u^2$ have the same sign, or when one or both are zero and the mixed derivative is not, but a change of variables will then be needed in order to display the saddle function as one for which the unmixed second derivatives have opposite signs (see Exercise 1.2.3).

The *implicit function theorem* is an important tool for showing that certain equations have solutions, and hence for giving explicit expression to quantities which depend on those solutions, including upper or lower bounds. A proof of the theorem in any finite number of dimensions is given by, for example, Apostol (1974, Theorem 13.7). Here we are working in the x, u plane. The conventional version of the theorem then states that if a C^1 function $f[x,u]$ is used to define a curve in the plane by $f[x,u]=0$, then in the neighbourhood of any point on the curve where $\partial f/\partial x \neq 0$, the curve may also be expressed as $x = x(u)$ in terms of a unique C^1 function $x(u)$ such that $(\mathrm{d}x/\mathrm{d}u)\partial f/\partial x + \partial f/\partial u = 0$. The result is a local one in the first instance, although a global result can often be built onto the local one by fitting together solutions separated by isolated places where $\partial f/\partial x = 0$. The conics provide simple examples, e.g. $f[x,u] = x^2 - 4u$.

Theorem 1.1
Any C^2 saddle function $L[x,u]$ of the type satisfying (1.10) *has the following properties:*

(a) The path defined on the surface $L = L[x,u]$ by $\partial L/\partial x = 0$ can be described as $L = J[u]$ in terms of a strictly convex function $J[u]$.

(b) The path defined on the surface $L = L[x,u]$ by $\partial L/\partial u = 0$ can be described as $L = K[x]$ in terms of a strictly concave function $K[x]$.

When a saddle point exists it has the following properties.

(c) A saddle point is unique.

(d) Any value of $J[u]$ will provide an upper bound to the height of the saddle point. Any value of $K[x]$ will provide a lower bound to the height of the saddle point.

(e) Minimization of $J[u]$ will determine the saddle point. Maximization of $K[x]$ will also determine the saddle point.

Proof

(a) In the implicit function theorem we choose $f[x,u]=\partial L/\partial x$. Since $\partial^2 L/\partial x^2 \neq 0$, the equation $\partial L/\partial x = 0$ can be solved as a unique C^1 function $x=x(u)$ whose slope satisfies

$$\frac{\partial^2 L}{\partial x^2}\frac{dx}{du}+\frac{\partial^2 L}{\partial x \partial u}=0. \tag{1.11}$$

The corresponding path on the saddle function has height

$$L[x(u),u]=J[u] \quad \text{(say)}$$

such that, by the chain rule,

$$\frac{dJ}{du}=\frac{\partial L}{\partial u}, \quad \frac{d^2 J}{du^2}=\frac{\partial^2 L}{\partial u^2}-\left(\frac{\partial^2 L}{\partial x \partial u}\right)^2\left(\frac{\partial^2 L}{\partial x^2}\right)^{-1}. \tag{1.12}$$

Hence $d^2 J/du^2 > 0$ by (1.10), and $J[u]$ is a strictly convex function.

(b) The roles of x and u in the implicit function theorem are reversed. Since $\partial^2 L/\partial u^2 \neq 0$, the equation $\partial L/\partial u = 0$ can be solved as a unique C^1 function $u(x)$. The corresponding path on the saddle function has height

$$L[x,u(x)]=K[x] \quad \text{(say)}$$

such that

$$\frac{dK}{dx}=\frac{\partial L}{\partial x}, \quad \frac{d^2 K}{dx^2}=\frac{\partial^2 L}{\partial x^2}-\left(\frac{\partial^2 L}{\partial x \partial u}\right)^2\left(\frac{\partial^2 L}{\partial u^2}\right)^{-1}. \tag{1.13}$$

Hence $d^2 K/dx^2 < 0$ by (1.10), and $K[x]$ is a strictly concave function.

(c) The slopes of any two curves implicitly defined in the x, u plane by $\partial L/\partial x = constant$ and $\partial L/\partial u = constant$ are given by (1.11) and its complement. Their product is $(\partial^2 L/\partial u^2)(\partial^2 L/\partial x^2)^{-1} < 0$ at every intersection. Therefore, by continuity, any two such curves can intersect at most once. The reader may sketch a plan view. In particular, a saddle point is unique. We shall give a different style of uniqueness proof later using (1.29), as in Theorem 1.7.

(d) The strict convexity of $J[u]$ and the strict concavity of $K[x]$ established in (a) and (b), respectively, may be rephrased as

$$J[u_+]-J[u_-]-(u_+-u_-)J'[u_-]>0$$

for any u_+ and u_-, and

$$K[x_+]-K[x_-]-(x_+-x_-)K'[x_+]>0$$

for any x_+ and x_-, in the notation of (1.4) and (1.6). Choosing u_- as the saddle point value, where $\mathrm{d}J/\mathrm{d}u=0$ by $(1.12)_1$, and u_+ arbitrary, gives the upper bound. Choosing x_+ as the saddle point value, where $\mathrm{d}K/\mathrm{d}x=0$ by $(1.13)_1$, and x_- arbitrary, gives the lower bound.

(e) For the present purpose we will regard it as self-evident that a strictly convex C^2 function of one variable which has zero slope at some point will attain its minimum there. Applying this to $J[u]$ with $(1.12)_1$, and to $-K[x]$ with $(1.13)_1$, gives the result. □

The path defined on the surface $L=L[x,u]$ by $\partial L/\partial x=0$ will not generally lie in a single vertical plane (except when $\partial L/\partial x$ is linear, for example when $L[x,u]$ is quadratic). The height L of this path may be studied as a function $\phi[s]$, say, of distance s measured along its projection $x=x(u)$ onto the horizontal plane. If we use s instead of u as the parameter, via an assumed piecewise C^2 function $u(s)$, we have $\phi[s]=L[x(u(s)),u(s)]$ and therefore

$$\frac{\mathrm{d}\phi}{\mathrm{d}s}=\frac{\mathrm{d}u}{\mathrm{d}s}\frac{\mathrm{d}J}{\mathrm{d}u},\quad \frac{\mathrm{d}^2\phi}{\mathrm{d}s^2}=\frac{\mathrm{d}^2u}{\mathrm{d}s^2}\frac{\mathrm{d}J}{\mathrm{d}u}+\left(\frac{\mathrm{d}u}{\mathrm{d}s}\right)^2\frac{\mathrm{d}^2J}{\mathrm{d}u^2}.$$

For linear $\partial L/\partial x$, u is proportional to s so that $\mathrm{d}^2u/\mathrm{d}s^2\equiv 0$ and $\mathrm{d}^2\phi/\mathrm{d}s^2>0$ at every point of the path, and $\phi[s]$ is strictly convex. Otherwise, e.g. for nonquadratic $L[x,u]$, $\mathrm{d}^2\phi/\mathrm{d}s^2$ is positive *at* the saddle point because $\mathrm{d}J/\mathrm{d}u=0$ there, but $\mathrm{d}^2\phi/\mathrm{d}s^2$ may take either sign away from the saddle point because neither $\mathrm{d}^2u/\mathrm{d}s^2$ nor $\mathrm{d}J/\mathrm{d}u$ need be zero there (compare the first term on the right in (1.8)). Hence $\phi[s]$ need not be strictly convex in general.

By contrast $J[u]$ *is* strictly convex, as we proved. This function describes the projection of the surface path $\partial L/\partial x=0$ onto the L,u plane. This projection procedure illustrates the remark after (1.8), that a nonlinear change of independent variable can destroy $(u\to s)$ or create $(s\to u)$ convexity.

Similar remarks apply to the other path $\partial L/\partial u=0$, when its height $\psi[s]$, say, is expressed via an assumed piecewise C^2 parametrization $x(s)$, so that $\psi[s]=L[x(s),u(x(s))]$ and

$$\frac{\mathrm{d}\psi}{\mathrm{d}s}=\frac{\mathrm{d}x}{\mathrm{d}s}\frac{\mathrm{d}K}{\mathrm{d}x},\quad \frac{\mathrm{d}^2\psi}{\mathrm{d}s^2}=\frac{\mathrm{d}^2x}{\mathrm{d}s^2}\frac{\mathrm{d}K}{\mathrm{d}x}+\left(\frac{\mathrm{d}x}{\mathrm{d}s}\right)^2\frac{\mathrm{d}^2K}{\mathrm{d}x^2}.$$

This path has a strictly concave projection $L=K[x]$ onto the L,x plane, but $\psi[s]$ is not necessarily concave in general.

(iv) Equivalence problems

The purpose of Theorem 1.1 is to put into focus the kind of result which may often be aimed at, but less frequently achieved in such explicit form. The hypotheses (1.10) are very strong, and we shall weaken them and the C^2 assumption later on. We shall aim, for example, to get theorems about upper and lower bounds which do not, of themselves, depend on the availability of an implicit function theorem. Thus we separate the question of proving the bounds from that of utilizing them, and also from the converse question of establishing minimum and maximum principles. We begin to introduce these distinctions here.

Consider the following problems I, II and III, all generated by the same given C^1 function $L[x,u]$. The problems can be posed regardless of whether $L[x,u]$ is a saddle function.

I. Let x_0, u_0 denote any point which satisfies both of the equations

$$\frac{\partial L}{\partial x}=0 \quad (\alpha), \qquad \frac{\partial L}{\partial u}=0 \quad (\beta). \tag{1.2 bis}$$

Let $\{x_0, u_0\}$ denote the set of such points. Find this set.

II. Among the set $\{x_\alpha, u_\alpha\}$ of points which satisfy (1.2α) but not necessarily (1.2β), find the set of *minimizers* of L, say $\{\hat{x}_\alpha, \hat{u}_\alpha\}$. We can write

$$L[\hat{x}_\alpha, \hat{u}_\alpha] = \min L[x_\alpha, u_\alpha].$$

III. Among the set $\{x_\beta, u_\beta\}$ of points which satisfy (1.2β) but not necessarily (1.2α), find the set of *maximizers* of L, say $\{\hat{x}_\beta, \hat{u}_\beta\}$. We can write

$$L[\hat{x}_\beta, \hat{u}_\beta] = \max L[x_\beta, u_\beta].$$

We shall say that any two of problems I, II and III are *equivalent* if and only if the set of solution points is the same for both problems. All three problems are equivalent if all three solution sets are the same.

In general, of course, no solution might exist to some or all of these problems without further specification of $L[x,u]$. It is known, however, that a continuous function on a closed and bounded set in a finite dimensional space attains its extreme values there (see Apostol, 1974, Theorems 3.31 and 4.28). Such extrema may be on the boundary, or internal. In many cases which we shall consider, however, the domain of $L[x,u]$ extends to infinity so that the hypothesis for the quoted existence result will not apply; then existence of solutions may have to be regarded as an implicit assumption which is necessary in order that a statement about solutions may have content.

For the particular C^2 (*a fortiori* C^1) saddle functions satisfying (1.10), Theorem 1.1(c) shows that there is at most one solution x_0, u_0 of I, namely the saddle point. In the suffix notation introduced above, the results of Theorem 1.1(d) and (e) can be written as a *pair of simultaneous upper and lower bounds*

$$J[u_\alpha] = L[x_\alpha(u_\alpha), u_\alpha] \geqslant L[x_0, u_0] \geqslant L[x_\beta, u_\beta(x_\beta)] = K[x_\beta]$$

generalizing (1.3); minimization of the strictly convex upper bound shows that II will have a unique solution $\hat{x}_\alpha, \hat{u}_\alpha$ at x_0, u_0, and maximization of the strictly concave lower bound shows that III will have a unique solution $\hat{x}_\beta, \hat{u}_\beta$ at x_0, u_0. Thus the three problems are equivalent for a function which satisfies the strong hypothesis (1.10) and has a saddle point. We have already seen this equivalence explicitly for the hyperbolic paraboloid.

In general, it may be self-evident that the set $\{x_0, u_0\}$ belongs to both sets $\{\hat{x}_\alpha, \hat{u}_\alpha\}$ and $\{\hat{x}_\beta, \hat{u}_\beta\}$, but equivalence requires one to prove the converses. Thus a true *minimum principle* will prove that every minimizer $\hat{x}_\alpha, \hat{u}_\alpha$ satisfies (1.2β); and a true *maximum principle* will prove that every maximizer $\hat{x}_\beta, \hat{u}_\beta$ satisfies (1.2α). Then we shall have a pair of *simultaneous extremum principles.* It is usually harder to prove such converses than it is to prove upper and lower bounds; an alternative proof of the latter is given in Theorem 1.4.

Another question of equivalence is introduced by the following problem IV, also posed in terms of a given C^1 function $L[x, u]$ which, for this purpose, need not be a saddle function.

IV. *Stationary value problem.* Find the set $\{\mathring{x}, \mathring{u}\}$, say, of points which are such that

$$L[x, u] - L[\mathring{x}, \mathring{u}] = o(\Delta),$$

where $\Delta = [(x - \mathring{x})^2 + (u - \mathring{u})^2]^{\frac{1}{2}}$, for all points x, u in a sufficiently small domain D enclosing $\mathring{x}, \mathring{u}$ within the interior of D. Any such point $\mathring{x}, \mathring{u}$ defines *a stationary point* of $L[x, u]$, and $L[\mathring{x}, \mathring{u}]$ is the corresponding stationary value.

In infinite dimensions, with $L[x, u]$ as a functional, IV would correspond to the classical starting point in the calculus of variations, where the fundamental problem is to prove that IV is equivalent to I, which we express as $\{\mathring{x}, \mathring{u}\} \equiv \{x_0, u_0\}$. In the current three dimensional context this equivalence is almost immediate for suitable D. Suppose there is a straight join from $\mathring{x}, \mathring{u}$ to every x, u in D (such a D is called star-shaped; in particular it could be convex). The first mean value theorem assures the existence of $\bar{x}, \bar{u}$ (say) on every such straight join, such that

$$L[x, u] - L[\mathring{x}, \mathring{u}] = (x - \mathring{x})\frac{\partial L}{\partial x}\bigg|_{\bar{x},\bar{u}} + (u - \mathring{u})\frac{\partial L}{\partial u}\bigg|_{\bar{x},\bar{u}}.$$

Dividing by Δ and then letting $\Delta \to 0$ along every straight join shows that IV implies

$$\xi \frac{\partial L}{\partial x}\bigg|_{\mathring{x},\mathring{u}} + \eta \frac{\partial L}{\partial u}\bigg|_{\mathring{x},\mathring{u}} = 0 \tag{1.14}$$

for every unit vector ξ, η pointing into D from $\mathring{x}, \mathring{u}$. In other words, we may say that the gradient of $L[x, u]$ at $\mathring{x}, \mathring{u}$ is orthogonal to every unit vector emanating from that point. This implies $\partial L/\partial x = \partial L/\partial u = 0$ at $\mathring{x}, \mathring{u}$, so that $\{\mathring{x}, \mathring{u}\} \subseteq \{x_0, u_0\}$. Every stationary point therefore satisfies (1.2). Conversely, any solution of (1.2) has the property $\xi(\partial L/\partial x)|_{x_0,u_0} + \eta(\partial L/\partial u)|_{x_0,u_0} = 0$, so that $\{x_0, u_0\} \subseteq \{\mathring{x}, \mathring{u}\}$ by comparing with (1.14). Hence $\{\mathring{x}, \mathring{u}\} \equiv \{x_0, u_0\}$. Therefore problems I and IV are equivalent, and we call this fact a *stationary principle*.

We anticipated the definition of a stationary point in IV, and the property $\{x_0, u_0\} \subseteq \{\mathring{x}, \mathring{u}\}$, when we said after (1.3) that points satisfying (1.2) would be called stationary points. The proof here that $\{\mathring{x}, \mathring{u}\} \subseteq \{x_0, u_0\}$ shows that there can be no stationary points (in the sense of the definition in IV) which do not satisfy (1.2).

The question of whether I is equivalent to IV is different from the questions of whether I is equivalent to II or III.

A stationary principle is more conventionally called a variational principle, in the sense that exploration by variation over a whole set of points ξ, η or D is involved in the determination of $\{\mathring{x}, \mathring{u}\}$ from (1.14) or from the defining equation in IV. Variational inequalities are defined in the same spirit (e.g. see Exercise 1.3.5 and §3.11).

An account of the calculus of variations which gives an explicit role to the notion of convexity is that of Troutman (1983). A treatment of variational inequalities has been given by Kinderlehrer and Stampacchia (1980).

Exercises 1.1

1. Prove that, when $F[u]$ is a C^1 function of a single scalar variable u with gradient $F'[u]$, the following definitions of a strictly convex function are equivalent. The method required is included in the proof of Theorem 1.2(a).
   (i) $F[u_+] - F[u_-] - (u_+ - u_-)F'[u_-] > 0 \ \forall$ distinct u_+ and u_-;
   (ii) $F[\lambda u_+ + (1-\lambda)u_-] < \lambda F[u_+] + (1-\lambda)F[u_-] \ \forall$ distinct u_+ and u_- and $\forall$ real λ in $0 < \lambda < 1$.
2. Verify (1.7). Prove that any piecewise C^2 function $F[u]$ satisfies

$$F[u_+] - F[u_-] - (u_+ - u_-)F'[u_-] = \tfrac{1}{2}(u_+ - u_-)^2 F''[\bar{u}]$$

for *some* value $\bar{u}$ between u_- and u_+ (and for *some* value of F'' among the interval of values possible if $\bar{u}$ is a point where F'' exhibits a finite jump).

3. Prove that the product of the principal curvatures of any C^2 function $L[x, u]$ referred to rectangular cartesian coordinates is

$$\left[\frac{\partial^2 L}{\partial x^2}\frac{\partial^2 L}{\partial u^2} - \left(\frac{\partial^2 L}{\partial x \partial u}\right)^2\right]\left[1 + \left(\frac{\partial L}{\partial x}\right)^2 + \left(\frac{\partial L}{\partial u}\right)^2\right]^{-2}.$$

4. For any C^2 function $L[x, u]$ of two scalar variables, prove from first principles that the sign of

$$\frac{\partial^2 L}{\partial x^2}\frac{\partial^2 L}{\partial u^2} - \left(\frac{\partial^2 L}{\partial x \partial u}\right)^2$$

is invariant when the independent variables are changed by any invertible linear transformation.

5. When the determinant in Exercise 4 is nonzero at every stationary point of $L[x, u]$, this function is called a Morse function. Establish that, in the neighbourhood of any particular stationary point of a sufficiently smooth Morse function, a diffeomorphism can be found to express the function exactly as one or other of the three 'canonical forms'

$$x^2 + u^2, \; -x^2 + u^2, \; -x^2 - u^2,$$

without remainder (refer to Poston and Stewart, 1976, p. 17). A diffeomorphism is an invertible C^1 change of variables.

6. (i) Show that every C^2 solution of Laplace's equation over a plane domain is saddle-shaped at every point.

   (ii) Let $z = x + iy$ be a complex variable, and $f[z]$ be a complex valued function of z. Use the Cauchy–Riemann equations for $f[z]$ to show that the two real components of the complex conjugate function $\overline{f[z]} = \bar{f}[\bar{z}]$ are not only saddle-shaped themselves, but are also the gradients of another solution of Laplace's equation which is therefore also saddle-shaped at every point. Illustrate with $f[z] = e^z$.

7. Construct examples to illustrate that, for a smooth landscape within a closed contour, the number of local maxima, minima and saddle points satisfies

$$\text{peaks} + \text{pits} - \text{passes} = 1.$$

(v) Quadratic example

Consider the general quadratic

$$L[x, u] = \tfrac{1}{2}ax^2 + hxu + \tfrac{1}{2}bu^2 + fx + gu + c \tag{1.15}$$

having constant coefficients. The stationary value problem I is

$$\frac{\partial L}{\partial x} = ax + hu + f = 0 \quad (\alpha), \qquad \frac{\partial L}{\partial u} = hx + bu + g = 0 \quad (\beta).$$

This has a unique solution if and only if $ab - h^2 \neq 0$, namely

$$x_0 = \frac{hg - bf}{ab - h^2}, \quad u_0 = \frac{hf - ag}{ab - h^2}.$$

The height of this stationary point is

$$L[x_0, u_0] = \frac{2fgh - ag^2 - bf^2}{2(ab - h^2)} + c.$$

Suppose first that

$$a < 0, \quad b > 0, \quad h \text{ unrestricted}, \tag{1.16}$$

Fig. 1.4. Plan view of features of the saddle function

$$\begin{aligned} L[x, u] &= -\tfrac{1}{2}x^2 + 2xu + u^2 \\ &= -\tfrac{1}{2}[x - (2 - \sqrt{6})u][x - (2 + \sqrt{6})u] \qquad \text{(asymptotes)} \\ &= -\tfrac{1}{5}(2x - u)^2 + \tfrac{3}{10}(x + 2u)^2 \end{aligned}$$

(orthogonal steepest convex and concave parabolas),

$$\frac{\partial L}{\partial x} = -x + 2u = 0, \quad J[u] = 3u^2,$$

(upper bounding parabolic section),

$$\frac{\partial L}{\partial u} = 2(x + u) = 0, \quad K[x] = -\tfrac{3}{2}x^2,$$

(lower bounding parabolic section).

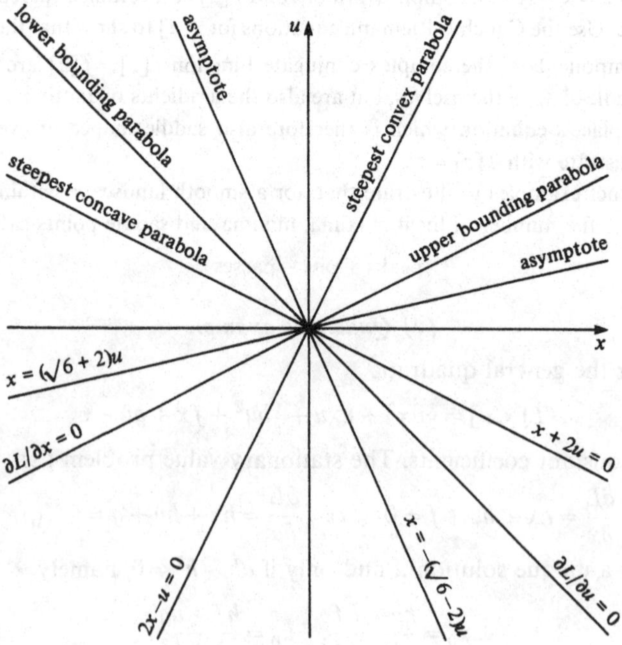

so that (1.15) is a saddle function of the particular type (1.10), and x_0, u_0 above is its unique saddle point.

The α-constraint alone is satisfied by $x_\alpha = -(f + hu_\alpha)/a$ with arbitrary u_α, and this defines a convex parabolic section

$$J[u_\alpha] = \tfrac{1}{2}\left(b - \frac{h^2}{a}\right)(u_\alpha - u_0)^2 + L[x_0, u_0] \tag{1.17}$$

of (1.15). This function can be used to illustrate (1.12). The problem II of minimizing these upper bounds is plainly solved by $\hat{x}_\alpha = x_0, \hat{u}_\alpha = u_0$. Hence I and II are equivalent.

The β-constraint alone is satisfied by $u_\beta = -(g + hx_\beta)/b$ with arbitrary x_β, and this defines a concave parabola

$$K[x_\beta] = \tfrac{1}{2}\left(a - \frac{h^2}{b}\right)(x_\beta - x_0)^2 + L[x_0, u_0] \tag{1.18}$$

illustrating (1.13). The problem III of maximizing these bounds is solved by $\hat{x}_\beta = x_0$, $\hat{u}_\beta = u_0$. Hence I and III are equivalent.

A plan view of a specific example of (1.15) with (1.16) is sketched in Fig. 1.4, where the minimizing and maximizing parabolic sections appear from above as straight lines. The reader may find it instructive to insert the hyperbolic level contours onto this diagram.

Suppose next that (1.15) is a saddle function in the sense of (1.9), so that

$$ab - h^2 < 0, \tag{1.19}$$

whether or not (1.16) holds. The saddle property can be alternatively verified by showing that (1.19) is necessary and sufficient for *some* particular cross-section $L = constant$ of the surface (1.15) to be written as a pair of real and nonparallel straight lines. These lines then serve as asymptotes for hyperbolae which are *other* sections $L = constant$. Therefore (1.19) guarantees that (1.15) is a hyperbolic paraboloid, and a linear change of variables could be found to rewrite (1.15) in the canonical version (1.1).

When $a \neq 0$ we see from $J[u_\alpha] = L[x_\alpha(u_\alpha), u_\alpha]$ that I and II are equivalent if and only if $b - h^2/a > 0$; with (1.9) this implies $a < 0$. If $a > 0$ with (1.19) the parabolic section is inverted, and I is equivalent to a maximization problem instead.

When $b \neq 0$ we see from $K[x_\beta] = L[x_\beta, u_\beta(x_\beta)]$ that I and III are equivalent if and only if $a - h^2/b < 0$; with (1.19) this implies $b > 0$. If $b < 0$ with (1.19), I is equivalent to a minimization problem instead.

Thus if, instead of (1.16),

$$0 < ab < h^2,$$

then I cannot be equivalent to both II and III. Instead, if (say) $a > 0$ and $b > 0$, the α- and β-sections define two different maximization problems which are each equivalent to I (Exercise 1.2.3 is an example). If $a < 0$ and $b < 0$, the two constraints define two different minimization problems which are each equivalent to I.

Put otherwise, suppose the coefficients in (1.15) are functions $a(\lambda)$, $b(\lambda)$, $h(\lambda)$ of a parameter λ whose evolution takes us successively through the three regimes

$$\text{(a)} \quad ab < 0 < h^2, \quad \text{(b)} \quad 0 < ab < h^2, \quad \text{(c)} \quad h^2 < ab. \tag{1.20}$$

The stationary point of $L[x, u]$ can be characterized by simultaneous upper and lower bounds in (a), and by either upper or lower bounds (but not both simultaneously) in (b) and (c), working throughout with the given choice of coordinates x and u; and if we work with changed coordinates we may recover simultaneous bounds in (b) but not in (c). The example $a = \lambda$, $b = \lambda^2$, $h = 1$ gives regime boundaries at $\lambda = 0$ and 1.

When $ab = 0$ and $h \neq 0$ I still has a unique solution. If $a = 0$ and $b > 0$ I and III are equivalent, but II is degenerate. If $b = 0$ and $a < 0$ I and II are equivalent, but III is degenerate. If $a = b = 0$ and $h \neq 0$ then

$$L[x, u] = hxu + fx + gu + c. \tag{1.21}$$

This bilinear saddle function satisfies (1.9) but not (1.10). It cannot, as it stands, illustrate nondegenerate extremum principles in the sense of II and III above. However, it provides a basic illustration of two new avenues for the construction of simultaneous extremum principles which can be developed, namely:

(a) change of variables designed to make a saddle function (already satisfying (1.9), but not (1.10)), satisfy (1.10) in the *new* variables. The simplest illustration is $L = xu$, which $x = x' + u'$, $u = -x' + u'$ converts to $L = -x'^2 + u'^2$. Clearly the new variables are along the steepest (x' and u') directions, and extremum principles expressed in terms of them are nondegenerate, as we have seen.

(b) introduction of inequality constraints, of which we give a very simple example in §1.2. These permit *non*stationary minima and/or maxima. Inequality constraints applied to (1.21) generate a linear programming problem, as we shall see in §1.6(i).

Exercises 1.2

1. Prove that $ab - h^2 < 0$ is necessary and sufficient for *some* particular cross-section $L =$ *constant* of the surface

$$L = \tfrac{1}{2}ax^2 + hxu + \tfrac{1}{2}bu^2 + fx + gu + c$$

to be expressible as a pair of real and nonparallel straight lines. The coefficients in L are constants.

2. Investigate in what sense the problem of minimizing

$$L[x,u] = hxu + \tfrac{1}{2}bu^2 + fx + gu + c \quad (b > 0)$$

subject to $\partial L/\partial x = 0$ is degenerate. Prove that the stationary value problem $\partial L/\partial x = \partial L/\partial u = 0$ is equivalent to maximizing L subject to the constraint $\partial L/\partial u = 0$.

3. Verify that

$$
\begin{aligned}
L &= x^2 + 3ux + 2u^2 \\
&= (x+2u)(x+u) && \text{(asymptotes)} \\
&= (x+\tfrac{3}{2}u)^2 - \tfrac{1}{4}u^2 && \text{(by completing the square)} \\
&= -\tfrac{1}{8}x^2 + \left(\frac{3}{2\sqrt{2}}x + \sqrt{2}u\right)^2 && \text{(by completing the other square)} \\
&= \left(\frac{3}{4\sqrt{2}} + \frac{1}{2}\right)(x+\sqrt{2}u)^2 - \left(\frac{3}{4\sqrt{2}} - \frac{1}{2}\right)(x-\sqrt{2}u)^2 && \text{(by inspection)} \\
&= \left(\frac{13}{36\sqrt{(10)}} + \frac{1}{9}\right)[(\sqrt{(10)}-1)x + 3u]^2 && \text{(steepest convex parabola)} \\
&\quad - \left(\frac{13}{36\sqrt{(10)}} - \frac{1}{9}\right)[(\sqrt{(10)}+1)x - 3u]^2 && \text{(steepest concave parabola).}
\end{aligned}
$$

Show that the last representation entails an *orthogonal* transformation from the original variables. Obtain the last representation (a) by calculating the bisectors of the asymptotes, and (b) by the standard computation of the eigenvectors of the Hessian. Draw a plan view of asymptotes, steepest convex and concave parabolas and the lines $\partial L/\partial x = 0$ and $\partial L/\partial u = 0$. Show that *each* of the last two sections provide lower bounds to the saddle point. Sketch the hyperbolic level contours.

(vi) A saddle quantity

Consider again a general single valued function $L[x,u]$ of two cartesian coordinates, over a rectangular domain whose sides are parallel to the axes. This time, however, we assume first only that $L[x,u]$ is continuous, and we give a definition that $L[x,u]$ be a saddle function which does not involve derivatives. Next we relate this definition to criteria expressed in terms of first derivatives, and then also second derivatives, when these are available.

Let x_+, u_+ and x_-, u_- denote any pair of points in the domain. We call them the 'plus' and 'minus' points respectively. They can each be joined to the point x_-, u_+ by straight paths which lie wholly within the domain of $L[x,u]$, so forming two consecutive sides of a rectangle parallel to the axes as shown in Fig. 1.5. Any plus and minus points with this property can be said to define a *rectangular subdomain* of $L[x,u]$. The two paths can be parametrized as the sets of points

$$x_\mu = \mu x_+ + (1-\mu)x_-, \quad u_\lambda = \lambda u_+ + (1-\lambda)u_-,$$

with $\mu = 0$ and $0 \leqslant \lambda \leqslant 1$, and then $\lambda = 1$ and $0 \leqslant \mu \leqslant 1$.

We shall went to be sure that *every* pair of distinct points in the domain of $L[x,u]$ defines a rectangular subdomain, for which it is necessary and sufficient that the whole domain of $L[x,u]$ be itself rectangular. This is why the hypothesis of a rectangular domain for $L[x,u]$ is adopted. In practice this will often be satisfied because the domain of $L[x,u]$ is either the whole x,u plane, or one of the half-spaces $x \geqslant 0$ or $u \geqslant 0$, or the quadrant $x \geqslant 0$ with $u \geqslant 0$.

We define $L[x,u]$ to be a *saddle function concave in x and convex in u* if the following properties (1.22a and b) both hold. For every fixed u_+, and for all distinct x_+ and x_-,

$$\mu L[x_+,u_+] + (1-\mu)L[x_-,u_+] \leqslant L[x_\mu,u_+] \tag{1.22a}$$

for all μ in $0 \leqslant \mu \leqslant 1$.
For every fixed x_-, and for all distinct u_+ and u_-,

$$\lambda L[x_-,u_+] + (1-\lambda)L[x_-,u_-] \geqslant L[x_-,u_\lambda] \tag{1.22b}$$

for all λ in $0 \leqslant \lambda \leqslant 1$.

Fig. 1.5. Rectangular subdomain in the x, u plane.

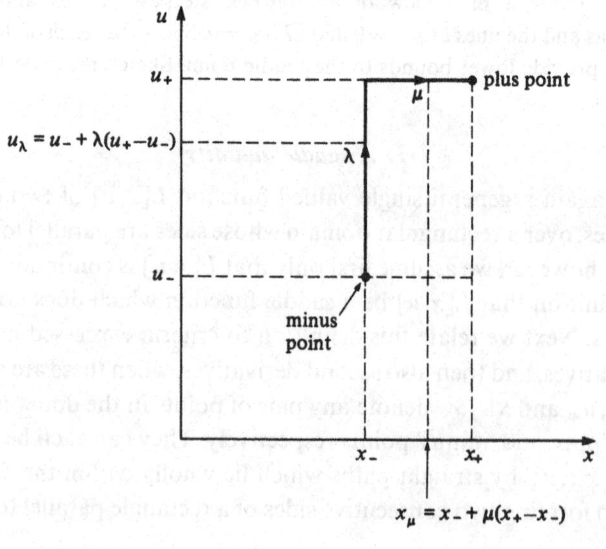

Such a function may be called either a *saddle function* or a *concave-convex function* for brevity, when it has been agreed that the choice of coordinates has been made once and for all, so that questions of coordinates change such as we discussed in relation to (1.9) and (1.19) (cf. Exercises 1.1.4 and 1.2.3) have already been settled. If strict inequalities hold in both (1.22a and b) when $0<\mu<1$ and $0<\lambda<1$, we say that $L[x,u]$ is a *strict* saddle function.

This definition of a saddle function parallels our initial definition of convexity in §1.1(ii), but when more smoothness is available it can be replaced by other criteria expressed in terms of derivatives. These latter criteria are more amenable to our purpose. To this end, when $L[x,u]$ is a C^1 function, and *a fortiori* when it is C^2 or piecewise C^2, we define the *saddle quantity*

$$S = L[x_+,u_+] - L[x_-,u_-] - (x_+ - x_-)\left.\frac{\partial L}{\partial x}\right|_{x_+,u_+} - (u_+ - u_-)\left.\frac{\partial L}{\partial u}\right|_{x_-,u_-}. \tag{1.23}$$

This quantity, or higher dimensional generalizations of it, will occur frequently in applications of those particular saddle functions which are concave in x and convex in u. We can regard S as a composition of the left sides of (1.4) and (1.6). It was proposed by Sewell (1969, equation (2.50)) as a basis for a very simple proof of upper and lower bounds, and we shall use S for this purpose here, in Theorem 1.4 and its later generalizations.

Theorem 1.2

(a) Let $L[x,u]$ be a C^1 function which is a saddle function in the sense that (1.22a *and* b) *hold for every pair of plus and minus points in the rectangular domain. This is equivalent to the property that*

$$S \geqslant 0 \tag{1.24}$$

for every pair of points in that domain.

(b) If $L[x,u]$ is a C^2 or piecewise C^2 function, the conditions that

$$\frac{\partial^2 L}{\partial x^2} \leqslant 0 \quad \textit{and} \quad \frac{\partial^2 L}{\partial u^2} \geqslant 0 \quad \textit{everywhere} \tag{1.25}$$

in the rectangular domain are sufficient for (1.24) *to hold. The strict inequalities* (1.10) *(also admitting isolated zeros) in place of* (1.25) *are sufficient to ensure that*

$$S > 0 \tag{1.26}$$

for every pair of points in the rectangular domain.

Proof

(a) We require an elaboration of the solution of Exercise 1.1.1. To prove that (1.22)⇒(1.24) we divide (1.22a) by $1-\mu$ and (1.22b) by λ, add the results to remove $L[x_-,u_+]$, and then let $\mu\to 1$ and $\lambda\to 0$. This gives (1.24).

To prove that (1.24)⇒(1.22), consider

$$\begin{aligned}
&\lambda L[x_-,u_+]+(1-\lambda)L[x_-,u_-]-L[x_-,u_\lambda]\\
&\qquad -\mu L[x_+,u_+]-(1-\mu)L[x_-,u_+]+L[x_\mu,u_+]\\
&\quad =\lambda\{L[x_-,u_+]-L[x_-,u_\lambda]\}+(1-\lambda)\{L[x_-,u_-]-L[x_-,u_\lambda]\}\\
&\qquad +\mu\{L[x_\mu,u_+]-L[x_+,u_+]\}+(1-\mu)\{L[x_\mu,u_+]-L[x_-,u_+]\}\\
&\quad \geqslant\{\lambda(u_+-u_\lambda)+(1-\lambda)(u_--u_\lambda)\}\left.\frac{\partial L}{\partial u}\right|_{x_-,u_\lambda}\\
&\qquad +\{\mu(x_\mu-x_+)+(1-\mu)(x_\mu-x_-)\}\left.\frac{\partial L}{\partial x}\right|_{x_\mu,u_+}
\end{aligned}$$

$=0$ by the definitions of u_λ and x_μ. The inequality in the proof is obtained by four particular uses of (1.24). The result establishes (1.22a and b) by making particular choices such as $\lambda=1$, any μ in $0\leqslant\mu\leqslant 1$ and then $\mu=0$, any λ in $0\leqslant\lambda\leqslant 1$.

Thus we have proved (1.22)⇔(1.24), which gives the equivalence in (a).

(b) On the two sides of the path shown in Fig. 1.5 we can define the functions

$$l(\lambda)=L[x_-,u_\lambda],\quad m(\mu)=L[x_\mu,u_+]$$

of the single scalars λ and μ, respectively, which are such that

$$\frac{d^2l}{d\mu^2}=(u_+-u_-)^2\frac{\partial^2L}{\partial {u_\lambda}^2},\quad \frac{d^2m}{d\mu^2}=(x_+-x_-)^2\frac{\partial^2L}{\partial {x_\mu}^2}\tag{1.27}$$

when these derivatives exist. If these are at least piecewise continuous on the respective paths it follows that

$$\begin{aligned}
S&=L[x_+,u_+]-L[x_-,u_+]-(x_+-x_-)\left.\frac{\partial L}{\partial x}\right|_{x_+,u_+}\\
&\qquad +L[x_-,u_+]-L[x_-,u_-]-(u_+-u_-)\left.\frac{\partial L}{\partial u}\right|_{x_-,u_-}\\
&=-\int_0^1\mu\frac{d^2m}{d\mu^2}d\mu+\int_0^1(1-\lambda)\frac{d^2l}{d\lambda^2}d\lambda
\end{aligned}\tag{1.28}$$

by two integrations by parts (compare (1.7)), one on each path as shown in Fig. 1.5. If the same starting pair of points had been labelled plus and minus in the reverse sense, another formula like (1.28) would have emerged, but with integration along the opposite pair of consecutive sides of the rectangle in Fig. 1.5.

The proofs are completed by inserting (1.25) or (1.10) into (1.27) and (1.28). At places of isolated discontinuity in the second derivatives (1.27) the hypothesized inequalities are to be applied to the derivatives evaluated from both sides. □

We call (1.26) a strict (and (1.24) a weak) *saddle inequality.*

By interchanging the role of the plus and minus points in the strict saddle inequality $S > 0$, and adding the two results, we see that this property of a strict saddle function implies

$$-(x_+ - x_-)\left(\frac{\partial L}{\partial x}\bigg|_{x_+,u_+} - \frac{\partial L}{\partial x}\bigg|_{x_-,u_-}\right) + (u_+ - u_-)\left(\frac{\partial L}{\partial u}\bigg|_{x_+,u_+} - \frac{\partial L}{\partial u}\bigg|_{x_-,u_-}\right) > 0 \tag{1.29}$$

for every pair of points in the domain. This is the generalization of (1.5), and it can be the basis of uniqueness proofs. For example, there could not be two saddle points each satisfying $\partial L/\partial x = \partial L/\partial u = 0$, one at the plus point and one at the minus point, because of the evident contradiction (compare the proof of (c) in Theorem 1.1).

1.2. Inequality constraints

(i) A single variable example

We now introduce another topic, which may appear at first to be quite different from the characterization of saddle points studied in §1.1. We shall soon demonstrate a strong connection between the two topics, however.

Let $J[u]$ be a C^1 function of a single scalar variable u, with gradient $J'[u]$. We regard $J[u]$ as given *ab initio* in this subsection and unconnected at this stage with anything in §1.1. It will be seen later (Exercises 1.5.6 and 7) how a particular example of the present $J[u]$ could be the $J[u]$ introduced in Theorem 1.1. The reason for using the same symbol here as in Theorem 1.1 is that $J[u]$ is minimized in each context, albeit over different regimes.

We wish to find conditions under which the following three problems are equivalent.

I. Find the set $\{u_0\}$, say, of points which satisfy the trio

$$\begin{aligned} J'[u] &\geqslant 0, \quad (\beta) \\ u &\geqslant 0, \quad (\alpha) \\ uJ'[u] &= 0. \end{aligned} \tag{1.30}$$

This is a pair of inequalities with an orthogonality condition. A trio with this structure connecting any two scalars, say u and w, can be summarized as the heavy two-segment line in Fig. 1.6 (at most one of the two quantities is positive). Such trios are often called 'complementarity conditions'. If $J[u]$ is quadratic, (1.30) can be called a linear complementarity problem.

II. Let $\{u_\alpha\}$ denote the set of points which satisfy (1.30α) but not necessarily the other two conditions in (1.30). Among these points find the set of minimizers of J, say $\{\hat{u}_\alpha\}$, so that

$$J[\hat{u}_\alpha] = \min J[u_\alpha]. \tag{1.31}$$

III. Let $\{u_\beta\}$ denote the set of points which satisfy (1.30β), but not necessarily the other two conditions in the trio. Among these points find the set of maximizers of $J - uJ'$, say $\{\hat{u}_\beta\}$, so that

$$J[\hat{u}_\beta] - \hat{u}_\beta J'[\hat{u}_\beta] = \max(J[u_\beta] - u_\beta J'[u_\beta]). \tag{1.32}$$

Equivalence will mean the same as in §1.1(iv), namely that the sets $\{u_0\}$, $\{\hat{u}_\alpha\}$ and $\{\hat{u}_\beta\}$ of solution points are the same. As compared with (1.2), $\{u_0\}$ is not here a set consisting exclusively of stationary points because the origin belongs to it even if $J'[0] > 0$; and the α and β conditions are each an inequality instead of an equation.

The *inverse function theorem* is another basic tool of analysis, like the implicit function theorem which is based upon it, and we shall summarize it for the case of finitely many dimensions in §1.5(i). Specialized to functions in the plane, the conventional inverse function theorem states that if the C^1 function $v[u]$ has finite nonzero slope at some point, then there exists, at least locally, a unique C^1 inverse function $u[v]$ also with finite nonzero slope, and such that $(dv/du)(du/dv) = 1$.

A strengthened version is available for a piecewise C^1 function $v[u]$

Fig. 1.6. Diagram of $w \geqslant 0$, $u \geqslant 0$, $uw = 0$.

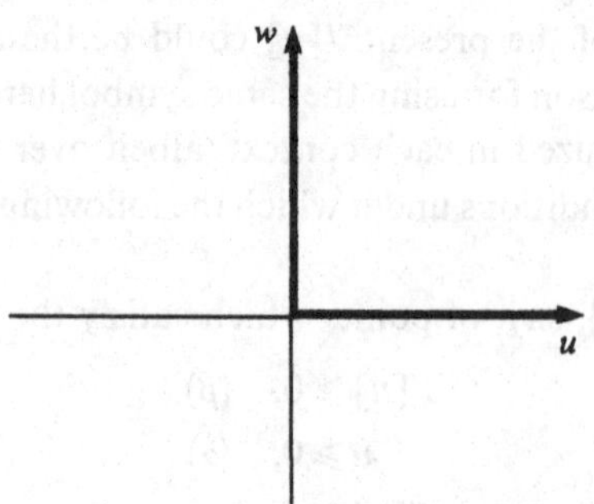

which is strictly monotonic, because the inverse function $u[v]$ is then also piecewise C^1 and strictly monotonic, and the result is globally valid as the sketch in Fig. 1.7 indicates. One-sided gradients apply at the corners, where there is a finite jump in slope. Even more strongly, the strict monotonicity is still preserved if we admit isolated locations where the slope of one function is zero and that of the inverse function infinite (as at the origin when $v = u^3$, $u = v^{1/3}$), and Fig. 1.7 indicates zero and infinite slopes too. We use this *strengthened inverse function theorem* in the proof of Theorem 1.3(b) following.

Theorem 1.3

(a) If $J[u]$ is a weakly convex C^1 function, then problems I *and* II *are equivalent.*

Fig. 1.7. Strengthened inverse theorem for piecewise C^1 strictly monotonic functions.

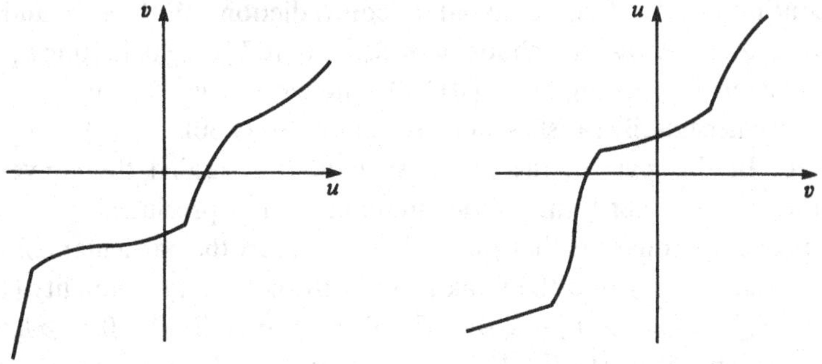

Fig. 1.8. Minimization of a weakly convex C^1 function on $u \geqslant 0$.

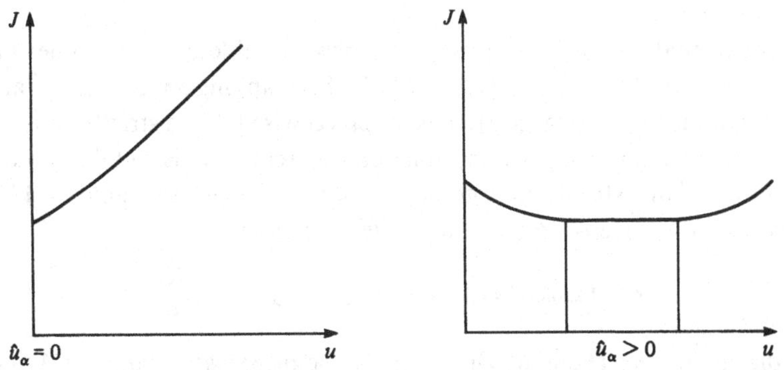

(*a*) $J'[\hat{u}_\alpha] > 0, \hat{u}_\alpha J'[\hat{u}_\alpha] = 0$

(*b*) $J'[\hat{u}_\alpha] = 0, \hat{u}_\alpha J'[\hat{u}_\alpha] = 0$

(b) If $J[u]$ is a strictly convex piecewise C^2 function, then problems I and III are equivalent. They are also equivalent to II a fortiori. The common solution of all three problems is unique.

Proof of (a)

Fig. 1.8 is a geometrical explanation of the expected result. There is a minimum which can be at the origin with nonnegative slope, or at an interior point (not necessarily unique) with slope zero. Minima with zero slope could also be at interior points and at the origin simultaneously.

To prove that $\{\hat{u}_\alpha\} \subseteq \{u_0\}$, note that $\bar{u}$ exists between each $\hat{u}_\alpha$ and any u_α such that

$$(u_\alpha - \hat{u}_\alpha)J'[\bar{u}] = J[u_\alpha] - J[\hat{u}_\alpha] \geqslant 0$$

by the first mean value theorem and the definition (1.31) of $\hat{u}_\alpha$. With an interior minimum $\hat{u}_\alpha > 0$ we can choose $u_\alpha > 0$ so that $u_\alpha - \hat{u}_\alpha$ has either sign. Then letting $u_\alpha \to \hat{u}_\alpha$ after division by $|u_\alpha - \hat{u}_\alpha|$, we must have $J'[\hat{u}_\alpha] = 0$ by the continuity of $J'[\bar{u}]$, to avoid a contradiction. With a boundary minimum $\hat{u}_\alpha = 0$ we cannot choose $u_\alpha < \hat{u}_\alpha$, so that $J'[\bar{u}] \geqslant 0$. Letting $u_\alpha \to 0$ after division by $|u_\alpha|$ implies $J'[0] \geqslant 0$. Therefore a global minimum of either type necessarily satisfies all three conditions (1.30), i.e. $\{\hat{u}_\alpha\} \subseteq \{u_0\}$ as required. In this part of the proof we have not needed the convexity hypothesis, and other local interior minima are not precluded.

To prove the converse, that $\{u_0\} \subseteq \{\hat{u}_\alpha\}$, we insert the particular choices $u_+ = u_\alpha$ and $u_- = u_0$ into the weak form of the convexity inequality (1.4). Hence $J[u_\alpha] - J[u_0] \geqslant (u_\alpha - u_0)J'[u_0] \geqslant 0$ because $u_0 J'[u_0] = 0$, $u_\alpha \geqslant 0$ and $J'[u_0] \geqslant 0$. This property $J[u_\alpha] \geqslant J[u_0]$ of every u_0, whether unique or not, is possessed by every $\hat{u}_\alpha$, since $J[u_\alpha] \geqslant J[\hat{u}_\alpha]$ by (1.31); hence $\{u_0\} \subseteq \{\hat{u}_\alpha\}$.

Hence $\{u_0\} \equiv \{\hat{u}_\alpha\}$, which proves (a).

Proof of (b)

To prove that $\{\hat{u}_\beta\} \subseteq \{u_0\}$, introduce a new variable v whose values are those of the gradient of $J[u]$, i.e. $v = J'[u]$. This mapping is piecewise C^1 and strictly monotonic (because $J[u]$ itself is piecewise C^2 and strictly convex), and by the strengthened inverse function theorem there is a globally valid piecewise C^1 and strictly monotonic inverse $u = u(v)$ such that $du/dv = 1/J''$.

We use this inverse to construct a new function

$$G[v] = vu(v) - J[u(v)] \quad \text{such that} \quad u = \frac{dG}{dv} \tag{1.33}$$

by the chain rule. Therefore $G[v]$ will also be piecewise C^2 and, by (1.5) in the form $(u_+ - u_-)(v_+ - v_-) > 0$, strictly convex. If there is an isolated zero of J'' it does not invalidate $u = \mathrm{d}G/\mathrm{d}v$, as we see by treating the latter as a coalescence of equal limits from either side.

This transformation converts problems III and I for $J[u]$ exactly into problems II and I for $G[v]$. The argument used to prove the first part of (a) can now be repeated verbatim in the new variables, to give $\{\hat{v}_\beta\} \subseteq \{v_0\}$. Strict monotonicity of $u(v)$ allows us to convert back to the original variables, giving $\{\hat{u}_\beta\} \subseteq \{u_0\}$.

To prove the converse, that $\{u_0\} \subseteq \{\hat{u}_\beta\}$, insert the particular choices $u_+ = u_0$ and $u_- = u_\beta$ into (1.4) for $J[u]$. Since $u_0 \geqslant 0$ and $J'[u_\beta] \geqslant 0$, we have $J[u_0] - J[u_\beta] + u_\beta J'[u_\beta] > u_0 J'[u_\beta] \geqslant 0$. This property $J[u_0] - u_0 J'[u_0] \geqslant J[u_\beta] - u_\beta J'[u_\beta]$ of u_0 is possessed by every $\hat{u}_\beta$, by (1.32); hence $\{u_0\} \subseteq \{\hat{u}_\beta\}$. Hence $\{u_0\} \equiv \{\hat{u}_\beta\}$.

The uniqueness of u_0 is established by the contradiction which follows from (1.5) for $J[u]$ by supposing that u_+ and u_- there are distinct solutions of (1.30). □

Fig. 1.9 is a geometrical illustration of the equivalence of problems I and III. Consider a piecewise C^2 function $J[u]$ whose strict convexity is assured by $J'' > 0$, with $J'' = 0$ permitted at isolated places. Then $(J - uJ')' = -uJ''$ has the sign of $-u$ except at those isolated places, and $(J - uJ')'' = -J'' - uJ'''$ can have any sign because J''' is unspecified. Therefore $J - uJ'$ is a decreasing function of u in $u \geqslant 0$ with a curvature which might oscillate in sign, as suggested in Fig. 1.9; i.e. it is monotonically decreasing except for isolated horizontal inflexions. The function is certainly stationary at $u = 0$, and it has the same value as $J[u]$ at $u = 0$ *and* where $J'[u] = 0$. Since

Fig. 1.9. Objective functions in problems II and III when $J[u]$ is strictly convex.

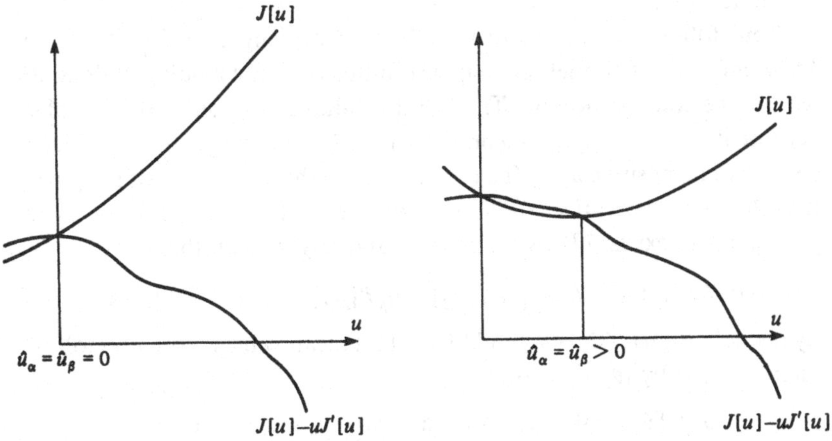

maximization is to be carried out over points where $J'[u] \geqslant 0$, Fig. 1.9 supports the equivalence of I and III.

The transformation from $J[u]$ to $G[v]$, with the properties

$$J+G=uv, \quad v=\frac{dJ}{du}, \quad u=\frac{dG}{dv}, \tag{1.34}$$

is our first encounter with what is called a *Legendre transformation.* $J[u]$ and $G[v]$ are called Legendre dual functions, or Legendre transforms of each other. In general these transformations are not confined either to convex or to C^1 functions, and we shall make a detailed study of them in Chapter 2. The example needed here possesses an unusual symmetry, in that a strictly convex and piecewise C^2 $J[u]$ leads to a strictly convex and piecewise C^2 $G[v]$, and vice versa. We shall see that it is more typical for there to be some asymmetry in convexity and in smoothness properties of Legendre dual functions, notwithstanding the symmetry in the defining equations, as in Exercise 1.3.1.

We may say that the strict convexity of $G[v]$, which follows by (1.5) from that of $J[u]$, can be illustrated by saying that the change of variable from u to v has the effect of smoothing out the wrinkles in $J[u]-uJ'[u]$ in Fig. 1.9 by the mechanism of (1.8).

It will be noticed that upper and lower bounds

$$J[u_\alpha] \geqslant J[u_0] \geqslant J[u_\beta]-u_\beta J'[u_\beta]$$

are much easier to prove than the full equivalences above, and require $J[u]$ only to be a weakly convex C^1 function. Exercise 1.3.1 illustrates why weak convexity is inadequate to ensure the equivalence of I and III. It shows that some nonunique solutions of III can have the property $\hat{u}_\beta J'[\hat{u}_\beta] \neq 0$ prohibited by I.

There follows an alternative method of proving that $\{\hat{u}_\beta\} \subseteq \{u_0\}$ in Theorem 1.3(b). This method may be omitted on a first reading. It does not require the construction of $G[v]$, but it is inhibited by isolated zeros of J'' because it requires $J''[u_\beta]>0$ or, if $J''[\hat{u}_\beta]=0$, that *some* even order higher derivative be positive at $\hat{u}_\beta$. To convey this method, suppose for simplicity that the $2m$th derivative $J^{(2m)}[u]$ is continuous for as large $m \geqslant 1$ as we please. There exists $\bar{u}$ between each $\hat{u}_\beta$ and any u_β such that

$$0 \leqslant J[\hat{u}_\beta]-\hat{u}_\beta J'[\hat{u}_\beta]-J[u_\beta]+u_\beta J'[u_\beta]=-(\hat{u}_\beta-u_\beta)\bar{u}J''[\bar{u}]$$

by the definition (1.32) of $\hat{u}_\beta$ and by the first mean value theorem. It follows after division by $|\hat{u}_\beta-u_\beta|$ that

$$\hat{u}_\beta J''[\hat{u}_\beta] \geqslant 0 \quad \text{or} \quad \leqslant 0 \quad \text{as} \quad u_\beta-\hat{u}_\beta \to 0+ \quad \text{or} \quad 0-$$

respectively. There also exists some $\bar{u}$ between each $\hat{u}_\beta$ and any u_β such that $J'[u_\beta] = J'[\hat{u}_\beta] + (u_\beta - \hat{u}_\beta)J''[\bar{u}]$, with $J''[\bar{u}] > 0$, except possibly at isolated locations where $J''[\bar{u}] = 0$. This property, with $J'[u_\beta] \geqslant 0$, permits only the limiting process $u_\beta - \hat{u}_\beta \to 0+$ in the case $J'[\hat{u}_\beta] = 0$. Then $\hat{u}_\beta \geqslant 0$ from above provided $J''[\hat{u}_\beta] > 0$. If $J'[\hat{u}_\beta] > 0$ we can choose $u_\beta - \hat{u}_\beta$ small enough to be of either sign, whence $\hat{u}_\beta = 0$ from above, again using $J''[\hat{u}_\beta] > 0$. Hence $\{\hat{u}_\beta\} \subseteq \{u_0\}$.

The case $J''[\hat{u}_\beta] = 0$ can also be incorporated as follows for certain strictly convex functions having extra smoothness at $\hat{u}_\beta$. The nth derivative of $J - uJ'$ is $(1-n)J^{(n)} - uJ^{(n+1)}$ for $n \geqslant 0$. Suppose $J^{(2m)}[\hat{u}_\beta] > 0$ for *some* $m > 1$. All odd order lower derivatives will be zero at $\hat{u}_\beta$ by the strict convexity. The proof of $\{\hat{u}_\beta\} \subseteq \{u_0\}$ then begins with a higher mean value theorem. Division by $|\hat{u}_\beta - u_\beta|^{2m-1}$ permits the same argument and conclusions as before, except that $J^{(2m)}[\hat{u}_\beta] > 0$ replaces $J''[\hat{u}_\beta] > 0$. The example $J[u] = \frac{1}{4}(u-a)^4$ with $a > 0$ illustrates this, and a sketch quickly shows that $u_0 = \hat{u}_\alpha = \hat{u}_\beta = a$. Exercise 1.3.2 illustrates a case for which *no* finite m exists with the above property and yet $\{\hat{u}_\beta\} \subseteq \{u_0\}$ is still true. This alternative method of proof fails in this example, but the method via the construction of $G[v]$ does not.

Suppose problem II is replaced by the simpler problem: minimize a given C^1 function $J[u]$ on the whole real line, without the restriction that u be positive. It may be asked: how should problems I and III be redefined so that an analogue of Theorem 1.3 applies? Let I be redefined by replacing the trio (1.30) by the single equation $J'[u] = 0$. Then Theorem 1.3(a) applies with u_α now unrestricted, so that the α subscript can now be dropped. The proof that $\{u_0\} \equiv \{\hat{u}\}$ applies *a fortiori* to establish the assumption used in proving Theorem 1.1(e). Thinking of $J'[u] = 0$ as labelled (β), let III be replaced by: maximize $J[u] - uJ'[u]$ subject to $J'[u] = 0$. This constraint has a unique solution when $J[u]$ is strictly convex, so although Theorem 1.3(b) now applies, the maximization is degenerate.

Exercises 1.3

1. Sketch the weakly convex piecewise C^2 function

$$J[u] = \begin{cases} \frac{1}{2}(u^2+1) , & u \geqslant 1, \\ u , & -1 \leqslant u \leqslant +1, \\ \frac{1}{2}(u^2+4u+1), & u \leqslant -1, \end{cases}$$

Show that maximizing points of $J[u] - uJ'[u]$ with respect to u, subject to $J'[u] \geqslant 0$, exist with the property $uJ'[u] \neq 0$. (This illustrates the need for strict rather than weak convexity in Theorem 1.3(b).) Explain why the change of variable $v = J'[u]$ is not uniquely invertible, but in spite of this construct and sketch the Legendre transform $G[v]$ defined by (1.33), and show that this is a strictly convex piecewise C^1 function with a unique minimum with respect to v at the slope discontinuity. This illustrates asymmetry in convexity and smoothness of $J[u]$ and $G[v]$.

2. Sketch the strictly convex C^2 function

$$J[u] = \begin{cases} \exp(-\tfrac{3}{2}) + k[u - 2(2/3)^{\frac{1}{2}}] + \tfrac{1}{3}[u - 2(2/3)^{\frac{1}{2}}]^3, & u \geqslant 2(2/3)^{\frac{1}{2}}, \\ \exp\{-[u - (2/3)^{\frac{1}{2}}]^{-2}\}, & 0 \leqslant u \leqslant 2(2/3)^{\frac{1}{2}}, \\ \exp(-\tfrac{3}{2}) - ku - \tfrac{1}{3}u^3, & u \leqslant 0, \end{cases}$$

where $k = 2(3/2e)^{\frac{3}{2}}$. Find the points where $J'' = 0$. Sketch the function $J[u] - uJ'[u]$ of u. Hence show that $u_0 = \hat{u}_\alpha = \hat{u}_\beta = (2/3)^{\frac{1}{2}}$, thus verifying that Theorem 1.3 includes this case even though every derivative of $J[u]$ is zero at $\hat{u}_\beta$. Sketch the Legendre transform $G[v]$ defined by (1.33), and verify that it is also strictly convex, notwithstanding the isolated locations where $J'' = 0$ and $du/dv = d^2G/dv^2 = \infty$.

3. Sketch the strictly convex C^2 function $J[u] = +(1 + u^2)^{\frac{1}{2}}$. Sketch its Legendre transform $G[v]$ having the properties (1.33), by showing that $G[v]$ is strictly convex, lies within the rectangle $-1 < v < +1$, $-1 \leqslant G < 0$ and has vertical tangents as $v \to \pm 1$.

4. Sketch non-convex C^1 functions $J[u]$ in $0 \leqslant u < \infty$ each of which has a global minimum point where

$$J' \geqslant 0, \quad u \geqslant 0, \quad uJ' = 0.$$

Consider both cases where the minimum occurs either (a) on the boundary or (b) in the interior.

5. Sketch nonconvex C^1 functions $J[u]$ in $0 \leqslant u < \infty$ which possess points (say u_1), either internal or on the boundary or both, which satisfy the variational inequality $(u - u_1)J'[u_1] \geqslant 0 \ \forall u$ in $0 \leqslant u < \infty$.

(ii) An inequality problem generated by a saddle function

Here we combine the ideas described for stationary value problems generated by saddle functions in §1.1(iv), with those for inequality problems generated by convex functions in the previous subsection. First we give an elementary discussion of one case in three dimensional space, and then in §1.3 we introduce generalizations to higher dimensions.

Let $L[x, u]$ be a given C^1 function which generates the following problems I, II and III. They are variants of those in §1.1(iv), motivated by the structure of those in §1.2(i). In the first instance, the problems can be posed regardless of whether $L[x, u]$ is a saddle function or not.

I. Let x_0, u_0 be any point which satisfies all the six conditions

$$\begin{aligned} \frac{\partial L}{\partial x} &\leqslant 0, \quad (\alpha) \\ x &\geqslant 0, \quad (\beta) \\ x\frac{\partial L}{\partial x} &= 0, \\ \frac{\partial L}{\partial u} &\geqslant 0, \quad (\beta) \\ u &\geqslant 0, \quad (\alpha) \\ u\frac{\partial L}{\partial u} &= 0. \end{aligned} \tag{1.35}$$

Find the set $\{x_0, u_0\}$, say, of such points.

Here x_0, u_0 will be a stationary point if both $x_0 > 0$ and $u_0 > 0$, but not necessarily otherwise. The set of conditions (1.35) has been divided into three subsets, one labelled (α), another labelled (β), and a pair of unlabelled orthogonality conditions. The reasons for this particular subdivision will become clear as we proceed. Representatives of (1.2), of (1.30) and of (1.35) in mechanics, for example, often require the (α) subset to consist entirely of kinematical conditions (e.g. compatibility) and the (β) subset to consist entirely of force conditions (e.g. equilibrium), or vice versa. The historical origin of trios of conditions like that in (1.30) and the two trios in (1.35) can be found independently in, for example, nonlinear programming and the plasticity of metals.

II. Let x_α, u_α be any point which satisfies (1.35α) but not necessarily the other four conditions in (1.35). Among the set $\{x_\alpha, u_\alpha\}$ of such points find the set $\{\hat{x}_\alpha, \hat{u}_\alpha\}$ which

$$\text{minimize } L - x\frac{\partial L}{\partial x}.$$

III. Among the set $\{x_\beta, u_\beta\}$ which satisfy (1.35β) but not necessarily the other four conditions, find the set $\{\hat{x}_\beta, \hat{u}_\beta\}$ which

$$\text{maximize } L - u\frac{\partial L}{\partial u}.$$

As before, these three problems will be said to be equivalent if and only if their solution sets $\{x_0, u_0\}$, $\{\hat{x}_\alpha, \hat{u}_\alpha\}$ and $\{\hat{x}_\beta, \hat{u}_\beta\}$ are the same. It will appear in §2.7 why it is no accident that the shape of the expressions in II and III closely resemble the form of a Legendre transformation (1.33).

The definition of a stationary value problem IV given in §1.1(iv) enters unchanged at this point, but it receives less emphasis in what follows because of the other possibilities permitted by the inequalities in (1.35). It is enough to recall, as we already have, that points which satisfy $\partial L/\partial x = \partial L/\partial u = 0$ are stationary, and that there are no other stationary points.

Subscripts $+, -, 0, \alpha, \beta$ will in future be used to denote values of *functions* of the coordinates, as well as coordinate values themselves. For example, the weak saddle inequality (1.24) with (1.23) can be rewritten

$$S = L_+ - L_- - (x_+ - x_-)\frac{\partial L}{\partial x}\bigg|_+ - (u_+ - u_-)\frac{\partial L}{\partial u}\bigg|_- \geqslant 0. \qquad (1.36)$$

Theorem 1.4 (Simultaneous upper and lower bounds)
Suppose that (1.35) *is generated by a single valued* C^1 *function* $L[x,u]$ *which is a weak saddle function on its domain in the sense that* (1.36) *holds for every pair of points there. Then*

$$L_\alpha - x_\alpha \frac{\partial L}{\partial x}\bigg|_\alpha \geqslant L_0 \geqslant L_\beta - u_\beta \frac{\partial L}{\partial u}\bigg|_\beta . \qquad (1.37)$$

Proof of upper bound.
The particular choices

$$x_+ = x_\alpha, u_+ = u_\alpha \quad \text{and} \quad x_- = x_0, u_- = u_0 \qquad (1.38)$$

inserted into (1.36) imply

$$L_\alpha - x_\alpha \frac{\partial L}{\partial x}\bigg|_\alpha - L_0 \geqslant - x_0 \frac{\partial L}{\partial x}\bigg|_\alpha + u_\alpha \frac{\partial L}{\partial u}\bigg|_0 \geqslant 0.$$

Proof of lower bound.
The particular choices

$$x_+ = x_0, u_+ = u_0 \quad \text{and} \quad x_- = x_\beta, u_- = u_\beta \qquad (1.39)$$

inserted into (1.36) imply

$$L_0 - L_\beta - u_\beta \frac{\partial L}{\partial u}\bigg|_\beta \geqslant - x_\beta \frac{\partial L}{\partial x}\bigg|_0 + u_0 \frac{\partial L}{\partial u}\bigg|_\beta \geqslant 0. \qquad \square$$

A simpler version of the same method of proof can be used for the stationary value problem (1.2), because there both $\partial L/\partial x|_\alpha = 0$ and $\partial L/\partial u|_\beta = 0$. The bounds (1.37) reduce to $L_\alpha \geqslant L_0 \geqslant L_\beta$, except that the hypothesis now is the weak saddle inequality (1.36), instead of the previous strict inequalities (1.10) which facilitated explicit solution of the respective constraints. The reader may find it instructive to compare the two different methods of proof given here and via Theorem 1.1. The proof of

Theorem 1.4 can also be adapted to two other problems, whose structure is a mix of (1.2) and (1.35) (see Exercise 1.4.1).

Theorem 1.4 can be viewed as showing that the set $\{x_0, u_0\}$ belongs to the set $\{\hat{x}_\alpha, \hat{u}_\alpha\}$ of minimizers and to the set $\{\hat{x}_\beta, \hat{u}_\beta\}$ of maximizers. We shall not prove here the converses, which are necessary for full equivalence and require stricter hypotheses. Instead we prefer to illustrate the equivalence by the following explicit example. The full proofs are given in a higher dimensional context in §1.5.

(iii) A graphical illustration of equivalence

Any solution point x_0, u_0 of (1.35) may be regarded as having one of the following four possible features.

(a) $\partial L/\partial x = \partial L/\partial u = 0$. This defines a stationary point of $L[x, u]$.

(b) $\partial L/\partial x = u = 0$.

(c) $x = \partial L/\partial u = 0$.

(d) $x = u = 0$.

The last three do not give stationary points of $L[x, u]$, but we shall see that they give stationary points of *other* functions.

Here we illustrate them by specific examples, which also serve to display the idea of equivalence explicitly for the three dimensional case, so extending the graphical illustrations for functions of one variable given in Figs. 1.8 and 1.9. Consider the four quadratic saddle functions

$$L[x, u] = -\tfrac{1}{2}x^2 + 2xu + u^2 + fx + gu, \tag{1.40}$$

defined by (a) $f = -2$, $g = -4$, (b) $f = +2$, $g = -1$, (c) $f = -2$, $g = -1$, (d) $f = -2$, $g = +1$. Their principal features are the same as those shown in plan view in Fig. 1.4, except that the saddle point is now shifted from the origin to the point $x = \frac{1}{3}(f - g)$, $u = -\frac{1}{6}(2f + g)$, as shown in Figs. 1.10α and β.

Consider problem I. The first three conditions in (1.35) consist of two inequalities and an orthogonality condition. This trio, for (1.40), is represented in Figs. 1.10α and β by the thick kinked line which includes part of the axis $x = 0$. The second trio of conditions in (1.35) is similarly represented by the other thick kinked line which includes part of the axis $u = 0$. The *intersection* of these two kinked lines for each given $L[x, u]$ in (1.40) is at the unique solution of (1.35). Figs. 1.10α and β display the four possible types (a)–(d) available to this solution.

Problem II requires the minimization of

$$L - x\frac{\partial L}{\partial x} = \tfrac{1}{2}x^2 + (u + \tfrac{1}{2}g)^2 - \tfrac{1}{4}g^2, \tag{1.41}$$

Fig. 1.10α. Equivalence of solutions of problems I and II generated by

$$L[x,u] = -\tfrac{1}{2}x^2 + 2xu + u^2 + fx + gu.$$

α region: $-x + 2u + f \leqslant 0, \quad u \geqslant 0.$

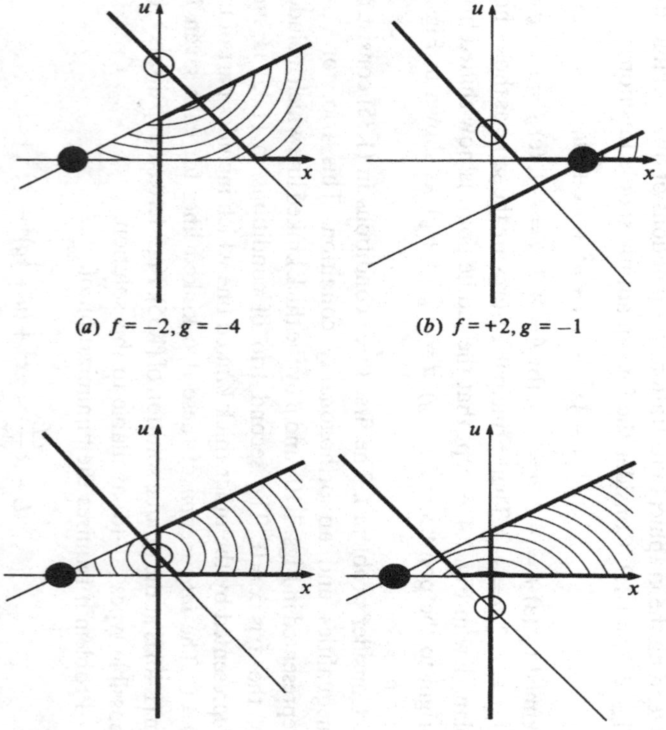

(a) $f = -2, g = -4$ (b) $f = +2, g = -1$ (c) $f = -2, g = -1$ (d) $f = -2, g = +1$

Fig. 1.10β. Equivalence of solutions of problems I and III generated by

$$L[x,u] = -\tfrac{1}{2}x^2 + 2xu + u^2 + fx + gu.$$

β region: $2x + 2u + g \geqslant 0, \quad x \geqslant 0.$

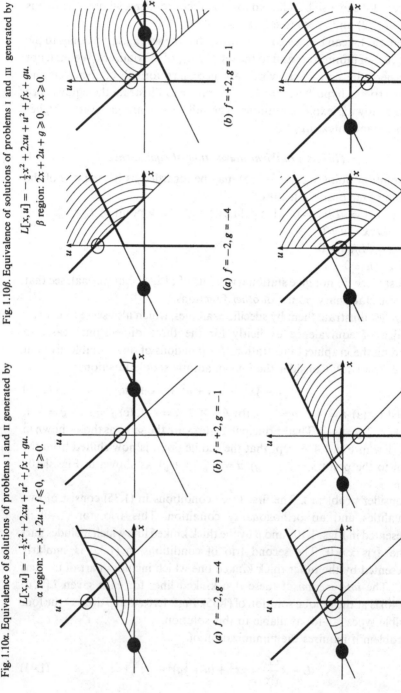

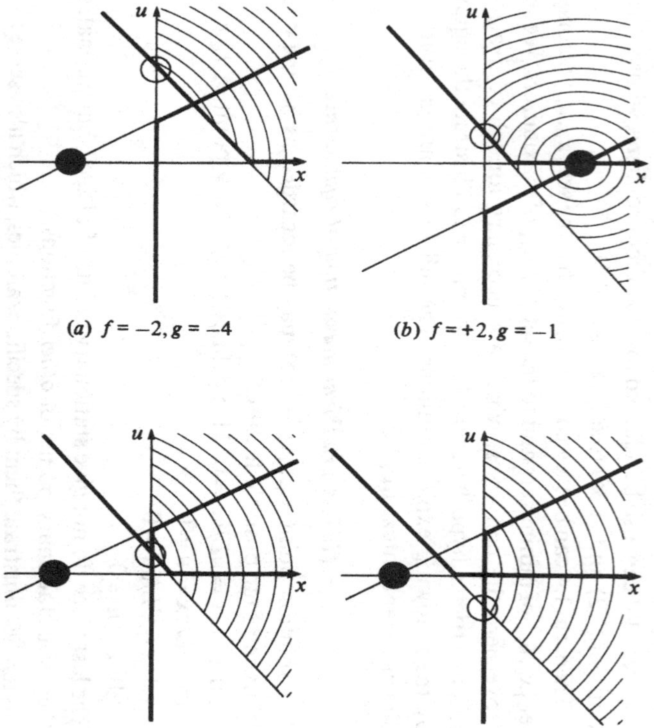

(a) $f = -2, g = -4$ (b) $f = +2, g = -1$ (c) $f = -2, g = -1$ (d) $f = -2, g = +1$

which is a convex (i.e. *bowl*-shaped) elliptical paraboloid, over the wedge-shaped α-region $0 \leqslant 2u \leqslant x - f$ defined by (1.35α) with (1.40). The bottom of the bowl is marked ○ at $x = 0$, $u = -\frac{1}{2}g$ in Figs. 1.10α and β. Fig. 1.10α shows those parts of the elliptical contours round this centre which also lie within the α-region. A study of these contours shows that the required minimizer is always *at* the above-mentioned solution of I, i.e. at the intersection of the kinked lines, for every type of solution (a)–(d). This proves equivalence of I with II in this example. Specifically, the curvature of the local contours is always *towards* the solution; types (a) and (d) have the solution on a straight edge part of the boundary to the α-region, with this α-edge *tangential* to the contours there; type (b) has the solution at a vertex on the α-boundary; type (c) has the solution interior to the α-region. These locations are summarized in Table 1.1. Every type of solution is therefore such that movement away from it in any direction into the α-region must be uphill on the bowl, so that the solution is the minimizing point for II. Evidently only type (c) is a *stationary* minimum, at the bottom of the bowl (1.41); as already stated, only type (a) is the stationary saddle point of (1.40).

Problem III requires the maximization of

$$L - u\frac{\partial L}{\partial u} = -\tfrac{1}{2}(x-f)^2 - u^2 + \tfrac{1}{2}f^2, \tag{1.42}$$

which is a concave (i.e. *dome*-shaped) elliptical paraboloid, over the β-region $x \geqslant 0$ with $2x \geqslant -(2u + g)$ defined by (1.35β) with (1.40). The top of the dome is marked ● at $x = f$, $u = 0$ in Figs. 1.10α and β. Fig. 1.10β shows those parts of the elliptical contours which also lie within the β-region. These show that I is equivalent to III, i.e. maximizing over the β-segment of the dome again finds the intersection of the kinked lines (again see Table 1.1). Every type of solution is such that movement away from it into the β-region must be downhill on the dome. Only type (b) is a *stationary* maximum of the dome (1.42).

Solution type (d), at the origin $x = u = 0$, is in fact the stationary saddle

Table 1.1. *Location of solution types for I–III with (1.40)*

	β-interior	β-edge	β-vertex
α-interior	—	—	(c) bowl-bottom
α-edge	—	(a) and (d)	—
α-vertex	(b) dome-top	—	—

point of another saddle function

$$L - x\frac{\partial L}{\partial x} - u\frac{\partial L}{\partial u} = \tfrac{1}{2}x^2 - 2xu - u^2, \tag{1.43}$$

which can be regarded as a *saddle function complementary* to (1.40). Every solution type (a)–(d), whether it be interior to or on the boundary of an α- or β-region, can therefore be regarded as the stationary point of at least one of the four functions (1.40)–(1.43).

The reader will find it helpful to repeat this study of equivalence for the cases in Exercise 1.4.1. Hyperbolic contours can there be used as an alternative to the ellipses in both or either of II and III. Also, this geometrical approach to the stationary value problem (1.2) can be compared with the analytical one given in §1.1(v).

Equivalence for *nonquadratic* saddle functions can also be studied by showing that the solution types (b), (c) and (d) are stationary points of a dome, bowl and complementary saddle respectively. Bowl and dome might no longer be convex and concave respectively (compare Fig. 1.9), but they can be rendered so if we change certain variables to gradient variables. This requires the Legendre transformation explored in detail in Chapter 2, and already broached in (1.33).

Exercises 1.4

1. Repeat the statement and proof of Theorem 1.4 with (1.35) replaced by

$$\frac{\partial L}{\partial x} = 0 \quad (\alpha), \quad \frac{\partial L}{\partial u} \geqslant 0 \quad (\beta), \quad u \geqslant 0 \quad (\alpha), \quad u\frac{\partial L}{\partial u} = 0;$$

and then by

$$\frac{\partial L}{\partial x} \leqslant 0 \quad (\alpha), \quad x \geqslant 0 \quad (\beta), \quad x\frac{\partial L}{\partial x} = 0, \quad \frac{\partial L}{\partial u} = 0 \quad (\beta).$$

2. Demonstrate geometrically the equivalence of problems I, II and III for the first problem in Exercise 1, by considering elliptical contours of $L - x(\partial L/\partial x)$ and $L - u(\partial L/\partial u)$ in the x, u plane, for the two saddle functions

$$L = -\tfrac{1}{2}x^2 + 2xu + u^2 + fx + gu,$$

with (a) $f = -2, g = -4$, (b) $f = +2, g = -1$.
Demonstrate the equivalence of I and II alternatively by considering the hyperbolic contours of L itself.

3. Is there any sense in which you would describe the function

$$L[x, u] = x^3 - 3xu^2$$

of two scalar variables as a saddle function? Sketch the function.

1.3. Transition to higher dimensions

(i) Notation

In the remainder of this chapter, and in Chapter 2, we shall deal with scalar functions $L[x_1, x_2, \ldots, x_n, u_1, u_2, \ldots, u_m]$ of $n+m$ real scalar variables. These variables are divided into two sets which we denote by

$$\begin{aligned} x &= \{x_1, x_2, \ldots, x_n\} = \{x_i : i = 1, \ldots, n\}, \\ u &= \{u_1, u_2, \ldots, u_m\} = \{u_\gamma : \gamma = 1, \ldots, m\}. \end{aligned} \tag{1.44}$$

Typical members of the sets will usually be written x_i or x_j, and u_γ or u_δ (*never* u_α or u_β; subscripts α and β are reserved for those u which satisfy α and β conditions respectively, as before). The definitions (1.44) allow us to write $L[x, u]$ as an abbreviation, designed to coincide with the usage of the preceding two sections if and only if $n = m = 1$.

The function $L[x, u]$ will always be single valued in this chapter, in accordance with a common convention for the word 'function'. In Chapter 2, however, this convention will be dropped to allow us to speak of multivalued functions which consist of two or more single valued parts.

It can be convenient, especially in quadratic expressions, to use matrix notation as an alternative to the set notation of (1.44). Column matrices are denoted by bold face symbols such as

$$\begin{aligned} \mathbf{x} &= [x_1 \quad x_2 \quad \cdots \quad x_n]^{\mathrm{T}}, \\ \mathbf{u} &= [u_1 \quad u_2 \quad \cdots \quad u_m]^{\mathrm{T}}. \end{aligned} \tag{1.45}$$

The superscript T will indicate the transpose of a matrix.

The theoretical structure of the infinite dimensional results which we begin to describe in Chapter 3 will be found to have many similarities with the finite dimensional theory. To emphasize these similarities and facilitate the comparison, we introduce an inner product notation here. Let $\mathbf{x}$ and $\mathbf{y}$ be any two $n \times 1$ matrices, with x and y as the alternative notation for the sets of their components, respectively. Let $\mathbf{u}$ and $\mathbf{v}$ be any two $m \times 1$ matrices, with u and v denoting the sets of their components. Define the two different *inner products*

$$(x, y) = \mathbf{x}^{\mathrm{T}}\mathbf{y} = x_i y_i, \quad \langle u, v \rangle = \mathbf{u}^{\mathrm{T}}\mathbf{v} = u_\gamma v_\gamma. \tag{1.46}$$

We shall use the convention of summation over the full range of a repeated suffix. It is easy to verify that (1.46) satisfies the standard axioms for an inner product (listed in §3.2(ii)). The central role of two distinct inner product spaces in theories of extremum principles was recognized by Noble and Sewell in a survey article (1972).

A C^1 function $L[x, u]$ is continuous and all its individual first partial derivatives $\partial L/\partial x_i$ and $\partial L/\partial u_\gamma$ exist uniquely. This does not mean that $L[x, u]$ could not be smoother; the point here is simply that there are a number of situations in which no mention is required of derivatives higher than the first, and so it is unnecessary to be specific about the continuity of such higher derivatives. We can regard these first derivatives as assembled into sets denoted by

$$\frac{\partial L}{\partial x} = \left\{\frac{\partial L}{\partial x_i} : i = 1, \ldots, n\right\}, \quad \frac{\partial L}{\partial u} = \left\{\frac{\partial L}{\partial u_\gamma} : \gamma = 1, \ldots, m\right\}, \tag{1.47}$$

or into column matrices denoted by

$$\frac{\partial L}{\partial \mathbf{x}} = \left[\frac{\partial L}{\partial x_1} \;\; \frac{\partial L}{\partial x_2} \;\; \cdots \;\; \frac{\partial L}{\partial x_n}\right]^{\mathrm{T}}, \quad \frac{\partial L}{\partial \mathbf{u}} = \left[\frac{\partial L}{\partial u_1} \;\; \frac{\partial L}{\partial u_2} \;\; \cdots \;\; \frac{\partial L}{\partial u_m}\right]^{\mathrm{T}}. \tag{1.48}$$

We call the sets (1.47) the x- and *u-gradients* of $L[x, u]$, anticipating the terminology for the gradients of functionals in §3.4(ii).

According to the definitions (1.46) we then have, in particular,

$$\begin{aligned} \left\langle x, \frac{\partial L}{\partial x}\right\rangle &= \mathbf{x}^{\mathrm{T}}\frac{\partial L}{\partial \mathbf{x}} = x_i\frac{\partial L}{\partial x_i}, \\ \left\langle u, \frac{\partial L}{\partial u}\right\rangle &= \mathbf{u}^{\mathrm{T}}\frac{\partial L}{\partial \mathbf{u}} = u_\gamma\frac{\partial L}{\partial u_\gamma}. \end{aligned} \tag{1.49}$$

(ii) Basic equivalence problems

By generalizing to higher dimensions the problems I, II and III in §1.2(ii) and IV in §1.1(iv), we define the following four problems, each generated by a given C^1 function $L[x, u]$ of the sets of variables in (1.44).

I. Find the set $\{x_0, u_0\}$, say, of points x_0, u_0 in the $n + m$ dimensional space which satisfy all the *governing conditions*

$$\begin{aligned} &\frac{\partial L}{\partial x} \leqslant 0, \quad (\alpha) \\ &x \geqslant 0, \quad (\beta) \\ &\left\langle x, \frac{\partial L}{\partial x}\right\rangle = 0, \\ &\frac{\partial L}{\partial u} \geqslant 0, \quad (\beta) \\ &u \geqslant 0, \quad (\alpha) \\ &\left\langle u, \frac{\partial L}{\partial u}\right\rangle = 0. \end{aligned} \tag{1.50}$$

Here the inequalities are intended to apply to each member of the respective sets, i.e. to every component in the corresponding column matrix, so that in (1.50) there are $2(n + m)$ component inequalities together with two orthogonality conditions. The braces {..} will denote various sets of $n+m$ dimensional points here and below. This is a compounded usage slightly different from the standard set notation in (1.44). Such braces will never denote an inner product; neither will square brackets, which are used to enclose the arguments of a function or functional.

II. Among the set $\{x_\alpha, u_\alpha\}$ of points which satisfy the α-conditions in I but not necessarily the other conditions there, find the set $\{\hat{x}_\alpha, \hat{u}_\alpha\}$ which minimize $L-(x, \partial L/\partial x)$, so that

$$L_\alpha - \left(x_\alpha, \frac{\partial L}{\partial x}\bigg|_\alpha \right) \geqslant \hat{L}_\alpha - \left(\hat{x}_\alpha, \frac{\widehat{\partial L}}{\partial x}\bigg|_\alpha \right). \tag{1.51}$$

III. Among the set $\{x_\beta, u_\beta\}$ of points which satisfy the β-conditions in I but not necessarily the other conditions there, find the set $\{\hat{x}_\beta, \hat{u}_\beta\}$ which maximize $L-\langle u, \partial L/\partial u\rangle$, so that

$$\hat{L}_\beta - \left\langle \hat{u}_\beta, \frac{\widehat{\partial L}}{\partial u}\bigg|_\beta \right\rangle \geqslant L_\beta - \left\langle u_\beta, \frac{\partial L}{\partial u}\bigg|_\beta \right\rangle. \tag{1.52}$$

As before, the equivalence of any two of the problems I–III will mean that they have a common set of solution points. The subscript notation introduced in (1.36) is now being extended to this higher dimensional case. In particular, the circumflex or 'hat' on L and its derivatives in (1.51) or (1.52) denotes evaluation at any solution point of II or III respectively, i.e. at any *minimizer* $\hat{x}_\alpha, \hat{u}_\alpha$ or *maximizer* $\hat{x}_\beta, \hat{u}_\beta$.

IV. *Stationary value problem.* Find the set $\{\mathring{x}, \mathring{u}\}$ of points which are stationary in the same sense as was defined in §1.1(iv), except that here $\Delta = [(x-\mathring{x}, x-\mathring{x}) + \langle u-\mathring{u}, u-\mathring{u}\rangle]^{\frac{1}{2}}$. We leave it as an exercise to show that this set of points is the same as that which satisfies

$$\frac{\partial L}{\partial x} = 0 \quad (\alpha), \qquad \frac{\partial L}{\partial u} = 0 \quad (\beta), \tag{1.53}$$

taken altogether. There are a total of $n+m$ scalar equations here, and they generalize (1.2) to $n+m$ dimensions.

The problem of solving (1.53) is a variant of problem I. Two other important variants are the *mixed problems*

$$\begin{aligned} \frac{\partial L}{\partial x} &= 0, \quad (\alpha) \\ \frac{\partial L}{\partial u} &\geqslant 0, \quad (\beta) \\ u &\geqslant 0, \quad (\alpha) \\ \left\langle u, \frac{\partial L}{\partial u} \right\rangle &= 0, \end{aligned} \tag{1.54}$$

and

$$\begin{aligned} \frac{\partial L}{\partial x} &\leqslant 0, \quad (\alpha) \\ x &\geqslant 0, \quad (\beta) \\ \left(x, \frac{\partial L}{\partial x}\right) &= 0, \\ \frac{\partial L}{\partial u} &= 0, \quad (\beta) \end{aligned} \tag{1.55}$$

which generalize the problems in Exercise 1.4.1. A corresponding minimizing problem II and maximizing problem III can be set down for each of (1.53)–(1.55), and the basic equivalence question posed in each case. The same notation for solution points of I, II and III will be used in each of the four variants.

We shall examine in detail the equivalence of I, II and III in the forms (1.50)–(1.52), giving upper and lower bounds in §1.4, and converses leading to equivalence in §1.5. We shall leave the reader to make the changes which are required in those proofs to handle the three variants (1.53)–(1.55). The methods are essentially the same so that §§*1.4 and 1.5 effectively handle all four cases simultaneously*. For the variants the most significant effect on the conclusions is that some of the minima and maxima turn out to be stationary with respect to either x or u (or both in (1.53)), which in general is not true for (1.50).

Notice in passing that since the α-conditions contain the equation $\partial L/\partial x = 0$ in (1.53) and (1.54), II may be then rewritten: minimize L subject to the α-constraints, so that $L_\alpha \geqslant \hat{L}_\alpha$. We shall often find in applications of (1.53) and (1.54), however, that $L - (x, \partial L/\partial x)$ is a simpler starting expression than L, because it excludes any bilinear terms which there may be in L. Likewise, the β-conditions contain $\partial L/\partial u = 0$ in (1.53) and (1.55), and III then may be rewritten: maximize L subject to the β-constraints, so that $\hat{L}_\beta \geqslant L_\beta$. But $L - \langle u, \partial L/\partial u \rangle$ is often simpler in its details than L.

(iii) Definition of a saddle function in higher dimensions

It will be clear from §§1.1 and 1.2 that the answer to the basic question of equivalence will require that the C^1 function $L[x, u]$ be a saddle function in some sense. For the converse theorems in §1.5 we shall also require it to be C^2. Here we generalize the definition of a saddle function given in §1.1(vi) to higher dimensions.

At any fixed x, the u-domain of $L[x, u]$ is a *convex set* if, for every pair of points u_+ and u_- in the set, and for all scalars λ in $0 \leqslant \lambda \leqslant 1$, the points $u_- + \lambda(u_+ - u_-) = \lambda u_+ + (1-\lambda)u_-$ all belong to the set also. This means that the straight join of every pair of points in the set lies wholly within the set. Similarly, at any fixed u, the x-domain of $L[x,u]$ is a convex set if, for every x_+ and x_- in the set, $\mu x_+ + (1-\mu)x_-$ belongs to the set for all μ in $0 \leqslant \mu \leqslant 1$. We say that the domain of $L[x,u]$ is *rectangular* if, for every pair of points x_+, u_+ and x_-, u_- in the domain, and for all real scalars μ and λ satisfying $0 \leqslant \mu \leqslant 1$, $0 \leqslant \lambda \leqslant 1$, the two parameter family of points

$$x_\mu = \mu x_+ + (1-\mu)x_-, \quad u_\lambda = \lambda u_+ + (1-\lambda)u_- \tag{1.56}$$

all lie within the domain. This domain may be thought of as a rectangular product of two convex sets. (Subscripts μ and λ are used here in a different sense to the i and γ in (1.44) since, like α and β, they qualify the whole sets x and u rather than pick out individual scalar members of them.)

We can now define $L[x,u]$ to be a *saddle function concave in x and convex in u* if the property (1.22) given for two dimensions holds verbatim in the present $n+m$ dimensional case. The only change of substance is that a rectangular domain now means the product of two convex sets in n and in m dimensions as just described, rather than merely the product of two connected line intervals which it was before (cf. Fig. 1.5). As previously, this definition of a saddle function only requires $L[x,u]$ to be continuous and single valued, with nothing needing to be presumed about even its first derivatives.

Next, for a C^1 function, we introduce the $n+m$ dimensional generalization of the *saddle quantity* (1.23) as

$$S = L_+ - L_- - \left(x_+ - x_-, \frac{\partial L}{\partial x}\bigg|_+ \right) - \left\langle u_+ - u_-, \frac{\partial L}{\partial u}\bigg|_- \right\rangle \tag{1.57}$$

for any pair of points x_+, u_+ and x_-, u_- in the domain of $L[x,u]$. Then it follows that Theorem 1.2(a) can be repeated verbatim for this higher dimensional case. The proof employs the same steps as before, the only change being the introduction of inner products in appropriate places to handle terms whose structure is that of the last two in (1.57).

This equivalence justifies the use of property (a) in the list below as an *alternative definition* of a C^1 saddle function concave in x and convex in u.

In some cases we shall find it desirable to adopt weaker hypotheses on C^1 functions, in order to obtain certain important theorems. Properties (b)–(f) below list some of these hypotheses, which are still conveniently expressed in terms of S, even though it becomes no longer appropriate to use the term saddle function in every such case, as we indicate below.

(a) The C^1 function $L[x,u]$ is a strict or weak *saddle function, concave in x and convex in u*, if

$$S > 0 \quad \text{or} \quad S \geqslant 0, \tag{1.58}$$

respectively, for every pair of distinct points in the domain of $L[x,u]$.

(b) $S \geqslant 0$ for every distinct pair in $\{x_\alpha, u_\alpha\} \cup \{x_\beta, u_\beta\}$. The latter is the set of points which satisfy either the α-conditions or the β-conditions or both in problem I, but not necessarily the orthogonality conditions.

(c) $S \geqslant 0$ for every distinct pair in $\{x_\alpha, u_\alpha\}$.

(d) As (c), together with $S > 0$ whenever $x_+ \neq x_-$ there.

(e) $S \geqslant 0$ for every distinct pair in $\{x_\beta, u_\beta\}$.

(f) As (e), together with $S > 0$ whenever $u_+ \neq u_-$ there.

Hypotheses (b)–(f) entail pairs of distinct points chosen from regions of much more restricted extent than in (a). These regions will often not be rectangular, as was required in Theorem 1.2(a) and its higher dimensional generalization. For example, when $n = m = 1$ the α-region in (c) is a wedge in (1.50) (with curvilinear boundaries, unless $L[x,u]$ is quadratic as illustrated in Fig. 1.10α), a half-plane in (1.55), a line in (1.53) and a half-line in (1.54) (see Exercise 1.5.4). Outside such a wedge $L[x,u]$ may have any shape, and (c) does not require it to be saddle-shaped there. Likewise, off such a line $\partial L/\partial x = 0$ the function $L[x,u]$ need not be saddle-shaped, and all that (c) really requires is that a convex function (such as $J[u]$ in Theorem 1.1(a)) shall be defined by the cross-section $\partial L/\partial x = 0$ of the surface $L = L[x,u]$.

Interrelations between hypotheses (a)–(f), and some of the consequent theorems, are summarized in Fig. 1.11. The advantage of such hypotheses is that they do not require mention of second derivatives, and are phrased

Fig. 1.11. Chart of $S \geqslant 0$ hypotheses.

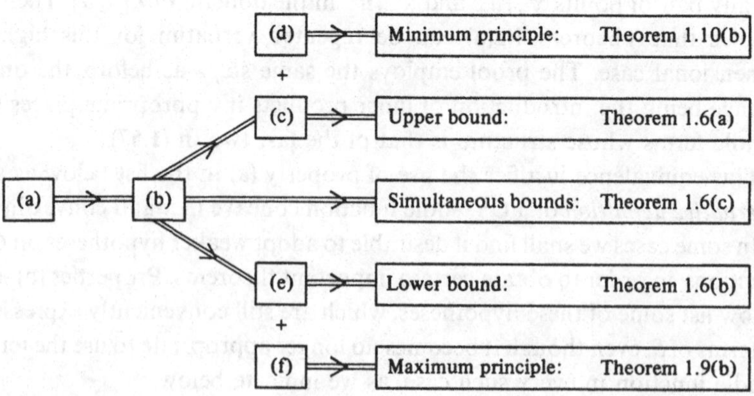

entirely in terms of (first) gradients and inner products; therefore proofs of similar structure can be expected in infinite dimensional inner product spaces later.

(iv) Second derivative hypotheses

We denote $n \times n$, $n \times m$, $m \times n$ and $m \times m$ matrices of second derivatives by

$$L_{xx} = \left[\frac{\partial^2 L}{\partial x_i \partial x_j}\right], \qquad L_{xu} = \left[\frac{\partial^2 L}{\partial x_i \partial u_\delta}\right],$$

$$L_{ux} = \left[\frac{\partial^2 L}{\partial u_\gamma \partial x_j}\right] = L_{xu}^{\mathrm{T}},\ L_{uu} = \left[\frac{\partial^2 L}{\partial u_\gamma \partial u_\delta}\right].$$

Theorem 1.5

If $L[x, u]$ is a C^2 function such that

$$-L_{xx} \text{ and } L_{uu} \text{ are positive (or nonnegative) definite} \tag{1.59}$$

everywhere in a rectangular domain, then $L[x, u]$ is a strict (or weak) saddle function in the sense of (1.58), concave in x and convex in u.

The proof follows closely that of Theorem 1.2(b), the main difference being that (1.27) is now replaced by

$$\begin{aligned} \frac{\mathrm{d}^2 l}{\mathrm{d}\lambda^2} &= (u_{\gamma_+} - u_{\gamma_-})(u_{\delta_+} - u_{\delta_-})\frac{\partial^2 L}{\partial u_\gamma \partial u_\delta}, \\ \frac{\mathrm{d}^2 m}{\mathrm{d}\mu^2} &= (x_{i_+} - x_{i_-})(x_{j_+} - x_{j_-})\frac{\partial^2 L}{\partial x_i \partial x_j}. \end{aligned} \tag{1.60}$$

We leave the reader to repeat the other steps. In Theorem 2.12 we shall invoke the result over every rectangular subdomain, which is a less restrictive hypothesis.

We now indicate the extension to higher dimensions of some ideas which were introduced in Theorem 1.1 and its proof. The condition $\partial L/\partial x = 0$ taken alone can be regarded as n scalar constraints between $n + m$ scalar variables, whose solution would in general be an m dimensional hypersurface in the $n + m$ dimensional space. We suppose that the solution has a parametric representation $x = x[u]$ such that (cf. (1.11)), when L_{xx} is nonsingular, the $n \times m$ matrix

$$[\partial x_i/\partial u_\delta] = -L_{xx}^{-1} L_{xu}$$

as a consequence of the implicit function theorem. We can then construct another function

$$J[u] = L[x[u], u] \quad \text{with} \quad \frac{\mathrm{d}J}{\mathrm{d}u} = \frac{\partial L}{\partial u}, \tag{1.61}$$

using the chain rule, whose $m \times m$ matrix of second derivatives, generalizing (1.12), is

$$J_{uu} = L_{uu} - L_{ux}L_{xx}^{-1}L_{xu}. \tag{1.62}$$

In §2.7 we extend this definition of $J[u]$ to incorporate multibranched solutions $x[u]$ of the trio $\partial L/\partial x \leqslant 0$, $x \geqslant 0$, $(x, \partial L/\partial x) = 0$, taken in place of $\partial L/\partial x = 0$ (cf. Exercise 1.5.7).

The condition $\partial L/\partial u = 0$ taken alone can similarly be supposed to have a parametric solution $u = u[x]$, when L_{uu} is nonsingular. We can use this to construct another function

$$K[x] = L[x, u[x]] \quad \text{with} \quad \frac{\mathrm{d}K}{\mathrm{d}x} = \frac{\partial L}{\partial x}, \tag{1.63}$$

whose $n \times n$ matrix of second derivatives, generalizing (1.13), is

$$K_{xx} = L_{xx} - L_{xu}L_{uu}^{-1}L_{ux}. \tag{1.64}$$

In §2.7 we extend this definition of $K[x]$ to incorporate solutions of the trio $\partial L/\partial u \geqslant 0$, $u \geqslant 0$, $\langle u, \partial L/\partial u \rangle = 0$, taken in place of $\partial L/\partial u = 0$.

For the special constraints of the type stated, we see that if one of the matrices in (1.59) is positive definite and the other nonnegative definite, the corresponding J_{uu} or $-K_{xx}$ is nonnegative definite. Fig. 1.12 summarizes

Fig. 1.12. Pointwise C^2 definiteness hypotheses and some consequences ($\Rightarrow$: 'implies', $\rightarrow$: 'is stronger than').

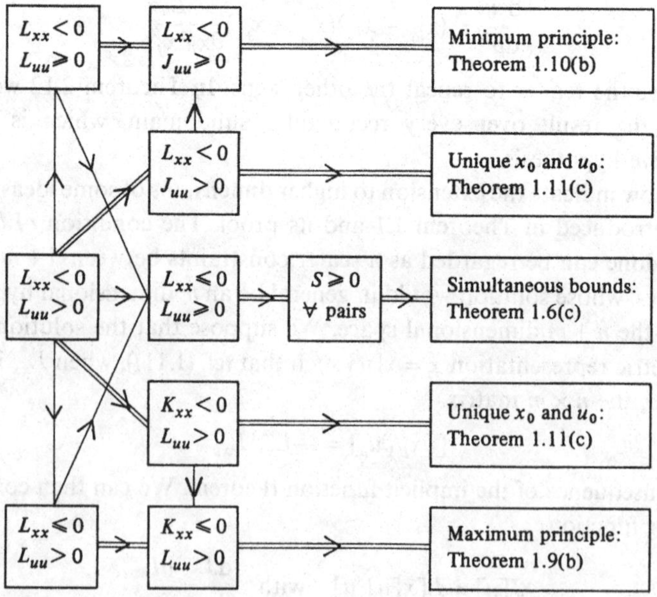

some interconnections between such pointwise hypotheses, and some theorems which they can lead to when they hold over a rectangular domain.

(v) Quadratic example

Let real matrices be given as follows: A is $n \times n$ and symmetric, B is $m \times m$ and symmetric, T is $m \times n$, T^{T} is the $n \times m$ transpose of T, $\mathbf{a}$ is $n \times 1$ and $\mathbf{b}$ is $m \times 1$. In the notation of matrices (1.45) consider

$$L[\mathbf{x},\mathbf{u}] = -\tfrac{1}{2}\mathbf{x}^{T}A\mathbf{x} + \mathbf{u}^{T}T\mathbf{x} + \tfrac{1}{2}\mathbf{u}^{T}B\mathbf{u} + \mathbf{a}^{T}\mathbf{x} + \mathbf{b}^{T}\mathbf{u}. \qquad (1.65)$$

According to the definitions (1.48) this has gradients

$$\frac{\partial L}{\partial \mathbf{x}} = -A\mathbf{x} + T^{T}\mathbf{u} + \mathbf{a}, \quad \frac{\partial L}{\partial \mathbf{u}} = T\mathbf{x} + B\mathbf{u} + \mathbf{b} \qquad (1.66)$$

in terms of which the reader can write out the three problems (1.50)–(1.52), and those for the other three variants (1.53)–(1.55). In arriving at $(1.66)_2$ we have used the transpose property that

$$\mathbf{x}^{T}T^{T}\mathbf{u} = \mathbf{u}^{T}T\mathbf{x} \quad \forall\mathbf{x} \quad \text{and} \quad \mathbf{u}. \qquad (1.67)$$

The objective functions in (1.51) and (1.52) are

$$L - \mathbf{x}^{T}\frac{\partial L}{\partial \mathbf{x}} = \tfrac{1}{2}\mathbf{x}^{T}A\mathbf{x} + \tfrac{1}{2}\mathbf{u}^{T}B\mathbf{u} + \mathbf{b}^{T}\mathbf{u}, \qquad (1.68)$$

and

$$L - \mathbf{u}^{T}\frac{\partial L}{\partial \mathbf{u}} = -\tfrac{1}{2}\mathbf{x}^{T}A\mathbf{x} - \tfrac{1}{2}\mathbf{u}^{T}B\mathbf{u} + \mathbf{a}^{T}\mathbf{x}. \qquad (1.69)$$

The saddle quantity (1.57) is

$$S = \tfrac{1}{2}(\mathbf{x}_{+} - \mathbf{x}_{-})^{T}A(\mathbf{x}_{+} - \mathbf{x}_{-}) + \tfrac{1}{2}(\mathbf{u}_{+} - \mathbf{u}_{-})^{T}B(\mathbf{u}_{+} - \mathbf{u}_{-}) \qquad (1.70)$$

for any pair of points $\mathbf{x}_{+},\mathbf{u}_{+}$ and $\mathbf{x}_{-},\mathbf{u}_{-}$ in the domain of $L[\mathbf{x},\mathbf{u}]$. This domain is the whole $n+m$ dimensional space and is therefore trivially rectangular. It follows either directly from (1.70), or from (1.59) since $L_{xx} = -A$, $L_{uu} = B$ here, that (1.65) is a weak saddle function, concave in x and convex in u, if

$$A \quad \text{and} \quad B \quad \text{are both nonnegative definite.} \qquad (1.71)$$

Then (1.68) and (1.69) are elliptic or cylindrical paraboloids, and

$$L - \mathbf{x}^{T}\frac{\partial L}{\partial \mathbf{x}} - \mathbf{u}^{T}\frac{\partial L}{\partial \mathbf{u}} = \tfrac{1}{2}\mathbf{x}^{T}A\mathbf{x} - \mathbf{u}^{T}T\mathbf{x} - \tfrac{1}{2}\mathbf{u}^{T}B\mathbf{u}$$
$$= -L + \mathbf{a}^{T}\mathbf{x} + \mathbf{b}^{T}\mathbf{u} \qquad (1.72)$$

is (as in (1.43)) another saddle function complementary to (1.65).

When A and B are positive definite, (1.68) and (1.69) are bowl- and dome-shaped elliptic paraboloids respectively, just as we found for the simpler examples (1.41) and (1.42); and so also are $J[u]$ and $K[x]$ since

$$J_{uu} = B + TA^{-1}T^{\mathrm{T}}, \quad -K_{xx} = A + T^{\mathrm{T}}B^{-1}T \tag{1.73}$$

from (1.62) and (1.64) and both are then positive definite.

The square and rectangular matrices A, B, T^{T} and T can also be regarded as operators, on the spaces X or U (say) of sets x or u respectively, with inner products (1.46), according to the scheme

$$A: X \to X, \quad B: U \to U,$$
$$T^*: U \to X, \quad T: X \to U.$$

Here and below we *rewrite* the (real) matrix T^{T} as T^* when we view it as an operator, to stimulate the comparison with more general adjoint operators in §3.3(ii). In this alternative notation we can rewrite (1.65) as

$$L[x, u] = -\tfrac{1}{2}(x, Ax) + \langle u, Tx\rangle + \tfrac{1}{2}\langle u, Bu\rangle + (a, x) + \langle b, u\rangle, \tag{1.74}$$

where a and b are the sets of n and m elements respectively of $\mathbf{a}$ and $\mathbf{b}$, and (1.67) becomes

$$(x, T^*u) = \langle u, Tx\rangle \quad \forall x \quad \text{and} \quad u. \tag{1.75}$$

This identifies the transpose T^{T} as our first example of the *adjoint* T^* of an operator, defined more generally in (3.10). We leave the reader to rewrite (1.68)–(1.70) in the alternative notation, as an exercise.

The sum $(x, y) + \langle u, v\rangle$ defines one possible inner product between the sets $\{x, u\}$ and $\{y, v\}$ in the $n + m$ dimensional space. An alternative inner product in that space is suggested by the form of (1.70) when A and B are positive definite, namely

$$(x, Ay) + \langle u, Bv\rangle. \tag{1.76}$$

This provides the point of connection between our approach and the hypercircle interpretation proposed by Synge (1957) for upper and lower bounds in linear problems (see §1.6(iv)).

(vi) Invariance in saddle definitions

The conditions (1.10) and their generalization (1.59) are *not* invariant with respect to all linear changes of the cartesian coordinates, and neither is the saddle quantity S of (1.23) and of (1.57); neither is the saddle property in (1.22) and in Exercise 1.5.5.

The advantages of using (1.58) as a definition of a C^1 saddle function concave in x and convex in u are

(a) in many applications where upper and/or lower bounds can be associated with a saddle function, the requisite decomposition of the variables into two groups corresponding to our x and u is present at the outset. In other words, no preliminary change of variables to recast a saddle function into a version which generates upper and lower bounds is needed;

(b) no mention of gradients higher than the first is required in (1.58), just as higher derivatives are not required in (1.50)–(1.55), nor in the proof of upper and lower bounds in §1.4 below.

The fortunate situation described in (a) does not apply in every context (e.g. it is not present in Exercise 1.2.3 or in §4.3). We then have to be ready to perform a preliminary change of variables. To see whether this is possible in principle we need to return to a definition of a saddle function which *is* invariant with respect to linear changes of coordinates, such as (1.9) (see Exercise 1.1.4) or a generalization of it. For example, it is obvious that surfaces lying in three dimensional space can often readily be recognized as being intrinsically either saddle-shaped or bowl-shaped or dome-shaped, and such shapes are independent of any particular set of cartesian coordinates used to describe them. In seeking a generalization of (1.9) to surfaces lying in $n + m + 1$ dimensional space, we know that the eigenvalues of the $(n + m) \times (n + m)$ Hessian matrix

$$\begin{bmatrix} L_{xx} & L_{xu} \\ L_{ux} & L_{uu} \end{bmatrix} \tag{1.77}$$

will be invariant with respect to linear changes of coordinates. The intrinsic local shape of such a surface could therefore be categorized in terms of the number of positive and of negative eigenvalues. For example, the intrinsic definition of a saddle function could be that (1.77) has both positive and negative eigenvalues (whether or not there should also be no zero eigenvalues we need not pursue here). The criterion (1.9) says that the product of the only two eigenvalues there must be negative.

Examining sign-definiteness of the quadratic from associated with (1.77) is an alternative way of categorizing $L[x, u]$ intrinsically. Positive-definite, negative-definite and indefinite correspond to bowl-, dome- and saddle-shaped functions respectively. Transformation to 'normal coordinates' (as they would be called in the linear vibration theory analogue) would associate the set of positive eigenvalues with a convex part of the surface and the set of negative eigenvalues with a concave part of the surface.

Another facet of invariance is the Morse lemma (see Poston and Stewart, 1976, p. 15) and its elaborations in singularity theory. The Morse lemma says, in effect, that if (1.77) has no zero eigenvalues at a stationary point of

$L[x,u]$, then a C^1 invertible change of variables exists which allows even nonquadratic $L[x,u]$ to be rewritten exactly (without any remainder term) as a sum of squares with coefficients ± 1. The number of positive (and therefore also of negative) coefficients is an intrinsic or characteristic or invariant property of the function, and it therefore categorizes the type of saddle function. In three dimensions (see Exercise 1.1.5) there can be only one such type, the hyperbolic paraboloid whose canonical form is (1.1).

Exercises 1.5

1. Prove that, when $F[u]$ is a C^1 function over an m dimensional convex set, with gradient $F'[u]=\{\partial F/\partial u_\gamma;\gamma=1,\ldots,m\}$, the following definitions of a weakly convex function are equivalent:
   (i) $F[u_+]-F[u_-]-\langle u_+-u_-,F'[u_-]\rangle \geqslant 0 \ \forall \ u_+$ and u_- in the set,
   (ii) $F[\lambda u_+ + (1-\lambda)u_-] \leqslant \lambda F[u_+]+(1-\lambda)F[u_-] \ \forall \ u_+$ and u_- in the set, and $\forall$ real λ in $0\leqslant\lambda\leqslant 1$.

   Supply the explicit details required to establish the generalization of Theorem 1.2(a) to the $n+m$ dimensional case.
2. Prove that, when $F[u]$ is a piecewise C^2 function over an m dimensional convex set such that F_{uu} is nonnegative definite everywhere, then $F[u]$ is a weak convex function in the sense of the previous question.
3. Prove that the bilinear function $L[\mathbf{x},\mathbf{u}]=\mathbf{u}^{\mathrm{T}}T\mathbf{x}+\mathbf{a}^{\mathrm{T}}\mathbf{x}+\mathbf{b}^{\mathrm{T}}\mathbf{u}$ of the $n\times 1$ and $m\times 1$ matrices $\mathbf{x}$ and $\mathbf{u}$, with given coefficients, makes the saddle quantity $S\equiv 0$.
4. In the case $n=m=1$, compute the saddle quantity S in (1.57) for
$$L[x,u]=\tfrac{1}{2}ax^2+xu+\tfrac{1}{2}u^2-x$$
   with given a. In the stationary value problem (1.53) show that:
   (i) $\left.\begin{array}{l}a<0\Leftrightarrow S>0\\ a=0\Leftrightarrow S\geqslant 0\end{array}\right\}$ $\forall$ distinct pairs of points;
   (ii) $a<0\Leftrightarrow S>0$ $\quad\forall$ distinct pairs in $\{x_\alpha,u_\alpha\}\cup\{x_\beta,u_\beta\}$;
   (iii) $\left.\begin{array}{l}a<1\Leftrightarrow S>0\\ a=1\Leftrightarrow S=0\end{array}\right\}$ $\forall$ distinct pairs in $\{x_\beta,u_\beta\}$;
   (iv) $\left.\begin{array}{l}a<0 \text{ or } a>1\Leftrightarrow S>0\\ a=0 \text{ or } a=1\Leftrightarrow S=0\end{array}\right\}$ $\forall$ distinct pairs in $\{x_\alpha,u_\alpha\}$;

   compare these hypotheses with (1.9) and (1.10).
5. If $L[x,u]$ is a C^1 function over a rectangular domain, use Exercise 1 to prove that the definition (1.22) of a saddle function concave in x and convex in u is equivalent to the statement that
$$\mu L[x_+,u_\lambda]+(1-\mu)L[x_-,u_\lambda]\leqslant L[x_\mu,u_\lambda]\leqslant \lambda L[x_\mu,u_+]+(1-\lambda)L[x_\mu,u_-]$$
   for every pair of points of type (1.56), in the $n+m$ dimensional case.
6. In the mixed problem (1.54), suppose that (1.54α) can be solved at the outset, to permit the elimination of x over the whole domain of $L[x,u]$ rather than over

$\{x_\alpha, u_\alpha\}$ only. Prove that (1.54) can then be reduced to the m dimensional generalization of (1.30).

7. Give a statement and explicit proof of the three dimensional Theorem 1.1 when the path in (a) is replaced by one defined by the trio $\partial L/\partial x \leqslant 0, x \geqslant 0, x\partial L/\partial x = 0$, and when the path in (b) is replaced by one defined by the trio $\partial L/\partial u \geqslant 0, u \geqslant 0$, $u\partial L/\partial u = 0$. Illustrate the revised (a) by plotting the graphs of $J[u]$ and its first two derivatives when $L = \frac{1}{2}ax^2 + xu$ for $a > 0$ and for $a < 0$.

1.4. Upper and lower bounds and uniqueness

(i) Introduction

Consider problems I–III in (1.50)–(1.52). Theorems 1.6 and 1.7 now following, and their simplified versions for the variants (1.53)–(1.55), are especially easy to establish by exploiting hypotheses selected from the list (1.58a–f) expressed in terms of the saddle quantity S in §1.3(iii). This viewpoint allows the theorems to be extended readily to infinite dimensions later. Recall that $L[x, u]$ here is a given single valued C^1 function, which need not be quadratic although it may be so.

(ii) Upper and lower bounds

Theorem 1.6

(a) Upper bound only

If $S \geqslant 0$ for every pair of distinct points in $\{x_\alpha, u_\alpha\}$, then

$$L_\alpha - \left(x_\alpha, \left.\frac{\partial L}{\partial x}\right|_\alpha \right) \geqslant L_0. \tag{1.78}$$

Also $\{x_0, u_0\} \subseteq \{\hat{x}_\alpha, \hat{u}_\alpha\}$, i.e. any solution of I, *whether unique or not, is a minimizer in* II.

(b) Lower bound only

If $S \geqslant 0$ for every pair of distinct points in $\{x_\beta, u_\beta\}$, then

$$L_0 \geqslant L_\beta - \left\langle u_\beta, \left.\frac{\partial L}{\partial u}\right|_\beta \right\rangle. \tag{1.79}$$

Also $\{x_0, u_0\} \subseteq \{\hat{x}_\beta, \hat{u}_\beta\}$, i.e. any solution of I *is a maximizer in* III.

(c) Simultaneous upper and lower bounds

If $S \geqslant 0$ for every pair of distinct points in $\{x_\alpha, u_\alpha\} \cup \{x_\beta, u_\beta\}$, then

$$L_\alpha - \left(x_\alpha, \left.\frac{\partial L}{\partial x}\right|_\alpha \right) \geqslant L_0 \geqslant L_\beta - \left\langle u_\beta, \left.\frac{\partial L}{\partial u}\right|_\beta \right\rangle. \tag{1.80}$$

Also $\{x_0, u_0\} \subseteq \{\hat{x}_\alpha, \hat{u}_\alpha\}$ and $\{x_0, u_0\} \subseteq \{\hat{x}_\beta, \hat{u}_\beta\}$, i.e. any solution of I *is both a minimizer in* II *and a maximizer in* III.

Proof

This is a direct adaptation of the two parts, taken separately and then together, of the proof of Theorem 1.4 in three dimensions but under weaker hypotheses. Only minor changes are required which the notation of §1.3(i) was devised to facilitate. We now use the definition of S in (1.57).

(a) Reference to (1.38) makes clear that only $\{x_\alpha, u_\alpha\}$ is actually deployed to prove the upper bound, since $\{x_0, u_0\} \subseteq \{x_\alpha, u_\alpha\}$. This is why a weaker hypothesis (1.58c) instead of (1.58b) ($S \geqslant 0$ over a smaller set of points) can be used to prove the individual bound in (1.78) than is required to prove the simultaneous bounds in (1.80). The quantity $\{x_0, \partial L/\partial x|_0\} = 0$ subtracted from L_0 shows that any x_0, u_0 satisfies the property (1.51) of the set of minimizers.

(b) Likewise (1.39) shows that only $\{x_\beta, u_\beta\}$ is actually used to prove the lower bound, since $\{x_0, u_0\} \subseteq \{x_\beta, u_\beta\}$. Again a (different) weaker hypothesis (1.58e) instead of (1.58b) is enough to prove the individual bound in (1.79) than is required to prove (1.80). Subtracting $\langle u_0, \partial L/\partial u|_0\rangle = 0$ from L_0 shows that any x_0, u_0 satisfies the property (1.52) of the set of maximizers.

(c) The hypothesis now is that $S \geqslant 0$ over a region just large enough to permit both choices (1.38) and (1.39) to be made in turn, giving the simultaneous bounds (1.80). In practice it may be easier to satisfy this hypothesis *a fortiori* by verifying $S \geqslant 0$ over the whole domain of $L[x, u]$ (we were content with that version in Theorem 1.4). The last result is just a combination of the last parts of (a) and (b). □

The two individual bounds in (1.78) and (1.79) are not necessarily valid simultaneously because their hypotheses are different; but each bound is stronger than the corresponding bound in the simultaneous pair (1.80), because the hypothesis for the latter is stronger ($S \geqslant 0$ over a larger set of points). Exercise 1.5.4 is an explicit illustration of the relative strengths of $S > 0$ over different sets of points. When the hypothesis for (1.80) fails, we can expect that those for (1.78) and (1.79) will be mutually exclusive; otherwise simultaneous bounds could be established indirectly, by (1.78) and (1.79) taken together, but not directly. It is the similarities in proof structure which make it economical to take the three parts of Theorem 1.6 together, and this was the viewpoint in the proofs of them given by Sewell (1982).

(iii) Example

It was easy to prove Theorem 1.6, given the saddle function viewpoint, because we deliberately separated that task from the usually harder ones of solving the α or β constraints to give explicit expression to $\{x_\alpha, u_\alpha\}$ or $\{x_\beta, u_\beta\}$ respectively. Theorem 1.6 applies to nonlinear problems, as well as to linear ones.

Straightforward illustrations of the above bounds, and of the minimum and maximum principles in §1.5, can be given in terms of the general quadratic function (1.65) of $n+m$ scalars. For this purpose we revert to the matrix notation. An upper bound will have the value of (1.68) subject to the α conditions, and a lower bound will have the value of (1.69) subject to the β conditions.

From (1.70) we see that the hypothesis of Theorem 1.6(c) is satisfied if

$$A \text{ and } B \text{ are both nonnegative definite,}$$

so that the simultaneous bounds (1.80) apply to all four problems (1.50) and (1.53)–(1.55). Explicit solution of the constraints is easier under other assumptions, however.

Consider the problem (1.53) with (1.65), i.e.

$$-A\mathbf{x} + T^{\mathrm{T}}\mathbf{u} + \mathbf{a} = 0 \quad (\alpha), \qquad T\mathbf{x} + B\mathbf{u} + \mathbf{b} = 0 \quad (\beta), \tag{1.81}$$

and also the problem (1.55) with (1.65), i.e.

$$\begin{aligned} -A\mathbf{x} + T^{\mathrm{T}}\mathbf{u} + \mathbf{a} \leqslant 0, \quad (\alpha) \\ T\mathbf{x} + B\mathbf{u} + \mathbf{b} = 0, \quad \mathbf{x} \geqslant 0, \quad (\beta) \\ \mathbf{x}^{\mathrm{T}}(-A\mathbf{x} + T^{\mathrm{T}}\mathbf{u} + \mathbf{a}) = 0. \end{aligned} \tag{1.82}$$

If the inverse B^{-1} of B exists, the set $\{\mathbf{x}_\beta, \mathbf{u}_\beta\}$ can be described explicitly as

$$\mathbf{u}_\beta = -B^{-1}(T\mathbf{x}_\beta + \mathbf{b}) \tag{1.83}$$

for all $\mathbf{x}_\beta$ in the case of (1.81), or for all $\mathbf{x}_\beta \geqslant 0$ in (1.82). Then from (1.70)

$$S = \tfrac{1}{2}(\mathbf{x}_{\beta+} - \mathbf{x}_{\beta-})^{\mathrm{T}}(A + T^{\mathrm{T}}B^{-1}T)(\mathbf{x}_{\beta+} - \mathbf{x}_{\beta-}) \tag{1.84}$$

for the difference of any pair of such $\mathbf{x}_\beta$, either in (1.81) or in (1.82). Therefore the conditions that

$$B^{-1} \text{ exists and } A + T^{\mathrm{T}}B^{-1}T \text{ is nonnegative definite} \tag{1.85}$$

are sufficient to satisfy the hypothesis of Theorem 1.6(b). Since $\partial L/\partial \mathbf{u}|_\beta = 0$ the lower bound in (1.79) can be evaluated explicitly in terms of (1.63) as $L_\beta = K[\mathbf{x}_\beta]$ for any $\mathbf{x}_\beta$ in (1.81), or for any $\mathbf{x}_\beta \geqslant 0$ in (1.82), where

$$K[\mathbf{x}] = -\tfrac{1}{2}\mathbf{x}^{\mathrm{T}}(A + T^{\mathrm{T}}B^{-1}T)\mathbf{x} + \mathbf{x}^{\mathrm{T}}(\mathbf{a} - T^{\mathrm{T}}B^{-1}\mathbf{b}) - \tfrac{1}{2}\mathbf{b}^{\mathrm{T}}B^{-1}\mathbf{b}. \tag{1.86}$$

Conditions (1.85) apply *a fortiori* if

$$B \text{ is positive definite} \quad \text{and} \quad A = 0,$$

and then the upper bound (1.78) is also available, because, from (1.70), $S \geqslant 0$

for every distinct pair of points. From (1.68) the upper bound is $\frac{1}{2}\mathbf{u}_\alpha{}^\mathrm{T} B\mathbf{u}_\alpha + \mathbf{b}^\mathrm{T}\mathbf{u}_\alpha$ for any $\mathbf{u}_\alpha$ satisfying $T^\mathrm{T}\mathbf{u}_\alpha + \mathbf{a} = 0$ in (1.81), or $T^\mathrm{T}\mathbf{u}_\alpha + \mathbf{a} \leqslant 0$ in (1.82).

(iv) Uniqueness of solution

The set of solutions x_0, u_0 of the governing conditions I has the following uniqueness properties.

Theorem 1.7

(a) Within any domain where $S > 0$ for every pair of points which have $x_+ \neq x_-$, there cannot be more than one x_0.

(b) Within any domain where $S > 0$ for every pair of points which have $u_+ \neq u_-$, there cannot be more than one u_0.

Proof

Within any domain where $S > 0$ for a set of pairs of distinct points, we can interchange the roles of plus and minus points, and add the resulting two inequalities $S > 0$ for each such pair. This gives

$$-\left(x_+ - x_-, \left.\frac{\partial L}{\partial x}\right|_+ - \left.\frac{\partial L}{\partial x}\right|_-\right) + \left\langle u_+ - u_-, \left.\frac{\partial L}{\partial u}\right|_+ - \left.\frac{\partial L}{\partial u}\right|_-\right\rangle > 0, \quad (1.87)$$

which is the generalization of (1.29).

If there were two distinct solutions of I, they could be taken as particular choices of the plus and minus points in (1.87). But the governing conditions, for example (1.50), applied to each would require the left side of (1.87) to be nonpositive (the same is true if (1.53), (1.54) or (1.55) is used in place of (1.50)).

Therefore there is a contradiction with any hypothesis for S which leads to the *strict* inequality in (1.87). By *reductio ad absurdum* it follows that x_0 (but not necessarily u_0) must be unique in case (a), and u_0 (but not necessarily x_0) must be unique in case (b). □

The smallest domains over which it might be realistic to expect to verify the hypotheses may be $\{x_\alpha, u_\alpha\}$ or $\{x_\beta, u_\beta\}$. We have illustrated the determination of $\{\mathbf{x}_\beta, \mathbf{u}_\beta\}$ in (1.83), and from (1.84) we see that the conditions that

$$B^{-1} \text{ exists and } A + T^\mathrm{T}B^{-1}T \text{ is positive definite} \quad (1.88)$$

are sufficient to ensure that $S > 0$ for every pair of points in $\{\mathbf{x}_\beta, \mathbf{u}_\beta\}$ which have $\mathbf{x}_{\beta+} \neq \mathbf{x}_{\beta-}$. Theorem 1.7(a) then generates unique $\mathbf{x}_0$, and not only for (1.81), but also for (1.82) because the extra β-inequality there implies a smaller $\{\mathbf{x}_\beta, \mathbf{u}_\beta\}$. For (1.81) we find

$$\mathbf{x}_0 = (A + T^\mathrm{T}B^{-1}T)^{-1}(\mathbf{a} - T^\mathrm{T}B^{-1}\mathbf{b}).$$

By back substitution we can see that $\mathbf{u}_0$ is also unique and find it, not because of Theorem 1.7(a) *per se*, but because of the way in which we are here able to make explicit the domain $\{\mathbf{x}_\beta, \mathbf{u}_\beta\}$ over which it applies.

(v) Embedding in concave/linear $L[x,u]$

Let $\phi[x]$ be a given scalar C^1 function of x, and let $v[x]$ be a set of m scalar C^1 functions of x. Suppose that the following optimization problem is posed: find the set of points which maximize $\phi[x]$ subject to the $m+n$ scalar constraints $v[x] \geqslant 0$ and $x \geqslant 0$.

We approach this by introducing another set u of m scalar variables, and constructing the C^1 function

$$L[x,u] = \phi[x] + \langle u, v[x] \rangle, \tag{1.89}$$

whose gradients are

$$\frac{\partial L}{\partial x} = \frac{\partial \phi}{\partial x} + \left\langle u, \frac{\partial v}{\partial x} \right\rangle, \quad \frac{\partial L}{\partial u} = v.$$

Here $\partial v/\partial x$ denotes the set of $n+m$ derivatives, typically $\partial v_\gamma/\partial x_i$. For this particular context we extend the meaning of the inner product symbol by defining $\langle u, \partial v/\partial x \rangle = u_\gamma \partial v_\gamma/\partial x$ as a set of n elements, typically $u_\gamma \partial v_\gamma/\partial x_i$. This set may be arranged as an $n \times 1$ column matrix when desired.

The function (1.89) generates the objective function as

$$L - \left\langle u, \frac{\partial L}{\partial u} \right\rangle = \phi[x] \tag{1.90}$$

and the following version of problem I from (1.50).

$$\begin{aligned} \frac{\partial \phi}{\partial x} + \left\langle u, \frac{\partial v}{\partial x} \right\rangle \leqslant 0, \quad u \geqslant 0, \quad (\alpha) \\ v[x] \geqslant 0, \quad x \geqslant 0, \quad (\beta) \\ \left\langle x, \frac{\partial \phi}{\partial x} + \left\langle u, \frac{\partial v}{\partial x} \right\rangle \right\rangle = \langle u, v \rangle = 0. \end{aligned} \tag{1.91}$$

From (1.90) and (1.91β) we can see that the given maximizing problem has thus been *embedded* as a particular version of problem III in our sequence of three problems (1.50)–(1.52). The set of maximizers could be written $\{\hat{x}_\beta\}$ with $\hat{u}_\beta$ arbitrary. The additional governing conditions which emerge in (1.91), namely (1.91α) and the orthogonality conditions, are usually called the Kuhn–Tucker conditions. The embedding procedure also induces an associated problem II, to minimize

$$L - \left\langle x, \frac{\partial L}{\partial x} \right\rangle = \phi - \left\langle x, \frac{\partial \phi}{\partial x} \right\rangle + \left\langle u, v - \left\langle x, \frac{\partial v}{\partial x} \right\rangle \right\rangle,$$

subject to (1.91α). This is much less simple than III, in which only x appears and not u.

In the original statement of the maximization problem, and in the process of embedding its constraints into an associated set of governing conditions (1.91), nothing has yet been supposed about the shape of $\phi[x]$ and $v[x]$ except that they be C^1. The purpose of the following theorem is to establish the relevance of (1.91) to the maximization problem in the sense that, under extra hypotheses on $\phi[x]$ and $v[x]$, any x which satisfies (1.91) is a maximizer of the original problem.

Theorem 1.8

Suppose that $\phi[x]$ and every member of $v[x]$ are each weakly concave functions over a common convex x-domain. Then for (1.89),

(a) $S \geqslant 0$ for every pair of distinct points in the rectangular domain which is the product of the convex x-domain with the region $u \geqslant 0$ (which is pyramidal in m dimensions);

(b) the simultaneous bounds (1.80) *apply, and in particular the stated concavities are sufficient for the lower bound property*

$$\phi[x_0] \geqslant \phi[x_\beta], \tag{1.92}$$

where x_0, u_0 is any solution of (1.91), *and x_β of* (1.91β); *also* $\{x_0\} \subseteq \{\hat{x}_\beta\}$.

Proof

(a) From (1.57), for any pair of plus and minus points in the domain of (1.89),

$$S = \phi_+ - \phi_- - \left(x_+ - x_-, \frac{\partial \phi}{\partial x}\bigg|_+ \right) + \left\langle u_+, v_+ - v_- - \left(x_+ - x_-, \frac{\partial v}{\partial x}\bigg|_+ \right) \right\rangle.$$

Evidently u_- is absent. The assumed concavities (cf. (1.6)) show that $S \geqslant 0$ for every pair of points having $u_+ \geqslant 0$. Note that we could not expect to allow $u < 0$ (unless all the functions in $v[x]$ are linear, in which case the coefficient of u_+ vanishes – see the linear programing example §1.6(i)).

(b) By (1.91α) $u_+ \geqslant 0$ holds for the two choices $u_+ = u_\alpha$ and $u_+ = u_0$ actually needed (again see (1.38) and (1.39)) to establish the bounds (1.80). The last part is immediate, as in the proof of Theorem 1.6(b). □

Exercises 1.6

1. Prove that the conditions that

$$A^{-1} \text{ exists and } B + TA^{-1}T^{\mathrm{T}} \text{ nonnegative definite}$$

are sufficient to satisfy the hypothesis of Theorem 1.6(a) for (1.81) and also for (1.54) with (1.65). Show that the upper bound in (1.78) can be evaluated explicitly in terms of (1.61) as $L_\alpha = J[\mathbf{u}_\alpha]$ for any $\mathbf{u}_\alpha$ in (1.81), and for any $\mathbf{u}_\alpha \geqslant 0$ in (1.54) with (1.65), where

$$J[\mathbf{u}] = \tfrac{1}{2}\mathbf{u}^{\mathrm{T}}(B + TA^{-1}T^{\mathrm{T}})\mathbf{u} + \mathbf{u}^{\mathrm{T}}(\mathbf{b} + TA^{-1}\mathbf{a}) + \tfrac{1}{2}\mathbf{a}^{\mathrm{T}}A^{-1}\mathbf{a}.$$

2. When the conditions in Exercise 1 are satisfied *a fortiori* because

$$A \text{ is positive definite} \quad \text{and} \quad B = 0,$$

show that the lower bound in (1.79) is also available, as $-\tfrac{1}{2}\mathbf{x}_\beta^{\mathrm{T}}A\mathbf{x}_\beta + \mathbf{a}^{\mathrm{T}}\mathbf{x}_\beta$ for any $\mathbf{x}_\beta$ satisfying $T\mathbf{x}_\beta + \mathbf{b} = 0$ in (1.81), or $T\mathbf{x}_\beta + \mathbf{b} \geqslant 0$ in (1.54) with (1.65).
   Show that simultaneous bounds are still available if

$$A \text{ is nonnegative definite} \quad \text{and} \quad B = 0,$$

and that the expression of the lower bound, but not that of the upper bound, is unaffected.
3. Write out the variant of Theorem 1.8 which is required to handle the problem of finding the set of points which maximize $\phi[\mathbf{x}] = -\tfrac{1}{2}\mathbf{x}^{\mathrm{T}}A\mathbf{x} + \mathbf{a}^{\mathrm{T}}\mathbf{x}$ subject to $T\mathbf{x} + \mathbf{b} \geqslant 0$.
4. Prove that the quadratic (1.65) has the property

$$2L = \mathbf{x}^{\mathrm{T}}\frac{\partial L}{\partial \mathbf{x}} + \mathbf{u}^{\mathrm{T}}\frac{\partial L}{\partial \mathbf{u}} + \mathbf{a}^{\mathrm{T}}\mathbf{x} + \mathbf{b}^{\mathrm{T}}\mathbf{u}.$$

Hence show that any solution $\mathbf{x}_0, \mathbf{u}_0$ of any one of the four problems (1.50) or (1.53)–(1.55) then gives to L the value

$$L_0 = \tfrac{1}{2}\mathbf{a}^{\mathrm{T}}\mathbf{x}_0 + \tfrac{1}{2}\mathbf{b}^{\mathrm{T}}\mathbf{u}_0.$$

5. Suppose that the optimization problem at the beginning of §(v) had been to find the set of points which maximize $\phi[x]$ subject to $v[x] = 0$. Prove that, if $\phi[x]$ and $v[x]$ are weakly concave over a common convex domain, any x which satisfies

$$\frac{\partial \phi}{\partial x} + \left\langle u, \frac{\partial v}{\partial x} \right\rangle = 0 \quad \text{and} \quad v[x] = 0$$

with some $u \geqslant 0$ is a maximizer of the original problem.
   When this maximization problem is replaced by a stationary value problem, with the convexity hypotheses dropped, the role of u becomes that of a Lagrange multiplier in the classical sense (see §1.5(iv)), and (1.89) may be called an augmented Lagrangian.

1.5. Converse theorems: extremum principles

(i) Introduction

In Theorem 1.6 we established, under various hypotheses of the type $S \geqslant 0$, that if $L - (x, \partial L/\partial x)$ has a minimum value subject to the α-constraints, then any solution of the governing conditions for problem I will be a minimizer; and that if $L - \langle u, \partial L/\partial u \rangle$ has a maximum value subject to the β-constraints, then any solution of I will be a maximizer.

This does not prove, however, that every solution of the constrained minimization problem, or of the constrained maximization problem, will necessarily satisfy I. It is important to be sure that no minimizer, and that no maximizer, lies outside the set of solutions x_0, u_0 of the conditions governing I. In this section we give hypotheses which allow us to prove that all minimizers satisfy I, and similarly for all maximizers. We call such results *converse theorems*, and we prove consequent theorems stating when problems I, II and III are equivalent. As in §1.4 we give the details for the problem governed by (1.50) and, except for Theorem 1.14, we leave the other variants as an exercise for the reader.

The method of proof required for fully nonlinear generating functions $L[x, u]$ turns out to be rather easier than that required for partly linear $L[x, u]$. The latter require a representation theorem which includes Farkas' lemma. For convenience we make a sharp distinction between the two cases. We assume in §1.5(ii) that $L[x, u]$ is C^2, rather than just C^1 as §1.4 required.

The reader may choose to omit the proofs in this section on a first reading, without undue disadvantage. The results given here will be referred to for particular purposes in §§1.6–1.9, but only very occasionally subsequently. Nevertheless they convey the importance of converse theorems in the theory at large. Generalizations to the infinite dimensional case of the following converse Theorems 1.9–1.14, whether or not they are available in the literature, require additional ideas which are outside the scope of this book. Perhaps the most familiar such result is the fundamental lemma of the calculus of variations, of which we give only a simple illustration in Theorem 3.8. The purpose of the present section is to indicate the theoretical importance of converse theorems by making some tangible proofs available in finite dimensions. By contrast, generalizations to the infinite dimensional case of the results in §1.4 on upper and lower bounds and uniqueness will be proved in Chapter 3, and used in Chapters 4 and 5.

We recall at this point the essence of the elementary *inverse function theorem* (e.g. see Apostol, 1974, Theorem 13.6). It states that if the C^1 function $v = v[u]$ taking an open set in $\mathbb{R}^m$ into another open set in $\mathbb{R}^m$ has a nonzero Jacobian (the determinant of the $m \times m$ matrix of partial derivatives $\partial v_\gamma/\partial u_\delta$) at some point, then in a neighbourhood surrounding that point there exists a unique inverse function $u = u[v]$ which is also

C^1, and which has the property

$$\frac{\partial v_\gamma}{\partial u_\delta}\frac{\partial u_\delta}{\partial v_\varepsilon}=\delta_{\gamma\varepsilon},$$

where $\delta_{\gamma\varepsilon}$ is the Kronecker delta. This is a local result. The inverse need not be unique globally, even when the Jacobian is nonzero at every point. For example, a negative Jacobian is possessed by the $v[u]:\mathbb{R}^2\to\mathbb{R}^2$ formed by the gradients of any function which is saddle-shaped at every point in the sense of (1.9). The global nonuniqueness is illustrated by the first derivatives (cf. Exercise 1.1.6)

$$v_1=\mathrm{e}^{u_1}\cos u_2,\quad v_2=-\mathrm{e}^{u_1}\sin u_2$$

of the saddle function $F[u_1,u_2]=\mathrm{e}^{u_1}\cos u_2$. The Jacobian here is $-\mathrm{e}^{2u_1}<0$ everywhere, but the inverse

$$u_1=\ln(v_1{}^2+v_2{}^2)^{\frac{1}{2}},\quad u_2=-\tan^{-1}\frac{v_2}{v_1}$$

has infinitely many u_2 values, differing by multiples of π, for any given v_1 and v_2.

In the course of proving Theorem 1.9(a) we shall establish *global* uniqueness of the inverse function $u=u[v]$ when its square matrix of partial derivatives is sign definite, rather than merely having a nonzero Jacobian. Under the hypothesis of a strictly convex (rather than saddle-shaped) function whose derivatives are $v[u]$, this is a global extension of the inverse function theorem which is appropriate to our purpose in Theorems 1.9(a), 1.10(a) and 1.11(a).

(ii) Extremum principles for nonlinear $L[x,u]$

Theorem 1.9 (Maximum principle)

(a) Converse to Theorem 1.6(b)

Let the C^2 function $L[x,u]$ be strictly convex with respect to u, over a rectangular domain which contains $\{x_\beta,u_\beta\}$. Then any maximizer in III *is also a solution of* I, *i.e.*

$$\{\hat{x}_\beta,\hat{u}_\beta\}\subseteq\{x_0,u_0\}.$$

(b) Maximum principle: equivalence theorem

When the hypothesis of (a) *holds and, in addition, $S\geqslant 0$ for every pair of distinct points in $\{x_\beta,u_\beta\}$, problems* I *and* III *are equivalent because they have the same set of solutions, i.e.*

$$\{x_0,u_0\}\equiv\{\hat{x}_\beta,\hat{u}_\beta\}.$$

(c) Uniqueness

When (a) and (b) hold, if the part u_0 of a solution of I *exists, it will be unique.*

If $S>0$ for every pair of distinct points in $\{x_\beta, u_\beta\}$, and if both parts x_0, u_0 of a solution of I *exists, they will be unique.*

Proof

(a) A rectangular domain is, by the definition given in relation to (1.56), the product of a convex x-domain and a convex u-domain. For every fixed x, and every distinct pair of u-values in the u-domain, the strict convexity with respect to u implies

$$\left\langle u_+ - u_-, \left.\frac{\partial L}{\partial u}\right|_+ - \left.\frac{\partial L}{\partial u}\right|_- \right\rangle > 0. \tag{1.93}$$

We introduce a new m dimensional variable v by the C^1 gradient mapping

$$v = v[x,u] = \frac{\partial L}{\partial u}. \tag{1.94}$$

(This notation is consistent with the $v[x]$ in (1.89), which is linear u). At every fixed x, the inverse mapping

$$u = u[x,v] \tag{1.95}$$

will be C^1 by the inverse function theorem. It must be a unique function of v globally, because if any pair of distinct u-points (i.e. with $u_+ \neq u_-$) existed in the convex u-domain corresponding to the same v-point (i.e. $v_+ = v_-$), this would contradict (1.93). For this reason (1.93) could be regarded as a definition of strict monotonicity of the gradient mapping (1.94).

We now use (1.95) to construct a new function

$$G[x,v] = \langle v, u[x,v]\rangle - L[x, u[x,v]]. \tag{1.96}$$

By the chain rule this will have the properties

$$u = \frac{\partial G}{\partial v}, \quad -\frac{\partial L}{\partial x} = \frac{\partial G}{\partial x}, \tag{1.97}$$

so that $G[x,v]$ will also be C^2. We have thereby constructed an m dimensional version of the Legendre transformation, first introduced for $m=1$ in (1.33), where the purpose was similar to our present one. The method of construction produces a single valued function $G[x,v]$ over the domain in x, v space into which the rectangular x, u domain has been mapped by the change of variable. Since the rectangular domain contains

$\{x_\beta, u_\beta\}$, by hypothesis, $G[x,v]$ is defined over a domain which contains the region $x \geqslant 0$, $v \geqslant 0$ (which we shall call a pyramid in $n+m$ dimensions).

Conditions (1.50) can now be rewritten

$$\frac{\partial G}{\partial x} \geqslant 0, \quad \frac{\partial G}{\partial v} \geqslant 0, \quad (\alpha)$$
$$x \geqslant 0, \quad v \geqslant 0, \quad (\beta) \qquad (1.98)$$
$$\left\langle x, \frac{\partial G}{\partial x} \right\rangle = \left\langle v, \frac{\partial G}{\partial v} \right\rangle = 0.$$

These conditions can be regarded as a higher dimensional generalization of (1.30), with the roles of (α) and (β) interchanged. Since

$$G = \left\langle u, \frac{\partial L}{\partial u} \right\rangle - L$$

in value, problem III in (1.52) can now be rewritten as follows: find the point(s) $\hat{x}_\beta$, $\hat{v}_\beta$ which minimize $G[x,v]$ among the points x_β, v_β satisfying (1.98β), so that

$$G[x_\beta, v_\beta] \geqslant G[\hat{x}_\beta, \hat{v}_\beta]. \qquad (1.99)$$

This is a generalization, from one to $n+m$ independent variables, of problem II in §1.2(i). The method used to prove the first part of Theorem 1.3(b), by converting it to the first part of the proof of Theorem 1.3(a) there, can be adapted to our present problem as follows.

The first mean value theorem for a function of $n+m$ variables, applied to $G[x,v]$, implies that there is at least one point $\bar{x}$, $\bar{v}$ (say) on the straight join of any $\hat{x}_\beta$, $\hat{v}_\beta$ to any x_β, v_β such that

$$\left\langle x_\beta - \hat{x}_\beta, \left.\frac{\partial G}{\partial x}\right|_{\bar{x},\bar{v}} \right\rangle + \left\langle v_\beta - \hat{v}_\beta, \left.\frac{\partial G}{\partial v}\right|_{\bar{x},\bar{v}} \right\rangle \geqslant 0, \qquad (1.100)$$

by (1.99).

If the minimizing point(s) exist only in the interior of the pyramid (so that $\hat{x}_\beta > 0$, $\hat{v}_\beta > 0$), straight joins to sufficiently close x_β, v_β will exist in every direction. Then we can choose $x_\beta \to \hat{x}_\beta$ and $v_\beta \to \hat{v}_\beta$ from every direction, so that by the continuity of the gradients in (1.100),

$$\frac{\partial G}{\partial x} = \frac{\partial G}{\partial v} = 0 \quad \text{at} \quad \hat{x}_\beta, \hat{v}_\beta. \qquad (1.101)$$

Every such minimizing point $\hat{x}_\beta$, $\hat{v}_\beta$ therefore satisfies *all* the conditions (1.98). It follows that every corresponding maximizer $\hat{x}_\beta$, $\hat{u}_\beta$ in III will also be a solution of I.

If $\hat{x}_\beta \geqslant 0$, $\hat{v}_\beta \geqslant 0$ with equality in (say) $r > 0$ of these $n+m$ scalar inequalities, then such a minimizing point is on the boundary of the pyramid; in fact on a face if $r = 1$, on an edge if $1 < r < n + m$, and at the vertex if $r = n + m$. Straight joins to arbitrarily close x_β, v_β on opposite sides of $\hat{x}_\beta$, $\hat{v}_\beta$ will exist parallel to $n + m - r$ axes. The foregoing limiting argument applied to (1.100) now implies that the $n + m$ dimensional gradient of G at $\hat{x}_\beta$, $\hat{v}_\beta$ has $n + m - r$ zero components and r nonnegative components, such that *all* the conditions (1.98) are again satisfied. Every corresponding maximizer $\hat{x}_\beta$, $\hat{u}_\beta$ in III will also be a solution of I.

Summarizing, regardless of whether $\hat{x}_\beta$, $\hat{v}_\beta$ is in the interior or on the boundary of the pyramid (1.98β), we have shown that

$$\{\hat{x}_\beta, \hat{u}_\beta\} \subseteq \{x_0, u_0\},$$

which proves (a).

(b) Combining the result just proved with that in Theorem 1.6(b) establishes the equivalence, regardless of whether or not the common solution set contains a unique solution point.

(c) When (a) and (b) hold, strict concavity in x is not insisted upon, and there might be $x_+ \neq x_-$, $u_+ = u_-$ for which the expression in (1.87) is zero, but Theorem 1.7(b) applies. Under the stronger final hypothesis Theorem 1.7(a) applies as well. □

As an example of this theorem we carry the discussion of §1.4(iii) further, reverting again to the matrix notation for

$$L[\mathbf{x}, \mathbf{u}] = -\tfrac{1}{2}\mathbf{x}^T A\mathbf{x} + \mathbf{u}^T T\mathbf{x} + \tfrac{1}{2}\mathbf{u}^T B\mathbf{u} + \mathbf{a}^T\mathbf{x} + \mathbf{b}^T\mathbf{u}. \qquad (1.65\text{ bis})$$

The hypothesis of Theorem 1.9(a) is satisfied *a fortiori* if

$$B \text{ is positive definite.}$$

The mapping (1.94) is

$$\mathbf{v} = T\mathbf{x} + B\mathbf{u} + \mathbf{b}$$

and manifestly now has the globally valid inverse

$$\mathbf{u} = B^{-1}(\mathbf{v} - T\mathbf{x} - \mathbf{b}),$$

so that the new function (1.96) in this case is

$$G[\mathbf{x}, \mathbf{v}] = \tfrac{1}{2}(\mathbf{v} - T\mathbf{x} - \mathbf{b})^T B^{-1}(\mathbf{v} - T\mathbf{x} - \mathbf{b}) + \tfrac{1}{2}\mathbf{x}^T A\mathbf{x} - \mathbf{a}^T\mathbf{x}.$$

The mean value theorem argument following (1.99) then applies to prove Theorem 1.9(a).

This can be illustrated most explicitly in the two cases of (1.81) and (1.82), in both of which $\mathbf{v}_\beta = 0$ because $\partial L/\partial \mathbf{u}|_\beta = 0$. In terms of (1.86) we notice that $G[\mathbf{x}, 0] = -K[\mathbf{x}]$ (this property is explained from another viewpoint in (2.100), where $G[\mathbf{x}, \mathbf{v}] = -k[\mathbf{x}, \mathbf{v}]$ and $\mathbf{v} = \partial L/\partial \mathbf{u}$). The problem (1.99) is therefore equivalent to the maximization of $K[\mathbf{x}_\beta]$, for all $\mathbf{x}_\beta$ in (1.81) and for all $\mathbf{x}_\beta \geqslant 0$ in (1.82). The mean value argument proves that the set of maximizers $\{\hat{\mathbf{x}}_\beta\}$ satisfy the governing conditions, as do the corresponding $\hat{\mathbf{u}}_\beta = -B^{-1}(T\hat{\mathbf{x}}_\beta + \mathbf{b})$.

Turning now to the maximum principle in Theorem 1.9(b), the extra hypothesis required by this was illustrated in (1.85). Therefore if

$$B \text{ is positive definite and } A + T^{\mathrm{T}}B^{-1}T \text{ is nonnegative definite} \tag{1.102}$$

the maximum principle applies to both (1.81) and (1.82). That is, those $\hat{\mathbf{x}}_\beta$ which maximize the $K[\mathbf{x}_\beta]$ of (1.86) are the same as those $\mathbf{x}_0$ which satisfy the governing conditions. If both

$$B \text{ and } A + T^{\mathrm{T}}B^{-1}T \text{ are positive definite,}$$

then Theorem 1.9(c) shows that $\mathbf{x}_0$ is unique as well as $\mathbf{u}_0$.

Theorem 1.10 (Minimum principle)

(a) Converse to Theorem 1.6(a)

Let the C^2 function $L[x,u]$ be strictly concave with respect to x, over a rectangular domain which contains $\{x_\alpha, u_\alpha\}$. Then any minimizer in II *is also a solution of* I, *i.e.*

$$\{\hat{x}_\alpha, \hat{u}_\alpha\} \subseteq \{x_0, u_0\}.$$

(b) Minimum principle: equivalence theorem

When the hypothesis of (a) *holds and, in addition $S \geqslant 0$ for every pair of distinct points in $\{x_\alpha, u_\alpha\}$, problems* I *and* II *are equivalent because they have the same set of solutions, i.e.*

$$\{x_0, u_0\} \equiv \{\hat{x}_\alpha, \hat{u}_\alpha\}.$$

(c) Uniqueness

When (a) *and* (b) *hold, if the part x_0 of a solution of* I *exists, it will be unique. If $S > 0$ for every pair of distinct points in $\{x_\alpha, u_\alpha\}$, and if both parts x_0, u_0 of a solution of* I *exists, they will be unique.*

The reader will be able to prove Theorem 1.10 by the obvious rephrasing of the proof of Theorem 1.9. An example is provided by carrying the discussion of Exercise 1.6.1 further, under the extra assumption that A be positive definite to satisfy the hypothesis of Theorem 1.10(a).

Theorem 1.11 (Simultaneous maximum and minimum principles)

Let the C^2 function $L[x,u]$ satisfy $S > 0$ for every pair of distinct points in a rectangular domain which includes $\{x_\alpha, u_\alpha\} \cup \{x_\beta, u_\beta\}$.

(a) Converse to Theorem 1.6(c)

Any minimizer in II *is also a solution of* I, *and so is any maximizer in* III, *i.e.*

$$\{\hat{x}_\alpha, \hat{u}_\alpha\} \subseteq \{x_0, u_0\} \quad \textit{and} \quad \{\hat{x}_\beta, \hat{u}_\beta\} \subseteq \{x_0, u_0\}.$$

(b) Simultaneous extremum principles: equivalence theorem

Problems I, II *and* III *are equivalent because they have the same set of solutions, i.e.*

$$\{x_0, u_0\} \equiv \{\hat{x}_\alpha, \hat{u}_\alpha\} \equiv \{\hat{x}_\beta, \hat{u}_\beta\}.$$

(c) Uniqueness

If such a common solution exists, it is unique.

The hypothesis of Theorem 1.11 ensures that $L[x, u]$ be a strict saddle function in the sense of (1.58). The reader will again be able to supply the proofs, which require just a combination of the proofs of Theorems 1.9 and 1.10.

The saddle quantity (1.23) for the general quadratic (1.15) in just two scalar variables x and u is

$$S = -\tfrac{1}{2}a(x_+ - x_-)^2 + \tfrac{1}{2}b(u_+ - u_-)^2.$$

When $a < 0$ and $b > 0$ the hypothesis of Theorem 1.11 is satisfied. An explicit illustration of these simultaneous maximum and minimum principles can now be recognized in our demonstration in §1.2(iii), for the particular quadratics (1.40), of the common solutions of problems I, II and III. Since $a = -1$ and $b = 2$ there, Theorem 1.11 guarantees the equivalence of the three problems in advance, as depicted in Figs. 1.10α and 1.10β.

This simple example can also be used to illustrate explicitly the weaker hypotheses of Theorem 1.9(a) or of Theorem 1.10(a), and the reader may find it helpful to work out the full details leading to the maximum principle in one case (as we have already indicated in $n + m$ dimensions after the proof of Theorem 1.9), and to the minimum principle in the other case.

(iii) Maximum principle by embedding

We now return to the problem of maximizing $\phi[x]$ subject to $v[x] \geqslant 0$ and $x \geqslant 0$, previously posed at the beginning of §1.4(v). We wish to prove here that its maximizing set $\{\hat{x}_\beta\} \subseteq \{x_0\}$, where $\{x_0\}$ is the x-component of the set of solutions of (1.91). This will be a converse to the last part of Theorem 1.8(b), hence giving a genuine maximum principle, as distinct from the lower bound property (1.92).

We still introduce u and $L[x, u] = \phi[x] + \langle u, v[x] \rangle$ to embed the given maximization problem as problem III, associated with (1.91) as problem

I. The converse Theorem 1.9 is not available, however, because this $L[x,u]$ is not strictly convex in u. An inversion like (1.95) is no longer available to eliminate the constraint $v[x]\geqslant 0$. A different approach is needed, and in (1.114) below we follow Kuhn and Tucker (1951) by employing Farkas' lemma, proved next.

Theorem 1.12 (Representation of a given vector a)
Let a, T and T^* indicate given $n\times 1$, $m\times n$ and $n\times m$ matrices written in the set notation of (1.74) *and* (1.75).

(*a*) $(a,x)=0$ $\forall x$ *such that* $Tx=0$ (1.103)

if and only if $a=T^*u$ *for some* u. (1.104)

(*b*) *Farkas' lemma* (1902)

$(a,x)\geqslant 0$ $\forall x$ *such that* $Tx\geqslant 0$ (1.105)

if and only if $a=T^*u$ *for some* $u\geqslant 0$. (1.106)

Proof
Sufficiency of (1.104) or (1.106) requires only adjointness (1.75),

$$\text{i.e.}\quad (a,x)=(x,T^*u)=\langle u,Tx\rangle=0 \text{ in (a)}, \quad \geqslant 0 \text{ in (b)}.$$

Necessity requires the following ideas. Let $\hat{u}$ denote a minimizer of the nonnegative weakly convex function $J[u]=\frac{1}{2}(T^*u-a,\,T^*u-a)$, over all u for case (a), or over all $u\geqslant 0$ for case (b). Of course, such $\hat{u}$ are not necessarily the same in the two cases. With $J'[u]=T(T^*u-a)$, the method used in the first part of the proof of Theorem 1.3(a) can be generalized to imply

$$T(T^*\hat{u}-a)\geqslant 0,\quad \hat{u}\geqslant 0,\quad \langle \hat{u},T(T^*\hat{u}-a)\rangle=0 \tag{1.107}$$

in case (b), or $T(T^*\hat{u}-a)=0$ in case (a). That is, we replace u_α there by u here, let $\bar{u}$ denote a point on the straight join of $\hat{u}$ to u such that

$$\langle u-\hat{u},T(T^*\bar{u}-a)\rangle=J[u]-J[\hat{u}]\geqslant 0,$$

divide by $\|u-\hat{u}\|$ and then let $u\to\hat{u}$ from all admissible directions. Next we note that

$$(a,T^*\hat{u}-a)=-(T^*\hat{u}-a,T^*\hat{u}-a)\leqslant 0$$

in both cases (a) and (b), since $(1.107)_3$ applies to both. Thus $(a,T^*\hat{u}-a)<0$ if *no* representation of the form (1.104) in (a) or (1.106) in (b) were available. But this implies a contradiction of (1.103) in (a) and (1.105) in (b), because $T^*\hat{u}-a$ is a possible choice of x in that it satisfies $Tx=0$ or $Tx\geqslant 0$ in the respective cases, as we saw above. □

Duffin *et al.* (1967) give a geometrical demonstration of Farkas' lemma.

Rockafellar (1970, §22) gives a different proof. The above proof is based on that of Iri (1969, p. 261). A particular infinite dimensional example of (a) is quoted by Temam (1979, p. 14). We shall use Theorem 1.12(b) in the proof of Theorem 1.13(a), and Theorem 1.12(a) in the proof of Theorem 1.14(a). We shall also use Theorem 1.12(a) to prove Theorem 2.19; in passing we notice that the latter Theorems imply that the following four matrix statements are equivalent, where R is a given rectangular matrix, perhaps different in size from T, but such that $TR^{\mathrm{T}} = 0$.

1. $\mathbf{a} = T^{\mathrm{T}}\mathbf{u}$ for some $\mathbf{u}$.
2. $\mathbf{a}^{\mathrm{T}}\mathbf{x} = 0 \ \forall \ \mathbf{x}$ such that $T\mathbf{x} = 0$.
3. $R\mathbf{a} = 0$.
4. $\mathbf{x} = R^{\mathrm{T}}\mathbf{c}$ for some column matrix $\mathbf{c}$, which may have a different number of entries from either $\mathbf{x}$ or $\mathbf{u}$.

Theorem 1.13 (Converse to Theorem 1.8)
Let $\{\hat{x}_\beta\}$ denote the set of points which maximize the given scalar C^1 function $\phi[x]$, subject to the $m+n$ given scalar C^1 constraints $v[x] \geqslant 0$ and $x \geqslant 0$.

(a) Without yet supposing that $\phi[x]$ and $v[x]$ are concave, every $\hat{x}_\beta$ will satisfy (1.91) *in association with some value(s) u_0 of u, provided certain consistency requirements specified in the proof below hold; i.e.*

$$\{\hat{x}_\beta\} \subseteq \{x_0\}.$$

In terms of the generating function (1.89) *the result is that every such maximizer of problem* III *is a solution of* I.

(b) Maximum principle: equivalence
When (a) *holds and, in addition, $\phi[x]$ and $v[x]$ are weakly concave over a convex x-domain, then*

$$\{\hat{x}_\beta\} \equiv \{x_0\}$$

for any fixed $\hat{u}_\beta \in \{u_0\}$. That is, problems III *and* I *are equivalent at least to the extent of having the same set of solution values of x.*

(c) Uniqueness theorem
When, over a convex x-domain, $\phi[x]$ is strictly concave and the m functions in $v[x]$ are weakly concave, there cannot be more than one common solution $\hat{x}_\beta = x_0$ of problems III *and* I *for any fixed u_0. If $\phi[x]$ and the components of $v[x]$ are all weakly concave, except that one component of $v[x]$ is strictly concave, then, with a positive corresponding component of any fixed u_0, there cannot be more than one common solution $\hat{x}_\beta = x_0$.*

Proof

(a) The first mean value theorem for a function of n variables, with $\phi[\hat{x}_\beta] \geqslant \phi[x_\beta]$, implies that there is at least one point $\bar{x}$ (say) on the straight join of any $\hat{x}_\beta$ to any x_β such that

$$\left(\hat{x}_\beta - x_\beta, \left.\frac{\partial \phi}{\partial x}\right|_{\bar{x}}\right) \geqslant 0, \tag{1.108}$$

provided this straight join lies within $\{x_\beta\}$, the set of points in the x-space defined by the constraints $v[x_\beta] \geqslant 0$, $x_\beta \geqslant 0$.

First consider maximizing points with the property $v[\hat{x}_\beta] > 0$. By arguments exactly similar to those following (1.101), but applied now to (1.108) and for an n dimensional pyramid instead of an $n+m$ dimensional one, we reach the following conclusions. If $\hat{x}_\beta > 0$ then

$$\left.\frac{\partial \phi}{\partial x}\right|_{\hat{x}_\beta} = 0, \tag{1.109}$$

but if $\hat{x}_\beta \geqslant 0$ with equality in some of these n scalar inequalities, then

$$\left.\frac{\partial \phi}{\partial x}\right|_{\hat{x}_\beta} \leqslant 0, \quad \hat{x}_\beta \geqslant 0, \quad \left(\hat{x}_\beta, \left.\frac{\partial \phi}{\partial x}\right|_{\hat{x}_\beta}\right) = 0. \tag{1.110}$$

In either case conditions (1.91) are all satisfied by choosing x_0 to be $\hat{x}_\beta$ and $u_0 = 0$. Since u_β is arbitrary, so is $\hat{u}_\beta$.

Next consider maximizing points with the property $v[\hat{x}_\beta] = 0$. We can regard this as m scalar equations without loss of generality, since if $\hat{x}_\beta$ turned out to be internal to a constraint boundary, that part of the constraint might just as well have been absent from the outset (as we can see, for example, from the absence of v from (1.109) and (1.110)). Let d belong to the cone of unit n-vectors pointing from $\hat{x}_\beta$ in the direction of arbitrarily close members of $\{x_\beta\}$, i.e.

$$d = \lim_{x_\beta \to \hat{x}_\beta} \frac{x_\beta - \hat{x}_\beta}{\| x_\beta - \hat{x}_\beta \|},$$

where the limit is taken along a straight join from x_β to $\hat{x}_\beta$, and $\|x\| = (x, x)^{\frac{1}{2}}$. By (1.108) and the continuity of $\partial\phi/\partial x$, we have the downhill property

$$\left(d, -\left.\frac{\partial \phi}{\partial x}\right|_{\hat{x}_\beta}\right) \geqslant 0. \tag{1.111}$$

By the first mean value theorem applied to each of the m inequalities in $0 \leqslant v[x_\beta] = v[x_\beta] - v[\hat{x}_\beta]$, followed by another limiting argument using the continuity of $\partial v/\partial x$, we have the admissibility property

$$\left(d, \left.\frac{\partial v}{\partial x}\right|_{\hat{x}_\beta}\right) \geqslant 0. \tag{1.112}$$

Also, if (say the last) s components of $\hat{x}_\beta$ are zero $(0 \leqslant s \leqslant n)$, the corresponding s components of d must be nonnegative: in matrix notation

$$\mathbf{d}^{\mathrm{T}} I_s \geqslant 0 \quad \text{where} \quad I_s = \underbrace{\begin{bmatrix} 0 \\ \hline 1 \quad 0 \\ \quad \ddots \\ 0 \quad 1 \end{bmatrix}}_{s} \Bigg\} n. \tag{1.113}$$

There are a total of $m+1+s$ homogeneous linear inequalities in (1.111)–(1.113), restricting the n components (not all zero) of d. There is no advance guarantee that these inequalities are self-consistent for arbitrarily chosen C^1 functions $\phi[x]$ and $v[x]$. To be certain of this consistency, a so-called 'constraint qualification' must be satisfied by $\phi[x]$ and $v[x]$. Consistency holds *a fortiori* if one of the inequalities holds for every d which satisfies the remaining $m+s$ inequalities: in particular (following Kuhn and Tucker, 1951, Theorem 1), Farkas' lemma shows that (1.111) holds for all d satisfying (1.112) and (1.113) if and only if the n equations

$$-\left.\frac{\partial \phi}{\partial x}\right|_{\hat{x}_\beta} = \left\langle \hat{u}_\beta, \left.\frac{\partial v}{\partial x}\right|_{\hat{x}_\beta} \right\rangle + \underbrace{\{0,\ldots,0,u_{m+1},\ldots,u_{m+s}\}}_{n} \tag{1.114}$$

hold for some set $\{u_1,\ldots,u_m,u_{m+1},\ldots,u_{m+s}\}$ of nonnegative scalars, where we have written $\hat{u}_\beta = \{u_1,\ldots,u_m\}$ to be consistent with (1.44). It follows from (1.114) that $(1.91\alpha)_1$ and its associated orthogonality condition, and therefore all of (1.91), are satisfied by choosing x_0 to be the particular type of $\hat{x}_\beta$ being considered here, and by choosing u_0 to be whatever set of values of the 'multipliers' $\hat{u}_\beta$ appears in (1.114).

This completes the proof of (a) under the constraint qualification that (1.111) holds for all d satisfying (1.112) and (1.113).

(b) In (a) we proved $\{\hat{x}_\beta\} \subseteq \{x_0\}$. The extra hypothesis that $\phi[x]$ and $v[x]$ are weakly concave over a convex x-domain means that $S \geqslant 0$ for every pair of points over the convex x-domain at any fixed u_0, as can be seen by adapting the proof of Theorem 1.8. From that Theorem it follows that $\{x_0\} \subseteq \{\hat{x}_\beta\}$. Hence

$$\{x_0\} \equiv \{\hat{x}_\beta\}$$

for any fixed $\hat{u}_\beta \in \{u_0\}$.

(c) A uniqueness proof of the type following (1.87) will be possible if and only if the hypotheses ensure that the quantity

$$-\left\langle x_+ - x_-, \left.\frac{\partial L}{\partial x}\right|_+ - \left.\frac{\partial L}{\partial x}\right|_- \right\rangle + \left\langle u_+ - u_-, \left.\frac{\partial L}{\partial u}\right|_+ - \left.\frac{\partial L}{\partial u}\right|_- \right\rangle$$

$$= -\left(x_+ - x_-, \frac{\partial \phi}{\partial x}\bigg|_+ - \frac{\partial \phi}{\partial x}\bigg|_- \right) + \langle u_+ - u_-, v_+ - v_- \rangle$$
$$- \left(x_+ - x_-, \left\langle u_+, \frac{\partial v}{\partial x}\bigg|_+ \right\rangle - \left\langle u_-, \frac{\partial v}{\partial x}\bigg|_- \right\rangle \right) > 0 \qquad (1.115)$$

for suitable pairs of distinct points. The simplest way to be sure of this is when

$$x_+ \neq x_-, \quad u_+ = u_- \geqslant 0, \qquad (1.116)$$

Fig. 1.13(*a*) $\phi[x_1, x_2] = x_1^3, \quad v[x_1, x_2] = 1 - x_1^2 - x_2,$

$$[d_1 d_2]\begin{bmatrix} -2 & 0 \\ -1 & 1 \end{bmatrix} \geqslant 0 \Rightarrow [d_1 d_2]\begin{bmatrix} -3 \\ 0 \end{bmatrix} \geqslant 0.$$

(*b*) $\phi[x_1, x_2] = x_1, \quad v[x_1, x_2] = (1 - x_1)^3 - x_2,$

$$[d_1 d_2]\begin{bmatrix} -1 & 0 & 0 \\ 0 & -1 & 1 \end{bmatrix} \geqslant 0 \Rightarrow \begin{bmatrix} d_1 \\ d_2 \end{bmatrix} = \begin{bmatrix} -1 \\ 0 \end{bmatrix}.$$

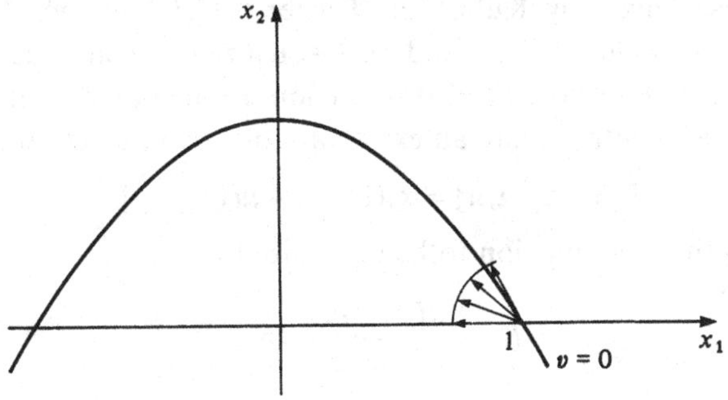

(*a*)

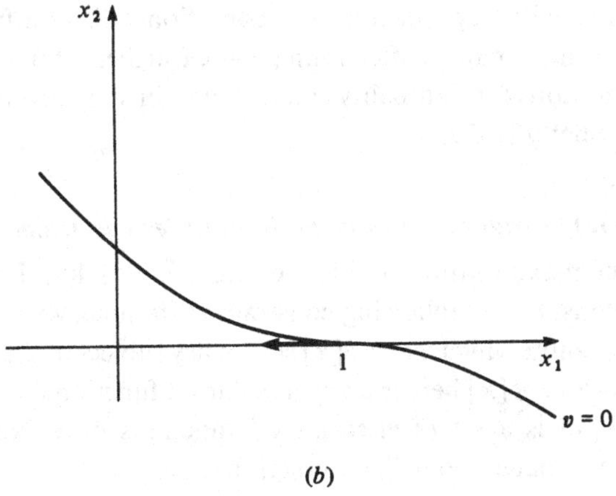

(*b*)

with strictly concave $\phi[x]$ and weakly concave $v[x]$. If there could be two distinct parts x_0 associated with the same part u_0 of a solution x_0, u_0 of I, we could choose these as the plus and minus points in (1.116), and so establish a contradiction. Hence x_0 is unique.

If one of the m common u-components in (1.116) were strictly positive, and if the corresponding component of $v[x]$ were strictly concave, then (1.115) would hold even if $\phi[x]$ were only weakly concave. Again not more than one x_0 can be associated with a u_0 of this particular type.

□

Fig. 1.13(a) shows a simple example ($n = 2, m = s = 1$) where the downhill property (1.111) is satisfied by every unit vector in the cone satisfying (1.112) and (1.113). It needs to be noted, however, that in general the $m + 1 + s$ inequalities (1.111)–(1.113) can be consistent even if only one unit vector d can be found to satisfy them. It is not necessary that one of them be a consequence of the remaining $m + s$, and in particular not necessary that (1.111) be a consequence of (1.112) with (1.113). Fig. 1.13(b) shows a case (given by Kuhn and Tucker (*op. cit.*)) in which (1.111)–(1.113) have a single solution, and no inequality is a consequence of the others. We could try to embed this maximization problem into an augmented space by introducing an extra variable λ and considering

$$L[x_1, x_2, \lambda, u] = x_1(1 - \lambda) + uv[x_1, x_2],$$

designed so that the function to be maximized is

$$L - \lambda\frac{\partial L}{\partial \lambda} - u\frac{\partial L}{\partial u} = x_1,$$

as before; but the associated addition to (1.91β) is $\partial L/\partial\lambda \geqslant 0$, i.e. $x_1 \leqslant 0$, implying $x_1 = 0$ which is too restrictive. In fact there can be other guarantees of consistency which lead to conditions different from (1.114), such as those usually named after John (1948). Further information about necessary conditions for optimality can be found in standard texts such as Bazaraa and Shetty (1979).

(iv) A stationary value problem under constraint

In place of the maximization problem discussed in §1.4(v), Exercise 1.6.5 and §1.5(iii), consider the following constrained stationary value problem. Find the set of points which make $\phi[x]$ stationary subject to the constraints $v[x] = 0$. As before, $\phi[x]$ here is a given scalar C^1 function of the set x of n scalars, and $v[x]$ is a set of m scalar C^1 functions of x. No concavity assumptions are made about these function here.

To define this problem precisely we must give a definition of what we mean by 'stationary subject to' such constraints. We shall say that $\mathring{x}$ is a *stationary point* of $\phi[x]$ *subject to the m scalar constraints* in $v[x]=0$ if

$$\phi[\mathring{x}]-\phi[x]=o(\|\mathring{x}-x\|) \tag{1.117}$$

for each x satisfying $v[x]=0$ in a sufficiently small convex neighbourhood of $\mathring{x}$. For brevity we call any such $\mathring{x}$ a *constrained stationary point*. This definition is consistent with the stationary value problems IV defined when there are no constraints in §§1.1(iv) and 1.3(ii). Let $\{\mathring{x}\}$ denote the set of all $\mathring{x}$ here.

The aim of the next theorem is to make this *constrained stationary value problem* more tangible by showing that it is equivalent to the problem of finding points x_0 (say) which satisfy both equations

$$\frac{\partial\phi}{\partial x}+\left\langle u,\frac{\partial v}{\partial x}\right\rangle=0 \quad (\alpha), \qquad v[x]=0 \quad (\beta), \tag{1.118}$$

with some finite u (say u_0). Previously we called this the version $\partial L/\partial x=\partial L/\partial u=0$ of problem I generated by the $L[x,u]=\phi[x]+\langle u,v[x]\rangle$ in (1.89). Let $\{x_0\}$ be the set of such points x_0.

The equivalence established in Theorem 1.14(c) may be expressed in conventional language by saying that the set of stationary points under the constraint can be found by the *Lagrange multiplier method*, i.e. by solving (1.118). In passing it is interesting to notice that the Lagrange multipliers are said by Oravas and McLean (1966) to have been used by Euler in 1732, four years before Lagrange was born.

Theorem 1.14 (Lagrange multiplier theorem)

(a) There exists some choice u_0 of the so-called multipliers u such that each constrained stationary point $\mathring{x}$ satisfies both (1.118α) *and* (1.118β), *i.e.*

$$\{\mathring{x}\}\subseteq\{x_0\}.$$

(b) Each solution x_0 of (1.118) *for which some u_0 exists is also a stationary point $\mathring{x}$ of $\phi[x]$ under the constraint* (1.118β), *i.e.*

$$\{x_0\}\subseteq\{\mathring{x}\}.$$

(c) $\{\mathring{x}\}\equiv\{x_0\}$.

Proof

(a) Let x_β denote any solution of (1.118β), so that $v[x_\beta]=0$, and x_β is written in place of x in (1.117). The first mean value theorem for a function of n variables implies that there is at least one point $\bar{x}$ (say) on the straight join of any $\mathring{x}$ to any x_β which are sufficiently close in the sense indicated after (1.117), such that

$$\left(\mathring{x} - x_\beta, \left.\frac{\partial \phi}{\partial x}\right|_{\bar{x}}\right) = o(\|\mathring{x} - x_\beta\|). \tag{1.119}$$

Let d this time (contrast the proof of Theorem 1.13(a)) be any unit vector of the type

$$d = \lim_{x_\beta \to \mathring{x}} \frac{x_\beta - \mathring{x}}{\|x_\beta - \mathring{x}\|}.$$

(This takes the place of the nonzero vector dx, du used for $n = m = 1$ in Theorem 1.1). Then by the continuity of $\partial\phi/\partial x$, the constrained stationary property in the form (1.119) implies

$$\left(d, \left.\frac{\partial \phi}{\partial x}\right|_{\mathring{x}}\right) = 0. \tag{1.120}$$

We also have $v[x_\beta] = v[\mathring{x}] = 0$. Hence, by the first mean value theorem applied to each of the m scalar equations in $0 = v[\mathring{x}] - v[x_\beta]$, and another limiting argument using the continuity of $\partial v/\partial x$, we find

$$\left(d, \left.\frac{\partial v}{\partial x}\right|_{\mathring{x}}\right) = 0 \tag{1.121}$$

for all the above d.

We have therefore shown that (1.120) applies for all d satisfying (1.121). We can therefore invoke the equality version of Farkas' lemma given in Theorem 1.12(a), to conclude that some u_0 exists to admit the representation

$$\left.\frac{\partial \phi}{\partial x}\right|_{\mathring{x}} = -\left\langle u_0, \left.\frac{\partial v}{\partial x}\right|_{\mathring{x}}\right\rangle. \tag{1.122}$$

Comparison with (1.118α) establishes that every $\mathring{x}$ has the properties of x_0, i.e.

$$\{\mathring{x}\} \subseteq \{x_0\}.$$

(b) Equations (1.118) are equivalent to the requirement that $L[x, u] = \phi[x] + \langle u, v[x]\rangle$ be stationary for unconstrained x and u. If we impose (1.118β) on this problem, which is in fact already satisfied by every x_0, we recover the problem of making $\phi[x]$ stationary subject to $v[x] = 0$. Hence

$$\{x_0\} \subseteq \{\mathring{x}\}$$

trivially.

(c) This is established by taking (a) and (b) together. □

Note that Theorem 1.14 requires no convexity or concavity properties of $\phi[x]$ or of $v[x]$. Note particularly that it is not restricted to a minimum or to a maximum property of $\phi[x]$ under the constraint, contrary to what is sometimes suggested in the literature. The conditions (1.118) themselves cannot distinguish between any maximum or minimum property.

There is an elementary particular illustration of Theorem 1.14 which also gives a pointer to certain procedures which are often adopted in more general situations. Suppose that a C^2 scalar function $\phi[x_1, x_2]$ of two scalar variables is given, and we wish to find the set C (say) of points where it is stationary subject to the single scalar constraint $\partial\phi/\partial x_1 = 0$. According to Theorem 1.14(c), C is the set of points which satisfy

$$u\frac{\partial^2\phi}{\partial x_1{}^2} = 0, \quad \frac{\partial\phi}{\partial x_2} + u\frac{\partial^2\phi}{\partial x_1 \partial x_2} = 0, \quad \frac{\partial\phi}{\partial x_1} = 0, \tag{1.123}$$

with some finite scalar u, from (1.118).

Therefore C will contain points such that

$$\frac{\partial^2\phi}{\partial x_1{}^2} \neq 0, \quad \frac{\partial\phi}{\partial x_1} = \frac{\partial\phi}{\partial x_2} = 0 \quad \text{with } u = 0 \tag{1.124}$$

and also points such that

$$\frac{\partial^2\phi}{\partial x_1{}^2} = \frac{\partial\phi}{\partial x_1} = 0, \quad \frac{\partial\phi}{\partial x_2} = -u\frac{\partial^2\phi}{\partial x_1 \partial x_2} \quad \text{with any finite } u. \tag{1.125}$$

In the neighbourhood of any point where $\partial^2\phi/\partial x_1{}^2$ and $\partial^2\phi/\partial x_1 \partial x_2$ are not both zero the implicit function theorem, as described just before Theorem 1.1, shows that the constraint $\partial\phi/\partial x_1 = 0$ may be solved locally as either $x_1(x_2)$ or $x_2(x_1)$ with slope $\mathrm{d}x_1/\mathrm{d}x_2$ which satisfies

$$\frac{\partial^2\phi}{\partial x_1{}^2}\mathrm{d}x_1 + \frac{\partial^2\phi}{\partial x_1 \partial x_2}\mathrm{d}x_2 = 0. \tag{1.126}$$

If $\partial^2\phi/\partial x_1{}^2 \neq 0$ this leads to a new function $\psi[x_2] = \phi[x_1(x_2), x_2]$ such that $\mathrm{d}\psi/\mathrm{d}x_2 = \partial\phi/\partial x_2$ in that neighbourhood. Therefore those points of C which satisfy (1.124) make $\psi[x_2]$ stationary in the unconstrained sense. On the other hand, if $\partial^2\phi/\partial x_1 \partial x_2 \neq 0$ (1.126) leads to a new function $\theta[x_1] = \phi[x_1, x_2(x_1)]$ such that

$$\frac{\mathrm{d}\theta}{\mathrm{d}x_1} = -\frac{\partial\phi}{\partial x_2}\frac{\partial^2\phi}{\partial x_1{}^2} \bigg/ \frac{\partial^2\phi}{\partial x_1 \partial x_2}$$

in that neighbourhood. Hence those points of C which satisfy either (1.124) or (1.125) with $\partial^2\phi/\partial x_1 \partial x_2 \neq 0$ make $\theta[x_1]$ stationary in the unconstrained sense. The overall conclusion is a *direct* reconciliation of the concept of a constrained stationary point via the Lagrange multiplier approach with the concept of an unconstrained stationary point, in so far as the implicit function theorem is available. A refinement of the latter would be required at points where both $\partial^2\phi/\partial x_1{}^2$ and $\partial^2\phi/\partial x_1\partial x_2$ are zero.

Let U denote the set of unconstrained stationary points of $\phi[x_1, x_2]$, i.e. those which satisfy $\partial\phi/\partial x_1 = \partial\phi/\partial x_2 = 0$. It is clear from (1.124) and (1.125) that, in general, C is larger than U. This is also evident if we imagine moving along a constrained path $\theta[x_1]$ on $\phi[x_1, x_2]$. Inclined directions on the surface will be found crossing such a path at some of the level points of the latter. Put otherwise, if $\partial\phi/\partial x_1 = 0$ is imposed as a constraint on the previously unconstrained problem of finding U, the ensuing constrained stationary value problem will determine C which, in general, contains more points than U.

This difference between C and U becomes unimportant when $\phi[x_1, x_2]$ is linear in x_1, say $\phi = x_1 f[x_2] + g[x_2]$. Then C contains no points of type (1.124), and those of type (1.125) require

$$f[x_2] = 0, \quad (x_1 + u)\frac{\mathrm{d}f}{\mathrm{d}x_2} + \frac{\mathrm{d}g}{\mathrm{d}x_2} = 0. \tag{1.127}$$

If the solution value of u is arbitrarily deemed to be zero, as seems conventional, we have $C = U$.

A related illustration of Theorem 1.14, returning now to the general context of $L[x, u] = \phi[x] + \langle u, v[x]\rangle$ in (1.89), is to find the set of x-points where this $L[x, u]$ is stationary under the constraint $\partial L/\partial u = 0$. Some change of notation is required. Let λ denote an m dimensional multiplier. Adapting (1.118), we have to solve the problem of setting to zero the gradients of the functional $L + \langle\lambda, v[x]\rangle = \phi[x] + \langle u + \lambda, v[x]\rangle$ of x, u and λ. Arbitrarily setting the solution value of λ to zero leaves (1.118) as it stands. A simple example in the notation of (1.74) is the problem of making $\frac{1}{2}(x, x)$ stationary subject to $Tx + b = 0$. From (1.118) the constrained stationary points are those which satisfy the constraint in conjunction with $x = T^*u$ for some u. Such u must therefore satisfy $TT^*u + b = 0$.

Exercises 1.7

1. Construct the statement and proofs of Theorems 1.6–1.11 and 1.13 for each of the variants (1.53)–(1.55) in place of (1.50).
2. Let $L[x, u]$ be a C^1 function defined over the $n + m$ dimensional pyramid $x \geqslant 0$, $u \geqslant 0$. Let $\check{x}, \check{u}$ be any point satisfying

$$L[\check{x}, u] \geqslant L[\check{x}, \check{u}] \geqslant L[x, \check{u}].$$

This is Kuhn and Tucker's (1951) definition of a saddle *point*, as the simultaneous solution of a minimizing problem over $u \geqslant 0$, and of a maximizing problem over

$x \geqslant 0$. It need not be a stationary point, and therefore is *not* the same as our definition of a saddle point (§1(iii)) which is required to be stationary. For example, in Figs. 1.10α and β all four solution points are examples of $\check{x}, \check{u}$, but only one of them is stationary.

(a) Prove that $\{\check{x}, \check{u}\} \subseteq \{x_0, u_0\}$, the set of solutions of (1.50).

(b) Prove that if $S \geqslant 0$ for all distinct pairs of points in the pyramid,

$$L[x_0, u] \geqslant L[x_0, u_0] \geqslant L[x, u_0]$$

and therefore

$$\{x_0, u_0\} \subseteq \{\check{x}, \check{u}\}.$$

(c) Hence state hypotheses under which the two extremizing problems taken together are equivalent to (1.50).

3. Explore how, in seeking the proof that $\{\hat{x}_\beta, \hat{u}_\beta\} \subseteq \{x_0, u_0\}$ for a piecewise C^2 function $L[x, u]$ which is *weakly* convex in u at each x, the methods of proof used for Theorems 1.9(a) and 1.13(a) may have to be amalgamated. Consider specifically a piecewise C^2 function $L[x, u]$ made up of a part which is strictly convex in u smoothly joined to a part which is linear in u (cf. Exercise 1.3.1).
4. Show that the set of points C defined by (1.124) and (1.125) for the case

$$\phi[x_1, x_2] = \tfrac{1}{3}x_1{}^3 + x_1x_2 - x_2{}^2 - x_2.$$

consists of two intersections of a line with a parabola (for $u = 0$), together with the origin (for $u = -\frac{1}{2}$). Find the corresponding functions $\psi[x_2]$ and $\phi[x_1]$.

5. For $\phi[x_1, x_2] = x_1x_2 - \frac{1}{2}x_2{}^2 + x_1 - 3x_2$, show that C is the line $x_2 = -1$, $x_1 = 2 + u$ for any u.
6. Unconstrained stationary points can be classified in terms of the *order of contact* made there with the horizontal. If x is a single scalar, for example, the function x^{n+1} may be said to have nth order of contact, or $(n+1)$ point contact, with the horizontal at $x = 0$, because its derivative has n coincident zeros there. Functions of several variables may have different order of contact along different paths through a particular stationary point. A close connection between this classification and the local description of bifurcation paths has been described by Sewell (1968*a*, *b*). Show by elementary methods that the function of two variables

$$\phi[x_1, x_2] = \tfrac{1}{2}ax_1{}^4 + bx_1{}^2x_2 + \tfrac{1}{2}cx_2{}^2$$

has, at the origin, first order contact with the horizontal plane in every direction except along a certain curved path where the contact is of order three. Assume $ac - b^2 \neq 0 \neq abc$.

1.6. Examples of links with other viewpoints

(i) Linear programming

Using the matrix notation of §1.3(v) as an alternative to our set notation, consider the bilinear function

$$L[\mathbf{x}, \mathbf{u}] = \mathbf{u}^{\mathrm{T}} T\mathbf{x} + \mathbf{a}^{\mathrm{T}}\mathbf{x} + \mathbf{b}^{\mathrm{T}}\mathbf{u} \tag{1.128}$$

which results by putting $A = B = 0$ in (1.65). Since $\partial L/\partial \mathbf{x} = T^{\mathrm{T}}\mathbf{u} + \mathbf{a}$ and $\partial L/\partial \mathbf{u} = T\mathbf{x} + \mathbf{b}$, the three problems (1.50)–(1.52) whose equivalence is sought can be stated as follows.

$$\text{I.}\quad \begin{aligned} &T^{\mathrm{T}}\mathbf{u} + \mathbf{a} \leqslant 0, \quad \mathbf{u} \geqslant 0, \quad (\alpha) \\ &T\mathbf{x} + \mathbf{b} \geqslant 0, \quad \mathbf{x} \geqslant 0, \quad (\beta) \\ &\mathbf{x}^{\mathrm{T}}(T^{\mathrm{T}}\mathbf{u} + \mathbf{a}) = \mathbf{u}^{\mathrm{T}}(T\mathbf{x} + \mathbf{b}) = 0. \end{aligned} \tag{1.129}$$

II. Minimize $\mathbf{b}^{\mathrm{T}}\mathbf{u}$ subject to (1.129α).

III. Maximize $\mathbf{a}^{\mathrm{T}}\mathbf{x}$ subject to (1.129β).

Each of the optimization problems is a so-called linear programming problem; i.e. a linear function is to be extremized subject to a system of linear inequalities. The variables separate out, as they do also in the other three variants obtained from (1.53)–(1.55). It is therefore enough to denote the sets of solutions of the α and β subsets, considered in turn, as $\{\mathbf{u}_\alpha\}$ and $\{\mathbf{x}_\beta\}$, respectively, since $\mathbf{u}_\beta$ and $\mathbf{x}_\alpha$ are both arbitrary. Problems of existence of such solutions, and in particular of minimizers $\{\hat{\mathbf{u}}_\alpha\}$ for II and maximizers $\{\hat{\mathbf{x}}_\beta\}$ for III, are nontrivial. Discussions of existence are given by Noble and Daniel (1977) and by Bazaraa and Shetty (1979, p. 212), for example.

The bilinear character of (1.128) means that $S \equiv 0$, as we can see from (1.70) (cf. our discussion of the rectangular hyperboloid (1.21)). Therefore the hypothesis of Theorem 1.6(c) holds trivially, and the simultaneous upper and lower bounds (1.80) apply, i.e.

$$\mathbf{b}^{\mathrm{T}}\mathbf{u}_\alpha \geqslant \mathbf{b}^{\mathrm{T}}\mathbf{u}_0 = \mathbf{a}^{\mathrm{T}}\mathbf{x}_0 \geqslant \mathbf{a}^{\mathrm{T}}\mathbf{x}_\beta. \tag{1.130}$$

The converse Theorem 1.11 does *not* apply. However, an alternative approach is available through the embedding method described in §§1.4(v) and 1.5(iii) with the particular choices

$$\phi[\mathbf{x}] = \mathbf{a}^{\mathrm{T}}\mathbf{x} \quad \text{and} \quad \mathbf{v}[\mathbf{x}] = T\mathbf{x} + \mathbf{b}$$

in (1.89). In passing, note that both bounds (1.130) can also be regarded as a consequence of a version of Theorem 1.8 in which the requirement $\mathbf{u} \geqslant 0$ can be dropped because $\mathbf{v}[\mathbf{x}]$ is a linear function.

For an appropriate converse of the lower bound we now look to Theorem 1.13. As the proof shows, the location of possible solutions $\hat{\mathbf{x}}_\beta$ of the maximizing problem depends on whether the given data $\mathbf{a}, T$ and $\mathbf{b}$ satisfy certain consistency conditions. If $T\hat{\mathbf{x}}_\beta + \mathbf{b} > 0$ and if the last s components $(0 \leqslant s \leqslant n)$ of $\hat{\mathbf{x}}_\beta$ are to be zero, then by (1.110) the last s components of $\mathbf{a}$ must be negative and its first $n - s$ components must be zero. If $T\hat{\mathbf{x}}_\beta + \mathbf{b} = 0$ we require

$$-\mathbf{d}^{\mathrm{T}}\mathbf{a} \geqslant 0, \quad \mathbf{d}^{\mathrm{T}}T^{\mathrm{T}} \geqslant 0, \quad \mathbf{d}^{\mathrm{T}}\mathbf{I}_s \geqslant 0 \tag{1.131}$$

from (1.111)–(1.113) When the first of (1.131) is required to be a consequence of the second and third we obtain the result stated in Theorem 1.13(a) and (b), including a maximum principle for linear programming. The hypothesis of the uniqueness Theorem 1.13(c) does not apply in this case.

We leave the reader to write out a complementary version of Theorem 1.13, with x and u interchanged, and to apply it to give a converse of the upper bound in (1.130). The result includes a minimum principle for linear programming.

(ii) Quadratic programming

The optimization of a quadratic function subject to linear inequality constraints is called a quadratic programming problem. The subject is discussed at length by, for example, Boot (1964) and by Bazaraa and Shetty (1979).

Such problems emerge naturally from our general approach. It can be seen by specializing (1.50)–(1.55) to (1.65)–(1.68) that

$$L[\mathbf{x},\mathbf{u}] = -\tfrac{1}{2}\mathbf{x}^{\mathrm{T}}A\mathbf{x} + \mathbf{u}^{\mathrm{T}}T\mathbf{x} + \tfrac{1}{2}\mathbf{u}^{\mathrm{T}}B\mathbf{u} + \mathbf{a}^{\mathrm{T}}\mathbf{x} + \mathbf{b}^{\mathrm{T}}\mathbf{u} \qquad (1.65\,bis)$$

generates the minimization of

$$\tfrac{1}{2}\mathbf{x}_\alpha{}^{\mathrm{T}}A\mathbf{x}_\alpha + \tfrac{1}{2}\mathbf{u}_\alpha{}^{\mathrm{T}}B\mathbf{u}_\alpha + \mathbf{b}^{\mathrm{T}}\mathbf{u}_\alpha \qquad (1.132)$$

among the solutions of a system of linear α constraints, and also generates the maximization of

$$-\tfrac{1}{2}\mathbf{x}_\beta{}^{\mathrm{T}}A\mathbf{x}_\beta - \tfrac{1}{2}\mathbf{u}_\beta{}^{\mathrm{T}}B\mathbf{u}_\beta + \mathbf{a}^{\mathrm{T}}\mathbf{x}_\beta \qquad (1.133)$$

among the solutions of a system of linear β constraints.

Examples in which the hypotheses of the bounding Theorem 1.6 are satisfied were worked out explicitly in §1.4(iii). In particular it was illustrated there how the β constraints could be solved explicitly to express the lower bound (1.133) as a function $K[\mathbf{x}_\beta]$ given by (1.86). Exercise 1.6.1 displays the expression $J[\mathbf{u}_\alpha]$ obtained for the upper bound (1.132) after a similar solution of α constraints.

Table 1.2. *Definiteness hypotheses for some quadratic programming results*

$A>0$ and $B+TA^{-1}T^{\mathrm{T}}\geqslant 0$	$\Rightarrow$ Minimum principle, Theorem 1.10(b)
$B+TA^{-1}T^{\mathrm{T}}\geqslant 0$	$\Rightarrow$ Upper bound, Theorem 1.6(a)
$A\geqslant 0$ and $B\geqslant 0$	$\Rightarrow$ Simultaneous bounds, Theorem 1.6(c)
$A+T^{\mathrm{T}}B^{-1}T\geqslant 0$	$\Rightarrow$ Lower bound, Theorem 1.6(b)
$B>0$ and $A+T^{\mathrm{T}}B^{-1}T\geqslant 0$	$\Rightarrow$ Maximum principle, Theorem 1.9(b)

From these examples, or from (1.62) and (1.64), we see that

$$L_{xx} = -A, L_{uu} = B, \quad J_{uu} = B + TA^{-1}T^{\mathrm{T}}, \quad K_{xx} = -(A + T^{\mathrm{T}}B^{-1}T).$$

It is helpful to reconstruct Figs. 1.11 and 1.12 for the case of quadratic $L[x,u]$. A simplified version is shown in Table 1.2, although it must be remembered that $J[u]$ and $K[x]$ do not always both exist. Sometimes only one of them exists, depending on which constraint is being solved. For example, we showed in §§1.4(iii) and 1.5(ii) that, for the problems governed by (1.81) or (1.82) a maximum principle applies if the hypothesis (1.102) is used in place of the weaker (1.85), the latter being sufficient to justify a lower bound $K[\mathbf{x}_\beta]$.

Exercise 1.6.1 may be similarly strengthened, to obtain a minimum principle there from Theorem 1.10(b). Included within the last case, with A positive definite and $B = 0$, is the equivalence of the problem

$$\begin{aligned} -A\mathbf{x} + T^{\mathrm{T}}\mathbf{u} + \mathbf{a} &= 0, \quad \mathbf{u} \geqslant 0, \quad (\alpha) \\ T\mathbf{x} + \mathbf{b} &\geqslant 0, \qquad\qquad\quad (\beta) \\ \mathbf{u}^{\mathrm{T}}(T\mathbf{x} + \mathbf{b}) &= 0 \end{aligned} \tag{1.134}$$

to the minimization of (1.132) subject to (1.134α). Conditions (1.134) illustrate (1.54). That (1.133) subject to (1.134β), with $B = 0$, provides a lower bound when A is nonnegative definite is established by Theorem 1.6(c), and Theorem 1.13(b) describes circumstances in which the maximum principle applies. A mechanical illustration of (1.134) and the associated bounds is given in §1.7.

When A and B are both positive definite in (1.65) the hypothesis of Theorem 1.11 is satisfied. Simultaneous maximum and minimum principles then apply, for each of the four cases (1.50) and (1.53)–(1.55).

(iii) Decomposition of a nonhomogeneous linear problem

Suppose we have to solve the real matrix equation

$$P\mathbf{x} = \lambda\mathbf{x} + \mathbf{a} \tag{1.135}$$

for the $n \times 1$ matrix $\mathbf{x}$, where $\mathbf{a}$ is a given $n \times 1$ matrix, λ is a given scalar, and P is a given symmetric nonnegative definite $n \times n$ matrix, so that $\mathbf{x}^{\mathrm{T}}P\mathbf{x} \geqslant 0$ for all $\mathbf{x}$. There exists a (non-unique) decomposition $P = T^{\mathrm{T}}T$, where T is an $m \times n$ matrix with transpose T^{T} if there are $n - m$ zero eigenvalues of P (see Exercise 1.8.1). Hence (1.135) can be written

$$\begin{aligned} \lambda\mathbf{x} + T^{\mathrm{T}}\mathbf{u} + \mathbf{a} &= 0, \quad (\alpha) \\ T\mathbf{x} + \mathbf{u} &= 0, \quad (\beta) \end{aligned} \tag{1.136}$$

where $\mathbf{u}$ is an auxiliary $m \times 1$ matrix variable. These equations are an example of the stationary value problem (1.53) generated by the gradients of

$$L[\mathbf{x},\mathbf{u}] = \tfrac{1}{2}\lambda\mathbf{x}^{\mathrm{T}}\mathbf{x} + \mathbf{u}^{\mathrm{T}}T\mathbf{x} + \tfrac{1}{2}\mathbf{u}^{\mathrm{T}}\mathbf{u} + \mathbf{a}^{\mathrm{T}}\mathbf{x}. \qquad (1.137)$$

We can always satisfy (1.136β) with arbitrarily chosen $\mathbf{x}_\beta$, so that $\mathbf{u}_\beta = -T\mathbf{x}_\beta$, and the consequent value of L is

$$L_\beta = \tfrac{1}{2}\lambda\mathbf{x}_\beta{}^{\mathrm{T}}\mathbf{x}_\beta - \tfrac{1}{2}\mathbf{x}_\beta{}^{\mathrm{T}}P\mathbf{x}_\beta + \mathbf{a}^{\mathrm{T}}\mathbf{x}_\beta. \qquad (1.138)$$

If $\lambda \neq 0$, we can always satisfy (1.136α) with arbitrarily chosen $\mathbf{u}_\alpha$ and consequent $\mathbf{x}_\alpha = -(T^{\mathrm{T}}\mathbf{u}_\alpha + \mathbf{a})/\lambda$. Then

$$L_\alpha = -\frac{1}{2\lambda}(T^{\mathrm{T}}\mathbf{u}_\alpha + \mathbf{a})^{\mathrm{T}}(T^{\mathrm{T}}\mathbf{u}_\alpha + \mathbf{a}) + \tfrac{1}{2}\mathbf{u}_\alpha{}^{\mathrm{T}}\mathbf{u}_\alpha.$$

We can make do with the knowledge that T exists, and avoid the need to determine T explicitly, if for $\mathbf{u}_\alpha$ we choose the $\mathbf{u}_\beta$, i.e. $\mathbf{u}_\alpha = -T\mathbf{x}_\beta$ for arbitrary $\mathbf{x}_\beta$. Then

$$L_\alpha = L_\beta - \frac{1}{2\lambda}(P\mathbf{x}_\beta - \lambda\mathbf{x}_\beta - \mathbf{a})^{\mathrm{T}}(P\mathbf{x}_\beta - \lambda\mathbf{x}_\beta - \mathbf{a}). \qquad (1.139)$$

If $\lambda < 0$ $L[\mathbf{x},\mathbf{u}]$ is a strict saddle function concave in $\mathbf{x}$ and convex in $\mathbf{u}$, since $S > 0$ for every pair of distinct points in the $\mathbf{x},\mathbf{u}$ space. Theorem 1.11 applies, and the simultaneous upper and lower bounds provided by (1.80) become

$$L_\alpha \geqslant \tfrac{1}{2}\mathbf{a}^{\mathrm{T}}\mathbf{x}_0 \geqslant L_\beta, \qquad (1.140)$$

where $\mathbf{x}_0$ is the unique solution of (1.135). In this case the bounds effectively provide a simple estimate for the component of the solution $\mathbf{x}_0$ in the direction of $\mathbf{a}$, with an error which does not exceed $-1/2\lambda$ times the square of the residual, by (1.139).

The saddle quantity (1.70) with $A = -\lambda I$ and $B = I$ becomes, for every pair of points which satisfy $\mathbf{u} = -T\mathbf{x}$,

$$S = \tfrac{1}{2}(\mathbf{x}_+ - \mathbf{x}_-)^{\mathrm{T}}P(\mathbf{x}_+ - \mathbf{x}_-) - \tfrac{1}{2}\lambda(\mathbf{x}_+ - \mathbf{x}_-)^{\mathrm{T}}(\mathbf{x}_+ - \mathbf{x}_-).$$

It follows that $S > 0$ for every pair of distinct points in the set $\{\mathbf{x}_\beta, \mathbf{u}_\beta\}$ satisfying (1.36β) if

$$\lambda < \min \frac{\mathbf{x}_\beta{}^{\mathrm{T}}P\mathbf{x}_\beta}{\mathbf{x}_\beta{}^{\mathrm{T}}\mathbf{x}_\beta}, \qquad (1.141)$$

where the minimum is taken with respect to every $\mathbf{x}_\beta \neq 0$. The calculation of this minimum is equivalent to a determination of the smallest eigenvalue of P. Whatever sign λ may have, if it is less than this smallest eigenvalue we know that Theorems 1.6(b) and 1.9 hold. The lower bound holds, i.e.

$$\tfrac{1}{2}\mathbf{a}^{\mathrm{T}}\mathbf{x}_0 \geqslant L_\beta,$$

and the equivalence proved in Theorem 1.9 justifies the associated maximizing principle. The upper bound in (1.140) need no longer hold because (1.141) does not insist that $\lambda < 0$. In fact when P is positive definite (rather than nonnegative definite), all the eigenvalues of the Rayleigh quotient on the right in (1.141) are positive, so that (1.141) is consistent with some $\lambda > 0$.

Whenever $\lambda \neq 0$ we also find from (1.70) that for every pair of points which satisfy (1.136α),

$$S = \tfrac{1}{2}(\mathbf{u}_+ - \mathbf{u}_-)^{\mathrm{T}}(\mathbf{u}_+ - \mathbf{u}_-) - \frac{1}{2\lambda}(\mathbf{u}_+ - \mathbf{u}_-)^{\mathrm{T}}Q(\mathbf{u}_+ - \mathbf{u}_-).$$

Here $Q = TT^{\mathrm{T}}$ is an $m \times m$ symmetric and positive definite matrix whose eigenvalues are the same as the positive ones of the original nonnegative definite P (see Exercise 1.8.1).

When $\lambda > 0$ it follows that $S > 0$ for every pair of distinct points in the set $\{\mathbf{x}_\alpha, \mathbf{u}_\alpha\}$ if

$$\lambda > \max \frac{\mathbf{u}_\alpha{}^{\mathrm{T}} Q \mathbf{u}_\alpha}{\mathbf{u}_\alpha{}^{\mathrm{T}} \mathbf{u}_\alpha}, \tag{1.142}$$

where the maximum is taken with respect to every $\mathbf{u}_\alpha \neq 0$. The calculation of this maximum is equivalent to a determination of the largest eigenvalue of Q, and therefore of P. Hence when λ exceeds the largest eigenvalue of P, we know that Theorem 1.6(a) holds, together with a converse which is a variant of Theorem 1.10 (in which (1.137) is strictly convex with respect to $\mathbf{x}$ at every fixed $\mathbf{u}$, because $\lambda > 0$ now). The outcome this time is that the upper bound still holds in (1.140), but not necessarily the lower bound, i.e.

$$L_\alpha \geqslant \tfrac{1}{2}\mathbf{a}^{\mathrm{T}}\mathbf{x}_0.$$

The equivalence proved by the variant of Theorem 1.10 justifies the associated minimizing principle.

When $\lambda \neq 0$ we can regard (1.136) as a decomposition of

$$Q\mathbf{u} = \lambda\mathbf{u} - T\mathbf{a},$$

which itself is a result of premultiplying (1.135) by T. Note that (1.141) and (1.142) do *not* hold simultaneously.

An infinite dimensional version of (1.139) is given in (3.130). If P has a smallest eigenvalue in an infinite dimensional problem, it often does not have a largest eigenvalue.

(iv) Hypercircle

We illustrate this for the finite dimensional case by reverting to the inner product and operator notation for matrices described at the end of §1.3(v).

The general quadratic generating function is now written

$$L[x,u] = -\tfrac{1}{2}(x, Ax) + (x, T^*u) + \tfrac{1}{2}\langle u, Bu\rangle + (a,x) + \langle b,u\rangle \tag{1.74 bis}$$

with gradients

$$\frac{\partial L}{\partial x} = -Ax + T^*u + a, \quad \frac{\partial L}{\partial u} = Tx + Bu + b. \tag{1.143}$$

Consider the stationary value problem (1.53). Then solutions of the constraints considered separately have the properties

$$0 = -Ax_\alpha + T^*u_\alpha + a, \quad 0 = Tx_\beta + Bu_\beta + b. \tag{1.144}$$

Any solution of both together is denoted by x_0, u_0. The following four equations include relations between the corresponding values of L.

$$\begin{aligned} L_\alpha - L_0 &= \tfrac{1}{2}(x_\alpha - x_0, A(x_\alpha - x_0)) + \tfrac{1}{2}\langle u_\alpha - u_0, B(u_\alpha - u_0)\rangle, \\ L_0 - L_\beta &= \tfrac{1}{2}(x_0 - x_\beta, A(x_0 - x_\beta)) + \tfrac{1}{2}\langle u_0 - u_\beta, B(u_0 - u_\beta)\rangle, \\ L_\alpha - L_\beta &= \tfrac{1}{2}(x_\alpha - x_\beta, A(x_\alpha - x_\beta)) + \tfrac{1}{2}\langle u_\alpha - u_\beta, B(u_\alpha - u_\beta)\rangle, \\ 0 &= (x_\alpha - x_0, A(x_\beta - x_0)) + \langle u_\alpha - u_0, B(u_\beta - u_0)\rangle. \end{aligned} \tag{1.145}$$

When A and B are positive definite, upper and lower bounds $L_\alpha \geqslant L_0 \geqslant L_\beta$ are predicted by (1.145) or by Theorem 1.6(c), and simultaneous extremum principles are assured by the converse Theorem 1.11. We can give a geometrical interpretation in an $n+m$ dimensional space having the inner product $\tfrac{1}{2}(.,A.) + \tfrac{1}{2}\langle .,B.\rangle$ given in (1.76). For then (see Fig. 1.14) any pair of approximating points x_α, u_α and x_β, u_β, each being required to satisfy only one of (1.144), will lie at opposite ends of a diameter of length $(L_\alpha - L_\beta)^{\frac{1}{2}}$ of a 'hypercircle', somewhere upon whose circumference the actual solution x_0, u_0 will lie. The right angle subtended there is expressed by $(1.145)_4$. The other two sides of the triangle have lengths $(L_\alpha - L_0)^{\frac{1}{2}}$ and $(L_0 - L_\beta)^{\frac{1}{2}}$. Pythagoras' theorem becomes $L_\alpha - L_\beta = (L_\alpha - L_0) + (L_0 - L_\beta)$. In terms of our saddle quantity (cf. (1.70)) the distance between any pair of points in this particular space is $S^{\frac{1}{2}}$. The inner

Fig. 1.14. Hypercircle.

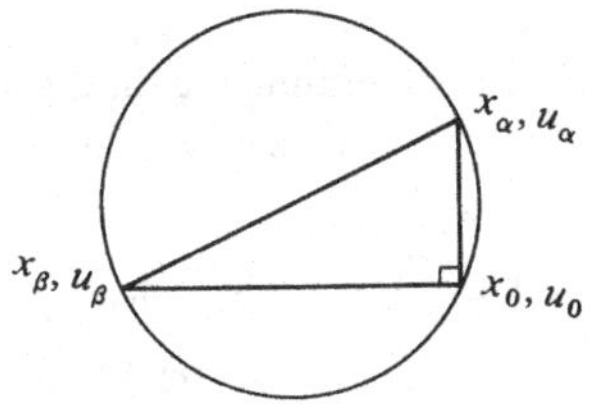

product notation here suggests corresponding formulae for linear stationary value problems in infinite dimensional spaces, for example as described in the text by Synge (1957).

Exercises 1.8

1. Let P be a real symmetric nonnegative definite $n \times n$ matrix whose eigenvectors assembled as columns make up the orthogonal matrix U, and whose positive eigenvalues are $\lambda_1^2, \ldots, \lambda_m^2$. Prove that $P = T^T T$ where $T^T = U\Lambda$ and Λ is any one of the 2^m $n \times m$ matrices

$$\Lambda = \begin{bmatrix} \pm\lambda_1 & & 0 \\ 0 & \ddots & \pm\lambda_m \\ \hline & 0 & \end{bmatrix}.$$

Hence prove that $TT^T = \text{diag}[\lambda_1^2 \cdots \lambda_m^2]$.

2. (i) Let R be a real symmetric indefinite $n \times n$ matrix. Find rectangular matrices T_1 and T_2 with transposes T_1^T and T_2^T such that

$$R = T_1^T T_1 - T_2^T T_2.$$

(ii) Hence decompose the problem $\mathbf{Rx} = \lambda\mathbf{x} + \mathbf{a}$ (given λ and $\mathbf{a}$) into a stationary value problem generated by a suitable generalization $L[\mathbf{x}, \mathbf{u}_1, \mathbf{u}_2]$ of (1.137). Hence show that extremum principles apply if λ lies outside the spectrum of eigenvalues of R. Some practical algorithms for spectral factorization of certain matrices have been described by Vanderwalle and Dewilde (1975).

3. Describe upper and lower bounds for the problem $P\mathbf{x} + \partial F/\partial\mathbf{x} = 0$ when $F[\mathbf{x}]$ is a strictly convex function and P is a symmetric nonnegative definite matrix. Repeat for the problem $P\partial F/\partial\mathbf{x} + \mathbf{x} = 0$

4. When x and u are scalars, show that $L[x, u] = x(u-1)$ generates problems (1.50)–(1.52) whose solution points are

	(1.50)	(1.51)	(1.52)
x	0	arbitrary	0
u	$\geqslant 1$	$\geqslant 1$	arbitrary

What can be said about partial equivalence of the three problems? Compare this example with Theorem 1.13.

5. Prove that the *ab initio* problem of minimizing the distance from the origin to the point x_1, x_2 subject to $x_1 + x_2 \geqslant 4$, $x_1 \geqslant 0$, $x_2 \geqslant 0$ can be regarded as problem (1.52) generated by

$$L[x_1, x_2, u] = -\tfrac{1}{2}(x_1^2 + x_2^2) + u(x_1 + x_2 - 4).$$

Show directly that all 12 problems (1.50)—(1.52), and (1.53)–(1.55) with their variants of II and III, have the same unique solution in x_1, x_2, u space (note that Theorem 1.10(c) ensures unique x_1 and x_2 but not u).

1.7. Initial motion problems

(i) Mechanical background

The remaining sections of this chapter may be read independently of each other. Here we use some standard notation of classical dynamics. Consider a system of particles whose typical position vector **r** may be expressed, e.g. via cartesian components with respect to a fixed reference frame, as a given function $\mathbf{r}[q_i, t]$ of n generalized coordinates q_i and the time t.

Let kinematical constraints be present which require the generalized velocities $\dot{q}_i$ to satisfy the m equations

$$A_{\gamma j}\dot{q}_j + B_\gamma = 0,$$

where the $m \times n$ $A_{\gamma j}[q_i, t]$ and the m $B_\gamma[q_i, t]$ are given C^1 functions of the indicated $n+1$ variables. The constraints may be holonomic or non-holonomic. In this section we use the convention that a repeated Latin or Greek suffix is to be summed over its full range of (respectively) n or m values; an isolated suffix is to receive its full range of values in turn.

The n equations of motion are

$$\frac{\mathrm{d}}{\mathrm{d}t}\left(\frac{\partial T}{\partial \dot{q}_j}\right) - \frac{\partial T}{\partial q_j} - Q_j - u_\gamma A_{\gamma j} = 0$$

if the constraints are workless in the usual sense implied by the derivation from D'Alembert's principle (a clear description of these preliminaries is provided by, for example, Synge (1960, §46)). Here the u_γ are m Lagrangian multipliers associated with the constraints, the generalized forces $Q_j[q_i, \dot{q}_i, t]$ are assigned functions of the $2n+1$ indicated variables, and the kinetic energy T has the particular form

$$T[q_i, \dot{q}_i, t] = \tfrac{1}{2}\dot{q}_i\dot{q}_j a_{ij}[q_k, t] + \dot{q}_i b_i[q_k, t] + c[q_k, t],$$

implied by the rheonomic assumption $\mathbf{r}[q_i, t]$. Here c, the n b_i, and the symmetric $n \times n$ matrix elements a_{ij} are given functions of the n q_k and t. We shall assume that the matrix of

$$a_{ij} \text{ is positive definite.}$$

This can be motivated in the first instance by the scleronomic case in which $\mathbf{r}[q_i]$ does not contain t, and so $T = \tfrac{1}{2}\dot{q}_i\dot{q}_j a_{ij}[q_k]$.

The foregoing $n+m$ equations of constraint and motion set the classical problem of determining the $n+m$ functions $q_i[t]$, $u_\gamma[t]$ for all time, subject to suitable starting conditions. Here we wish to pose different problems, such as the following.

(ii) Initial motion problem

At a given instant, say $t=0$, the value of the generalized displacements q_i and velocities $\dot{q}_i$ are regarded as known. The kinematical constraints are assumed to be sustained through $t=0$, in the sense that the time-differentiated version of the above constraint equations holds at $t=0$. What are the values at the given instant of the generalized accelerations $\ddot{q}_i[0]$ and the multipliers $u_\gamma[0]$ (or at least of the reactions $u_\gamma A_{\gamma j}$)? The $n+m$ governing equations of this problem are

$$\begin{aligned} a_{ij}\ddot{q}_j - A_{\gamma i}u_\gamma &= z_i, \quad (\alpha) \\ A_{\gamma j}\ddot{q}_j &= s_\gamma, \quad (\beta) \end{aligned} \qquad (1.146)$$

where we have denoted known data at $t=0$ by

$$z_i = \frac{\partial T}{\partial q_i} + Q_i - \dot{a}_{ij}\dot{q}_j - \dot{b}_i, \quad s_\gamma = -\dot{A}_{\gamma j}\dot{q}_j - \dot{B}_\gamma.$$

This is a problem of linear algebra whose complete resolution will depend on further assumptions about the values of $A_{\gamma i}$ at $t=0$.

(iii) Generating saddle function

The $\ddot{q}_i$ can be identified with our previous x_i, and (1.146) can be regarded as a stationary value problem (1.53)

$$\frac{\partial L}{\partial \ddot{q}_i} = 0 \quad (\alpha), \quad \frac{\partial L}{\partial u_\gamma} = 0 \quad (\beta),$$

generated by the saddle function

$$L[\ddot{q}_i, u_\gamma] = -\tfrac{1}{2}a_{ij}\ddot{q}_i\ddot{q}_j + u_\gamma A_{\gamma i}\ddot{q}_i + z_i\ddot{q}_i - s_\gamma u_\gamma, \qquad (1.147)$$

which is strictly concave in the $\ddot{q}_i$ and otherwise bilinear. This L is manifestly not the classical Lagrangian which is a function of the $q_i, \dot{q}_i$ and t. Nevertheless, in choosing L as the label throughout this book for our generating functions, we are acknowledging the influential role of the structure of classical mechanics, as a parent of new theories beyond the mechanical realm.

We can now use (1.147) to generate problems defined by (1.50), (1.54) and (1.55), in addition to (1.53); and then to read off uniqueness theorems and simultaneous extremum principles characterizing them from Theorems 1.6–1.14. Of these other three possible cases, we show next that (1.54) does have mechanical significance.

(iv) Unilateral constraints

We now pose a different initial motion problem to that of §(ii) above. The values of the q_i and $\dot{q}_i$ are again regarded as known and satisfying $A_{\gamma j}\dot{q}_j + B_\gamma = 0$ at $t = 0$, but in place of (1.146) we seek to determine the current values of the $\ddot{q}_i$ and u_γ from

$$\left.\begin{aligned} a_{ij}\ddot{q}_j - A_{\gamma i}u_\gamma &= z_i, \quad (\alpha)\\ A_{\gamma j}\ddot{q}_j &\geqslant s_\gamma, \quad (\beta)\\ u_\gamma &\geqslant 0, \quad (\alpha)\\ u_\gamma[A_{\gamma j}\ddot{q}_j - s_\gamma] &= 0. \end{aligned}\right\} \quad (1.148)$$

These can be written down following (1.54) with (1.147). This problem, and the ensuing simultaneous extremum principles, were described in the present generality, including the nonholonomic case, by Sewell (1969). The mechanical reasoning is that unless $u_\gamma > 0$, the corresponding kinematical constraint need no longer be sustained through $t = 0$, but can experience incipient breakdown in the sense of a jump from zero to a positive value of $\mathrm{d}(A_{\gamma j}\dot{q}_j + B_\gamma)/\mathrm{d}t$, provided $u_\gamma = 0$. The constraint is passive in the sense that $u_\gamma < 0$ is impossible.

(v) Examples

In the simplest situations this mechanical reasoning means that a reactive push ($u_\gamma > 0$) but not pull ($u_\gamma < 0$) is possible, but $u_\gamma = 0$ at surfaces where separation is incipient. For example, following Moreau (1966*a*) who

Fig. 1.15. Mackie's problem.

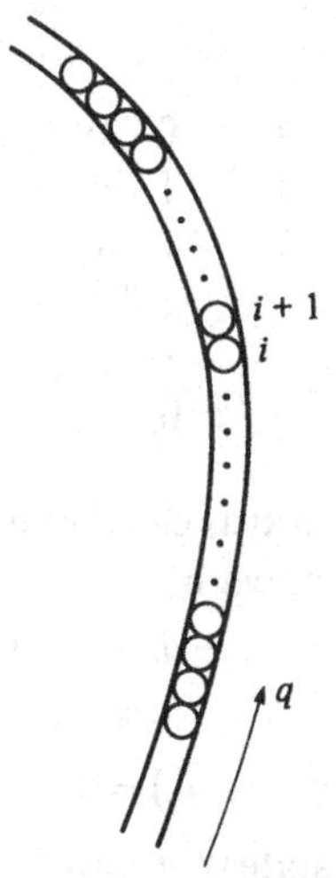

introduced such 'unilateral' constraints in the holonomic case, there may be m functions $f_\gamma[q_i, t]$ such that

$$f_\gamma \geqslant 0, \quad u_\gamma \geqslant 0, \quad u_\gamma f_\gamma = 0$$

for all time, and therefore in particular

$$\dot{u}_\gamma f_\gamma + u_\gamma \dot{f}_\gamma = 0, \quad \ddot{u}_\gamma f_\gamma + 2\dot{u}_\gamma \dot{f}_\gamma + u_\gamma \ddot{f}_\gamma = 0.$$

We suppose that $f_\gamma = \dot{f}_\gamma = 0$ at $t = 0$, implying $(1.148)_4$ in the form $u_\gamma \ddot{f}_\gamma = 0$ and $(1.148)_2$ as $\ddot{f}_\gamma \geqslant 0$. Then $A_{\gamma i} = \partial f_\gamma / \partial q_i$, $B_\gamma = \partial f_\gamma / \partial t$ from $\ddot{f}_\gamma = 0$.

Mackie (1977) has examined the particular problem of n perfectly smooth, rigid, identical spherical balls fitting exactly into the interior of a curved circular tube, as shown in Fig. 1.15. The radius of curvature of the tube is assumed large compared with the radius of a ball, and the tube is planar. The tube is fixed in a vertical plane and the balls are supported at rest in contact with one another. At $t = 0$ the support is removed so that the balls begin to fall. The problem is to find the initial acceleration of each ball, and in particular which balls remain in contact and which start to separate from each other. Let q_i be distance measured along the curved tube to the ith ball, and let u_i be the compressive reaction between the ith and $(i+1)$th balls immediately after release. Formally setting $u_0 = u_n = 0$ we have a situation where $m = n - 1$. Mackie's approximation for the equations of motion is

$$m\ddot{q}_i - u_{i-1} + u_i = z_i, \quad (\alpha)$$

where z_i is the weight of the ith ball resolved along the tube in the q-direction, and m is the mass of a ball. It follows that the $(n-1) \times n$ constraint matrix is

$$[A_{\gamma i}] = \begin{bmatrix} -1 & 1 & 0 & 0 & \cdot & \cdot & & \cdot & \cdot \\ 0 & -1 & 1 & 0 & \cdot & \cdot & & \cdot & \cdot \\ 0 & 0 & -1 & 1 & \cdot & \cdot & & \cdot & \cdot \\ \cdot & \cdot & \cdot & \cdot & \cdot & \cdot & & \cdot & \cdot \\ \cdot & \cdot & \cdot & \cdot & \cdot & \cdot & & 1 & 0 \\ 0 & 0 & 0 & 0 & & & & -1 & 1 \end{bmatrix}.$$

Since the balls cannot move towards each other, but can only either stay in contact or start to move apart, we have

$$\ddot{q}_{i+1} - \ddot{q}_i \geqslant 0, \quad (\beta)$$

$$u_i \geqslant 0, \quad (\alpha)$$

$$u_i(\ddot{q}_{i+1} - \ddot{q}_i) = 0,$$

for each $i = 1, \ldots, n-1$. These state that if any pair of adjacent balls begin to separate their relative acceleration may be positive and their mutual reaction zero. If they stay together their compressive reaction may be positive but their relative acceleration will be zero. We have thus given an illustration of (1.148) with $s_\gamma = 0$. The generating function (1.147) in the case of three balls is

$$L = -\tfrac{1}{2}m\left[\left(\ddot{q}_1 - \frac{z_1}{m}\right)^2 + \left(\ddot{q}_2 - \frac{z_2}{m}\right)^2 + \left(\ddot{q}_3 - \frac{z_2}{m}\right)^2\right]$$

$$+\frac{1}{2m}({z_1}^2 + {z_2}^2 + {z_3}^2) + u_1(-\ddot{q}_1 + \ddot{q}_2) + u_2(-\ddot{q}_2 + \ddot{q}_3).$$

(vi) Simultaneous extremum principles

Transcribing notation identifies (1.148) with (1.134), and so the maximum and minimum problems here are illustrations of the two quadratic programming problems mentioned there. Explicitly, the mechanical results are as follows.

Theorem 1.7(a) ensures uniqueness of the $\ddot{q}_i$ part, but not the u_γ part, of the solution of (1.146) and of (1.148). Moreau (1966a) claims uniqueness of the u_γ by other methods. The solution value L_0 of (1.147) in both problems is the solution value of

$$L_0 = \tfrac{1}{2}\ddot{q}_i(z_i - A_{\gamma i}u_\gamma).$$

Theorem 1.6(c) shows that this is bounded above and below by (1.80).

The upper bound has the values of

$$L - \ddot{q}_i\frac{\partial L}{\partial \ddot{q}_i} = \tfrac{1}{2}a_{ij}\ddot{q}_i\ddot{q}_j - s_\gamma u_\gamma,$$

subject to the (α) conditions. Theorem 1.10 adapted to the variants (1.53) or (1.54) shows that the problem of minimizing this quantity is equivalent to solving (1.146) or (1.148) respectively. If a_{ij}^{-1} is the typical element of the inverse (not the inverse of the typical element) of the matrix a_{ij}, we can write the upper bound as

$$\tfrac{1}{2}a_{ij}^{-1}(z_i + A_{\gamma i}u_\gamma)(z_j + A_{\delta j}u_\delta) - s_\gamma u_\gamma.$$

This nonhomogeneous quadratic bound can be further improved by minimization with respect to arbitrary u_γ in the 'bilateral' case (1.146), and with respect to $u_\gamma \geqslant 0$ in the unilateral case (1.148).

The lower bound to L_0 has the values of

$$L - u_\gamma\frac{\partial L}{\partial u_\gamma} = -\tfrac{1}{2}a_{ij}\ddot{q}_i\ddot{q}_j + z_i\ddot{q}_i,$$

subject to the (β) conditions. Under the conditions of Theorem 1.13(b) adapted to (1.148) the problem III of maximizing this quantity is equivalent to solving (1.148) at least in respect of the $\ddot{q}_i$ part of the solution. For the bilateral problem the maximization is carried out with respect to all $\ddot{q}_i$ satisfying equations (1.146β). Maximizing this lower bound is equivalent to minimizing

$$\tfrac{1}{2}a_{ij}(\ddot{q}_i - a_{ik}^{-1}z_k)(\ddot{q}_j - a_{jl}^{-1}z_l).$$

This quantity corresponds to what has been called (in the bilateral case) the 'compulsion' (Truesdell and Toupin, 1960, §237) or 'constraint' (Whittaker, 1937, §105) or square of the 'curvature' (Synge, 1960, §85) of the motion. Thus our maximum principle is a generalization, to include nonholonomic and unilateral constraints, of what is often called 'Gauss' principle of least constraint'. Our minimum principle above is usually omitted from standard texts altogether.

(vii) Cavitation

Cavitation in a continuous medium can be regarded as a breakdown of the kinematical constraint of continuity, either internally or at a boundary. In an incompressible medium the inception of cavitation can be modelled as occurring where the pressure drops to a nominated vaporization value, via an inequality like (1.148β) in terms of the time-derivative of the equation of continuity. In this way infinite dimensional versions of the foregoing upper and lower bounds have been proved by Sewell (1969) for a wide class of media with internal bilateral and unilateral constraints (Exercise 5.4.5), generalizing work on the incompressible perfect fluid by Moreau (1967)– see also Mackie (1969).

Another infinite dimensional problem having a similar theoretical structure arises in a rigid perfectly plastic body subjected to dynamic loading. Reddy (1981) has obtained simultaneous extremum principles for the stresses and accelerations in this context by following the approach described above, i.e. of identifying the central role of a saddle-shaped generating functional. He summarizes the previous literature in his context (cf. Exercise 5.4.5).

1.8. Geometric programming

(i) Introduction

In this section and the next, which are due in substance to Noble (1982, private communication), we show how some applied optimization problems fall within the theoretical framework for constructing extremum principles which has been described above. Sometimes a special twist or change of variable is needed at the outset before, for example, a generating saddle function can be discerned. Once that is done, however, and the given problem is recognized as governed by one of the cases (1.50) or (1.53)–(1.55), the other three variants offer the possibility of immediate new results if the requisite inequalities on the variables are appropriate to the applied context, as we saw in §1.7.

The term 'geometric programming' was employed by Duffin *et al.* (1967) to describe certain optimization problems in engineering design and economics which they handle via the inequality relating arithmetic and geometric means. We treat these problems differently, as an illustration of §1.4(v) and Exercise 1.6.5.

(ii) Primal problem

We are required to minimize a 'total cost' function

$$\sum_{i=1}^{m} c_i t_1^{\alpha_{i1}} t_2^{\alpha_{i2}} \cdots t_r^{\alpha_{ir}}$$

of r positive 'design variables' t_j $(j = 1, \ldots, r)$, where the $m \times r$ exponents α_{ij} are given constants, and the m coefficients c_i are given positive constants.

A strict inequality constraint $t > 0$ can always be accommodated automatically by working with a new variable $\tau = \ln t$ which may have either sign, so that $t = e^{\tau}$. The structure of the present cost function makes it especially appropriate to introduce r τ_j, all unrestricted in sign, such that $\tau_j = \ln t_j$ and $t_j = e^{\tau_j}$. Let $\xi_i = \sum_{j=1}^{r} \alpha_{ij}\tau_j$ and rewrite the primal problem in terms of $1 + m$ scalars ϕ and v_i as:

$$\text{maximize} \qquad \phi[\xi_i] = -\sum_{i=1}^{m} c_i e^{\xi_i}$$

$$\text{subject to} \qquad v_i[\xi_i, \tau_j] = \xi_i - \sum_{j=1}^{r} \alpha_{ij}\tau_j = 0.$$

(iii) Saddle function

We can introduce m Lagrangian multipliers u_i and, following (1.89), construct the generating function

$$L[\xi_i, \tau_j; u_i] = -\sum_{i=1}^{m} \left[c_i e^{\xi_i} - u_i \left(\xi_i - \sum_{j=1}^{r} \alpha_{ij}\tau_j \right) \right]. \qquad (1.149)$$

We can identify this with the variables x and u of our general theory (cf. (1.44) with $n = m + r$ here) by defining

$$x = \{\xi_1, \ldots, \xi_m, \tau_1, \ldots, \tau_r\}, \quad u = \{u_1, \ldots, u_m\}.$$

Each e^{ξ_i} is strictly convex, so that we can regard (1.149) as a non-quadratic saddle function $L[x,u]$ consisting of a weakly concave part $\phi[x]$ added to a bilinear part $\langle u, v[x]\rangle$.

This is enough to assure us that the simultaneous bounds (1.80) apply, whichever one of the four variants of governing conditions we ultimately adopt. The upper bound has the values of

$$L-\sum_{i=1}^{m}\xi_i\frac{\partial L}{\partial \xi_i}-\sum_{j=1}^{r}\tau_j\frac{\partial L}{\partial \tau_j}=\sum_{i=1}^{m}c_i(\xi_i-1)e^{\xi_i}$$

among the solutions of the (α) conditions. The lower bound has the values of

$$L-\sum_{i=1}^{m}u_i\frac{\partial L}{\partial u_i}=-\sum_{i=1}^{m}c_ie^{\xi_i}$$

among the solutions of the (β) conditions.

(iv) Governing conditions

The primal problem can now be identified as that of maximizing this lower bound, subject to the (β) conditions, provided the latter consist only of $v_i[\xi_i,\tau_j]=0$. Since we have no inherent reason to impose any $\xi_i\geqslant 0$ or $\tau_j\geqslant 0$ *a priori*, we are led to choose (1.53) in preference to (1.55). This argument embeds the primal problem into our general scheme governed by the following stationary value problem:

$$\begin{aligned}
\frac{\partial L}{\partial \xi_i}&=u_i-c_ie^{\xi_i}=0, \quad (\alpha)\\
\frac{\partial L}{\partial \tau_j}&=-\sum_{i=1}^{m}u_i\alpha_{ij}=0, \quad (\alpha)\\
\frac{\partial L}{\partial u_i}&=\xi_i-\sum_{j=1}^{r}\alpha_{ij}\tau_j=0. \quad (\beta)
\end{aligned} \qquad (1.150)$$

The quantity L_0 bounded by (1.80) is therefore the value of (1.149) appropriate to any solution of this system of $2m+r$ equations; by elimination of the u_i and τ_j we could write

$$L_0=-\sum_{i=1}^{m}c_ie^{\xi_i}\quad\text{such that}\quad\sum_{i=1}^{m}c_ie^{\xi_i}\alpha_{ij}=0.$$

(v) Dual problem

By eliminating the ξ_i at the outset, we can say that the primal problem seeks to maximize

$$-\sum_{i=1}^{m}c_i\exp\left(\sum_{j=1}^{r}\alpha_{ij}\tau_j\right),$$

directly with respect to the arbitrary τ_j. An associated maximum principle, along the lines of Theorem 1.13(b) but for the stationary value problem, would need to guarantee that the set of solutions for the τ_j obtained by this direct procedure is the same as the set of values of the τ_j which satisfy (1.150).

Our main purpose here, however, is to infer a *dual problem* to be associated with the primal problem. We define the dual problem to be that of minimizing the foregoing upper bound subject to (1.150α). All $u_i > 0$ under these constraints, which allow the u_i to be eliminated in favour of the ξ_i which are unrestricted in sign. Then the dual problem

$$\text{minimizes} \quad \sum_{i=1}^{m} c_i(\xi_i - 1)e^{\xi_i}$$

$$\text{subject to} \quad \sum_{i=1}^{m} c_i\alpha_{ij}e^{\xi_i} = 0 \quad (j = 1, \ldots, r).$$

An associated minimum principle would require a variant of Theorem 1.10 allowing also for the fact that (1.149) is strictly concave only with respect to *part* of x (the ξ_i part): the principle would guarantee that the set of solutions for the ξ_i obtained by minimization is the same as the set of values of the ξ_i which satisfy (1.149).

The 'dual' terminology is conventional in this context. In terms of the duality introduced in Chapter 2 it is justified in §2.7(i).

(vi) Alternative version of the dual problem

Under the constraints $u_i = c_i e^{\xi_i}$ we have all $u_i > 0$ so that $\xi_i = \ln(u_i/c_i)$. Introducing m normalized variables

$$\delta_i = \frac{u_i}{\Delta} \quad \text{with} \quad \Delta = \sum_{i=1}^{m} u_i > 0$$

the dual problem can be rewritten:

$$\text{minimize} \quad \Delta(\ln\Delta - 1) + \Delta\ln\prod_{i=1}^{m}\left(\frac{\delta_i}{c_i}\right)^{\delta_i}$$

with respect to varying $\delta_i > 0$ and $\Delta > 0$ subject to

$$\sum_{i=1}^{m}\delta_i = 1, \quad \sum_{i=1}^{m}\delta_i\alpha_{ij} = 0 \quad (j = 1, \ldots, r).$$

After performing the minimization with respect to Δ, there remains the problem of

$$\text{maximizing} \quad \left(\frac{c_1}{\delta_1}\right)^{\delta_1}\left(\frac{c_2}{\delta_2}\right)^{\delta_2}\cdots\left(\frac{c_m}{\delta_m}\right)^{\delta_m} \tag{1.151}$$

with respect to the m δ_i, subject to the constraints just stated. This last version is the form of the dual problem reached by Duffin *et al.* (1967) from arguments based on the geometric mean. Note how different in appearance it is from our version of the dual obtained by constructing the saddle function (1.149). Many concrete applications of geometric programming are described in the book of Duffin *et al.* (*op. cit.*).

(vii) Inequality constraints

We described briefly a variant of the foregoing primal and dual in which inequality constraints also appear. The same $m \times r$ exponents α_{ij} and the same m $c_i > 0$ are given, as before, but the cost function is defined differently. Let $p + 1$ positive integers m_k $(k = 0, 1, 2, \ldots, p)$ be given in any way such that $1 \leqslant m_0 < m_1 < m_2 \cdots < m_p = m$. Each of these is the highest number of a set M_k (say) of consecutive positive integers starting with $s_k = m_{k-1} + 1$ (with $s_0 = 1$), so filling the number line from 1 to m as illustrated in Fig. 1.16.

The primal problem is now to minimize the function

$$\sum_{i \in M_0} c_i t_1^{\alpha_{i1}} t_2^{\alpha_{i2}} \ldots t_r^{\alpha_{ir}}$$

of $t_j > 0$ $(j = 1, \ldots, r)$ subject also to

$$\sum_{i \in M_k} c_i t_1^{\alpha_{i1}} t_2^{\alpha_{i2}} \cdots t_r^{\alpha_{ir}} \leqslant 1 \quad \text{for} \quad k = 1, 2, \ldots, p.$$

Introducing the variables τ_j and ξ_i used earlier, these p inequalities become

$$\sum_{i \in M_k} c_i e^{\xi_i} \leqslant 1 \quad \text{for} \quad k = 1, 2, \ldots, p.$$

We require an extra p Lagrangian multipliers λ_k (say) leading to the new generating function

$$L[\xi_i, \tau_j; u_i, \lambda_k]$$

$$= -\sum_{i \in M_0} c_i e^{\xi_i} + \sum_{i=1}^{m} u_i\left(\xi_i - \sum_{j=1}^{r} \alpha_{ij}\tau_j\right) - \sum_{k=1}^{p} \lambda_k\left(\sum_{i \in M_k} c_i e^{\xi_i} - 1\right). \qquad (1.152)$$

Correspondence with our general theory is effected by the identifications

$$x = \{\xi_1, \ldots, \xi_m, \tau_1, \ldots, \tau_r\}, \quad u = \{u_1, \ldots, u_m, \lambda_1, \ldots, \lambda_p\}$$

and we find that the saddle quantity satisfies $S \geqslant 0$ when every $\lambda_k \geqslant 0$. The

Fig. 1.16. Sequence of sets M_k $(k = 0, 1, \ldots, p)$.

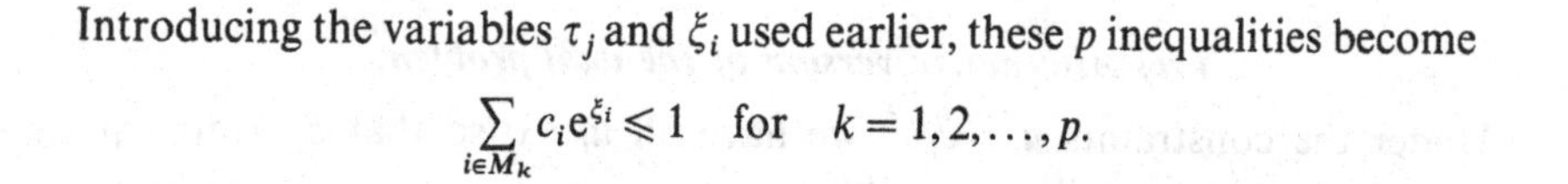

upper bound will have the values of

$$L-\sum_{i=1}^{m}\xi_i\frac{\partial L}{\partial \xi_i}-\sum_{j=1}^{r}\tau_j\frac{\partial L}{\partial \tau_j}$$

$$=\sum_{i\in M_0}c_i(\xi_i-1)e^{\xi_i}+\sum_{k=1}^{p}\lambda_k\left[1+\sum_{i\in M_k}c_i(\xi_i-1)e^{\xi_i}\right] \qquad (1.153)$$

among the solutions of an α subset of governing conditions still to be identified. The lower bound will have the values of

$$L-\sum_{i=1}^{m}u_i\frac{\partial L}{\partial u_i}-\sum_{k=1}^{p}\lambda_k\frac{\partial L}{\partial \lambda_k}=-\sum_{i\in M_0}c_ie^{\xi_i} \qquad (1.154)$$

among the solutions of a β subset still to be identified.

We can embed the new inequality constraints in the following set of governing conditions, which are deduced as a *combination* of (1.53) and (1.54).

$$\frac{\partial L}{\partial \xi_i}=\begin{cases}u_i-c_ie^{\xi_i}=0 & \text{if } i\in M_0,\\ u_i-\lambda_kc_ie^{\xi_i}=0 & \text{if } i\in M_k \quad (k\neq 0),\end{cases} \qquad (\alpha)$$

$$\frac{\partial L}{\partial \tau_j}=-\sum_{i=1}^{m}u_i\alpha_{ij}=0, \qquad (\alpha)$$

$$\frac{\partial L}{\partial u_i}=\xi_i-\sum_{j=1}^{r}\alpha_{ij}\tau_j=0, \qquad (\beta) \qquad (1.155)$$

$$\frac{\partial L}{\partial \lambda_k}=1-\sum_{i\in M_k}c_ie^{\xi_i}\geqslant 0, \qquad (\beta)$$

$$\lambda_k\geqslant 0, \qquad (\alpha)$$

$$\sum_{k=1}^{p}\lambda_k\frac{\partial L}{\partial \lambda_k}=\sum_{k=1}^{p}\lambda_k\left(1-\sum_{i\in M_k}c_ie^{\xi_i}\right)=0.$$

The primal problem is now seen to be that of maximizing (1.154) subject to (1.155β). We deduce from our theoretical framework that the dual problem is to minimize (1.153) subject to (1.155α).

To convert this dual into one announced by Duffin *et al.* (1967, pp. 78–99), it is enough to suppose that $\lambda_k>0$, since any zero λ_k will simply delete the corresponding term from (1.153). Then by $(1.155)_1$ $\xi_i=\ln u_i/c_i$ or $\ln u_i/c_i\lambda_k$ and (1.153) becomes

$$\sum_{i\in M_0}u_i\left(\ln\frac{u_i}{c_i}-1\right)+\sum_{k=1}^{p}\left[\lambda_k+\sum_{i\in M_k}u_i\left(\ln\frac{u_i}{c_i\lambda_k}-1\right)\right].$$

The stationary minimum of this with respect to the λ_k occurs when $\lambda_k=\sum_{i\in M_k}u_i$. This minimum can be expressed, in terms of $m+1+p$ new variables defined this time by

$$\delta_i = \frac{u_i}{\Delta}, \quad \Delta = \sum_{i \in M_0} u_i, \quad \gamma_k = \sum_{i \in M_k} \delta_i$$

as

$$\Delta\left[\ln \Delta - 1 + \sum_{i=1}^{m} \delta_i \ln \frac{\delta_i}{c_i} - \sum_{k=1}^{p} \gamma_k \ln \gamma_k\right].$$

Here $\Delta > 0$, and the stationary minimum of this expression with respect to Δ is

$$-\left[\prod_{k=1}^{p} \gamma_k^{\gamma_k}\right] \prod_{i=1}^{m} \left(\frac{c_i}{\delta_i}\right)^{\delta_i}. \tag{1.156}$$

Inserting the definitions of γ_k, this is the expression to be minimized with respect to the δ_i in the dual stated by Duffin *et al.* (*op. cit.*, p. 78 equation (6)). The remaining constraints are that

$$\text{all} \quad \delta_i > 0 \quad (i = 1, \ldots, m), \quad \sum_{i \in M_0} \delta_i = 1,$$

$$\sum_{i=1}^{m} \delta_i \alpha_{ij} = 0 \quad (j = 1, \ldots, r).$$

Again notice how different in *appearance* this dual is from the version (1.153) obtained by a systematic application of our procedure.

1.9. Allocation problem

(i) Introduction

The problem of allocating finite resources (such as money or human effort or natural energy) to competing activities, in a way which will maximize some chosen measure of the total return or output from those activities, is a familiar one in economics (fraught as it often is by political disagreement about what particular measure of the return should be chosen). Texts on optimization theory, such as Luenberger (1969) or Whittle (1971), often introduce the problem as follows.

Let there be n competing activities. Suppose that when x_i units of resource are allocated to the ith activity, the return from that activity is measured by a C^1 function $\phi_i[x_i]$ of one variable, for each $i = 1, \ldots, n$. Suppose that a 'law of diminishing returns' operates, defined as follows. Each $x_i \geqslant 0$. Each $\phi_i[x_i]$ is a monotonically increasing but concave function, beginning at $\phi_i[0] = 0$ and tending to a finite positive ceiling as $x_i \to b$. Here b is a fixed positive number representing the total available resources in the sense that $\sum_{i=1}^{n} x_i \leqslant b$.

(ii) Saddle function and governing conditions

The primal problem is therefore of the type posed in §1.4(v), namely:

$$\text{maximize} \quad \phi[x_i] = \sum_{i=1}^{n} \phi_i[x_i]$$

$$\text{subject to} \quad v[x_i] = b - \sum_{i=1}^{n} x_i \geqslant 0, \quad \text{all } x_i \geqslant 0.$$

We infer at once that this is an example of our general problem (1.52) generated in this case by the saddle function

$$L[x_i, u] = \sum_{i=1}^{n} \phi_i[x_i] + u\left(b - \sum_{i=1}^{n} x_i\right) \tag{1.157}$$

of type (1.89), where u here is a single scalar multiplier. The associated governing equations (1.50) become

$$\begin{aligned} &\frac{\partial L}{\partial x_i} = \frac{d\phi_i}{dx_i} - u \leqslant 0, \quad u \geqslant 0, \quad (\alpha) \\ &\frac{\partial L}{\partial u} = b - \sum_{i=1}^{n} x_i \geqslant 0, \quad x_i \geqslant 0, \quad (\beta) \\ &\sum_{i=1}^{n} x_i\left(\frac{d\phi_i}{dx_i} - u\right) = u\left(b - \sum_{i=1}^{n} x_i\right) = 0. \end{aligned} \tag{1.158}$$

We could have begun by requiring that all resources be actually used, so that $\sum_{i=1}^{n} x_i = b$ instead of the inequality. This would have required (1.55) instead of (1.50), and the sign of u could have been left unrestricted. But with the assumptions of increasing $\phi_i[x_i]$, the multiplier is automatically non-negative when $u \geqslant$ every $d\phi_i/dx_i$; so in that sense nothing new results by insisting that all resources be used.

(iii) Simultaneous extremum principles

Theorem 1.8 applies (*a fortiori* since $v[x_i]$ is linear). The lower bound has the values of

$$L - u\frac{\partial L}{\partial u} = \sum_{i=1}^{n} \phi_i[x_i], \tag{1.159}$$

as already anticipated, and is to be maximized subject to (1.158β). This is the primal problem. Theorem 1.13 provides circumstances in which the values of the x_i which maximize this lower bound are also the values which satisfy all of (1.158).

The upper bound has the values of

$$L - \sum_{i=1}^{n} x_i \frac{\partial L}{\partial x_i} = \sum_{i=1}^{n} \left[\phi_i[x_i] - x_i \frac{d\phi_i}{dx_i}\right] + ub. \tag{1.160}$$

We define the dual problem, for the above primal, to be that of minimizing this quantity subject to (1.158α). When every $\phi_i[x_i]$ is a strictly concave function, Theorem 1.10(a) and (b) holds, so that the dual problem is equivalent to that of solving (1.158). If some particular $\phi_i[x_i]$ is only weakly concave, a more detailed investigation of equivalence will be required.

With our special assumption that every $\phi_i[x_i]$ is increasing, we can minimize with respect to $u \geqslant$ every $\mathrm{d}\phi_i/\mathrm{d}x_i \geqslant 0$ from the outset. This leaves the problem of minimizing

$$\sum_{i=1}^{n}\left[\phi_i[x_i]-(x_i-b)\frac{\mathrm{d}\phi_i}{\mathrm{d}x_i}\right] \tag{1.161}$$

with respect to the x_i. We can easily simplify this expression when the $\phi_i[x_i]$ are strictly concave by introducing new gradient variables

$$y_i=\frac{\mathrm{d}\phi_i}{\mathrm{d}x_i},$$

as in (1.94). The n individual inverses $x_i = x_i[y_i]$ lead, via the Legendre transformation as in (1.93)–(1.97), to new strictly concave functions

$$\psi_i[y_i]=y_i x_i[y_i]-\phi_i[x_i[y_i]] \quad \text{with} \quad \frac{\mathrm{d}\psi_i}{\mathrm{d}y_i}=x_i.$$

We have then to minimize the strictly convex function

$$\sum_{i=1}^{n}[by_i-\psi_i[y_i]] \tag{1.162}$$

with respect to the $y_i \geqslant 0$.

2

Duality and Legendre transformations

2.1. Introduction

The main purpose of this chapter is to give a very explicit account of the Legendre transformation, including information which is not widely known. We work in finite dimensions only, and to that extent the treatment is elementary. Basic properties of the transformation are laid out in §2.2. Detailed illustrations are given in §§2.3 and 2.4 for functions of one and two variables respectively.

The transformation may be viewed as a conversion of one continuous scalar function into another. Under quite light restrictions it turns out that the transformation is reversible, and then we say that each function is the *dual* of the other, or that each is the Legendre transform of the other. The reversible Legendre transformation is also called the Legendre dual transformation. It is important for both practical and theoretical reasons.

On the practical side we can often expect to solve at least some of the constraints governing a problem, whether it be only a stationary value problem or, more strongly, a problem offering an upper bound to be evaluated prior to minimization. The Legendre transformation is one of the devices which can effect the solution of such partial constraints. This can be explained in the notation of the very particular examples of the transformation already used to facilitate the proofs of Theorems 1.3(b) and 1.9(a) (see also Exercises 1.3). The dual functions there, for example $J[u]$ and $G[v]$ in (1.34), have the properties $v = J'[u]$ and $u = G'[v]$ where the prime denotes the gradient. We shall see that these equations generalize to nonconvex and piecewise C^1 functions of many variables. Any solution u_0 of the stationary value problem $J'[u_0] = 0$ is evidently a gradient $G'[0] = u_0$ of $G[v]$ at the origin. Such a viewpoint embeds the problem of finding stationary points of $J[u]$ within the problem of finding the Legendre dual function $G[v]$. If $J[u]$ is single valued but not convex it may have several stationary points, and then $G[v]$ will be multivalued with several single valued parts lying above the origin, each with a different gradient there. Finally, if there are other variables present, such as x in $L[x, u]$ rather than u alone in $J[u]$, we may be faced with a partial constraint $\partial L/\partial u = 0$ to solve, as in (1.53) or (1.55), in place of $J'[u] = 0$.

Other reasons for the importance of the Legendre dual transformation include the symmetric way in which it organises details of physical theories, as we shall illustrate for plasticity in §2.6, for thermodynamics in §5.2, and for compressible fluid mechanics in §5.3(ii). In the physical examples which interest us most in this book, the dual functions have dimensions of energy.

Another aspect of this organizing feature stems from the fact that a function $L[x, u]$ can be used to generate three other Legendre dual functions, by holding fixed either x or u or neither. Such quartets of corresponding functions are studied in §2.5. The curvatures of corresponding parts of the functions in a quartet are found to be related in specific ways. Quadratic and nonquadratic examples of such quartets are given in §2.6 and §5.2.

In §2.7 we apply these ideas. In particular we show, in an important special case, how a saddle-shaped part of $L[x, u]$ maps into convex and concave functions having the values

$$L - \left\langle x, \frac{\partial L}{\partial x} \right\rangle \quad \text{and} \quad L - \left\langle u, \frac{\partial L}{\partial u} \right\rangle$$

of the respective upper and lower bounds associated with (1.50)–(1.55). We also discuss the solution of constraints. An indication is given in §2.7(ii) of how bifurcation theory can fit in to the theoretical structure. A changing control parameter may cause a single saddle-shaped part of $L[x, u]$ to evolve into several co-existing parts of different shapes. A generalized Hamiltonian structure is described in §2.7(iii). This is illustrated by two alternative approaches to network theory in §2.8, where energy functions less smooth than the generating functions of Chapter 1 are common.

In order to bring out these matters thoroughly, we make a fresh start in this chapter on the description of the Legendre transformation. Many facets present themselves, and it is rewarding to explore them in turn. It is illuminating to begin with continuous Legendre dual functions which need not be smooth. The transformation can be described both analytically and geometrically, and we find that a readiness to use either description can be helpful. At least one function in a dual pair often turns out to be multivalued, as illustrated above. A continuous multivalued function will be understood to consist of two or more *single valued parts* joined together regressively.

We do not aim to give an exhaustive coverage of what might be regarded as the standard aspects of duality theory, because these are usually expressed too much in the restricted context of convex functions to allow discussion of multivaluedness. Instead we discuss more recent topics, including an account of the role of singularities in the Legendre transformation, building upon a review article of Sewell (1982).

In this chapter we sometimes use the set notation of §1.3, as an alternative to the suffix notation, in order to suggest how certain formulae have infinite dimensional generalizations. The formulae for the *values* of the bounds is one of our principal examples of this, as Theorem 3.4 will show. However, additional issues arise in the infinite dimensional case and a fully coherent account of the generalization to that case is not available. Chapter 3 is therefore written in such a way that it can be read independently of Chapter 2 if desired. Thus we treat certain issues in Chapter 3 on their merits, using Chapter 2 as a guide to the structure but not as an essential prerequisite.

We hope to show that the Legendre transformation is both elegant and useful. It is not as widely appreciated by applied mathematicians as it deserves to be in spite of its long history, and it often lurks unacknowledged behind seemingly *ad hoc* manipulations or envelope constructions.

Early ideas of poles and polars are due to Apollonius. We notice that, according to Oravas and McLean (1966), Euler used the 'Legendre' transformation in 1770, 17 years before Legendre himself first used it in 1787. We shall use the conventional name, however, since to do otherwise would be to crusade for a lost cause.

2.2. Legendre transformation

(i) Introduction

We develop a definition of the Legendre transformation from scratch, in stages:

(a) as a correspondence between a point, called a *pole*, and a plane, called a *polar*;

(b) as a correspondence between two continuous surfaces, defined by allowing the pole to move on one such surface, and finding that the other is the envelope of the polar;

(c) by recognizing that each such surface may be composed of a succession of contiguous parts, called *branches*. A branch will be defined by requiring the sign of a quantity related to the curvature to be constant;

(d) we classify corresponding, or *dual*, pairs of individual branches as being either *regular*, or *singular* of various types, according to definitions to be given;

(e) finally a Legendre dual transformation is a correspondence between two continuous scalar functions which may each be composed of several such regular or singular branches.

As remarked in the Introduction, we have already noticed how explicit use of the Legendre transformation assisted our analysis in Chapter 1. Here we begin afresh. We describe the transformation using a non-committal notation which is not unduly loaded towards one particular application. Illustrations are given in the notation of their own context.

(ii) Duality between a point and a plane

Let x_i, X $(i=1,\ldots,n)$ be cartesian coordinates spanning an $n+1$ dimensional space. Let any particular point of the space be called the *pole*, with coordinates y_i, Y (say). Then the plane

$$X+Y=x_i y_i \tag{2.1}$$

is called the *polar*. The summation convention will again be used in this chapter. Poles and polars have a simple relationship with the paraboloid

$$2X=x_i^2. \tag{2.2}$$

Theorem 2.1

(a) When the pole is outside the paraboloid, the points of contact of all tangents drawn from pole to paraboloid will lie on the polar, as shown in Fig. 2.1(a).

(b) When the pole is inside the paraboloid, the polar lies outside it.

Proof

(a) The normal direction $-x_i$, 1 to the paraboloid at a tangency point will be orthogonal to the direction of the join of that point to the pole, i.e.

$$-(y_i-x_i)x_i+Y-X=0.$$

With (2.2) this implies (2.1).

(b) When the pole is inside the paraboloid $2Y>y_i^2$. Also $2X\geqslant x_i^2$ for any point of the polar inside or on the paraboloid. But the sum of these inequalities, with (2.1), implies $0>(x_i-y_i)^2$ and therefore a contradiction. □

Evidently when the pole is outside the paraboloid it is the apex of a tangent cone whose base is part of the polar, so that the polar passes through the paraboloid. Fig. 2.1(a) is a section containing the pole and the axis of the paraboloid.

A different section through the pole, perpendicular to the axis of the paraboloid would, when $y_i^2 > 2Y > 0$, express the relationship between poles and polars as a tangent construction to a sphere or, if $n = 2$, to a circle as in Fig. 2.1(*b*). The circle is the circle of Apollonius (Exercise 2.1.1).

Given a pole at a finite point, we can write down the polar. Given a polar of finite slope we can read off from (2.1) the coordinates of a finite pole. This correspondence is called a *duality* between points and planes. It may also be called a Legendre dual transformation between them. The correspondence is available algebraically, regardless of whether it is helpful to use the relationship with the paraboloid or sphere (or circle).

Fig. 2.1. (*a*) Pole outside the paraboloid, and its polar. (*b*) Section $X = Y$ through paraboloid, polar and pole.

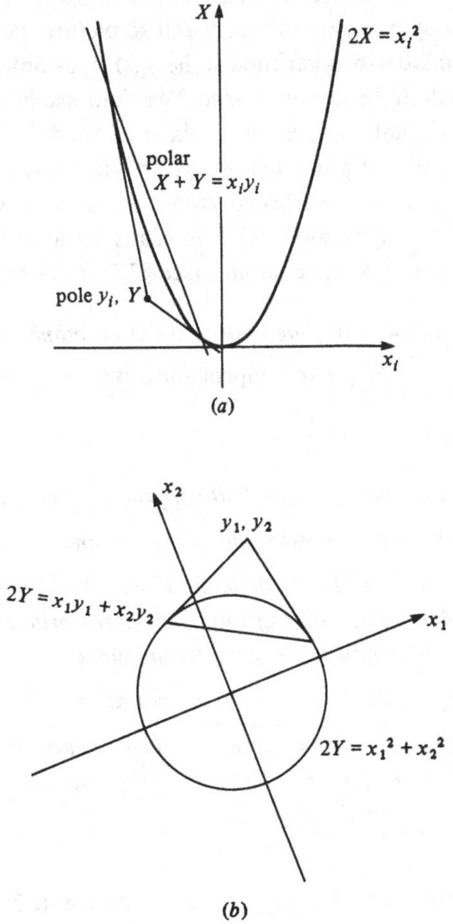

(iii) Polar reciprocation

Next we imagine the pole to move on a given continuous surface

$$Y = Y[y_i],$$

described by a given function $Y[y_i]$ of all the y_i, with smoothness to be specified. In response to this movement of the pole, the polar plane will change position and in general it will envelop another continuous surface

$$X = X[x_i],$$

described by another function $X[x_i]$. We justify this in specific cases below. The construction may be called *polar reciprocation.* It establishes a correspondence between the two surfaces, which can be called a Legendre dual transformation between the *dual functions* $Y[y_i]$ and $X[x_i]$.

According to convenience we can either introduce another $n+1$ dimensional space spanned by the y_i, Y, and so picture the surfaces in two different spaces; or we can superimpose the y_i, Y axes onto the x_i, X axes, and picture everything in the one space. We shall use both devices.

The intercept of dual surfaces on a plane of fixed $X = Y > 0$ may be described as the result of polar reciprocation with respect to a circle (or sphere). An example of this special case occurs in wave propagation problems, where the dispersion relation leads to a so-called slowness surface and, as its dual, a wave surface (see §3.5 of Sewell, 1982).

(iv) Plane locus of pole dual to point envelope of polar

As a first illustration of polar reciprocation we examine an important special case.

Theorem 2.2

(a) When the pole traverses any finite segment of the plane

$$Y = c_i y_i - C \text{ with given finite } c_i \text{ and } C, \tag{2.3}$$

the polars all pass through the single point $x_i = c_i$, $X = C$. This point is the envelope of the polars. The polar of this point is the original given plane.

(b) When the pole is any finite point on the plane

$$0 = c_i y_i - C \text{ with given finite } c_i \text{ (not all zero) and } C, \tag{2.4}$$

each polar contains a line which is parallel to the fixed direction c_i, C. If $n = 1$ such polars can only meet at infinity, but if $n > 1$ they can also intersect along a common line with the stated direction.

Proof

(a) By eliminating Y the polar (2.1) of any point of (2.3) is

$$X = (x_i - c_i)y_i + C.$$

This is satisfied by $x_i = c_i$, $X = C$ for any set of *finite* values of the n coordinates y_i. The envelope is defined by introducing the function

$$f(y_i) = (x_i - c_i)y_i + C - X \tag{2.5}$$

of the unconstrained parameters y_i, and eliminating them between the $n+1$ equations $\partial f/\partial y_i = 0$ and $f = 0$.

That the polar of the intersection point c_i, C is the plane (2.3) follows immediately from the symmetry of the definition (2.1). Fig. 2.2(*a*) is an illustration.

(b) Every polar (2.1) has normal direction $-y_i, 1$ which, by (2.4), is orthogonal to the direction c_i, C. Every polar therefore contains a line having this fixed direction. If $n = 1$ the polars are parallel lines which can only meet at infinity, as shown for $C \neq 0$ in Fig. 2.2(*b*). If $n > 1$, every pole on

Fig. 2.2. Singular envelopes for poles on a plane. (*a*) Single point envelope for poles on $Y = c_i\, y_i - C$. (*b*) Parallel polars for poles on $0 = cy - C$. (*c*) Common polar intersection for poles with the same $Y = C > 0$ on $C = c_1 y_1 + c_2 y_2$.

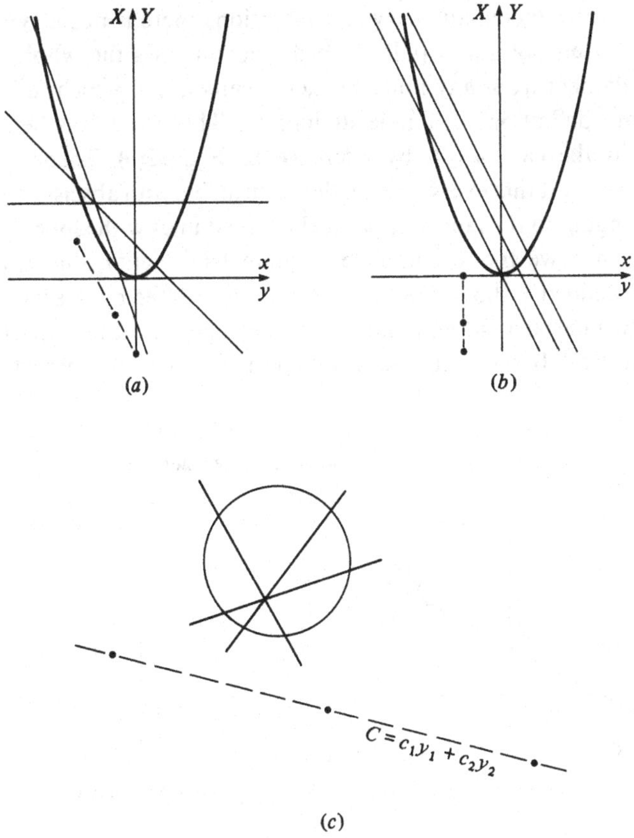

(2.4) having the same $Y = C$ has a polar which contains the line $(x_i - c_i)/c_i = X/C$ (n equations, no sum on i). This meets the plane $X = C$ in the point $x_i = 2c_i$, $X = C$. For $n = 2$ and $C > 0$ this is shown in Fig. 2.2(*c*) as a common intersection of polars which can be drawn by the tangent construction of Fig. 2.1(*b*). □

It will be noticed that our definition of the pole required it to be, by implication, at a *finite* point, since only then is the polar plane well defined as it stands in (2.1). Likewise the proof of Theorem 2.2(a) is confined to poles at finite points. The statement of Theorem 2.2(a) expresses a fully symmetric duality between finite points and finite segments of planes which have finite slope. We may call such a single point function

$$x_i = c_i, \quad X = C$$

an *accumulated singularity* of a Legendre transformation, corresponding to any finite segment of a plane (2.3) which itself will be an illustration of a *distributed singularity* of the transformation. Fig. 2.3 illustrates these notions for $n = 1$.

Without introducing auxiliary considerations there is not a symmetric duality between poles and polars which encompasses the whole infinite polar, as distinct from any finite segment thereof, and which induces an appropriate definition of a pole at infinity. This issue can be sharply focussed in the case $n = 1$, by reference to Fig. 2.2(*b*). Theorem 2.2(b) includes the result that every pole on the vertical line with abscissa C/c has a polar belonging to the family of parallel inclined lines with slope C/c. This suggests, when we try to complete a symmetric duality, that a pole at infinity be defined in the two opposite directions defined by this family, and that the two corresponding polars each be defined as at least part of the original vertical. In particular, these two parts may be half-lines which can

Fig. 2.3. Dual distributed and accumulated singularities.

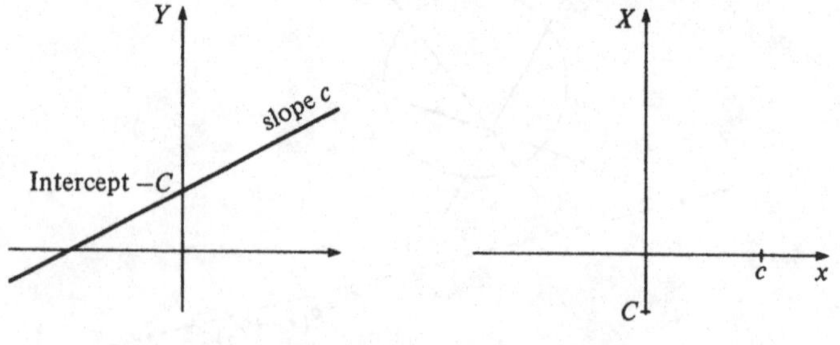

Finite straight segment $Y = cy - C$

Single point c, C

either oppose or overlay each other, as we show in §2.3(v) by introducing a limiting argument based on convexity considerations.

(v) Duality between regular branches

Let $Y[y_i]$ now be a C^2 function, at least locally. We define a *regular point* of $Y[y_i]$ to be a point where the determinant of its Hessian satisfies

$$\left|\frac{\partial^2 Y}{\partial y_i \partial y_j}\right| \neq 0 \quad \text{or} \quad \pm\infty. \tag{2.6}$$

We define a *regular branch* to be a contiguous set of regular points.

A *singular point* of a C^2 function is defined to be a point which is not regular. We describe a classification of various types of singular point in the next subsection.

Theorem 2.3

When a pole moves on a regular C^2 branch of a surface over a finite domain, its polar envelops a regular C^2 branch of another surface, also over a finite domain. When another pole moves on this second regular branch, its own polar envelops the original regular branch of the first surface.

Proof

Locally the envelope $X = X[x_i]$ of the polars (2.1) is defined analytically as described after (2.5), but with the nonlinear

$$f(y_i) = x_i y_i - Y[y_i] - X$$

in place of the linear (2.5); that is, by eliminating the y_i from

$$X = x_i y_i - Y[y_i] \quad \text{with} \quad 0 = x_i - \frac{\partial Y}{\partial y_i}. \tag{2.7}$$

By (2.6) the elementary inverse function theorem quoted in §1.5(i) applies, and this elimination can be carried out by inverting the C^1 gradient mapping $(2.7)_2$. Inserting this C^1 inverse $y_i = y_i[x_j]$ into $(2.7)_1$ gives the function

$$X[x_i] = x_i y_i[x_j] - Y[y_i[x_j]] \tag{2.8}$$

defining the envelope, with gradient

$$\frac{\partial X}{\partial x_i} = y_i \tag{2.9}$$

by the chain rule, so that $X[x_i]$ is C^2.

Using Latin suffixes to denote *second* derivatives of X and Y with respect to x_i and y_i, respectively, the inverse function theorem also gives

$$X_{ik} Y_{kj} = \delta_{ij}, \tag{2.10}$$

where δ_{ij} is the Kronecker delta. Taking determinants we have

$$|X_{ik}||Y_{kj}| = 1.$$

From this we see that (2.6) is equivalent to the same property for X_{ij}. The Legendre transformation from a regular C^2 branch of $Y[y_i]$ to a regular C^2 branch of $X[x_i]$ can therefore equally well be carried out in the reverse order. The finite gradients implied for one branch entail a finite domain for the dual branch, as $(2.7)_2$ and (2.9) show. □

Because of the envelope property in polar reciprocation, the Legendre transformation is often called a 'contact' transformation. This geometrical viewpoint also indicates that isolated locations where the curvature of $Y[y_i]$ changes discontinuously need not prevent the unique local inversion of $(2.7)_2$ which is required to construct $X[x_i]$. This permits the extension of Theorem 2.3 to the case of regular piecewise C^2 branches.

The variables y_i and x_i are called dual *active* variables in the Legendre transformation. The function $Y[y_i]$ often carries m additional parameters, say u_γ, which also appear in $X[x_i]$, so that

$$Y = Y[y_i, u_\gamma] \Leftrightarrow X = X[x_i, u_\gamma] \tag{2.11}$$

explicitly. Such u_γ are called *passive* variables in the Legendre transformation, and it is easy to show that

$$\frac{\partial Y}{\partial u_\gamma} + \frac{\partial X}{\partial u_\gamma} = 0. \tag{2.12}$$

This is another symmetric relation, to be added to the equations

$$X + Y = x_i y_i, \quad y_i = \frac{\partial X}{\partial x_i}, \quad x_i = \frac{\partial Y}{\partial y_i}, \tag{2.13}$$

obtained during the proof of Theorem 2.3. We recall that precursors of (2.13) were obtained under convexity hypotheses, which are stronger than (2.6), in (1.34) and (1.94)–(1.97).

Theorem 2.4

Dual regular C^2 branches $Y[y_i]$ and $X[x_i]$ are either (a) *both strictly convex* (b) *both strictly concave or* (c) *both saddle-shaped.*

Proof

Given an arbitrary n-tuple $\{\lambda_i\}$, we can construct in terms of it another arbitrary n-tuple $\{\mu_k\}$ such that

$$\mu_k = Y_{kj}\lambda_j, \quad \lambda_i = X_{ik}\mu_k, \tag{2.14}$$

by (2.6) and (2.10). There follows

$$\lambda_i \mu_i = Y_{ij}\lambda_i\lambda_j = X_{ij}\mu_i\mu_j. \tag{2.15}$$

Therefore (a) positive definiteness, (b) negative definiteness or (c) indefiniteness of Y_{ij} is transferred to X_{ij}, and vice versa, by (2.15). This is sufficient to establish the results, by recalling Theorem 1.5 and the remarks following (1.77). □

The elementary inverse function theorem, as applied to the gradient mapping $x_i = \partial Y/\partial y_i$ which emerges in $(2.7)_2$, includes the result that increments of coordinates and of gradients at a regular point are uniquely related, by $\mathrm{d}x_i = Y_{ij}\,\mathrm{d}y_j$ with (2.6). Notwithstanding this, however, we shall be led by (2.13) to construct regular branches of C^2 Legendre dual functions which contain *self-intersections*, for example in Fig. 2.16(*b*) when $n > 1$. We shall assume that the conflict which this type of local nonuniqueness appears to offer with the local uniqueness in the inverse function theorem can be resolved by identifying different single valued parts of the regular branch and dealing with them one at a time. In fact the standard inverse function theorem quoted in §1.5(i) can be strengthened in various particular circumstances, for example as we have already illustrated in Fig. 1.7, and as is also the case for the gradient mapping derived from Figs. 2.5 and 2.15 below.

(vi) Classification of singularities

The last two subsections described two particular types of pairs of dual functions, namely finite points and finite segments of planes in §(iv), and regular branches of C^2 functions in §(v). Associated singularities were defined on page 100 and after (2.6) respectively. It is convenient to see how the latter fit into a brief classification of singularities of the Legendre transformation. This is summarized in Fig. 2.4, and explained in the following paragraphs. We return to the generality envisaged for polar reciprocation in §(iii) above, where the pole moves on any given continuous surface in $n+1$ dimensional space, and the polar envelops another continuous surface.

It is clear from Fig. 2.2(*b*) that if the pole moves parallel to the axis of the paraboloid, if only for a short distance, the envelope of the polar will have a point or points at infinity. This means that if the function $Y[y_i]$ has a vertical tangent, or if it has a finite vertical discontinuity, the corresponding points of the dual function $X[x_i]$ will be at infinity. We say that an *infinite singularity* is a place of infinite slope or location in $Y[y_i]$ or $X[x_i]$. We can expect them to occur in dual pairs, one of each type.

Next suppose that the pole moves on part of a given continuous surface where there is not an infinite singularity, but instead a finite gradient. More than one type of finite singularity can occur. If there is an isolated location

Fig. 2.4. Classification of dual pairs of singularities of continuous locus of pole $Y = Y[y_i]$ and envelope of polar $X = X[x_i]$.

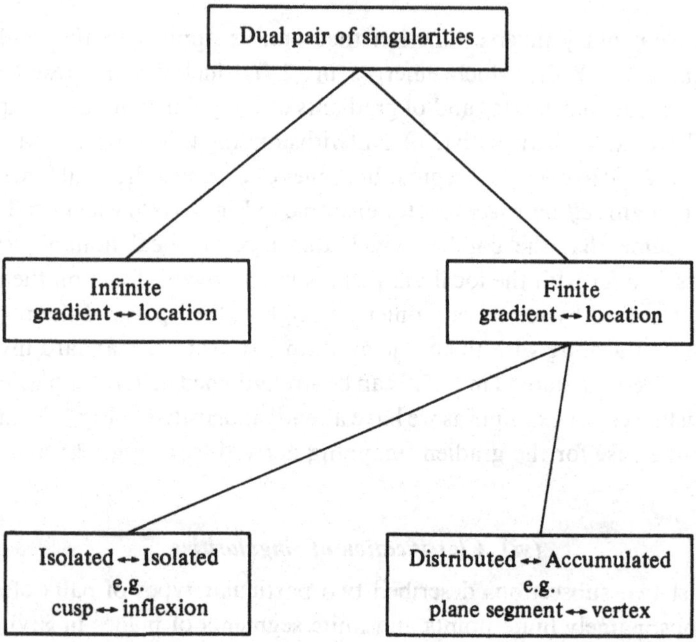

Fig. 2.5. Example of accumulated and distributed dual singularities.

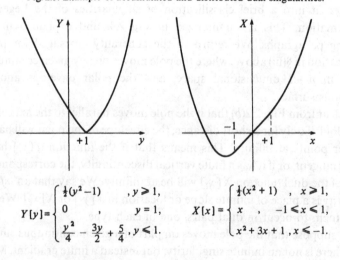

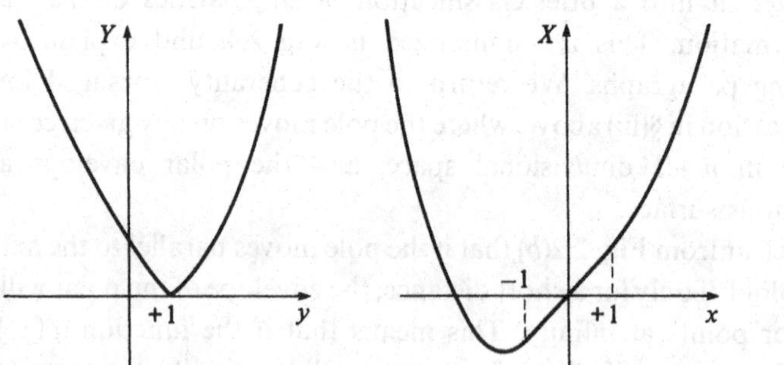

$$Y[y] = \begin{cases} \frac{1}{2}(y^2 - 1) & , y \geqslant 1, \\ 0 & , \quad y = 1, \\ \frac{y^2}{4} - \frac{3y}{2} + \frac{5}{4} & , y \leqslant 1. \end{cases} \qquad X[x] = \begin{cases} \frac{1}{2}(x^2 + 1) & , \quad x \geqslant 1, \\ x & , \quad -1 \leqslant x \leqslant 1, \\ x^2 + 3x + 1 & , x \leqslant -1. \end{cases}$$

(*a*) Accumulated at $y = 1$ (*b*) Distributed over $-1 \leqslant x \leqslant +1$

where the gradient has a nonzero finite jump, we say that there is an *accumulated singularity* at that location, by extending the terminology introduced after Theorem 2.2. We have in mind that the pole moves on a piecewise C^1 function which has vertices or edges at the accumulated singularities, and is C^2 or piecewise C^2 elsewhere, as in $Y[y_1, y_2] = -|y_1| + |y_2|$ which could be called a sharp saddle function. An example in one variable is shown in Fig. 2.5, where a strictly convex piecewise C^1 function $Y[y]$ containing an accumulated singularity maps into a weakly convex piecewise C^2 function $X[x]$ containing a distributed singularity. This is a variant of Exercise 1.3.1.

Another type of *finite singularity* will be said to occur on a C^2 function $Y[y_i]$ where (2.6) fails, i.e. where

$$|Y_{ij}| = 0 \quad \text{or} \quad |Y_{ij}| = \pm\infty. \tag{2.16}$$

In this sense we can think of a finite singularity as closing one or more regular branches. The origin in the case of y^4 or $y^{\frac{4}{3}}$ are examples. The second derivatives in (2.16) could also be one-sided in the case of a piecewise C^2 function.

Finite singularities where $|Y_{ij}| = 0$ will be called *isolated* or *distributed* in y-space according as the dimension d of the region so defined there satisfies $d < n$ or $d = n$ respectively.

The form of a distributed singularity is to be found by treating $|Y_{ij}| = 0$ as a differential equation or (if Y_{ij} has rank $r < n - 1$) system of differential equations to be integrated, and so a natural alternative name for a distributed singularity would be a *singular branch* of $Y[y_i]$. When $r = 0$ such integration is easily shown (Exercise 2.1.2) to imply that a plane segment is the only possible distributed singularity. This is consistent with the terminology first motivated after Theorem 2.2. When $0 < r < n$, however, a variety of nonplanar singular branches can be expected. For example, when $n = 2$ and $r = 1$ and under the special assumption that the variables y_1 and y_2 are separable, it is straightforward to show by direct integration that the only singular branches possible must be proportional with arbitrary sign to one of the forms

$$Y[y_1, y_2] = \begin{cases} \exp(b_1 y_1 + b_2 y_2), & (2.17) \\ (b_1 y_1 + c_1)^a (b_2 y_2 + c_2)^{1-a}, & (2.18) \end{cases}$$

where a, b_1, b_2, c_1, c_2 are any constants. Dropping the separability assumption, any C^2 function of $b_1 y_1 + b_2 y_2$ satisfies $|Y_{ij}| = 0$; so also does $(b_1 {y_1}^a + b_2 {y_2}^a)^{1/a}$ for any $a \neq 0$. When $n > 2$ and $r \geqslant 1$ other singular branches can be inferred by symmetric generalizations of these forms (Exercise

2.1.3). The location and value of the accumulated singular functions $X[x_1, x_2]$ which are dual to (2.17) and (2.18) are worked out in §2.4(ii).

Taking the hint from this, and from the special example exhibited in Fig. 2.3, we can seek to interpret a general accumulated singularity of (say) $X[x_i]$ as defined above as the Legendre dual of a (not necessarily planar) distributed singularity of $Y[y_i]$. When $|X_{ik}||Y_{kj}| = 1$ holds at a finite singularity, as at least a one-sided limit approached from dual regular branches, we can conclude that

$$|Y_{ij}| = 0 \Leftrightarrow |X_{ij}| = \pm\infty. \tag{2.19}$$

In this sense we can say that finite singularities of $Y[y_i]$ and $X[x_i]$ also occur in dual pairs.

Rather than pursue further generalities, it will be sufficient for our purpose to give some explicit examples in §§2.3 and 2.4. Like Fig. 2.5, these illustrate how distributed and accumulated singularities occur in dual pairs. They also illustrate how a finite singularity where $|X_{ij}| = \pm\infty$ can be isolated in x-space, and indicate how isolated finite singularities occur in dual pairs. An inflexion and cusp are simple examples of the latter, as also are $\frac{1}{4}x^4$ and $\frac{3}{4}y^{\frac{4}{3}}$ at the origin.

With the aid of the classification summarized in Fig. 2.4 we can now expect to construct a Legendre transformation starting from a quite complicated function, perhaps consisting of several regular branches joined by infinite or finite singularities, which may be distributed, accumulated or localized. The construction is not inhibited by the idea that the transformation may 'break down' at a singularity. Examples from physics are shown in Figs. 5.2 and 5.3.

Where an accumulated singularity joins together two regular branches of C^2 functions we may expect to evaluate a *subdifferential*, as defined in convex analysis, if the resulting composite function is indeed convex. (For example, at $y = 1$ in Fig. 2.5(a) the subdifferential is the set of all values $-1 \leqslant x \leqslant 1$ between the extreme gradients -1 and $+1$ evaluated from left and right respectively. Each particular x in this interval is called a *subgradient* at $y = 1$.) Some theories of duality are systematically expressed in terms of subdifferentials. We shall only need to mention them briefly, for example in relation to Fig. 2.15 below.

A self-intersection is not a singularity in the sense of the foregoing definitions.

(vii) Stability of singularities

There is a large body of literature stimulated by Whitney (1955), Arnol'd (1968, 1974), Thom (1969, 1975) and Zeeman (1977), often called catastrophe theory, which focusses on selected isolated singularities which are stable with respect to certain mathematical perturbations. This is the stimulus for the specific illustrations of isolated singularities in our next two sections. It may be that our distributed and accumulated singularities would be unstable unless those perturbations are severely constrained, but for our purposes such instability is a side issue at this stage. This is because we regard plane segments and vertices, for example, as helpful theoretical idealizations. In another context, witness that the experimental difficulty of observing a vertex in stress space on the plastic yield surface for a metal does not vitiate the usefulness of the vertex as an idealization in theoretical discussions.

The idea of stability of singularities in certain plane mappings is conveyed by the photograph in Fig. 2.25 below. A deep and extensive classification of certain 'Legendrian singularities' is summarized by Arnol'd *et al.* (1985).

(viii) Bifurcation set

This is an appropriate place to introduce some terminology which we shall need presently. A given function

$$V = V[y_i, u_\gamma], \tag{2.20}$$

say C^2 for simplicity to be specific, of n variables y_i and m assignable parameters u_γ, such as were introduced in (2.11), induces the stationary value problem to find the y_i from the n equations

$$\frac{\partial V}{\partial y_i} = 0 \tag{2.21}$$

for each set of parameter values. By allowing the parameters to vary we can define in the $n+m$ dimensional space an *equilibrium surface*, introduced by Sewell (1966) in a mechanical context where V has the interpretation of a potential energy. The equilibrium surface is thus the locus of all possible equilibriated combinations of 'behaviour' or 'configuration' (y_i) variables and 'control' parameters (u_γ). The latter could be, for example, assignable loads, dimensions, moduli or imperfections. The mechanical terminology can help the imagination, but is not imperative mathematically. In other contexts the term 'catastrophe manifold', as explained by Poston and Stewart (1978), is used in place of 'equilibrium surface'.

The *bifurcation set* is the set of points in the parameter space which is obtained by eliminating the y_i from (2.21) and

$$\left| \frac{\partial^2 V}{\partial y_i \partial y_j} \right| = 0. \tag{2.22}$$

In Fig. 2.10 we shall relate this elimination procedure to the elimination which was involved in the envelope calculation (2.7) of Legendre dual functions.

The original purpose of the equilibrium surface was to geometrize discussion of mechanical stability. Such stability can fail when the control parameters lie on the bifurcation set, and we shall see that this is associated with singularities of a Legendre transformation, which in turn set limits on the range of validity of an extremum principle.

Exercises 2.1

1. A plane section $X = Y > 0$ perpendicular to the axis of the paraboloid $2X = x_1{}^2 + x_2{}^2$ meets the latter in a circle, and the polar of y_1, y_2, Y in the line $2Y = x_1y_1 + x_2y_2$. Establish the connections of these intercepts with the circle of Apollonius. The latter is the locus of a point such that the ratio of its distances from two fixed points is constant (e.g. see Durell, 1920, Theorem 9). Show that this ratio is $[(y_1{}^2 + y_2{}^2)/2Y]^{\frac{1}{2}}$.
2. When the rank of the Hessian $\partial^2 Y/\partial y_i \partial y_j$ is zero over an extended region, for general $n > 1$, prove that the function $Y[y_i]$ of the n y_i is necessarily a plane segment.
3. Verify that

$$Y[y_1, y_2, y_3] = \begin{cases} \exp(b_1y_1 + b_2y_2 + b_3y_3), \\ (b_1y_1 + c_1)^{a_1}(b_2y_2 + c_2)^{a_2}(b_3y_3 + c_3)^{a_3}, \end{cases}$$

where the coefficients and exponents are constants and $a_1 + a_2 + a_3 = 1$, are distributed singular branches in the same sense as (2.17) and (2.18), but with $r = 1$ and $r = 2$ respectively $(0 < r < n = 3)$. Do the same for

$$(b_1y_1{}^a + b_2y_2{}^a + \cdots + b_ny_n{}^a)^{1/a}$$

for any $a \neq 0$.

4. If $Y[y]$ is a weakly convex piecewise C^2 function of a single variable (contrast the $J[u]$ of Theorem 1.3(b)), prove that its Legendre dual $X[x]$ is a strictly convex piecewise C^1 function.

2.3. Legendre transformations in one active variable, with singularities

(i) Inflexion and cusp as dual isolated singularities

The function

$$Y[y, u] = \tfrac{1}{3}y^3 + uy \tag{2.23}$$

is a C^∞, and therefore C^2, function of two scalar variables. At each fixed u the cubic is a single valued function made up of two regular branches, where

$\partial^2 Y/\partial y^2 > 0$ over $y > 0$ and $\partial^2 Y/\partial y^2 < 0$ over $y < 0$, joined smoothly and progressively by the isolated finite singularity at the inflexion point $y = 0$ itself. There are no infinite singularities. The gradient equation $x = \partial Y/\partial y$ is

$$x = y^2 + u, \tag{2.24}$$

which can be inverted locally on each regular branch in turn. The pair of inverses

$$y = \pm(x-u)^{\frac{1}{2}} \tag{2.25}$$

over $x > u$ are the gradients $y = \partial X/\partial x$ of respective branches of the dual function

$$X[x,u] = \pm\tfrac{2}{3}(x-u)^{\frac{3}{2}}, \tag{2.26}$$

found by eliminating y from $X + Y = xy$ with $x = \partial Y/\partial y$.

This is a two valued function which can be regarded as made up of two single valued regular branches, whereon $\partial^2 X/\partial x^2$ is finite for $x > u$, joined smoothly but regressively by a cusp point at $x = u$. This cusp point is an isolated and, in the sense of $(2.19)_2$, finite singularity, where $\partial^2 X/\partial x^2 = \pm\frac{1}{2}(x-u)^{-\frac{1}{2}}$ jumps from $-\infty$ to $+\infty$. The function (2.26) is therefore piecewise C^2 over $x \geqslant u$. Fig. 2.6 shows two cases, in each of which B denotes the pair of dual isolated singularities closing one pair AB and another pair BC of single valued dual regular branches.

Fig. 2.6. Inflexion and cusp as dual isolated singularities

$Y[y,u] = \frac{1}{3}y^3 + uy, \quad X[x,u] = \pm\frac{2}{3}(x-u)^{\frac{3}{2}}.$

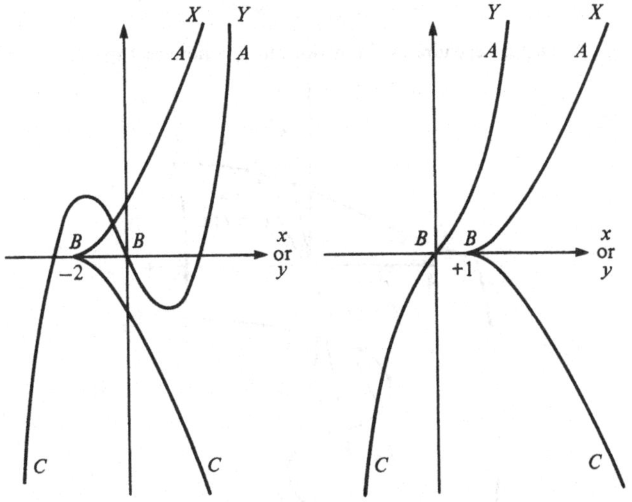

(a) $u = -2$ (b) $u = +1$

Starting from (2.26), the reader can easily verify that the double valuedness presents no obstacle to carrying through the Legendre transformation analytically in the reverse order.

Geometrically it is possible to perceive the pair of dual functions (2.23) and (2.26) simultaneously in the same three dimensional picture, and for every value of u, by a superposition of axes different to that of Fig. 2.6, as follows. We can write (2.23) either as

$$\tfrac{1}{3}y^3 + uy - Y = 0 \tag{2.27}$$

or, by substituting $Y = xy - X$ which relates the coordinates of pole and polar, as

$$\tfrac{1}{3}y^3 - (x-u)y + X = 0. \tag{2.28}$$

Allowing u to vary, and using axes of y, u, Y superposed on axes of y, $-(x-u)$, $-X$, we can display either (2.27) or (2.28) as the same folded surface exhibited in Fig. 2.7. The surface can be regarded as made up of straight lines, on each of which $y = constant$, and we show one of them.

Using the y, u, Y axes every cross-section $u = constant$ of the surface is a member of the family of cubics (2.23). The Legendre dual (2.26) of every such cubic is found by projecting the surface parallel to the y-axis onto the x–u, X plane. The perceived fold-line in the surface projects into the curve (2.26).

This is because the pair of equations (2.7) from which y is eliminated to reach (2.26) by the envelope viewpoint is exactly the same as the pair of equations from which y has to be eliminated to perform the above

Fig. 2.7. Legendre dual functions in superimposed spaces.

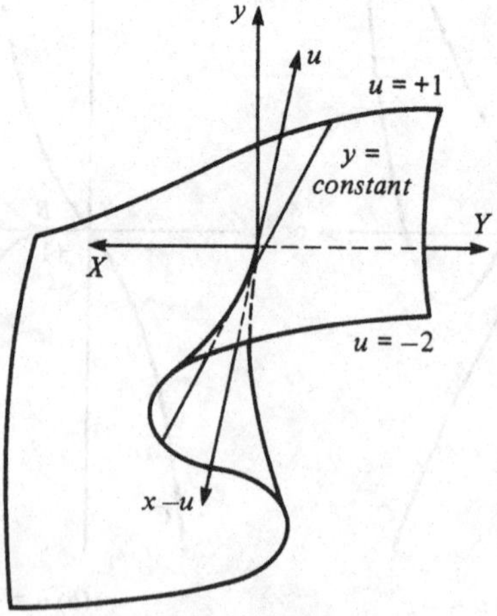

projection. It is straightforward to verify that the latter pair are the equations $\partial V/\partial y = \partial^2 V/\partial y^2 = 0$ defining, by (2.21) and (2.22), the bifurcation set for the so-called 'cusp catastrophe potential' in the form

$$V[y, x-u, X] = \tfrac{1}{12}y^4 - \tfrac{1}{2}(x-u)y^2 + Xy, \qquad (2.29)$$

treating $x-u$ and X as assignable control parameters for this purpose. The surface in Fig. 2.7 is thus the ubiquitous equilibrium surface associated with this quartic potential. (For a mechanical example, it has been known since 1969 that in the buckling of elastic struts, Fig. 2.7 relates the equilibrium values of load $(x-u)$, imperfection (X) and amplitude (y) of deflection in the fundamental mode, cf. Fig. 4 of Sewell (1966).)

This unified view in a single three dimensional diagram of pairs of Legendre dual functions applies more generally (see Theorem 2.6). The photograph in Fig. 2.8 (from Sewell (1978*a*)) of a string model of the surface (2.27) or (2.28), taken parallel to the y-axis and therefore in the direction of the projection, shows the surface made up of straight lines (the strings, on each of which $y = constant$, at intervals of $\tfrac{1}{4}$ in $-2 \leqslant y \leqslant +2$) which are seen to envelop the cusped curve (2.26). Cubic cross-sections $u = \tfrac{2}{3}$ and $u = -2$ in (2.27), where the strings pass through upper and lower end planes

Fig. 2.8. Strings $y = constant$ on surface $\tfrac{1}{3}y^3 - (x-u)y + X = 0$, enveloping $X = \pm\tfrac{2}{3}(x-u)^{\frac{3}{2}}$ in the $x-u$, X plane.

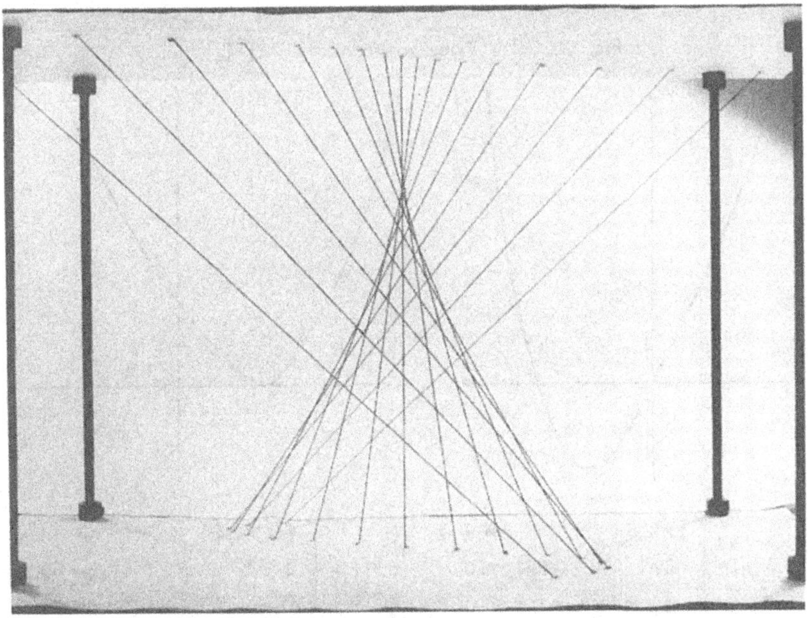

respectively, are also visible. If we regard any point on the surface as a pole with coordinates y, u, Y, its own polar in the previous sense is the string which passes through that same point but is regarded as a line (2.28) in the $x - u, X$ plane with given y and u. In this sense *a pole lies on its own polar.* The photograph displays the enveloping property of these polars.

(ii) Dual distributed and accumulated singularities

In the previous example the Legendre transformation applied to the C^2 cubic over $-\infty < y < \infty$ preserved both the isolated and the C^1 character of the singularity, but induced a jump in the (one-sided) second derivative of the dual function at the cusp.

Consider now the example of Fig. 2.9, in which $Y[y]$ is a C^2 function consisting of two regular branches joined by a linear segment. The reader may find it a useful algebraic exercise to verify that the dual function $X[x]$ is as stated there, namely a piecewise C^1 function consisting of two C^2 regular branches joined by an accumulated singularity across which there is a jump in (one-sided) gradient. This vertex has mapped from the finite linear branch of $Y[y]$. In other words, even though $Y[y]$ is C^2, its singular branch means that $X[x]$ is not even C^1 because an accumulated singularity in the form of a vertex is induced.

Another example was shown in Fig. 2.5, this time in which a piecewise C^2 function with linear segment is dual to a piecewise C^1 function with vertex.

Fig. 2.9. C^2 $Y[y]$ dual to piecewise C^1 $X[x]$.

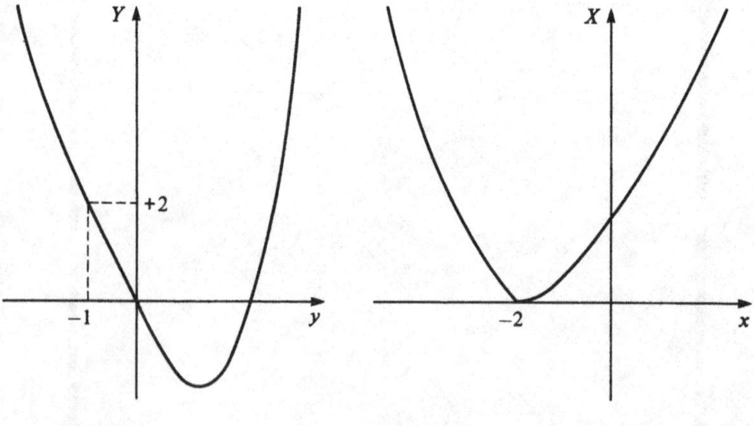

$$Y[y] = \begin{cases} \tfrac{1}{3}y^3 - 2y, & y \geqslant 0 \\ -2y, & -1 \leqslant y \leqslant 0 \\ -\tfrac{1}{3}(y+1)^3 - 2y, & y \leqslant -1 \end{cases} \qquad X[x] = \begin{cases} \tfrac{2}{3}(x+2)^{\frac{3}{2}}, & x \geqslant -2 \\ 0, & x = -2 \\ \tfrac{2}{3}(-x-2)^{\frac{3}{2}} + (-x-2), & x \leqslant -2 \end{cases}$$

To illustrate a different choice of viewpoint, we started there with $Y[y]$ as the function with the vertex, to be viewed as an accumulated singularity, and then used the gradient inversions of the Legendre transformation to construct the stated $X[x]$ having a linear segment dual to the vertex.

(iii) Some general theorems

Here we generalize the connections between Legendre dual functions, bifurcation sets, and their representation in a single three dimensional diagram, which we illustrated in the specific example of §(i) above. The results were given by Sewell (1982).

Theorem 2.5

Let $\phi[y]$ be a given function, supposed C^3 for simplicity, of one scalar variable with gradient $\phi'[y]$. Then the Legendre dual of the function

$$Y[y] = \phi'[y] + uy \tag{2.30}$$

for any fixed u, has the same shape as the bifurcation set in $x - u, X$ space of the potential function

$$V[y, x-u, X] = \phi[y] - \tfrac{1}{2}(x-u)y^2 + Xy, \tag{2.31}$$

where $x - u$ and X are treated as assignable parameters.

If $\phi'[y]$ contains $m - 1$ passive variables in addition to y, the same result applies in the $m + 1$ dimensional space spanned by $x - u, X$ and the additional $m - 1$ parameters.

Proof

According to (2.7) the dual function in the $x - u, X$ plane is found by eliminating y from the pair of equations

$$X = (x-u)y - \phi'[y], \quad 0 = x - u - \phi''[y]$$

defining the envelope property.

But exactly the same equations also follow from

$$\frac{\partial V}{\partial y} = 0 \quad \text{and} \quad \frac{\partial^2 V}{\partial y^2} = 0$$

applied to (2.31). The result follows from the definition of bifurcation set given in relation to (2.21) and (2.22).

The last part is immediate as soon as the $m - 1$ extra parameters are written out explicitly. □

Theorem 2.6

The surface drawn in three dimensions as

$$\phi'[y] + uy - Y = 0 \tag{2.32}$$

or

$$\phi'[y] - (x - u)y + X = 0 \tag{2.33}$$

on superimposed y, u, Y *and* $y, -(x-u), -X$ *axes respectively has the following property. The Legendre dual of every function* $Y[y]$ *defined by any cross-section* $u = constant$ *of* (2.32) *is found, as a bifurcation set of* (2.31), *by projection of* (2.33) *parallel to the y-axis onto the* $x-u$, X *plane.*

The proof is a geometrical rephrasing of the previous theorem.

The surface (2.32) or (2.33) is made up of straight lines on each of which $y = constant$. For a given pole y, u, Y satisfying (2.32), the polar (2.33) passes through that pole in the superimposed three dimensional diagram, and is perceived as the projection of that line onto the $x-u, X$ plane. The alternative envelope and projection viewpoints entail the elimination of y from

$$X = y\phi''[y] - \phi'[y], \quad x - u = \phi''[y],$$

as we see from above. These Theorems are all illustrated in §(i) and Figs. 2.6–2.8 above for the case $\phi[y] = y^4/12$.

We leave the reader to explore the extent to which the C^3 assumption in Theorem 2.5 may be relaxed, to incorporate the presence of distributed and accumulated singularities in the three dimensional picture involved in Theorem 2.6, and as illustrations to represent the curves of Figs. 2.5 and 2.9 in this way (see Exercise 2.2.3).

Exercises 2.2

1. For the functions of Fig. 2.9, plot
   (i) $-X$ as a function $f[y] = Y - y\,\mathrm{d}Y/\mathrm{d}y$ of y, and verify that this is a piecewise C^2 function having nonisolated (distributed) maxima, with respect to all y, at the points induced by the linear branch of $Y[y]$;
   (ii) $-Y$ as a function $g[x] = X - x\,\mathrm{d}X/\mathrm{d}x$ of x, and verify that this has a discontinuity induced by the vertex in $X[x]$, and an inflexion elsewhere.
   Show that $f[y]$ and $g[x]$ are *not* Legendre dual functions.
2. Perform similar calculations to those of Exercise 1 for the functions of Fig. 2.5. Show in particular that the linear branch of $X[x]$ induces nonisolated maxima of $g[x] = X - x\,\mathrm{d}X/\mathrm{d}x$ with respect to $\mathrm{d}X/\mathrm{d}x \geqslant 0$ at points which do not satisfy $x\,\mathrm{d}X/\mathrm{d}x = 0$. Contrast Theorem 1.3(b). Compare Exercise 1.3.1.
3. (i) Determine the function $\phi[y]$ which is required to identify the $Y[y]$ of (2.30) with the $Y[y]$ of Fig. 2.9, and specify its smoothness.
   (ii) Construct an actual string model of the three dimensional surface (2.32) with the $\phi[y]$ which you have found.
   (iii) Verify that its projection parallel to the y-axis reveals the dual function $X[x+2]$ in Fig. 2.9. Verify that the strings can be regarded as polars through their own pole.

4. Repeat Exercise 3 for the functions of Fig. 2.5.
5. Show that, for given $a \neq 1$ or 0, every regular branch of the dual of x^a is proportional to y^b, where $(1/a)+(1/b)=1$.

(iv) Ladder for the cuspoids

The significance of Theorem 2.5 is that bifurcation sets of (2.31) have already been determined in the catastrophe theory literature for certain particular functions $\phi[y]$, and therefore the associated Legendre duals of (2.30) can simply be read off from this known information. In particular this is true when $\phi[y]$ is one of the polynomials

$$\phi[y] = \frac{y^{p+1}}{p(p+1)} + b_{p-1}\frac{y^{p-1}}{(p-2)(p-1)} + b_{p-2}\frac{y^{p-2}}{(p-3)(p-2)} + \cdots$$
$$+ \frac{b_4}{12}y^4 + \frac{b_3}{6}y^3 \qquad (2.34)$$

for any positive integer $p \geqslant 2$, where the coefficients are given fixed values of passive variables. When quadratic and linear terms are added, as in the case of (2.30) and (2.31), the resulting polynomial is sometimes called a *cuspoid* potential. The coefficient of the penultimate highest degree term can always be transformed to zero by a change of y-origin, and we have supposed without loss of generality that this has already been done. Fixing the highest coefficient as indicated leaves $p-2$ assignable coefficients in a cuspoid potential of degree p.

***Theorem* 2.7** (Ladder for the cuspoids)
The Legendre dual of the cuspoid of degree p, in a $(p-1)$ *dimensional space, has the same shape as the bifurcation set of the cuspoid of degree* $p+1$.

Proof
This is an immediate consequence of applying the last part of Theorem 2.5 to the example (2.34), since there $m = p-2$. □

The result was originally established in a different way by Sewell (1977a) using explicit scaling arguments, and expressed as a 'ladder for the cuspoids'. The first few steps of this mnemonic are indicated in Fig. 2.10. In the left column is the list of names which have become conventional for the cuspoid potentials of degree 3 (fold), 4 (cusp), 5 (swallowtail) and 6 (butterfly). In the right column are surfaces lying in spaces of dimension 2 (our (2.26) in the $x-u, X$ space implied by Fig. 2.6), 3 (our Fig. 2.11(b) below) and 4, each obtainable in two different ways – as a Legendre dual function by stepping up the ladder or as a bifurcation set by stepping down it. Recall that the Legendre dual $X[x]$ of (2.30) is actually always of the form $X[x-u]$.

Computer pictures of bifurcation sets for the cases $p = 3$ to $p = 8$ in (2.31) with (2.34) have been published by Woodcock and Poston (1974), by regarding them as surfaces ruled by lines $y = constant$. They are complicated surfaces in coefficient spaces, being typically multivalued and composed of several regular branches, sometimes self-intersecting, but joining with continuous value and gradient at isolated singularities, which may be cusped edges ('ribs') of regression. Nevertheless they may be considered known, especially as regards their qualitative structure. In that sense the corresponding Legendre dual functions are therefore also known by reading them off the ladder.

The explicit calculation in §(i) above can now be seen to illustrate the first

Fig. 2.10. Ladder for the cuspoids

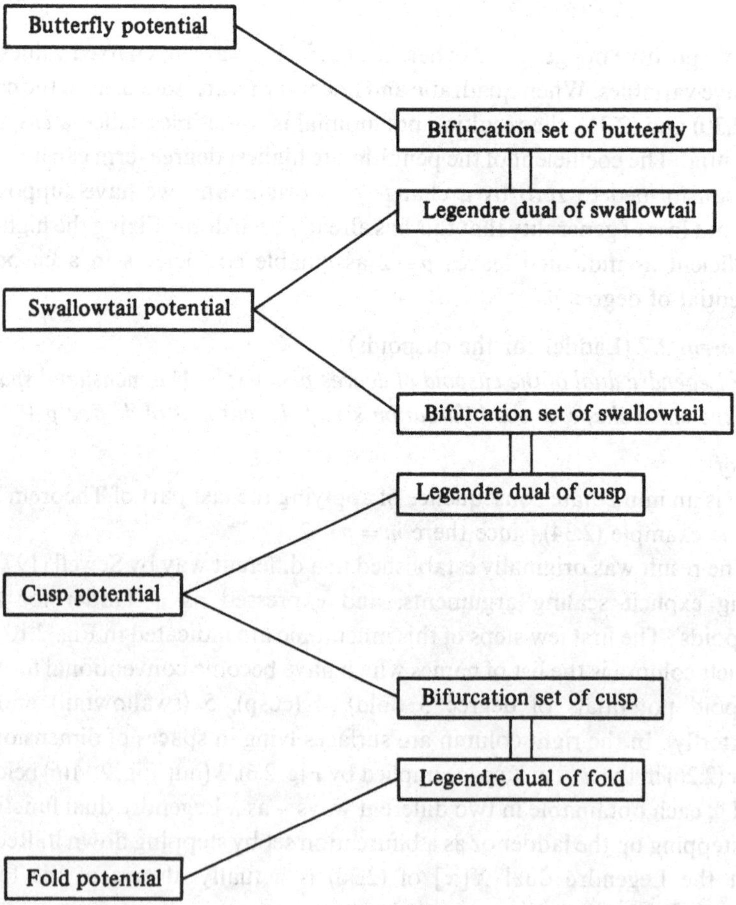

result in the ladder, that the Legendre dual of the cubic fold potential has the same shape as the bifurcation set of the quartic cusp potential. In a compressible fluids example in §5.3(ii) and Fig. 5.3(*a*) and (*b*) we shall be guided by the second result in the ladder, that the Legendre dual $X = X[x-u, b]$ of the quartic cusp potential

$$Y[y, u, b] = \tfrac{1}{4}y^4 + \tfrac{1}{2}by^2 + uy \tag{2.35}$$

for any fixed u has the same shape as the known bifurcation set of the quintic swallowtail potential

$$V[y, b, x-u, X] = \tfrac{1}{20}y^5 + \tfrac{1}{6}by^3 - \tfrac{1}{2}(x-u)y^2 + Xy. \tag{2.36}$$

Although there is not a simple algebraic expression for $X[x-u, b]$ such as we found in (2.26), qualitative knowledge of the result via (2.36) facilitates the following direct calculation based on (2.35).

From (2.35) we can easily plot $Y - uy = \tfrac{1}{4}y^4 + \tfrac{1}{2}by^2$ in three dimensions as a function of one active variable (y) and one passive variable (b), as shown

Fig. 2.11. Quartic (2.35) and its swallowtail dual function.

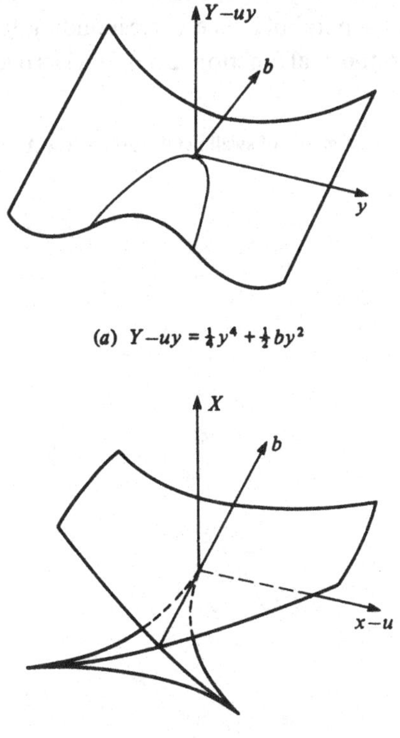

(*a*) $Y - uy = \tfrac{1}{4}y^4 + \tfrac{1}{2}by^2$

(*b*) Dual $X = X[x-u, b]$ of (*a*)

in Fig. 2.11(*a*). This is a single valued surface made up of symmetric quartic cross-sections for each fixed b. For $b < 0$ each quartic section contains two inflexions joining three regular branches; for $b > 0$ each quartic section is one regular branch with no inflexion. The gradient equation $(2.7)_2$ to be inverted is here $x = \partial Y/\partial y$ for each b, i.e.

$$x - u = y^3 + by. \tag{2.37}$$

The inversion of this is controlled by the inflexions which are isolated singularities where $\partial^2 Y/\partial y^2 = 0$, i.e.

$$0 = 3y^2 + b. \tag{2.38}$$

The sign of $\partial^2 Y/\partial y^2$ changes as y traverses the inflexion at any given $b < 0$, and so we can expect from the simpler example of Fig. 2.6 that the dual of each inflexion will be a cusp. We therefore expect the dual function

$$X = X[x - u, b] \tag{2.39}$$

to consist, for $b < 0$, of three regular branches joining at two cusps where $\partial^2 X/\partial x^2$ changes sign (by (2.11)); and for $b > 0$, of just one regular branch.

When b is allowed to vary (2.38) shows that the locus of inflexions on Fig. 2.11(*a*) lies over a parabola. The corresponding locus of cusps is an edge of regression on the dual function (2.39), and lies over a curve found by

Fig. 2.12. String model of swallowtail surface $X = X\ [x-u, b]$.

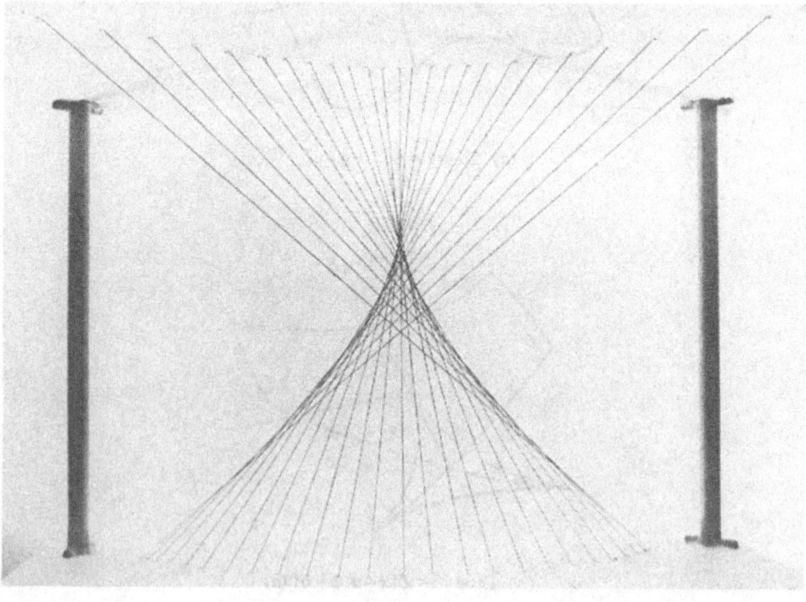

eliminating y from (2.37) and (2.38), namely

$$x - u = \mp 2\left(-\frac{b}{3}\right)^{\frac{3}{2}}.$$

This curve *itself* has a cusp at $x = u, b = 0$. The corresponding point on the dual function (2.39) itself, now shown in three dimensions in Fig. 2.11(*b*), is called the *swallowtail point*. The dual function can be drawn as a succession of cross-sections $b = constant$. It is symmetric with respect to the plane $x - u = 0$ in Fig. 2.11(*b*).

Many more details were described by Sewell (1977a), including the photograph of a string model of $X = X[x - u, b]$ shown in Fig. 2.12. The photograph is taken looking along the negative X-axis. The b-axis is vertically upward, and floor and ceiling of the model are the planes $b = -5$ and $b = 2.5$ respectively. On each string y is constant, and this value varies at intervals of 0.1 in the range $-1.2 \leqslant y \leqslant 1.2$ to generate successive strings. Thus the model displays the dual function (2.39) as parametrized by y according to the equations

$$X = \tfrac{3}{4}y^4 + \tfrac{1}{2}by^2, \quad x - u = y^3 + by$$

specified in (2.7).

(v) Half-line dual of a pole at infinity

We now return to the issue raised at the end of §2.2(iv), namely of how to define a polar to be associated with a pole at infinity. We give two examples which indicate the kind of auxiliary considerations required to handle this question.

Let c and C be two fixed numbers, and u be another assignable parameter

Table 2.1. *Families of parabolas and cubics and their duals*

$Y[y, u]$	$cy - C + \tfrac{1}{2}uy^2$	$cy - C + \tfrac{1}{3}uy^3$
$x = \dfrac{dY}{dy}$	$c + uy$	$c + uy^2$
$y = \dfrac{dX}{dx}$	$\dfrac{x - c}{u}$	$\pm\left(\dfrac{x - c}{u}\right)^{\frac{1}{2}}$
$X[x, u]$	$\dfrac{(x - c)^2}{2u} + C$	$\pm\dfrac{2(x - c)^{\frac{3}{2}}}{3u^{\frac{1}{2}}} + C$

which can be regarded as generating a family $Y[y,u]$ of quadratics or cubics as listed in Table 2.1. For each u, except $u=0$, these are regular functions of y excepting also that the cubics have an isolated singularity at their inflexion.

Treating u as a passive variable, the application of

$$x=\frac{\mathrm{d}Y}{\mathrm{d}y},\quad y=\frac{\mathrm{d}X}{\mathrm{d}x},\quad xy=X+Y, \tag{2.40}$$

from (2.13) for $n=1$, shows that the respective families $X[x,u]$ of Legendre

Fig. 2.13. Duals of parabolas and cubics for $C=c=1$ and their limits as $u\to 0$.

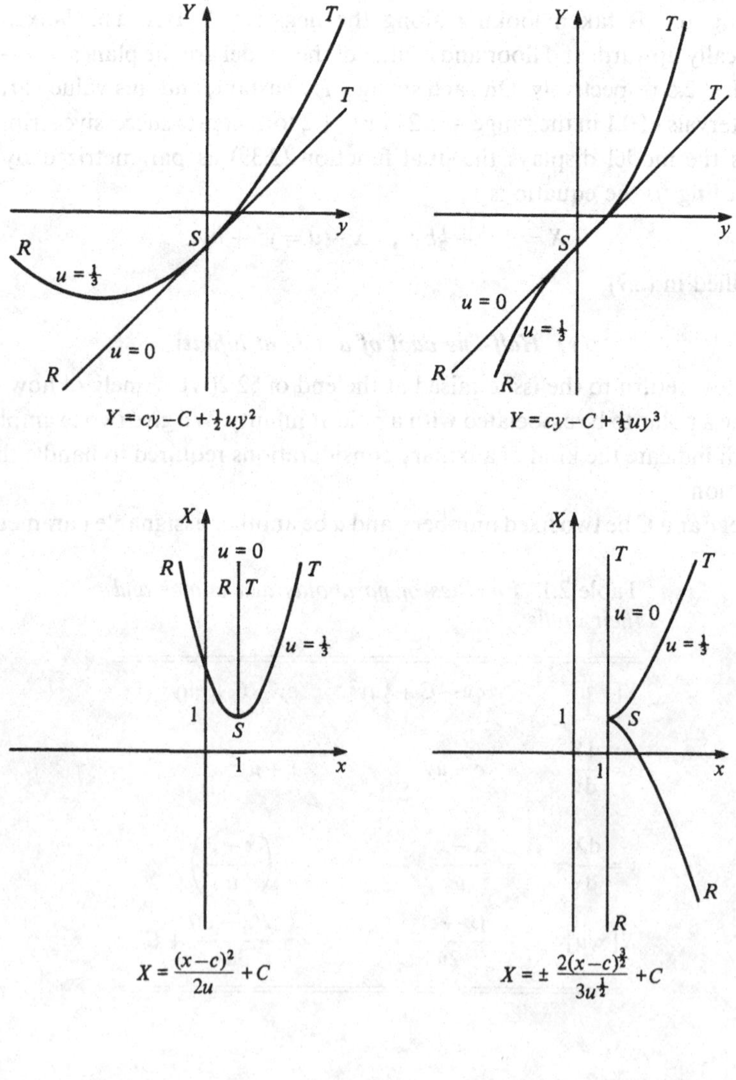

dual functions are those listed in Table 2.1. For $c = C = 1$ and $u = \frac{1}{3}$ these functions are all shown in Fig. 2.13. A parabola maps into a parabola. A cubic maps into the cusped curve of §2.3(i). Half-curves labelled RS and ST map into half-curves with the same labels.

In the limit $u \to 0$, from above or below, every $Y[y, u]$ becomes the straight line $Y = cy - C$ also shown in Fig. 2.13. Regarded as a distributed singularity, we already know that every finite segment of this line has a dual accumulated at the single point $x = c$, $X = C$. This is at the nose of the dual parabola, or at the cusp of the dual to the cubic.

Consider now what should be the dual of the *whole* line $Y = cy - C$, including points at infinity in both directions. We treat separately the two half-lines where $y \geqslant 0$ and $y \leqslant 0$. Fig. 2.13 indicates that, when such a half-line is viewed as the limit of a convex curve, the dual is the limit of another convex curve, namely the vertically upward half-line

$$x = c, \quad X \geqslant C.$$

On the other hand, when an inclined half-line is viewed as the limit of a concave curve, the dual is the limit of another concave curve, namely the vertically downward half-line

$$x = c, \quad X \leqslant C.$$

Half-lines labelled RS and ST in Fig. 2.13 map into half-lines with the same labels. The limit of the duals of the parabolas is two half-lines overlaying each other; the limit of the duals of the cubics is two half-lines opposing each other.

Put otherwise, when a pole is deemed to be at infinity in a direction specified by an inclined half-line, the polar may be chosen as either an upward or a downward vertical half-line. The choice depends on the auxiliary consideration of whether limits are approached via a family of convex or of concave curves.

(vi) Response curves and complementary areas

The gradient equations $x_i = \partial Y / \partial y_i$ and $y_i = \partial X / \partial x_i$ in (2.13) can each represent response relations in physical contexts, such as stress–strain laws in continuum mechanics or voltage–current characteristics in electrical network problems. From this viewpoint the x_i and y_i in a given context will have different *a priori* physical definitions, such as the components of one of the various precise tensor measures of stress and strain, respectively, or their rates of change; or like the analogues of voltage and current in an element of a transportation or economic network. The scalar functions

$Y[y_i]$ and $X[x_i]$ themselves are interpretable as measures of energy or complementary energy.

When only one active variable is present the response relations, now $(2.40)_1$ and $(2.40)_2$, can be drawn as *response curves* $x = x[y]$ and $y = y[x]$ in an x, y plane. So long as such a response curve leaving the origin remains monotonic, $(2.40)_3$ written as $X[x] + Y[y] = xy$ has the interpretation that

Fig. 2.14. Areas under a response curve with isolated singularity.

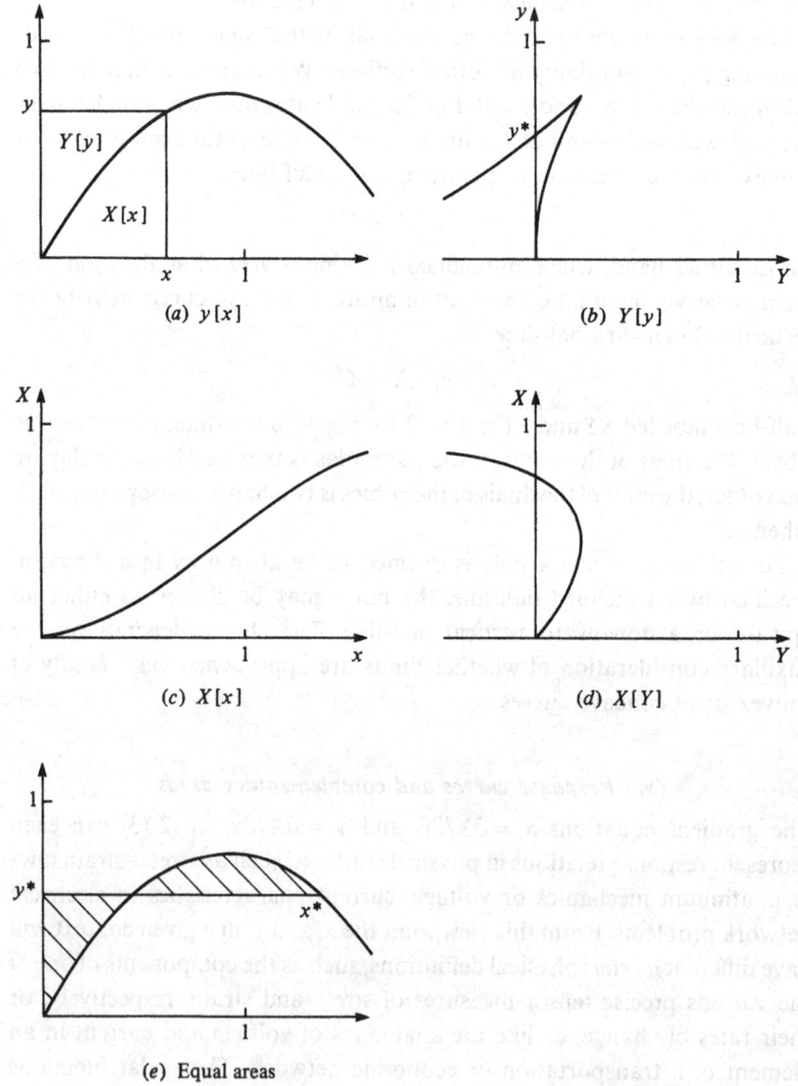

the rectangle between the axes, the origin, and the point x, y is divided by the response curve into two *complementary areas* $Y[y]$ and $X[x]$ lying between the curve and the axes, as shown in Fig. 2.14(*a*).

This area interpretation can be extended through a finite singularity, for example where $\mathrm{d}y/\mathrm{d}x = 0$, if we add the convention that, when the ordinate is falling, areas subtended onto the ordinate axis are counted as negative. Fig. 2.14, adapted from Sewell (1980), illustrates a response curve on which $\mathrm{d}y/\mathrm{d}x$ decreases monotonically from positive values through an isolated zero to negative values. At the maximum y the area $X[x]$ passes through an inflexion, and the complementary area $Y[y]$ passes through a cusp. The foregoing sign convention then comes into operation, allowing $Y[y]$ to start falling and even to pass through zero at a later point x^*, y^* (say) on the response curve, where $X = x^*y^*$ in value. The two shaded areas in Fig. 2.14(*e*) are then equal.

The area values X and Y are themselves related by another function $X[Y]$, as illustrated in Fig. 2.14(*d*), which is such that

$$\frac{\mathrm{d}X}{\mathrm{d}Y} = \frac{y}{x}\frac{\mathrm{d}x}{\mathrm{d}y}. \tag{2.41}$$

This slope is unity at the origin, for any $y[x]$ which passes through the origin. Fig. 2.14 is calculated from part of $y = 7x/4 - x^2$. As another example, a continuous response curve which contains a straight horizontal segment will subtend area functions $X[x]$ having a straight inclined segment and $Y[y]$ having a vertex. Of course the labels on the axes may all be interchanged if it is desired to present $X[x]$ as the multivalued function instead of $Y[y]$, to be in line with some of our other diagrams; but this is not necessary.

A well-known example of Fig. 2.14(*a*) is found in the one-dimensional engineering theory of a bar in tension, where y is nominal stress (current load per unit initial cross-sectional area) and $x + 1$ is uniform stretch (longitudinal after/before length ratio). Evidently the associated strain energy and complementary energy are examples of Legendre dual functions having isolated singularities in the form of the inflexion and cusp respectively in Figs. 2.14(*c*) and (*b*).

Certain problems of crack propagation give rise to load (y)-deflection (x) response curves which have a pointed maximum instead of the smooth one shown in Fig. 2.14(*a*), and even a saw-tooth characteristic (see Fig. 8.1 of Atkins and Mai (1985)). If the two slopes calculated from either side of such a vertex are finite, there is again an inflexion and a cusp in the area curves $X[x]$ and $Y[y]$ respectively.

Response curves in a plane diagram are not necessarily confined to one-dimensional theories, since they may relate scalar invariants of vectors or tensors from a three dimensional theory. We give an example in Fig. 5.3(*a*) and (*b*) from compressible inviscid fluid mechanics, where an isolated 'sonic singularity' is present on a curve relating fluid *speed* (x) and the magnitude (y) of the mass flow vector.

(*vii*) *Nondecreasing response characteristics*

Response curves derived from energies having distributed and accumulated singularities are not uncommon in applications. In particular, nondecreasing single step and multistep response curves often describe the voltage–current characteristics in branches of electrical networks (e.g., see Iri 1969, Rockafellar 1982). We illustrate upper and lower bounds associated with these in §2.8. (The physical branch meant here is unrelated, of course, to the mathematical branches defined after (2.6) and (2.16).) The step can arise from the presence of a diode in the branch. Linear resistors have linear voltage–current characteristics. Combinations of current and voltage sources, diodes and linear resistors in a single branch can result in a zig–zag

Fig. 2.15. Multistep response characteristic and energies.

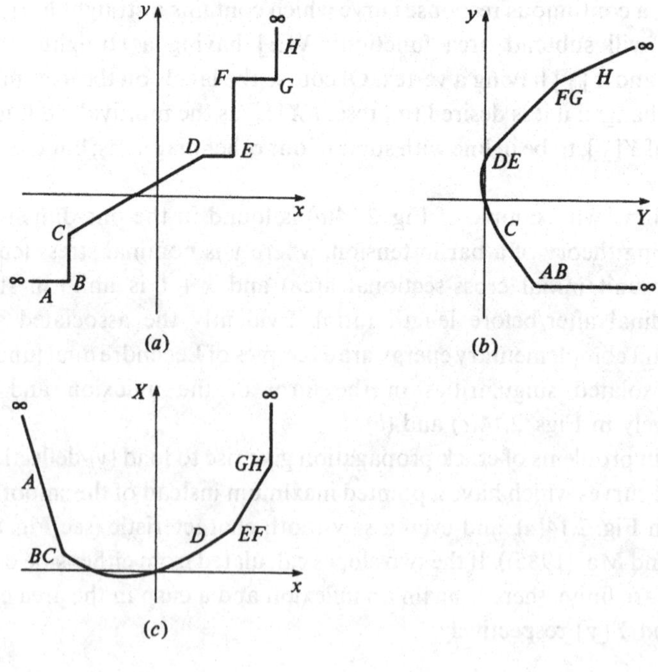

characteristic, examples $y = y[x]$ of which might be as in Fig. 2.15(*a*) or Fig. 2.30. The associated energies in Fig. 2.15(*b*) and (*c*) each consist of a parabolic arc *CD* joined continuously to distributed and accumulated singularities. This is an example in which the branch characteristic is completed by producing the finite segments *BA* and *GH* to infinity, parallel to the respective axes. The corresponding energies are completed by upward half-lines, from the point *AB* on *Y* and from the point *GH* on *X*. This is consistent with regarding them as the limits of convex curves in the sense indicated by Fig. 2.13.

In the context of numerical analysis, nodes and finite elements might represent accumulated and distributed singularities, respectively. Then a function approximated by nodes would have a dual approximated by finite elements (cf. Fig. 2.15(*b*) and (*c*)).

Formulae describing dual functions $Y[y]$ and $X[x]$ of the type shown in Fig. 2.15(*b*) and (*c*) can be constructed as follows. Let $Y = \Psi[y]$ and $X = \Phi[x]$ denote the parts *BCDEFG* in Figs. 2.15(*b*) and (*c*) respectively; these have the properties $x = \mathrm{d}\Psi/\mathrm{d}y$ and $y = \mathrm{d}\Phi/\mathrm{d}x$, allowing for the finite distributed and accumulated singularities. The end pieces *A* and *H* are then incorporated as follows. Let x_B, y_B and x_G, y_G be the coordinates of *B* and *G*, respectively, in Fig. 2.15(*a*). Define

$$\phi[x] = \begin{cases} \Phi[x], & x_B \leqslant x \leqslant x_G, \\ y_B(x - x_B) + \Phi[x_B], & x \leqslant x_B, \end{cases}$$

$$X[x] \geqslant \phi[x], \quad x \leqslant x_G, \quad (X[x] - \phi[x])(x - x_G) = 0,$$

$$\psi[y] = \begin{cases} x_G(y - y_G) + \Psi[y_G], & y \geqslant y_G, \\ \Psi[y], & y_G \geqslant y \geqslant y_B, \end{cases}$$

$$Y[y] \geqslant \psi[y], \quad y \geqslant y_B, \quad (Y[y] - \psi[y])(y - y_B) = 0.$$

Thus the trio of two inequalities with an orthogonality condition, of the type which we first met in (1.30), reappears here. We do not define $X[x]$ for $x > x_G$, or $Y[y]$ for $y < y_B$. Approaches to convex functions which follow Rockafellar (1970) find technical advantages in extending the domain of, for example, a function such as $X[x]$ by assigning the value $+\infty$ to it where $x > x_G$. We do not use this approach here.

The characteristic itself in Fig. 2.15(*a*) can now be described as

$$y \geqslant \frac{\mathrm{d}\phi}{\mathrm{d}x}, \quad x \leqslant x_G, \quad \left(y - \frac{\mathrm{d}\phi}{\mathrm{d}x}\right)(x - x_G) = 0,$$

or alternatively as

$$x \leqslant \frac{\mathrm{d}\psi}{\mathrm{d}y}, \quad y \geqslant y_B, \quad \left(x - \frac{\mathrm{d}\psi}{\mathrm{d}y}\right)(y - y_B) = 0.$$

Idealized elements of frameworks and structures made of rigid/perfectly plastic material have stress–strain characteristics not unlike the branch characteristic of a pair of diodes. Elastic/perfectly plastic material has an added linear element like the electrical resistor. More is said about plastic materials in §§2.6(iii), 5.8 and 5.10, and a simple example of Fig. 2.15 is given in Fig. 2.23.

By now it will be evident that the inverse function theorem $y = y[x] \Leftrightarrow x = x[y]$ relating two scalars can be strengthened even more than was done in Fig. 1.6. It can be adapted, for example, to Fig. 2.15(*a*) with the inverses expressed as in (2.40). This requires that the gradients of $X[x]$ and $Y[y]$ at the vertices in Fig. 2.15(*b*) and (*c*) be regarded as the jumps in the one-sided gradients there. Such jumps are also called subdifferentials.

2.4. Legendre transformations in two active variables, with singularities

(i) Umbilics, illustrating dual isolated singularities

The two variable polynomials

$$Y[y_1, y_2] = y_1{}^2 y_2 \pm y_2{}^3 + u y_1{}^2 \qquad (2.42)$$

parametrized by u are, when linear terms are added, the so-called 'universal unfoldings' of the hyperbolic (+ sign alternative) and elliptic (− sign) umbilics. To find the Legendre dual function $X[x_1, x_2]$ of either, with the value of u fixed, requires the simultaneous inversion of two equations $x_i = \partial Y/\partial y_i$ by $(2.7)_2$, namely of

$$x_1 = 2y_1(y_2 + u), \quad x_2 = y_1{}^2 \pm 3y_2{}^2, \qquad (2.43)$$

and substitution of the result into

$$X = x_1 y_1 + x_2 y_2 - Y = y_1{}^2(2y_2 + u) \pm 2y_2{}^3. \qquad (2.44)$$

The linear terms omitted from (2.42) will not affect the form of the result but, just as we found for one active variable (Theorem 2.5, e.g. (2.26)), they will merely shift the x_i-origin.

It is easiest to explain the calculation of the dual function if we have a three dimensional picture of the result already before us. Therefore, for any fixed $u < 0$, we show in Fig. 2.16(*a*) the single valued C^∞ function (2.42 +), i.e. for the hyperbolic case; and in Fig. 2.16(*b*) the corresponding dual function $X[x_1, x_2]$, which is found to contain cusped edges of regression and a self-intersection line, thus implying multivaluedness in the composite global function. Corresponding individual regular branches on the two surfaces are labelled *A*, *B* and *C*. We justify this diagram with the following extracts from detailed arguments given by Sewell (1978*b*) for the calculation of the Legendre duals of both hyperbolic and elliptic umbilics.

From $(2.16)_1$ the Legendre transformation has an isolated singularity where the Hessian matrix

$$\left[\frac{\partial^2 Y}{\partial(y_1, y_2)}\right] = 2\begin{bmatrix} y_2 + u & y_1 \\ y_1 & \pm 3y_2 \end{bmatrix} \tag{2.45}$$

has zero determinant, i.e.

$$\pm 3y_2(y_2 + u) = y_1{}^2. \tag{2.46}$$

Fig. 2.16. (*a*) Hyperbolic umblic potential $Y[y_1, y_2] = y_1{}^2 y_2 + y_2{}^3 + uy_1{}^2$ with singular hyperbola, for fixed $u < 0$. (b) Legendre dual function $X[x_1, x_2]$, for $u < 0$.

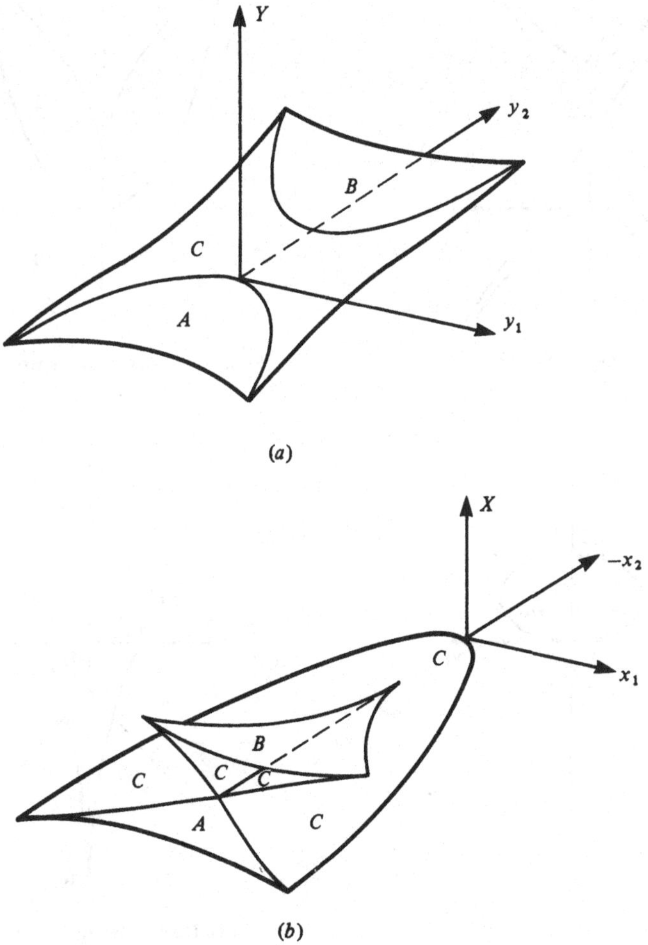

When $u \neq 0$ this singularity traces a hyperbola or ellipse in the y_1, y_2 plane, as shown for $u < 0$ in Fig. 2.17(*a*) and (*c*), which degenerates to a pair of asymptotes or a single point respectively if $u = 0$.

The mapping of the singularity onto the x_1, x_2 plane, as shown for $u < 0$ in Fig. 2.17(*b*) and (*d*), can be described explicitly by inserting the parametric version of (2.46) into (2.43). Thus as a parameter θ ranges through $0 \leqslant \theta < 2\pi$, the isolated singularity in the hyperbolic case is the

Fig. 2.17. Plan view of isolated singularity for umbilics ($u < 0$), showing corresponding θ values, and multiplicities (bold face) of $X[x_1, x_2]$.

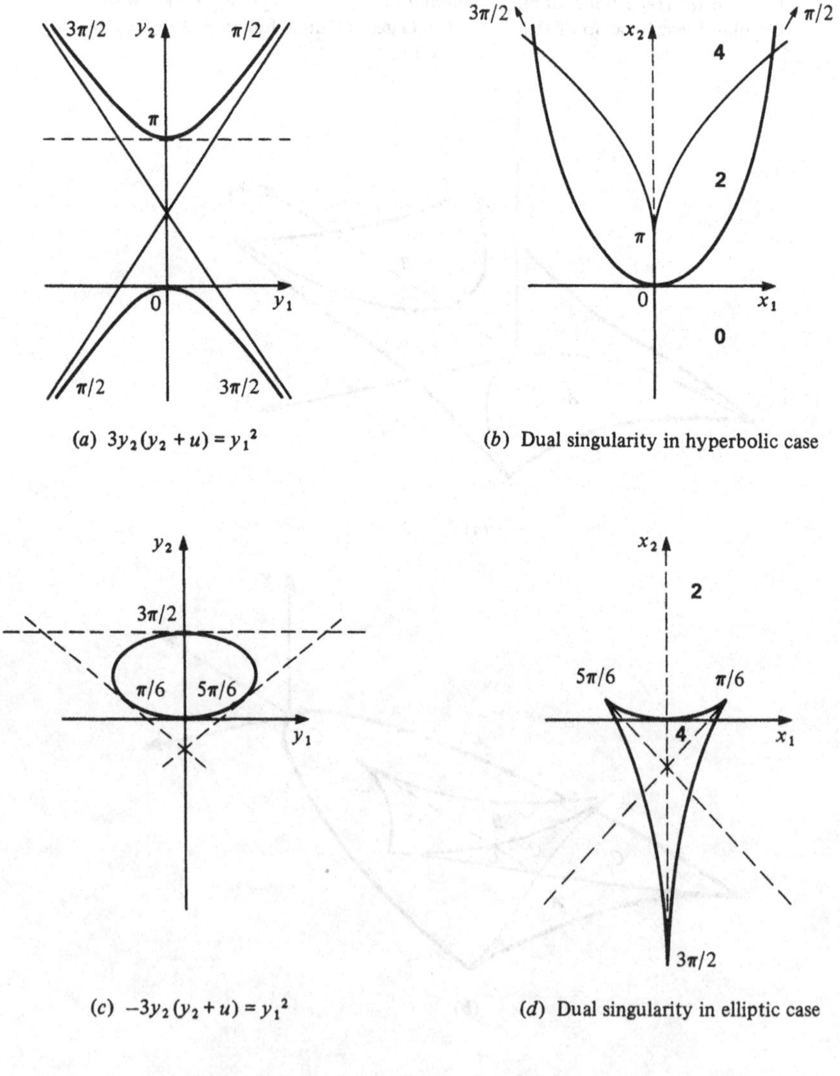

(*a*) $3y_2(y_2 + u) = y_1^2$

(*b*) Dual singularity in hyperbolic case

(*c*) $-3y_2(y_2 + u) = y_1^2$

(*d*) Dual singularity in elliptic case

curve traced out by

$$y_1 = \frac{\sqrt{3}}{2} u \tan\theta, \quad y_2 = \tfrac{1}{2} u (\sec\theta - 1),$$

or

$$x_1 = \frac{\sqrt{3}}{2} u^2 \tan\theta (1 + \sec\theta), \quad x_2 = \tfrac{3}{2} u^2 \sec\theta (\sec\theta - 1) \geqslant 0. \qquad (2.47+)$$

In the elliptic case the singularity is

$$y_1 = \frac{\sqrt{3}}{2} u \cos\theta, \quad y_2 = \tfrac{1}{2} u (\sin\theta - 1),$$

or

$$x_1 = \frac{\sqrt{3}}{2} u^2 \cos\theta (1 + \sin\theta), \quad x_2 = \tfrac{3}{2} u^2 \sin\theta (1 - \sin\theta). \qquad (2.47-)$$

In either case the height of the singularity is given by $Y[\theta]$ or $X[\theta]$ from (2.42) or (2.44) respectively, thus completing a parametric description of the singularity as a space curve in three dimensions.

For almost every value of θ the eigenvector of (2.45) is transverse to the hyperbola/ellipse (2.46 $\pm$), implying a local cubic-like inflexional trace on $Y[y_1, y_2]$, which maps into a local cusp-like trace on $X[x_1, x_2]$, as we would expect from Fig. 2.6. This explains why the singularity in x_1, x_2, X space emerges as a cusped edge of regression, as shown in Fig. 2.16. At exceptional values of θ, however, namely $\theta = \pi$ in the hyperbolic case and $\theta = \pi/6, 5\pi/6, 3\pi/2$ in the elliptic case, the eigenvector of (2.45) is tangential to the conic singularity (2.46 $\pm$), and these points map into swallowtail points (cf. Fig. 2.11(*b*)) of $X[x_1, x_2]$. The latter look like cusps in plan view, but they also terminate a line of self-intersection (shown dashed in Fig. 2.17(*b*) and (*d*)) which has mapped from the corresponding tangents shown dashed in Fig. 2.17(*a*) and (*c*)).

The foregoing singularities are the boundaries of regular branches. In the hyperbolic case there are three regular branches, labelled A, B and C in Fig. 2.16. It is easily verified by examining the sign-definiteness of (2.45) that, for $Y[y_1, y_2]$ with $u < 0$, A is strictly concave, B is strictly convex and C is saddle-shaped. It follows from Theorem 2.4 that the corresponding branches A, B, C of $X[x_1, x_2]$ with $u < 0$ have the same concave, convex and saddle shapes respectively. Moreover, the latter C contains one self-intersection line. In the elliptic case there are only two regular branches, corresponding to the inside and outside of the ellipse (2.46 $-$). The inside branch is concave if $u < 0$ and convex if $u > 0$, and it maps onto a single valued regular branch of $X[x_1, x_2]$ whose boundary has the shape of a hypocycloid. The outside branch of $Y[y_1, y_2]$ is saddle-shaped, and maps onto a saddle-shaped branch of $X[x_1, x_2]$ which contains three self-intersection lines instead of one.

These juxtapositions of self-intersection lines, regular branches and singularities, with the requirement of continuity of $X[x_1, x_2]$ and its (possibly regressive) slope, imply that this dual function must be multi-valued, with multiplicities (written boldface on Fig. 2.17) 0, 2, or 4 in the hyperbolic case and 2 or 4 in the elliptic case.

We leave the reader to construct the three dimensional diagrams, like Fig. 2.16 for the hyperbolic cases $u \geqslant 0$, and for the elliptic case. It is also possible to construct paper models of all these surfaces, introducing pleats in suitable directions to enforce convexities or concavities, and using staples to hold together regular branches meeting along a common cusped edge. Those parts of the two surfaces which map into each other may be painted in the same colour.

On the conic singularity (2.46 $\pm$) the rank of (2.45) is one if $u \neq 0$, but zero if $u = 0$. In this way either umbilic organizes the passage of a Legendre transformation through an isolated compound singularity of multiplicity two. This is not the only way in which a double zero can appear in the singularity of a Legendre transformation, but it is one of the simplest. Arnol'd (1974) described it in terms of the evolution of a wave-front. Other typical ways are to be found via four more umbilics listed in the statement of the so-called Thom's Theorem (see Poston and Stewart, 1978, p. 122), the associated singularities also having the attribute of stability mentioned in §2.2(vii). The way to use these results, in an applied context where there are more active variables than the multiplicity of the isolated singularity (when this multiplicity is one or two), is to seek smooth changes of variable which will sometimes identify one or two active variables contributing essentially to the singularity.

(ii) Accumulated duals of nonplanar distributed singularities

The distributed singular function

$$Y[y_1, y_2] = \exp(b_1 y_1 + b_2 y_2) \tag{2.17 bis}$$

induces, by $x_i = \partial Y / \partial y_i$, the dual active variables

$$x_1 = b_1 Y, \quad x_2 = b_2 Y \tag{2.48}$$

and therefore a dual singularity accumulated over the straight line

$$\frac{x_1}{b_1} = \frac{x_2}{b_2}.$$

The height X of the dual accumulated singular function over this line is

$$\begin{aligned} X &= x_1 y_1 + x_2 y_2 - Y \\ &= (\ln Y - 1)Y. \end{aligned} \tag{2.49}$$

We can regard (2.48) and (2.49) as a parametric description of the accumulated singularity, parametrized by $Y > 0$.

The distributed singular function

$$Y[y_1, y_2] = (b_1 y_1 + c_1)^a (b_2 y_2 + c_2)^{1-a} \tag{2.18 bis}$$

(with $a \neq 0, 1$ to retain two variables) induces the dual active variables

$$x_1 = ab_1 t^{1-a}, \quad x_2 = (1-a)b_2 t^{-a}, \quad t = \frac{b_2 y_2 + c_2}{b_1 y_1 + c_1} \tag{2.50}$$

and therefore a dual singularity accumulated over the curve

$$\left(\frac{x_1}{ab_1}\right)^{-a} = \left(\frac{x_2}{(1-a)b_2}\right)^{1-a},$$

which can be parametrized by t. The height of the dual accumulated singular function over this curve is

$$X = -\frac{c_1}{b_1} x_1 - \frac{c_2}{b_2} x_2. \tag{2.51}$$

The possibility is now available of joining regular branches via distributed singular functions which are nonplanar such as (2.17) or (2.18), or via their accumulated duals (above). This discussion offers a view of how to calculate the duals of certain piecewise C^1 functions of several variables, by regarding the gradient discontinuities as accumulated singularities whose duals will be distributed singularities of some form, not necessarily planar.

Exercises 2.3

1. Sketch a three dimensional view, in y_1, y_2, Y space, of the special forms taken by hyperbolic and elliptic umbilic functions (2.42 $\pm$) when $u = 0$.
2. Deduce the Legendre dual functions $X[x_1, x_2]$ of the hyperbolic and elliptic umbilics when $u = 0$. Sketch a three dimensional view of them, and identify corresponding regular branches. Determine which pairs of branches are convex, concave or saddle-shaped.
3. Repeat Exercises 1 and 2 for the elliptic case with $u > 0$ and with $u < 0$. Hence describe the evolution of $X[x_1, x_2]$ as u passes through zero.
4. Repeat Exercise 2 for the hyperbolic case with $u > 0$. Hence describe the evolution of $X[x_1, x_2]$ as u passes through zero.
5. Show that (2.49) also applies to the first function $Y[y_1, y_2, y_3]$ in Exercise 2.1.3,

and that the domain of the accumulated dual function is a straight line in three dimensional space.

6. Show that the domain of the accumulated dual $X[x_1, x_2, x_3]$ of the second function in Exercise 2.1.3 is the surface

$$\left(\frac{x_1}{a_1 b_1}\right)^{a_1}\left(\frac{x_2}{a_2 b_2}\right)^{a_2}\left(\frac{x_3}{a_3 b_3}\right)^{a_3} = 1.$$

2.5. Closed chain of Legendre transformations

(i) Indicial notation

We now explore the effect of interchanging the roles of active and passive variables. For this purpose we return to the general dimensional case envisaged in (2.11)–(2.13). We first recapitulate the primary equations describing a Legendre transformation.

Given a function

$$Y = Y[y_i, u_\gamma] \tag{2.52}$$

of the y_i ($i = 1,\ldots,n$) and the u_γ ($\gamma = 1,\ldots,m$), having (at least one-sided) gradients

$$x_i = \frac{\partial Y}{\partial y_i}, \quad v_\gamma = \frac{\partial Y}{\partial u_\gamma}, \tag{2.53}$$

the Legendre dual function in a transformation with the y_i and x_i active and the u_γ passive is the function

$$X = X[x_i, u_\gamma] \tag{2.54}$$

of the x_i and the u_γ, having the gradients

$$y_i = \frac{\partial X}{\partial x_i}, \quad -v_\gamma = \frac{\partial X}{\partial u_\gamma} \tag{2.55}$$

and values such that

$$X + Y = x_i y_i. \tag{2.56}$$

We have given a very explicit discussion of the role of singularities in the last three sections. Broadly speaking, this background, now taken as read, is seen to allow either function $Y[y_i, u_\gamma]$ or $X[x_i, u_\gamma]$ to be a piecewise C^1 multivalued function, composed of regular and singular branches joined together. There will be some asymmetry of gradient continuity between the two functions, in that a C^1 join at an isolated singularity may be progressive in one function and regressive in its dual (e.g. Fig. 2.6); and a C^1 join at a distributed singularity may map onto an accumulated singularity in the form of a finite-angled corner at a vertex or edge on the dual function (e.g. Fig. 2.5). Given such qualifications, we can regard (2.52)–(2.56) as a global analytical statement of a Legendre transformation, no longer inhibited by local difficulties with the inversion procedure at singularities.

We now seek to invert $(2.53)_2$ rather than $(2.53)_1$, by treating the u_γ and v_γ as active and the y_i as passive. The result can be expressed in terms of another global dual function, say

$$W = W[y_i, v_\gamma], \tag{2.57}$$

having at least one-sided gradients

$$-x_i = \frac{\partial W}{\partial y_i}, \quad u_\gamma = \frac{\partial W}{\partial v_\gamma} \tag{2.58}$$

and values such that

$$Y + W = u_\gamma v_\gamma. \tag{2.59}$$

A further possibility is to seek the global inversion of $(2.55)_2$ with the x_i as passive, or of $(2.58)_1$ with the v_γ as passive. Either of these last two inversions leads to the same new global dual function

$$Z = Z[x_i, v_\gamma] \tag{2.60}$$

with gradients

$$-y_i = \frac{\partial Z}{\partial x_i}, \quad -u_\gamma = \frac{\partial Z}{\partial v_\gamma} \tag{2.61}$$

and values such that

$$X + Y + Z + W = 0. \tag{2.62}$$

Fig. 2.18. Closed chain of Legendre transformations relating global functions.

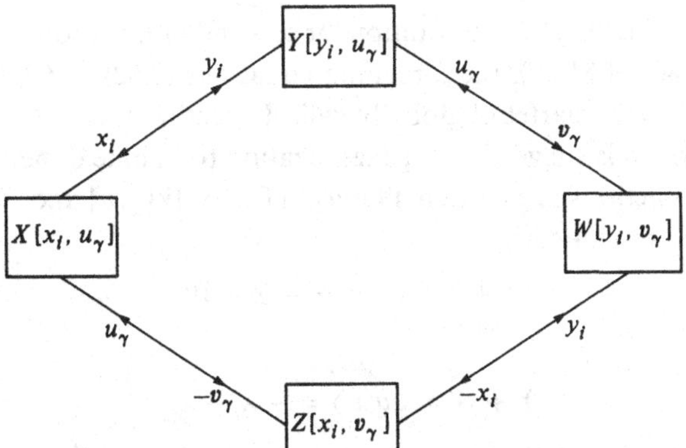

We can thereby construct a closed chain of four Legendre transformations, which may all be global under the qualifications mentioned. A mnemonic summarizing these interconnections is the *diamond diagram* of Fig. 2.18. At each corner of the diamond is written a globally valid and possibly multivalued function which can generate, by inversion of its gradient equations, either of two Legendre transformations. Each link in the diagram represents a Legendre transformation admitting isolated, distributed or accumulated singularities. The dual active variables in each link are written next to the arrows. With these conventions a quartet of Legendre transformations can be read off directly from the diamond diagram. Note that the presence of two minus signs in the diagram is intrinsic.

Furthermore, two successive links in the chain may be applied simultaneously with care. Thus by treating all $n+m$ variables as active, with none passive, it may be possible to go directly from $X[x_i, u_\gamma]$ to $-W[y_i, v_\gamma]$, or from $Y[y_i, u_\gamma]$ to $-Z[x_i, v_\gamma]$, since

$$X - W = x_i y_i - u_\gamma v_\gamma, \quad Y - Z = x_i y_i + u_\gamma v_\gamma.$$

Fig. 2.18 provides a symmetrical standpoint, with the likely asymmetry in gradient continuity left to be implicit. Any corner of the diamond may be chosen, according to convenience, as a starting point from which to develop certain aspects of a physical theory. For example, by differentiating any one generating function we obtain two sets of equations relating four sets of variables. Supplementary information will be required from other sources to complete a fully posed algebraic problem. Sometimes there is not a real need for more than one or two of the four alternative starting points, but in other problems (e.g. thermodynamics – see §5.2) all four have their uses.

(ii) Inner product space notation

It is worth noting that, in the notation for sets and inner products described in (1.44)–(1.46) of §1.3(i), the foregoing equations (2.52)–(2.62) describing the essentials of a quartet of globally valid Legendre transformations take on a form which is suggestive of generalization to infinite dimensions. The generating functions are written $Y[y,u]$, $X[x,u]$, $W[y,v]$ and $Z[x,v]$ and their values are related by

$$X + Y = (x, y) = -Z - W \tag{2.63}$$

and

$$Y + W = \langle u, v \rangle = -X - Z. \tag{2.64}$$

The gradient equations in, for example, the transformation (2.52)–(2.56) from $Y[y,u]$ to $X[x,u]$ become

$$\frac{\partial Y}{\partial y}=x, \quad \frac{\partial X}{\partial x}=y, \quad \frac{\partial Y}{\partial u}=v=-\frac{\partial X}{\partial u}. \tag{2.65}$$

From this viewpoint, any pair of given globally valid functions $Y[y,u]$ and $X[x,u]$ would be said to be related by a Legendre transformation provided only that their values satisfy $(2.63)_1$ and that their gradients satisfy (2.65). The possible presence of singularities of various kinds, and of multivaluedness, is accepted from the outset.

(iii) Convex and saddle-shaped branches

The local shapes of two dual branches in a single Legendre transformation are related in characteristic ways. We gave examples of this in Theorem 2.4. The object now is to give further explicit examples, and to extend such results to describe the transmutation of shape around particular quartets of branches in a closed chain of Legendre transformations.

Let us first recall how the definition of a convex function, and of a saddle function, depends on how much smoothness one wishes to assume. For example, convexity of a continuous function of one variable can be defined (Exercise 1.1.1(ii)) but, over a domain where first derivatives exist uniquely, an equivalent definition (1.4) expressed in terms of gradients is also available, and the later type of definition is especially useful for our purposes in this book. When second derivatives are also available, sufficient conditions for convexity can be expressed in terms of them via formulae such as (1.7). The multidimensional generalization led, via the idea of rectangular subdomains (Fig. 1.5) in two dimensions, to the eventual definition of an $n+m$ dimensional saddle function in (1.58) in terms of first derivatives, and to sufficient conditions for it in terms of second derivatives.

It is instructive to maintain this distinction between two levels of smoothness in the discussion of transmutation properties which follows. First we give results which depend only on the existence of first derivatives, which can therefore be expressed entirely in the inner product notation for $n+m$ variables, as in the previous subsection. Secondly we give associated results which use second derivatives.

Theorem 2.8

Let x_+, u_+ and x_-, u_- be any two points in the domain of the global and possibly multivalued $X[x,u]$, and let X_+ and X_- each be any one of the corresponding values of this function. Let y_+, u_+ and y_-, u_- be corresponding points in the domain of $Y[y,u]$, according to $y=\partial X/\partial x$, and let Y_+ and Y_- be *corresponding values of* $Y=(x,y)-X$. *Then*

$$X_+ - X_- - \left(x_+ - x_-, \left.\frac{\partial X}{\partial x}\right|_-\right) - \left\langle u_+ - u_-, \left.\frac{\partial X}{\partial u}\right|_+\right\rangle$$
$$= Y_- - Y_+ - \left(y_- - y_+, \left.\frac{\partial Y}{\partial y}\right|_+\right) - \left\langle u_- - u_+, \left.\frac{\partial Y}{\partial u}\right|_+\right\rangle. \quad (2.66)$$

Proof
This is an immediate consequence of the stated properties of the global transformation (2.52)–(2.56), in the inner product notation (2.63) and (2.65). □

The formula (2.66) was given by Noble and Sewell (1972, equation (6.1)) for the case of single valued functions. The reader will find it helpful to verify that the formula is valid even when x_+, u_+, X_+ and x_-, u_-, X_- are on different branches of $X[x,u]$ (Exercise 2.4.1). The multivaluedness of one of the dual functions and its gradient does not invalidate the formula. The Theorem could be stated in reverse order, i.e. with y_+, u_+, Y_+ and y_-, u_-, Y_- as given, with x_+, u_+, X_+ and x_-, u_-, X_- computed from them. In particular it is instructive to check (Exercise 2.4.2) that (2.66) still holds where there is a self-intersection of one of the functions, as in Fig. 2.16(*b*) for the dual of the hyperbolic umbilic. This example also illustrates that distinct values of y_+ and y_- do not necessarily map into distinct values of x_+ and x_-.

The saddle definition (1.58) was in terms of any pair of plus and minus points in the domain of $L[x,u]$, and $L[x,u]$ was *single* valued over that domain (otherwise the sign of S could be arbitrarily chosen by reversing the values of L_+ and L_- for a pair of points such that $x_+ - x_-$ and $u_+ - u_-$ are almost zero). To avoid this difficulty when applying the definition to branch C of $X[x_1, x_2]$ in Fig. 2.16(*b*) evidently requires the selection of single valued parts of C, which can be individually bounded by the self-intersection line.

Our next result invokes an extended definition of convexity, which is in the same spirit as the saddle definition (1.58), in that we no longer wish to insist (as in Exercise 1.5.1(i)) that the domain of a convex function is itself convex. The latter requirement, although a desirable starting point, would ultimately be too restrictive, as can be seen from the example of branch B of $X[x_1, x_2]$ in Fig. 2.16(*b*). This branch is defined over a domain which is *concave*, being interior to a cusp-shaped boundary, but the branch itself is convex according to a sufficient criterion in terms of second derivatives over every convex subdomain.

We define $Y[y, u]$ to be strictly (or weakly) *jointly convex* in y and u considered as a single set of variables if the expression on the right of (2.66) is positive (or nonnegative) for every distinct pair of points in the domain of $Y[y, u]$, supposed single valued over that domain.

Theorem 2.9
A single valued part of $X[x, u]$ which is strictly (or weakly) saddle-shaped in the sense of (1.58), *but convex in x and concave in u, maps into a strictly (or weakly) jointly convex part of $Y[y, u]$, provided every distinct pair of x, u points maps into a distinct pair of y, u points and vice versa.*

Proof
Given the qualifications already made, the result expresses the fact that when the left side of (2.66) is positive (or nonnegative) so is the right side, and vice versa. □

The result was established by Noble and Sewell (1972, §6) and is of some basic importance in describing the influence of passive variables (u here) on transmutation of the shape of generating functions in a Legendre transformation (with x and y active here). We now approach the result from a more local viewpoint, incorporating second derivatives. This provides an alternative and in some respects more explicit account than the foregoing discussion in terms of first derivatives (only) evaluated at pairs of points.

Individual second derivatives will here be denoted by

$$X_{ij} = \frac{\partial^2 X}{\partial x_i \partial x_j}, \quad X_{i\gamma} = \frac{\partial^2 X}{\partial x_i \partial u_\gamma}, \quad X_{\gamma\delta} = \frac{\partial^2 X}{\partial u_\gamma \partial u_\delta}$$

for $X[x_i, u_\gamma]$, and for other functions similarly (cf. §1.3(iv) and (2.10)).

Theorem 2.10
Dual C^2 branches of $Y[y_i, u_\gamma]$ and $X[x_i, u_\gamma]$, *which are regular in the sense of* (2.6) *with respect to the active variables, with the u_γ passive, satisfy*

$$X_{i\gamma} = -X_{ik}Y_{k\gamma}, \quad Y_{i\gamma} = -Y_{ik}X_{k\gamma}, \tag{2.67}$$

$$X_{\gamma\delta} + Y_{\gamma\delta} = -X_{\gamma k}Y_{k\delta} = -X_{\delta k}Y_{k\gamma} \tag{2.68}$$

$$= X_{ij}Y_{i\gamma}Y_{j\delta} = Y_{ij}X_{i\gamma}X_{j\delta}. \tag{2.69}$$

Also, for an arbitrary $(n+m)$-tuple $\{\lambda_i\ \lambda_\gamma\}$, we can construct in terms of it another arbitrary $(n+m)$-tuple $\{\mu_i\ \lambda_\gamma\}$ such that

$$\mu_i = Y_{ik}\lambda_k + Y_{i\gamma}\lambda_\gamma, \quad \lambda_i = X_{ik}\mu_k + X_{i\gamma}\lambda_\gamma \tag{2.70}$$

and

$$Y_{ij}\lambda_i\lambda_j + 2Y_{i\gamma}\lambda_i\lambda_\gamma + Y_{\gamma\delta}\lambda_\gamma\lambda_\delta = X_{ij}\mu_i\mu_j - X_{\gamma\delta}\lambda_\gamma\lambda_\delta. \tag{2.71}$$

Proof

To establish (2.67) and (2.68) is an exercise in the use of implicit functions, which we leave to the reader. It is an extension of the arguments needed to establish (2.10), i.e. $X_{ik}Y_{kj} = \delta_{ij}$. Formulae (2.69) are an immediate combination of (2.67) and (2.68). We need (2.10) and (2.67) to get $(2.70)_2$ from $(2.70)_1$, and the arbitrariness of μ_i follows from that of λ_i by the regularity postulate. Formula (2.71), which is an extension of (2.15), is a consequence of (2.10) and (2.67)–(2.70), and is valid without requiring any sign-definiteness hypothesis. □

Theorem 2.11

Under the conditions of Theorem 2.10 *the two requirements (a) that the* $(n+m)\times(n+m)$ *Hessian*

$$\begin{bmatrix} Y_{ij} & Y_{i\delta} \\ Y_{\gamma j} & Y_{\gamma\delta} \end{bmatrix} \text{ is positive definite,} \tag{2.72}$$

and (b), that the $n \times n$ *and* $m \times m$ *Hessians*

$$X_{ij} \text{ and } -X_{\gamma\delta} \text{ are each positive definite,} \tag{2.73}$$

are equivalent.

Proof

Condition (2.72) holds, by definition, if and only if the left side of (2.71) is positive for arbitrary choices of $\{\lambda_i\ \lambda_\gamma\}$. But this implies that the right side of (2.72) is positive for arbitrary choices of $\{\lambda_i\ \lambda_\gamma\}$, and therefore also for arbitrary choices of $\{\mu_i\ \lambda_\gamma\}$ by the regularity property $|X_{ik}| \neq 0$ applied to (2.70). Choosing all $\mu_i = 0$ with all λ_γ arbitrary, and then all $\lambda_\gamma = 0$ with all μ_i arbitrary, establishes (2.73). A similar argument can be applied in the reverse order. □

The content of Theorems 2.10 and 2.11 was given by Sewell (1982).

Theorem 2.12

(a) A C^2 *part of* $X[x_i, u_\gamma]$ *which satisfies* (2.73) *in every rectangular subdomain is strictly saddle-shaped, strictly convex in the* x_i *and strictly concave in the* u_γ.

(b) It maps by a nonsingular Legendre transformation into a C^2 *part of* $Y[y_i, u_\gamma]$ *which satisfies* (2.72) *and is strictly jointly convex.*

Proof

(a) This is a consequence of Theorem 1.5, with $L[x, u]$ there replaced by $-X[x, u]$. In other words, (2.73) are sufficient conditions for the saddle-shaped property. They are the generalization of (1.10).

(b) This requires Theorem 2.11 and the fact that (2.72) are sufficient conditions for strict joint convexity. The latter fact entails an extension of the result in Exercise 1.5.2, in which the right side of (2.66) is expressed as a positive definite quadratic form over a convex subdomain of y, u space. □

Next we replace (2.73) by a weaker hypothesis. Let X_{ij}^{-1} denote the typical element of the inverse (*not* the inverse of the typical element) of the Hessian whose typical element is X_{ij}.

Theorem 2.13

(a) The following two pairs of hypotheses are equivalent:

$$X_{ij} \text{ is positive definite, } -X_{\gamma\delta}+X_{ij}^{-1}X_{i\gamma}X_{j\delta} \text{ nonnegative definite}$$

and

$$Y_{ij} \text{ is positive definite, } Y_{\gamma\delta} \text{ is nonnegative definite.} \tag{2.74}$$

(b) The intercept defined by the n equations $\partial X/\partial x_i = 0$ on the surface $X = X[x_i, u_\gamma]$ is a concave function of the u_γ when (2.74) *holds.*

Proof

(a) From (2.15) we see that positive definite X_{ij} is equivalent to positive definite Y_{ij}. From (2.10) $X_{ij}^{-1} = Y_{ij}$, so that $Y_{\gamma\delta} = -X_{\gamma\delta} + X_{ij}^{-1}X_{i\gamma}X_{j\delta}$ from (2.69).

(b) Positive definite X_{ij} and the implicit function theorem lead to a function of the u_γ with Hessian

$$X_{\gamma\delta} - X_{ij}^{-1}X_{i\gamma}X_{j\delta},$$

by (1.62), which is nonpositive definite by (2.74). □

Theorem 2.13 holds *a fortiori* if both matrices in either of the hypotheses (2.74) are positive definite, and then the intercept function in (b) is strictly concave.

A case of special interest may be described in terms of a pair of Legendre dual functions $\psi[x_i]$ and $\phi[q_i]$, each of n scalar variables only and such that

$$\frac{\partial\psi}{\partial x_i} = q_i, \quad \frac{\partial\phi}{\partial q_i} = x_i, \quad \psi + \phi = x_i q_i.$$

Then if $T_{\gamma i}$ is the typical element of a given $m \times n$ matrix, whose transpose has typical element $T_{i\gamma}$, the functions

$$\begin{aligned} X[x_i, u_\gamma] &= \psi[x_i] - u_\gamma T_{\gamma i} x_i, \\ Y[y_i, u_\gamma] &= \phi[y_i + T_{i\gamma}u_\gamma], \end{aligned} \tag{2.75}$$

are also Legendre duals satisfying (2.52)–(2.56) with passive u_γ. Thus the $n+m$ arguments of the function $Y[y_i, u_\gamma]$ actually appear only within the n combinations $y_i + T_{i\gamma}u_\gamma$ acting as the arguments of ϕ. Second derivatives are $X_{i\gamma} = -T_{i\gamma}$, $X_{\gamma\delta} \equiv 0$ and

$$X_{ij} = \psi_{ij}, \quad -X_{\gamma\delta} + X_{ij}^{-1}X_{i\gamma}X_{j\delta} = \psi_{ij}^{-1}T_{i\gamma}T_{j\delta} = Y_{\gamma\delta}.$$

Hence positive definite ψ_{ij} is sufficient to satisfy *both* hypotheses in each of the pairs (2.74). The concave function which Theorem 2.13(b) then predicts is $-\phi[T_{i\gamma}u_\gamma]$. We shall illustrate this example in the context of bifurcation theory in §2.7(iii), where ϕ and ψ may have the roles of potential energy and complementary energy in mechanical examples.

The equivalent hypotheses (2.72) and (2.73) are strong ones. This is evident in itself, and it also becomes clear by contrasting them with the weaker (2.74), which turn out to be more representative of some practical applications, for example in bifurcation theory as first mentioned. When the global functions $Y[y_i, u_\gamma]$ and $X[x_i, u_\gamma]$ do have a pair of corresponding branches which satisfy (2.72) and (2.73), it is sometimes convenient to call them *strong branches*.

By reversing the roles of active and passive variables we also obtain a strong branch of $Z[x_i, v_\gamma]$ from one of $X[x_i, u_\gamma]$ and a strong branch of $W[y_i, v_\gamma]$ from one of $Y[y_i, u_\gamma]$ (or from one of $Z[x_i, v_\gamma]$). With (2.73), which implies $(2.74)_1$ *and* the similar inequalities with the roles of the x_i and u_γ interchanged, we obtain a *strong quartet* of branches.

We see that either hypothesis (2.72) or (2.73) is sufficient for $Z[x_i, v_\gamma]$

Fig. 2.19. Strong quartet of branches of Fig. 2.18, showing transmission of strict convexities between regular branches.

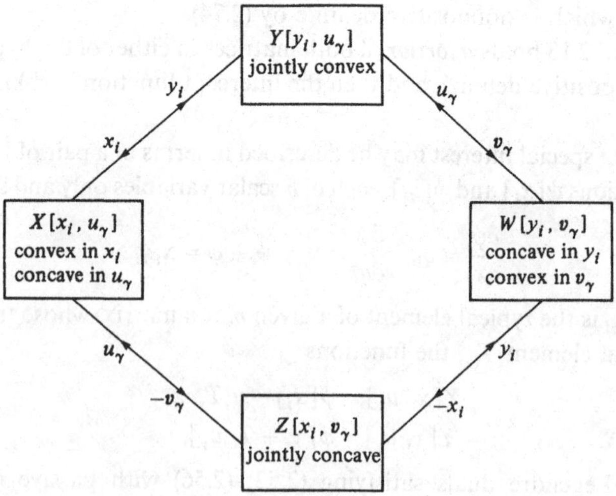

to be a jointly strictly concave function, and for $W[y_i, v_\gamma]$ to be another strict saddle function but concave in the y_i and convex in the v_γ. Hence we obtain the relations summarized in Fig. 2.19 (given by Noble and Sewell, 1972) between the shapes in a strong quartet, which is regular in *both* sets of variables and forms part of the global and possibly multi-branched functions in Fig. 2.18.

On the other hand hypotheses like (2.74) allow, for example, that $X[x_i, u_\gamma]$ could be weakly concave in the u_γ, even though strictly convex in the x_i. This is consistent with $|X_{\gamma\delta}| = 0$ and, in particular, with a linear dependence of $X[x_i, u_\gamma]$ on the u_γ. This can introduce distributed and accumulated singularities into those Legendre transformations where the u_γ are active variables.

The C^2 requirement of Theorems 2.12 and 2.13 can also be relaxed somewhat, for example to allow for certain piecewise C^2 functions.

As an indication of what can happen on branches where even (2.74) does not apply, and as a contrast to Theorem 2.11, we note the following result.

Theorem 2.14

Under the conditions of Theorem 2.10 *any set of three (possibly different) sign-definiteness hypotheses which apply to the three Hessians*

$$X_{ij}, X_{\gamma\delta} \quad \text{and} \quad X_{ij}^{-1} X_{i\gamma} X_{j\delta} - X_{\gamma\delta} \tag{2.76}$$

are transmitted collectively to the derivatives of $Y[y_i, u_\gamma]$. *In other words, these collective requirements are invariant in the Legendre transformation.*

Proof

From (2.10) we have $X_{ij}^{-1} = Y_{ij}$ and $Y_{ij}^{-1} = X_{ij}$. From (2.15) we see that any sign-definiteness property of X_{ij} applies also to its inverse Y_{ij}. From (2.69) it follows that

$$X_{\gamma\delta} = Y_{ij}^{-1} Y_{i\gamma} Y_{j\delta} - Y_{\gamma\delta} \quad \text{and} \quad X_{ij}^{-1} X_{i\gamma} X_{j\delta} - X_{\gamma\delta} = Y_{\gamma\delta}.$$

Taken as a *pair* considered collectively, these two $m \times m$ Hessians are invariant in the transformation, and therefore so are any pair of sign-definiteness hypotheses which may be applied to them. □

Various cases are included in the foregoing Theorem. The non-singular Hessian X_{ij} could be positive definite, negative definite or indefinite, and likewise Y_{ij} respectively. The pair of $m \times m$ Hessians displayed in the above proof could transmit the same sign, or a pair of different signs which switch round within the pair; and semidefiniteness can also be included since $X_{\gamma\delta}$ and $Y_{\gamma\delta}$ are not required to be nonsingular.

For example, if all three Hessians in (2.76) are positive definite for $X[x_i, u_\gamma]$, the same applies to the corresponding Hessians for $Y[y_i, u_\gamma]$ (Sewell, 1982). These conditions are sufficient for a special type of separate convexity. The special feature is provided by the third hypothesis ($Y_{\gamma\delta}$ positive definite) in (2.76) which, at least when $n = m = 1$, means that the function is a saddle function with respect to *another* choice of variables. This can be seen by comparison with (1.20b) and Exercise 1.2.3; see also Exercise 2.4.5. Exercise 2.4.6 illustrates other possibilities, including the transmission of indefiniteness.

Let us now denote matrices of second derivatives of $X[x_i, u_\gamma]$ alternatively by

$$N = [X_{ij}], \quad R = [X_{i\delta}], \quad R^{\mathrm{T}} = [X_{\gamma j}], \quad M = [X_{\gamma\delta}],$$

so that R^{T} is the $m \times n$ transpose of the $n \times m$ matrix R, N is $n \times n$ symmetric and M is $m \times m$ symmetric. We can then express the second derivatives of $Y[y_i, u_\gamma]$, $Z[x_i, v_\gamma]$ and $W[y_i, v_\gamma]$ in terms of those of $X[x_i, u_\gamma]$, as in Table 2.2, by repeated application of (2.67)–(2.69), for any quartet of branches where all the indicated inverses exist (and without any other sign definiteness hypotheses). The two alternatives presented for the second derivative of $W[y_i, v_\gamma]$ correspond to the two possible approaches to it from $X[x_i, u_\gamma]$ via $Y[y_i, u_\gamma]$ and then via $Z[x_i, v_\gamma]$ respectively. In the particular case when the functions are quadratic Table 2.2 is a table of their coefficients, but at this stage there is no need for the functions to be quadratic.

Theorem 2.15

(a) A C^2 branch $X[x_i, u_\gamma]$ satisfying

$$N, M \textit{ and } R^{\mathrm{T}}N^{-1}R - M \textit{ nonsingular} \tag{2.77}$$

generates a quartet of regular branches of $Y[y_i, u_\gamma]$, $Z[x_i, v_\gamma]$ and $W[y_i, v_\gamma]$.

(b) A C^2 branch satisfying (2.73), *i.e.*

$$N \textit{ and } -M \textit{ positive definite,} \tag{2.78}$$

generates a strong quartet of regular branches, as shown in Fig. 2.19.

Proof

(a) The hypothesis (2.77) permits three nonsingular Legendre transformations $Z \leftrightarrow X \leftrightarrow Y \leftrightarrow W$, and we can expect that since it is possible to go the long way round the circuit from Z to W in this way, it must be possible to go the short way $Z \leftrightarrow W$ in one step. The proof that the quartet

Table 2.2. *Matrices of second derivatives in a quartet of regular branches*

Derivatives	ij	$i\delta$	$\gamma\delta$
$X[x_i, u_\gamma]$	N	R	M
$Y[y_i, u_\gamma]$	N^{-1}	$-N^{-1}R$	$R^T N^{-1} R - M$
$W[y_i, v_\gamma]$	$N^{-1}R(R^T N^{-1} R - M)^{-1} R^T N^{-1}$ $- N^{-1} = (RM^{-1}R^T - M)^{-1}$	$N^{-1}R(R^T N^{-1} R - M)^{-1}$ $= (RM^{-1}R^T - N)^{-1} RM^{-1}$	$(R^T N^{-1} R - M)^{-1} = -M^{-1}$ $+ M^{-1}R^T(RM^{-1}R^T - N)^{-1} RM^{-1}$
$Z[x_i, v_\gamma]$	$RM^{-1}R^T - N$	RM^{-1}	M^{-1}

can be closed in this way requires a matrix proof that (2.77) imply $RM^{-1}R^{T}-N$ is nonsingular. We leave this as an exercise (Exercise 2.4.6).

(b) This is just a restatement in the matrix notation of a previous deduction from Theorem 2.11. Expressed otherwise (cf. Table 2.2) it is easy to verify that (2.73) imply $R^{T}N^{-1}R-M$ and $-RM^{-1}R^{T}+N$ are positive definite. □

Exercises 2.4

1. Verify that both sides of formula (2.66) have the value $\frac{1}{3}(5+2^{\frac{5}{2}})$ when $x_+=3$, $u_+=1$, $X_+=\frac{1}{3}2^{\frac{5}{2}}$ and $x_-=2$, $u_-=1$, $X_-=-\frac{2}{3}$ in the transformation (2.23)–(2.26), i.e. of Fig. 2.6(*b*).
2. Verify that for the hyperbolic umbilic (2.42+) and its dual, both sides of the formula (2.66) have the value zero when
$$y_{1+}=1,\quad y_{2+}=1,\quad u_+=-1,$$
$$y_{1-}=-1,\quad y_{2-}=1,\quad u_-=-1.$$
3. Prove Theorem 2.10.
4. Complete the explicit details of the proof of Theorem 2.12.
5. Verify that *separate* convexity of $Y[y,u]$, i.e. with respect to y and u separately (e.g. assured by positive definiteness of the two Hessians Y_{ij} and $Y_{\gamma\delta}$) is necessary but not sufficient for *joint* convexity.
6. For an $n\times n$ matrix N, an $m\times m$ matrix M, and an $n\times m$ matrix R, prove by direct matrix methods alone that N, M and $R^{T}N^{-1}R-M$ nonsingular imply $RM^{-1}R^{T}-N$ nonsingular.
7. Consider the following two quadratic examples generated by
   (a) $X[x_1,x_2,u]=\frac{1}{2}(x_1{}^2+x_2{}^2)+rx_1u+\frac{1}{2}u^2$
   where r is an assigned parameter ($n=2$, $m=1$ here);
   (b) $X[x,u]=\frac{1}{2}ax^2+hxu+\frac{1}{2}bu^2+fx+gu+c$
   where the coefficients are all assigned ($n=m=1$).
   In each example verify the formulae (2.67)–(2.71), calculate the other three generating functions in the regular cases, and compute Table 2.2. Use the results to illustrate Theorems 2.14 and 2.15, and also to illustrate Exercise 6 above.
8. In the examples of Exercise 7, show that when N, M and $R^{T}N^{-1}R-M$ are positive definite, it follows that $R^{T}M^{-1}R-N$ is positive definite in (b) but not in (a).
9. When the transformations from $Y[y,u]$ to $X[x,u]$ and $W[y,v]$ are both regular, prove that for $n=m=1$ the second derivatives satisfy
$$X_{xx}=\frac{1}{Y_{yy}}=\left[\frac{W_{yv}{}^2}{W_{vv}}-W_{yy}\right]^{-1},\quad X_{uu}=\frac{Y_{yu}{}^2}{Y_{yy}}-Y_{uu}=\left[\frac{W_{yv}{}^2}{W_{yy}}-W_{vv}\right]^{-1}$$
$$W_{vv}=\frac{1}{Y_{uu}}=\left[\frac{X_{xu}{}^2}{X_{xx}}-X_{uu}\right]^{-1},\quad W_{yy}=\frac{Y_{yu}{}^2}{Y_{uu}}-Y_{yy}=\left[\frac{X_{xu}{}^2}{X_{uu}}-X_{xx}\right]^{-1}.$$

2.6. Examples of quartets of Legendre transformations

(i) Introductory examples

A basic elementary illustration of Fig. 2.19 when $n = m = 1$ is

$$\begin{aligned} X &= \tfrac{1}{2}(x^2 - u^2), & Y &= \tfrac{1}{2}(y^2 + u^2), \\ Z &= -\tfrac{1}{2}(x^2 + v^2), & W &= \tfrac{1}{2}(-y^2 + v^2), \end{aligned} \tag{2.79}$$

noted by Noble and Sewell (1972). This is a strong quartet of regular branches, i.e. all the convexities are strict and there are no singularities.

More structure can be illustrated by using the general quadratic

$$\tfrac{1}{2}ax^2 + hxu + \tfrac{1}{2}bu^2 + fx + gu + c$$

as one of the generating functions in Fig. 2.18. For example, the reader will find it instructive at this stage to carry out Exercise 2.4.7(b). With this quadratic $X[x, u]$ it is found (cf. Theorem 2.15) that all four transformations are regular if and only if $a \neq 0$, $b \neq 0$ and $ab - h^2 \neq 0$, that a strong quartet of regular branches exists if $a > 0$ and $b < 0$, and that a quartet of regular branches which is not a strong quartet is present if $0 < ab < h^2$. In each of these cases $X[x, u]$ is a saddle function after suitable change of variables (cf. Exercise 2.4.8, and recall (1.9) and (1.19)). Such results can be displayed geometrically in a three dimensional coefficient space spanned by axes along which a, b and h are measured, and separated into different regions by the planes $a = 0$ and $b = 0$ and the cone $ab - h^2 = 0$ in that space.

The functions (2.79) are called canonical Morse functions in the sense indicated in Exercise 1.1.5, i.e. as being the simplest form obtainable by smooth changes of variables (diffeomorphisms) applied to *ab initio* regular functions which need not even be quadratic. In applications, however, we have to be ready for the possibility that the diffeomorphisms required to reach the canonical Morse functions (in the regular case), or the elementary catastrophes (in the singular case) may take us unduly far from the natural physical variables. For this reason one needs to strike a balance between the *ab initio* expression of a theory in concrete cases, and the insight into singularities which a canonical expression can give.

Simple illustrations of quartets which contain singularities are also found by specializing the general quadratic of Exercise 2.4.7(b). For example, if $h \neq 0$ there are two regular transformations connecting

$$X = \tfrac{1}{2}x^2 + hxu - u, \quad Y = \tfrac{1}{2}(y - hu)^2 + u,$$

and (2.80)

$$W = \frac{1}{2h^2}[(v - 1)^2 + 2h(v - 1)y];$$

and two singular transformations converting the distributed singularities on X(linear in u) and on W(linear in y) into a singularity which accumulates in the x, v plane on the line $v + hx = 1$. The quantity Z is defined only on this line, having the *values* of $-\frac{1}{2}x^2$ there, but it is not defined as a function $Z[x, v]$ of two independent variables over the whole plane (unless by an arbitrary extension off the line, e.g. as a cylinder, which we do not pursue). This illustrates the limitations which qualify the construction of quartets of branches when singularities are present.

If $h = 0$ in the X and Y of (2.80) we have a single regular transformation connecting

$$X = \tfrac{1}{2}x^2 - u, \quad Y = \tfrac{1}{2}y^2 + u$$

and two singular transformations converting each into quantities with values

$$Z = -\tfrac{1}{2}x^2, \quad W = -\tfrac{1}{2}y^2, \tag{2.81}$$

respectively, and defined only on the line $v = 1$ in the x, v and y, v spaces respectively. It turns out that if we regard (2.81) as functions of one variable on the respective lines $v = 1$, they are related by a regular Legendre transformation.

Quadratic examples of quartets of regular branches in which $n > 1$ and $m > 1$ are provided when the matrices in Table 2.2 have assigned values, for example

$$X[\mathbf{x}, \mathbf{u}] = \tfrac{1}{2}\mathbf{x}^{\mathrm{T}}N\mathbf{x} + \mathbf{x}^{\mathrm{T}}R\mathbf{u} + \tfrac{1}{2}\mathbf{u}^{\mathrm{T}}M\mathbf{u}$$

in the matrix notation for $\mathbf{x}$ and $\mathbf{u}$ also used in (1.45). One can immediately write down $Y[\mathbf{y}, \mathbf{u}]$, $Z[\mathbf{x}, \mathbf{v}]$ and $W[\mathbf{y}, \mathbf{v}]$.

(ii) Strain energy and complementary energy density

Basic variables in the mechanics of solids and fluids are 3×3 symmetric matrices of stress components τ_{ij} and strain components e_{ij} at each point of the medium $(i, j = 1, 2, 3)$. Their precise mechanical definitions are not needed at this point. We rearrange the six independent components of such matrices as 6×1 column matrices with definitions such as

$$\begin{aligned} \boldsymbol{\tau} &= [\tau_{11} \quad \tau_{22} \quad \tau_{33} \quad 2^{\frac{1}{2}}\tau_{23} \quad 2^{\frac{1}{2}}\tau_{31} \quad 2^{\frac{1}{2}}\tau_{12}]^{\mathrm{T}} \\ &= [\tau_1 \quad \tau_2 \quad \tau_3 \quad \tau_4 \quad \tau_5 \quad \tau_6]^{\mathrm{T}}, \end{aligned}$$

with $\mathbf{e}$ defined similarly. The $2^{\frac{1}{2}}$ maintains bookkeeping in $(2.82)_3$.

Hill (1956*a*) drew attention to the fact that in several types of material a convex work function $U[\mathbf{e}]$ and a convex complementary work function $U_c[\boldsymbol{\tau}]$ exist such that an invertible stress–strain relation is expressed by a regular Legendre transformation

$$\boldsymbol{\tau} = \frac{\partial U}{\partial \mathbf{e}}, \quad \mathbf{e} = \frac{\partial U_c}{\partial \boldsymbol{\tau}}, \quad U + U_c = \boldsymbol{\tau}^T \mathbf{e} = \tau_{ij} e_{ij}, \tag{2.82}$$

and that this fact is central to the existence of extremum principles and uniqueness theorems in such materials.

For example, in classical elasticity we require a particular definition of $\boldsymbol{\tau}$ which is called true (Cauchy) stress, and which we shall denote by $\boldsymbol{\sigma}$ here and in §5.6. The generalized Hooke's law is a linear stress–strain relation

$$\boldsymbol{\sigma} = K\mathbf{e}, \quad \mathbf{e} = M\boldsymbol{\sigma}, \quad M = K^{-1},$$

where K is a 6×6 positive definite symmetric matrix of 21 given elastic moduli, M is its inverse matrix of so-called compliances, and $\mathbf{e}$ is standard small strain. Then the strain energy density and complementary energy density are

$$U[\mathbf{e}] = \tfrac{1}{2}\mathbf{e}^T K \mathbf{e}, \quad U_c[\boldsymbol{\sigma}] = \tfrac{1}{2}\boldsymbol{\sigma}^T M \boldsymbol{\sigma} \tag{2.83}$$

respectively. This is an example of a pair of Legendre dual functions. In isotropic compressible material there remain only two independent moduli, and

$$U[\mathbf{e}] = \mu(e_1{}^2 + e_2{}^2 + e_3{}^2 + e_4{}^2 + e_5{}^2 + e_6{}^2) + \tfrac{1}{2}\lambda(e_1 + e_2 + e_3)^2,$$
$$U_c[\boldsymbol{\sigma}] = \tfrac{1}{2}(1 + \nu)E^{-1}(\sigma_1{}^2 + \sigma_2{}^2 + \sigma_3{}^2 + \sigma_4{}^2 + \sigma_5{}^2 + \sigma_6{}^2)$$
$$- \tfrac{1}{2}\nu E^{-1}(\sigma_1 + \sigma_2 + \sigma_3)^2.$$

Here $E > 0$ is Young's modulus, ν is Poisson's ratio, $\mu = E/2(1 + \nu) > 0$ is the shear modulus and $3\lambda + 2\mu = E/(1 - 2\nu) > 0$ is the bulk modulus.

The idea underlying the expression of constitutive equations in the form (2.82) (where there are no passive variables) extends to cases where $\boldsymbol{\tau}$ and $\mathbf{e}$ have other interpretations; e.g. $\mathbf{e}$ could be strain rate as in a Newtonian fluid (cf. §5.7).

(iii) Incremental elastic/plastic constitutive equations

This example builds upon (2.82) and (2.83) and illustrates well how the role of passive variables comes into its own. We have in mind either a single metal crystal capable of glide or, alternatively, a material element of an elastic/plastic body, in a given state of stress and known current geometry. The precise constitutive equations have been elucidated by Hill (1966, 1967 respectively in the two cases). They relate additional increments $\dot{\boldsymbol{\tau}}$ of stress and $\dot{\mathbf{e}}$ of strain in quasistatic distortion. Real time effects are absent, by hypothesis, and it is then a standard convention to describe increments as quasistatic rates, hence the superposed dots (for it is easier to write $\dot{\boldsymbol{\tau}}$ and $\dot{\mathbf{e}}$

frequently than, say, $d\tau$ and de). Again we do not need to give precise mechanical definitions of $\dot{\tau}$ and $\dot{e}$ until we address boundary value problems (§5.10), because the analytical structure of the constitutive relations now to be described is common to several theories of plasticity. The review articles of Hill (1978) and Havner (1982) contain more specialist refinements.

Two sets $\{\lambda_\gamma\}$ and $\{\mu_\gamma\}$, each of $m \geqslant 1$ intermediate scalar variables, are first introduced. The λ_γ can be interpreted as magnitudes of certain shear rates, for example on m given slip directions in the single crystal, and the μ_γ are associated *slack* stress-rates. These λ_γ and μ_γ are eventually eliminated to leave constitutive relations between $\dot{\tau}$ and $\dot{e}$ alone, and so it is possible to think of the λ_γ and μ_γ as *internal variables* and $\dot{\tau}$ and $\dot{e}$ as *external variables.*

It is important to avoid premature elimination, however, if we wish to recognise that incremental elastic/plastic constitutive equations can be expressed in the form shown in Fig. 2.18, and hence that their analytical structure has much in common with other physical and mathematical situations. Thus Sewell (1974) showed that Fig. 2.20 is a succinct starting point for the description of elastic/plastic response, transcribing the notation for Legendre transformations which is associated with Fig. 2.18 in (2.52)–(2.62). We may generate two (sets of) equations relating the four (sets of) variables from any convenient corner of Fig. 2.20. It may be simplest to start from the given function $X[\dot{\tau}, \lambda_\gamma]$ which generates

$$\dot{e} = \frac{\partial X}{\partial \dot{\tau}}, \quad \mu_\gamma = \frac{\partial X}{\partial \lambda_\gamma}, \tag{2.84}$$

according to (2.55). Alternatively, any other corner of Fig. 2.20 which can be

Fig. 2.20. Elastic/plastic response expressed via Legendre transformations.

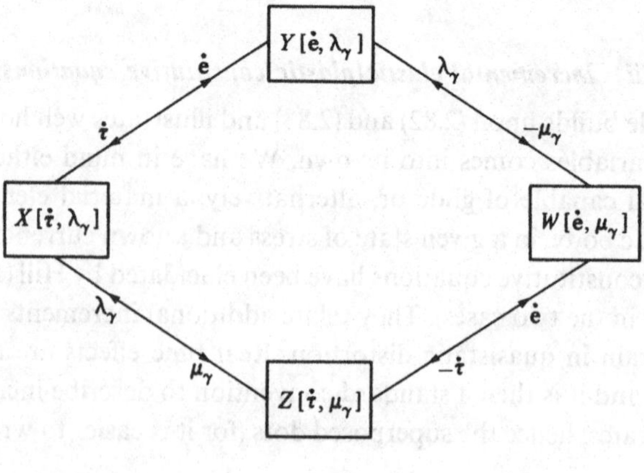

reached by Legendre transformations will generate an equivalent pair of equations.

Any such pair underdetermines the four variables, however, and we need supplementary information not contained in Fig. 2.20. This is provided by a trio of 'plastic flow conditions'

$$\lambda_\gamma \geqslant 0, \quad \mu_\gamma \leqslant 0, \quad \lambda_\gamma \mu_\gamma = 0 \tag{2.85}$$

having the structure of Fig. 1.6 (summation on repeated γ as in (1.46)), and acting as one 'equation'; and fourthly by assigning values to either $\dot{\mathbf{e}}$ or $\dot{\boldsymbol{\tau}}$.

The function $X[\dot{\boldsymbol{\tau}}, \lambda_\gamma]$ will always be assumed homogeneous of degree two in the $\dot{\boldsymbol{\tau}}$ and λ_γ taken together. This is sufficient to ensure that the real time effects are absent.

The most familiar example of Fig. 2.20 is generated by

$$X[\dot{\boldsymbol{\tau}}, \lambda_\gamma] = \tfrac{1}{2}\dot{\boldsymbol{\tau}}^{\mathrm{T}} M \dot{\boldsymbol{\tau}} + \dot{\boldsymbol{\tau}}^{\mathrm{T}} \boldsymbol{v}_\gamma \lambda_\gamma - \tfrac{1}{2} h_{\gamma\delta} \lambda_\gamma \lambda_\delta. \tag{2.86}$$

Here M is a given positive definite 6×6 matrix of elastic compliances. Each of the m given $\boldsymbol{v}_\gamma$ is a 6×1 rearrangement of the components $v_{ij\gamma}$ of a 3×3 symmetric matrix. This describes either the simple shear direction in the crystal, by $v_{ij\gamma} = \tfrac{1}{2}(n_i t_j + n_j t_i)_\gamma$ where t_i and n_i are given unit vector components of the slip direction and slip plane normal respectively, or the plastic yield surface normal in the continuum. The yield surface in stress space has a pyramidal vertex at the given stress state if $m > 1$, and is locally smooth if $m = 1$. The most commonly used yield surfaces are cylinders in principal stress space, equally inclined to the three positive axes, and with circular (Mises) or hexagonal (Tresca) cross-section. For plane stress their traces on the plane of zero third stress are shown in Fig. 2.21, together with

Fig. 2.21. Plane stress sections of common yield cylinders.

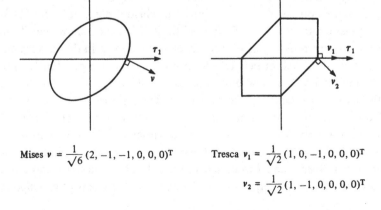

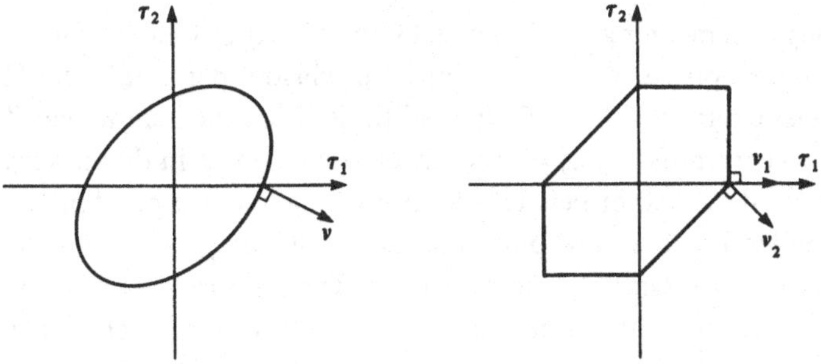

Mises $\boldsymbol{v} = \dfrac{1}{\sqrt{6}}(2, -1, -1, 0, 0, 0)^{\mathrm{T}}$

Tresca $\boldsymbol{v}_1 = \dfrac{1}{\sqrt{2}}(1, 0, -1, 0, 0, 0)^{\mathrm{T}}$

$\boldsymbol{v}_2 = \dfrac{1}{\sqrt{2}}(1, -1, 0, 0, 0, 0)^{\mathrm{T}}$

(the projections of) their normals at the uniaxial stress point. In our theory we do not need to know the shape of the yield surface other than locally at the current stress point. The $h_{\gamma\delta}$ is the typical entry in a given $m \times m$ so-called hardening matrix. From (2.84) and (2.86) we find that

$$\dot{\mathbf{e}} = M\dot{\boldsymbol{\tau}} + \boldsymbol{v}_\gamma\lambda_\gamma, \quad \mu_\gamma = \dot{\boldsymbol{\tau}}^{\mathrm{T}}\boldsymbol{v}_\gamma - h_{\gamma\delta}\lambda_\delta, \tag{2.87}$$
$$X = \tfrac{1}{2}(\dot{\boldsymbol{\tau}}^{\mathrm{T}}\dot{\mathbf{e}} + \mu_\gamma\lambda_\gamma) \text{ in value,}$$

before (2.85) are applied.

By performing the Legendre transformations directly, or by transcribing Table 2.2 since (2.86) is quadratic, we can find the other three generating functions in Fig. 2.20. Thus in terms of $K = M^{-1}$ and the $m \times m$ symmetric matrix with elements defined by

$$g_{\gamma\delta} = h_{\gamma\delta} + \boldsymbol{v}_\gamma{}^{\mathrm{T}}K\boldsymbol{v}_\delta,$$

we find

$$\begin{aligned} Y[\dot{\mathbf{e}}, \lambda_\gamma] &= \tfrac{1}{2}\dot{\mathbf{e}}^{\mathrm{T}}K\dot{\mathbf{e}} - \lambda_\gamma\boldsymbol{v}_\gamma{}^{\mathrm{T}}K\dot{\mathbf{e}} + \tfrac{1}{2}g_{\gamma\delta}\lambda_\gamma\lambda_\delta \\ &= \tfrac{1}{2}(\dot{\mathbf{e}}^{\mathrm{T}} - \lambda_\gamma\boldsymbol{v}_\gamma{}^{\mathrm{T}})K(\dot{\mathbf{e}} - \lambda_\delta\boldsymbol{v}_\delta) + \tfrac{1}{2}h_{\gamma\delta}\lambda_\gamma\lambda_\delta \\ &= \tfrac{1}{2}(\dot{\boldsymbol{\tau}}^{\mathrm{T}}\dot{\mathbf{e}} - \mu_\gamma\lambda_\gamma) \text{ in value.} \end{aligned} \tag{2.88}$$

If the hardening matrix is nonsingular, and $h_{\gamma\delta}^{-1}$ is the typical element of its inverse,

$$\begin{aligned} Z[\dot{\boldsymbol{\tau}}, \mu_\gamma] &= -\tfrac{1}{2}\dot{\boldsymbol{\tau}}^{\mathrm{T}}M\dot{\boldsymbol{\tau}} - \tfrac{1}{2}h_{\gamma\delta}^{-1}(\mu_\gamma - \boldsymbol{v}_\gamma{}^{\mathrm{T}}\dot{\boldsymbol{\tau}})(\mu_\delta - \boldsymbol{v}_\delta{}^{\mathrm{T}}\dot{\boldsymbol{\tau}}) \\ &= \tfrac{1}{2}(-\dot{\boldsymbol{\tau}}^{\mathrm{T}}\dot{\mathbf{e}} + \mu_\gamma\lambda_\gamma) \text{ in value.} \end{aligned} \tag{2.89}$$

If the matrix with elements $g_{\gamma\delta}$ is nonsingular, with $g_{\gamma\delta}^{-1}$ as the typical element of its inverse,

$$\begin{aligned} W[\dot{\mathbf{e}}, \mu_\gamma] &= -\tfrac{1}{2}\dot{\mathbf{e}}^{\mathrm{T}}K\dot{\mathbf{e}} + \tfrac{1}{2}g_{\gamma\delta}^{-1}(\boldsymbol{v}_\gamma{}^{\mathrm{T}}K\dot{\mathbf{e}} - \mu_\gamma)(\boldsymbol{v}_\delta{}^{\mathrm{T}}K\dot{\mathbf{e}} - \mu_\delta) \\ &= -\tfrac{1}{2}(\dot{\boldsymbol{\tau}}^{\mathrm{T}}\dot{\mathbf{e}} + \mu_\gamma\lambda_\gamma) \text{ in value.} \end{aligned} \tag{2.90}$$

The physical meaning of adding (2.85) to (2.87) is as follows. The strain-rate is decomposed by $(2.87)_1$ into an elastic part $M\dot{\boldsymbol{\tau}}$ (cf. (2.83)) and a plastic part $\boldsymbol{v}_\gamma\lambda_\gamma$. The flow conditions (2.85) then show that if any $\mu_\gamma < 0$ the corresponding $\lambda_\gamma = 0$ (purely elastic response in that mode), and that any mode $\lambda_\gamma > 0$ of plastic response entails a corresponding $\mu_\gamma = 0$.

The reader may find it helpful to repeat all the foregoing manipulations in the case $m = 1$ (smooth yield surface). Then we can drop the Greek suffixes altogether, and aside from $\lambda = \mu = 0$ which implies $\dot{\boldsymbol{\tau}}^{\mathrm{T}}\boldsymbol{v} = 0$, we have only the two possibilities that either $\mu = \dot{\boldsymbol{\tau}}^{\mathrm{T}}\boldsymbol{v} < 0$ with $\lambda = 0$, or $\lambda > 0$ with $\mu = \dot{\boldsymbol{\tau}}^{\mathrm{T}}\boldsymbol{v} - h\lambda = 0$. If $h \neq 0$ the use of (2.85) to eliminate the internal variables μ and λ in favour of the external variables $\dot{\boldsymbol{\tau}}$ and $\dot{\mathbf{e}}$ thus allows the constitutive equations to be expressed in terms of a *composite* stress-rate potential defined by

$$V_c[\dot{\tau}] = \begin{cases} -Z[\dot{\tau},0] & \text{if } \dfrac{\dot{\tau}^T\nu}{h} \geqslant 0, \\ X[\dot{\tau},0] & \text{if } \dot{\tau}^T\nu < 0 \end{cases} \tag{2.91}$$

as

$$\dot{\mathbf{e}} = \frac{\partial V_c}{\partial \dot{\tau}}. \tag{2.92}$$

The inverse of this can be expressed, if $g = h + \nu^T K \nu \neq 0$, in terms of a composite strain-rate potential defined by

$$V[\dot{\mathbf{e}}] = \begin{cases} -W[\dot{\mathbf{e}},0] & \text{if } \dfrac{\nu^T K \dot{\mathbf{e}}}{g} \geqslant 0, \\ Y[\dot{\mathbf{e}},0] & \text{if } \nu^T K \dot{\mathbf{e}} < 0 \end{cases} \tag{2.93}$$

with

$$\dot{\tau} = \frac{\partial V}{\partial \dot{\mathbf{e}}}. \tag{2.94}$$

The composite potentials have the common value $V = V_c = \frac{1}{2}\dot{\tau}^T\dot{\mathbf{e}}$, by inserting $\lambda\mu = 0$ in above, and hence

$$V + V_c = \dot{\tau}^T\dot{\mathbf{e}}. \tag{2.95}$$

This, with (2.92) and (2.94), shows that the composite potentials are piecewise C^2 functions which can be regarded as generating a Legendre transformation directly. Such composite potentials have been introduced *ab initio* by Hill (1962) and, in the terms here of Fig. 2.20 and its quartet of single valued functions, by Sewell (1982).

If $m > 1$ (yield surface vertex), the use of (2.85) to eliminate internal variables can again lead to composite stress-rate and strain-rate potentials, but with more components. If $m = 2$, for example, (2.91) with $\lambda_1 > 0, \lambda_2 > 0$ or with $\mu_1 < 0, \mu_2 < 0$ needs to be augmented by the two extra possibilities $\lambda_1 > 0,\ \mu_2 < 0$ and $\mu_1 < 0,\ \lambda_2 > 0$. The latter entail extra Legendre transformations with λ_1, μ_1 active and λ_2, μ_2 passive, and then vice versa, leading to new mixed dual functions $P[\dot{\tau}, \mu_1, \lambda_2]$ and $Q[\dot{\tau}, \lambda_1, \mu_2]$ (say). When these exist (2.92) will apply with the definition

$$V_c[\dot{\tau}] = \begin{cases} -Z[\dot{\tau},0,0] & \text{if } \lambda_1 > 0, \quad \lambda_2 > 0, \\ -P[\dot{\tau},0,0] & \text{if } \lambda_1 > 0, \quad \mu_2 < 0, \\ -Q[\dot{\tau},0,0] & \text{if } \mu_1 < 0, \quad \lambda_2 > 0, \\ X[\dot{\tau},0,0] & \text{if } \mu_1 < 0, \quad \mu_2 < 0. \end{cases} \tag{2.96}$$

In general there are $m!/(m-r)!r!$ ways of selecting r plastic modes $\lambda_\gamma > 0$, implying associated $\mu_\gamma = 0$, from the possible m. Hence the composite potentials, when they exist, will have $\sum_{r=0}^{m} m!/(m-r)!r!$ components.

Particular interest attaches to the case when all m $\lambda_\gamma > 0$, called *fully plastic loading*. If the determinant $|h_{\gamma\delta}| \neq 0$ we can eliminate the λ_γ from (2.87) to express $\dot{\mathbf{e}}$ in terms of $\dot{\boldsymbol{\tau}}$ only. When $|g_{\gamma\delta}| \neq 0$ also, the inverse of these last relations can also be written as $-\dot{\boldsymbol{\tau}} = \partial W/\partial \dot{\mathbf{e}}$ (cf. (2.58)$_1$ and Fig. 2.20) with all $\mu_\gamma = 0$, i.e.

$$\dot{\boldsymbol{\tau}} = K\dot{\mathbf{e}} - K\mathbf{v}_\gamma g_{\gamma\delta}^{-1}\mathbf{v}_\delta{}^{\mathrm{T}}K\dot{\mathbf{e}} \tag{2.97}$$

from (2.90). The coefficient of $\dot{\mathbf{e}}$ here is the 6×6 symmetric matrix of *tangent moduli* for full plastic loading in the incremental distortion. In the one dimensional stress/strain response curve for $m = 1$ sketched in Fig. 2.22 the slopes of the two segments are $K = E$, Young's modulus, and the tangent modulus $E(1 - E/(h + E)) = h/(1 + h/E)$.

Hypotheses on the hardening matrix $h_{\gamma\delta}$ control the further details. If $h_{\gamma\delta}$ is positive definite, then (2.86) is a strict saddle function and Fig. 2.20 is a single strong quartet as in Fig. 2.19. This is the most straightforward case, in which Theorems 2.11, 2.12 and 2.15(*b*) apply, the relation between $\dot{\boldsymbol{\tau}}$ and $\dot{\mathbf{e}}$ is unique, and dual extremum principles exist to characterize it (see Exercises 2.5.3 and 4). If $h_{\gamma\delta}$ is nonnegative definite we still have a regular pair of branches of $X[\dot{\boldsymbol{\tau}}, \lambda_\gamma]$ and $Y[\dot{\mathbf{e}}, \lambda_\gamma]$ and certain weakened extremum principles survive, but the uniqueness in the relation between $\dot{\boldsymbol{\tau}}$ and $\dot{\mathbf{e}}$ is lost, and because $|h_{\gamma\delta}|$ can be zero it may happen that Z is only defined on an accumulated singularity (cf. (2.80)). If $h_{\gamma\delta}$ is negative definite while $g_{\gamma\delta}$ is positive definite, the special type of separate convexity mentioned after Theorem 2.14 applies to both $X[\dot{\boldsymbol{\tau}}, \lambda_\gamma]$ and $Y[\dot{\mathbf{e}}, \lambda_\gamma]$, since then the three matrices in (2.76) are all positive definite. A simple

Fig. 2.22. One dimensional elastic/plastic response.

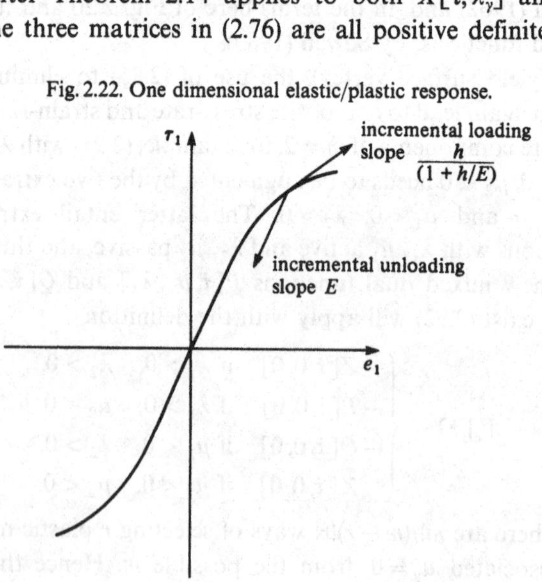

calculation with $m = 1$ and $h < 0$ (strain softening) will illustrate the multi-valuedness in the relation between $\dot{\mathbf{t}}$ and $\dot{\mathbf{e}}$ when $h_{\gamma\delta}$ is not positive definite.

A set of supplementary conditions different from (2.85) would be that all $\mu_\gamma = 0$ regardless of the sign of the λ_γ. When $g_{\gamma\delta}$ is positive definite it can be shown that the fictitious linear comparison solid so defined, with constitutive equation (2.97), has the following property (Hill, 1958, for $m = 1$; Sewell, 1972, for $m > 1$). A bifurcation in the real elastic/plastic solid cannot precede one in the more easily analyzed comparison solid, whose tangent moduli are the same for both loading and unloading (see Hill, 1978, for further details).

This comparison theorem rests upon an application to (2.90) of a property of any quadratic $W[\mathbf{y}, \mathbf{v}]$, say

$$W[\mathbf{y},\mathbf{v}] = -\tfrac{1}{2}\mathbf{y}^{\mathrm{T}}A\mathbf{y} + \mathbf{v}^{\mathrm{T}}T\mathbf{y} + \tfrac{1}{2}\mathbf{v}^{\mathrm{T}}B\mathbf{v} + \mathbf{a}^{\mathrm{T}}\mathbf{y} + \mathbf{b}^{\mathrm{T}}\mathbf{v},$$

whose gradients are $\mathbf{u} = T\mathbf{y} + B\mathbf{v} + \mathbf{b}$, $-\mathbf{x} = -A\mathbf{y} + T^{\mathrm{T}}\mathbf{v} + \mathbf{a}$ by (2.58). For any pair of arguments $\mathbf{y}_+$, $\mathbf{v}_+$ and $\mathbf{y}_-$, $\mathbf{v}_-$ with finite differences denoted by $\mathbf{y}_+ - \mathbf{y}_- = \Delta\mathbf{y}$, etc., we find (Sewell, 1974)

$$\Delta\mathbf{x}^{\mathrm{T}}\Delta\mathbf{y} - \Delta\mathbf{y}^{\mathrm{T}}A\Delta\mathbf{y} = -\Delta\mathbf{u}^{\mathrm{T}}\Delta\mathbf{v} + \Delta\mathbf{v}^{\mathrm{T}}B\Delta\mathbf{v}.$$

If B is nonnegative definite and if the plus and minus points both satisfy $\mathbf{u} \geqslant 0$, $\mathbf{v} \geqslant 0$, $\mathbf{u}^{\mathrm{T}}\mathbf{v} = 0$, we can write $A\mathbf{y} + \mathbf{a} = -\mathbf{x}_{\mathrm{c}}$ (say) so that

$$\Delta\mathbf{x}^{\mathrm{T}}\Delta\mathbf{y} \geqslant \Delta\mathbf{x}_{\mathrm{c}}{}^{\mathrm{T}}\Delta\mathbf{y}. \tag{2.98}$$

The result cited in the previous paragraph can be shown to follow from (2.98).

The basic analytical structure contained in (2.84) with (2.85) is not only transmitted up the size hierarchy from single metal crystals to polycrystalline aggregates as already indicated. It also grew independently in nonlinear programming (cf. §1.6(i) and (ii)). In addition some aspects of it reappear in a diversity of boundary value problems posed for three dimensional bodies (see Chapter 5). The engineering theory of plastic structures represents a fusion of these approaches, going from the continuum back to the discrete and exploiting explicit algorithms from the nonlinear programming literature (e.g. see the reviews by Maier, 1977; and Maier and Munro, 1982).

Exercises 2.5

1. Establish $Y[\dot{\mathbf{e}}, \lambda_\gamma]$ in (2.88). Show that it is strictly jointly convex when $h_{\gamma\delta}$ is positive definite. Establish $Z[\dot{\mathbf{t}}, \mu_\gamma]$ in (2.89) then and show that it is strictly jointly concave, as Fig. 2.19 requires.
2. Assemble the Table 2.2 in the notation of (2.86), and specialize it to the case $m = 1$.

3. When $h_{\gamma\delta}$ is positive definite and $\dot{\mathbf{e}}$ is regarded as assigned, show that (2.87) with (2.85) can be regarded as an illustration of problem (1.54) with

$$L[\dot{\mathbf{t}}, \lambda_\gamma] = \dot{\mathbf{t}}^{\mathrm{T}}\dot{\mathbf{e}} - X[\dot{\mathbf{t}}, \lambda_\gamma]$$

as a saddle function strictly concave in $\dot{\mathbf{t}}$ and strictly convex in the λ_γ. Hence show that the solution values of $\dot{\mathbf{t}}$ and the λ_γ are unique; and that (cf. Theorem 1.11) they will minimize

$$L - \dot{\mathbf{t}}^{\mathrm{T}}\frac{\partial L}{\partial \dot{\mathbf{t}}} = Y[\dot{\mathbf{e}}, \lambda_\gamma] \text{ among all } \lambda_\gamma \geqslant 0;$$

and that they will maximize

$$L - \lambda_\gamma \frac{\partial L}{\partial \lambda_\gamma} = \dot{\mathbf{t}}^{\mathrm{T}}\dot{\mathbf{e}} + Z[\dot{\mathbf{t}}, \mu_\gamma] \text{ among all } \dot{\mathbf{t}} \text{ such that } \mu_\gamma \leqslant 0,$$

treating the μ_γ as auxiliary parameters satisfying $(2.87)_2$ (Hill, 1966, Noble and Sewell, 1972). How are these results affected if $h_{\gamma\delta}$ is nonnegative definite?

4. When $\dot{\mathbf{t}}$ is assigned and $m = 1$ with $h_{\gamma\delta} = h > 0$, show by inserting $J[\lambda] = -X[\dot{\mathbf{t}}, \lambda]$ into Theorem 1.3 that any λ satisfying $(2.84)_2$ with (2.85) is unique and will

$$\begin{aligned} &\text{minimize} \quad -X[\dot{\mathbf{t}}, \lambda] \text{ among all } \lambda \geqslant 0; \text{ and} \\ &\text{maximize} \quad Z[\dot{\mathbf{t}}, \mu] \text{ among all } \mu \leqslant 0. \end{aligned}$$

5. In the case $m = 2$, and when the then 2×2 hardening matrix has the properties

$$h_{11} \neq 0, \quad h_{22} \neq 0, \quad h_{11}h_{22} - h_{12}{}^2 \neq 0$$

with any choice of signs, calculate $P[\dot{\mathbf{t}}, \mu_1, \lambda_2]$ and $Q[\dot{\mathbf{t}}, \lambda_1, \mu_2]$ and hence describe the composite potential (2.96) explicitly as a piecewise C^2 function of $\dot{\mathbf{t}}$. Display the domains on a plane spanned by the two scalar variables $\mathbf{v}_1^{\mathrm{T}}\dot{\mathbf{t}}$ and $\mathbf{v}_2^{\mathrm{T}}\dot{\mathbf{t}}$ for the various choices of sign.

6. Let M and r be given positive constants. Show that the function

$$X[x, \lambda] = \tfrac{1}{2}Mx^2 + \tfrac{1}{2}\lambda(x^2 - r^2)$$

of two scalar variables x and λ generates

$$Y[y, \lambda] = \tfrac{1}{2}(M + \lambda)^{-1}y^2 + \tfrac{1}{2}\lambda r^2$$

by a Legendre transformation which is regular if the passive $\lambda \neq -M$. What can be said about the dual functions if $\lambda = -M$? (cf. Fig. 2.3 and §2.3(vii)).

Then show that, with active $\lambda \neq -M$ and passive $y \neq 0$, another regular transformation generates

$$W[y, \mu] = \mp y(2\mu + r^2)^{\frac{1}{2}} + \tfrac{1}{2}M(2\mu + r^2).$$

What can be said about the dual functions if $y = 0$?

Next show that two singular transformations, starting with active λ and passive x in $X[x, \lambda]$ or with active y and passive μ in $W[y, \mu]$, convert any finite segment of those two linear functions into the same value of

$$Z = -\tfrac{1}{2}Mx^2 = -\tfrac{1}{2}M(2\mu + r^2)$$

accumulated at

$$\mu = \tfrac{1}{2}(x^2 - r^2) \quad \text{or} \quad -x = \mp(2\mu + r)^{\frac{1}{2}}$$

respectively.

Construct the diagram corresponding to Fig. 2.18 or 2.20 in the case when $\lambda > -M$, $y > 0$, $x > 0$, $\mu > -\frac{1}{2}r^2$. If x and y are thought of as prototype stress-like and strain-like variables $\frac{1}{2}(x^2 - r^2)$ has the role of a prototype yield function.

7. Consider the function $Y[y, \lambda]$ constructed in Exercise 6 for the case $\lambda > -M$. By adjoining all three conditions

$$\lambda \geqslant 0, \quad \mu \leqslant 0, \quad \lambda\mu = 0$$

to the pair

$$x = \frac{\partial Y}{\partial y}, \quad \mu = -\frac{\partial Y}{\partial \lambda}$$

generated by such $Y[y, \lambda]$, eliminate the internal variables λ and μ. Hence express the response relation between the external variables x and y as shown in Fig. 2.23(a).

Show that this relation can be written as

$$x = \frac{\mathrm{d}V}{\mathrm{d}y}$$

Fig. 2.23. Response function and composite potentials for Exercise 2.5.7.

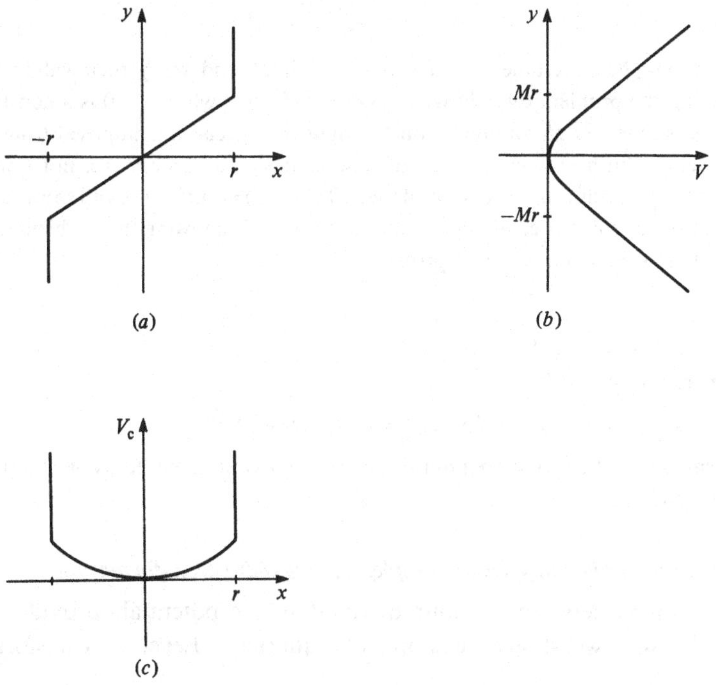

in terms of the function $V[y]$ shown in Fig. 2.23(*b*), which is a composite piecewise C^2 function of the type (2.93). Discuss the conventions which are needed to allow the response relation to be written alternatively as

$$y = \frac{dV_c}{dx}$$

in terms of the function $V_c[x]$ shown in Fig. 2.23(*c*). (This is a composite function of the type (2.91), augmented by vertical branches where $V_c[\pm r] > \frac{1}{2}Mr^2$. These are to be regarded as polars of poles at infinity on $V[y]$, invoking also the auxiliary convexity discussed in §2.3(v).)

8. Let $M > 0$ and $T \neq 0$ be given constants, and $f(x)$ be a given convex function of x. Show that the function

$$L[x; \lambda, u] = uTx - \tfrac{1}{2}Mx^2 - \lambda f(x)$$

of three scalar variables x, λ and u has a nonnegative saddle quantity S (see (1.57)) in $\lambda \geqslant 0$, so that it may be regarded as concave in x and convex in u and λ there.

For the governing conditions

$$\frac{\partial L}{\partial x} = 0, \quad \lambda \geqslant 0, \quad (\alpha)$$

$$\frac{\partial L}{\partial u} = 0, \quad \frac{\partial L}{\partial \lambda} \geqslant 0, \quad (\beta)$$

$$\lambda \frac{\partial L}{\partial \lambda} = 0,$$

write out the simultaneous extremum principles, and verify their validity by solving the problem directly when $f(x) = \frac{1}{2}(x^2 - r^2)$, where $r > 0$ is a constant.

9. Suppose the trio of conditions on the right in Exercise 8 is imposed from the outset on both extremum principles, so that supplementary (i.e. not strictly necessary) conditions are deployed there. Take $f(x) = \frac{1}{2}(x^2 - r^2)$, and show that λ can then be eliminated along the lines of Exercise 7, allowing the problem to be written as a *stationary* value problem (cf. (1.63))

$$\frac{\partial L}{\partial x} = 0 \quad (\alpha), \qquad \frac{\partial L}{\partial y} = \frac{\partial L}{\partial u} = 0 \quad (\beta),$$

generated by

$$L[x; y, u] = x(Tu - y) + V[y].$$

Write down the new extremum principles, and verify their equivalence to the previous ones.

(iv) Other physical examples of Legendre transformations

The relations between the four thermodynamic potentials provide perhaps the most well-known example of a quartet of Legendre transformations, which may contain several regular branches joined by at least isolated singularities. We give some details of this in §5.2.

Another physical example of Fig. 2.19 occurs in the semigeostrophic equations of meteorology, as examined in detail by Chynoweth and Sewell (1989). These equations model mid-latitude motions of the atmosphere on a sub-continental scale, and allow for the presence of weather fronts via discontinuities of velocity and temperature. Local cartesian coordinates describe the position of each parcel of air by a horizontal vector $\mathbf{x}$ and a vertical pseudo-height z defined in terms of pressure. The geopotential of the parcel or particle at each time t is expressed as a function $P[\mathbf{x}, z, t]$ having gradients equal to

$$\mathbf{M} = \frac{\partial P}{\partial \mathbf{x}}, \quad \theta = \frac{\partial P}{\partial z}. \tag{2.99}$$

Here $\mathbf{M}$ represents the horizontal momentum vector, and θ is a function of entropy called the potential temperature. Three other Legendre dual functions are defined as summarized in Fig. 2.24, by transcribing the notation of §2.5(i). The presence of a Legendre transformation in this context was explicitly recognized only recently by Chynoweth. Porter and Sewell (1988). An example of flow in a vertical plane allows $P[\mathbf{x}, z]$ to be the convex part of a swallowtail function like Fig. 2.11(b), with the atmospheric front located by the self-intersection where the slope discontinuities are jumps in $\mathbf{M}$ and θ, by (2.99). Fig. 2.24 facilitates alternative starting points, such as a single-valued saddle function $S[\mathbf{M}, z]$, instead of $P[\mathbf{x}, z]$, to be used for the construction of models. Numerical methods of solving the semigeostrophic equations can involve dual discretizations like those of Fig. 2.15, as described by Chynoweth and Sewell (1990).

Fig. 2.24. Legendre transformations in semigeostrophic theory.

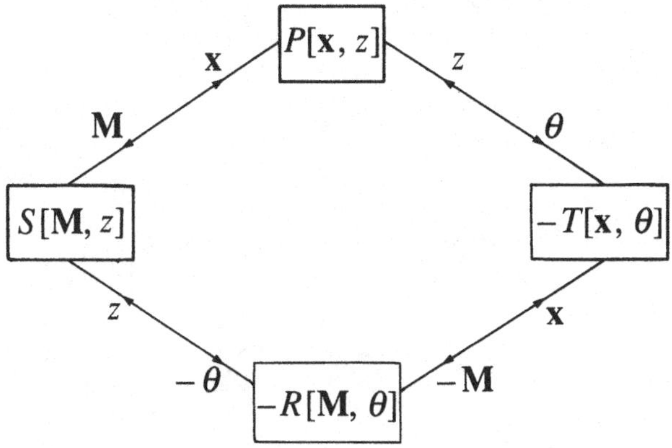

In the theory of singularities which are stable with respect to certain mathematical perturbations (see §2.2(vii)), it is known that the bifurcation sets for fold, cusp and swallowtail (defined after Theorem 2.7) represent the typical singularities of non-gradient mappings of $\mathbb{R}^3 \to \mathbb{R}^3$ (see Thorndike *et al.*, 1978), just as the fold and cusp are the only stable singularities of smooth mappings of the plane into the plane (Whitney, 1955). Therefore these three forms will appear more widely than as Legendre singularities, which can be viewed as singularities of gradient mappings.

For example, in the two or three dimensional kinematics of any continuous medium, the hodograph transformation between (say cartesian coordinates of) particle position x_i and velocity v_i can exhibit only the stated singularities whether or not the motion is irrotational. This is the underlying reason why cusps will appear frequently in applications to plane fluid flow. An early basic investigation of the singularities of plane mappings by Craggs (1948) was not restricted to gradient mappings, and so could have been used for rotational flows in spite of his restrictive title. His only example, however, was an application to irrotational plane flows. It is

Fig. 2.25. Singularities of a constrained mapping $\mathbb{R}^2 \to \mathbb{R}^2$.

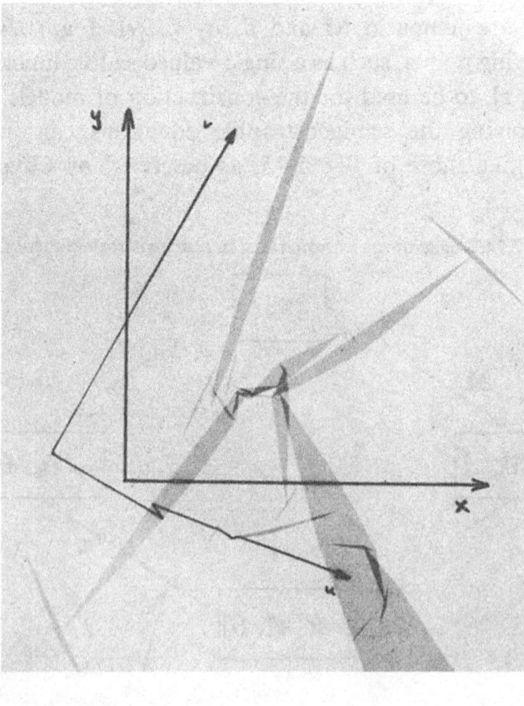

only for irrotational flow, i.e. $v_i = \partial\phi/\partial x_i$ whether plane or in three dimensions, that the hodograph transformation reduces to a Legendre transformation. Here $\phi[x_i]$ is the velocity potential. It is not an energy, and so will not have that role in the integrands of extremum principles.

(v) Stable singularities in a constrained plane mapping

Figure 2.25 shows an attempt to exhibit experimentally the Whitney result for $\mathbb{R}^2 \to \mathbb{R}^2$ quoted above. It is a photograph of an arbitrarily crumpled plastic sheet flattened under clear glass, which may also be displayed using an overhead projector (as proposed by Sewell, 1980, 1981). Cartesian u, v axes have previously been drawn on the plastic sheet, and x, y axes on the working plate of the projector. We see that a multivalued mapping of the x, y plane onto the u, v plane is produced, i.e. of the plate onto the plastic. The multiplicity of the mapping correlates with the shade of grey in the photograph, or on the projection screen. The singularities of the mapping appear as folds, which seem to run out in pairs at nonzero angles instead of cusps. The folds seem to be mostly straight rather than curved. The singularities are typical because the sheet was crumpled arbitrarily before flattening, and the reader may repeat the experiment in the lecture room. The difference from the Whitney result, i.e. nonzero angles instead of cusps, may lie in the fact that the plastic is almost inextensible, thus constraining the exhibited mapping to be less than arbitrary. There is some literature (e.g. Callahan, 1974: Robertson, 1977) on the pattern of creases in crumpled flattened paper. Fig. 2.25 gives some support to Robertson's claim that usually four straight folds run out together at the same point, making a pair of complementary angles (adding to π).

2.7. The structure of maximum and minimum principles

(i) Bounds and Legendre transformations

Using the inner product and set notation of §§1.3(i) and 2.5(ii), we bring together here some ideas from Chapters 1 and 2. A given function $-L[x,u]$ of $n+m$ scalar variables can be used as a particular example of $X[x,u]$ to generate, by Legendre transformations which may be globally valid, three other functions which we denote by

$$j\left[\frac{\partial L}{\partial x}, u\right] = L - \left(x, \frac{\partial L}{\partial x}\right), \quad k\left[x, \frac{\partial L}{\partial u}\right] = L - \left\langle u, \frac{\partial L}{\partial u}\right\rangle,$$
$$h\left[\frac{\partial L}{\partial x}, \frac{\partial L}{\partial u}\right] = -L + \left(x, \frac{\partial L}{\partial x}\right) + \left\langle u, \frac{\partial L}{\partial u}\right\rangle, \tag{2.100}$$

where independent variables are written on the left within square brackets and values are written on the right. The connections between the functions are displayed in Fig. 2.26. This is a particular example of Fig. 2.18, obtainable just by transcribing (2.52)–(2.62). Even if one of the four functions is single valued, at least one of the other three functions could be multivalued with regular and singular branches, as allowed for in §§2.5(i) and (ii). The functions (2.100) can be constructed in advance of the choice of whichever set of the governing conditions (1.50) or (1.53)–(1.55) are to apply. First and second derivatives with respect to passive variables on a regular branch are, by $(2.65)_3$ and (2.69),

$$\frac{\partial j}{\partial u}=\frac{\partial L}{\partial u},\quad j_{uu}=L_{uu}-L_{ux}L_{xx}^{-1}L_{xu},$$
$$\frac{\partial k}{\partial x}=\frac{\partial L}{\partial x},\quad k_{xx}=L_{xx}-L_{xu}L_{uu}^{-1}L_{ux}. \tag{2.101}$$

A particular branch of $k[x, \partial L/\partial u]$, denoted by $-G[x, v]$, was introduced in (1.96). At the end of §(iv) below it will be convenient to define $\bar{y}$ and $\bar{v}$ as shorthand symbols for $\partial L/\partial x$ and $\partial L/\partial u$ respectively, but to begin with we can be more explicit by desisting from this shorthand.

Simple illustrations of (2.100) can be generated from the quadratic

$$L[x,u]=\tfrac{1}{2}\lambda(x,x)+(x,T^*u)+\tfrac{1}{2}\langle u,u\rangle+(a,x), \tag{2.102}$$

which is just (1.137) rewritten in the operator notation. If $\lambda \neq 0$ we find

$$j\left[\frac{\partial L}{\partial x},u\right]=-\frac{1}{2\lambda}\left(\frac{\partial L}{\partial x}-T^*u-a,\frac{\partial L}{\partial x}-T^*u-a\right)+\tfrac{1}{2}\langle u,u\rangle \tag{2.103}$$

Fig. 2.26. Values of global functions generated by $L[x,u]$ via Fig. 2.18.

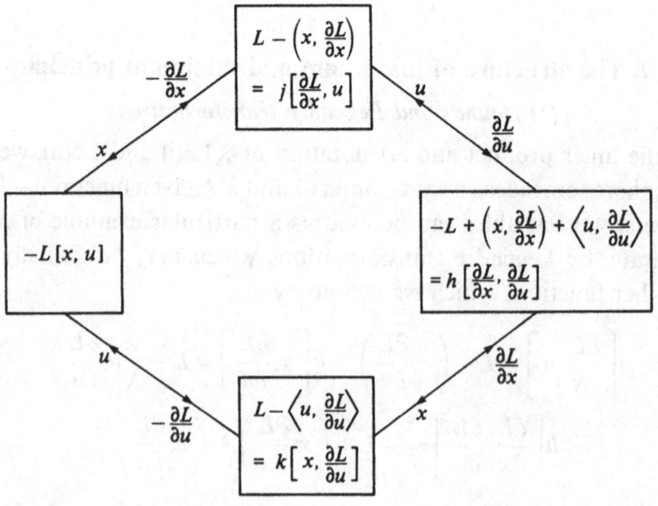

(if $\lambda = 0$ we can still define values of $j = \frac{1}{2}\langle u, u\rangle$); and for any λ

$$k\left[x, \frac{\partial L}{\partial u}\right] = \tfrac{1}{2}\lambda(x, x) + (a, x) - \tfrac{1}{2}\left\langle \frac{\partial L}{\partial u} - Tx, \frac{\partial L}{\partial u} - Tx \right\rangle. \qquad (2.104)$$

Recall that T is an $m \times n$ matrix whose transpose T^{T} is rewritten as T^* for consistency with the operator notation in use here (see (1.75)). When $\lambda < 0$, these $j[0, u]$ and $k[x, 0]$ are the bounds L_α and L_β calculated in (1.140).

Similar close relations between the functions in (2.100) and the bounds in (1.80) are valid much more generally, as we shall see in Table 2.3. However, we can see already by comparing Theorem 1.6 with Fig. 2.26 that, when appropriate hypotheses on $L[x, u]$ are satisfied, *the values of upper and lower bounds are the same as those of the associated Legendre dual functions.*

(ii) Bifurcation theory

Nonquadratic examples of $L[x, u]$ can illustrate aspects of bifurcation theory. A prototype can be described in terms of a simplified version of (2.75). Suppose that $V_c[x]$ and $V[q]$ are given functions, each of a single scalar variable, which are related by a Legendre dual transformation

$$V_c'[x] = q, \quad V'[q] = x, \quad V_c + V = xq, \qquad (2.105)$$

where the prime signifies differentiation in each case. Again we have in mind that, in mechanical contexts for example, $V[q]$ is the potential energy and $V_c[x]$ is the complementary energy. Examples include

$$V[q] = \tfrac{1}{3}q^3 + aq \quad \text{and} \quad V[q] = \tfrac{1}{4}q^4 + \tfrac{1}{2}bq^2 + aq,$$

where a and b are assignable parameters. These are the fold and cusp potentials (2.23) and (2.35) respectively, in transcribed notation. The corresponding $V_c[x]$ will be multivalued, as in Fig. 2.6 for the cusped function $\pm\frac{2}{3}(x-a)^{\frac{3}{2}}$, or the swallowtail $V_c[x]$ shown as a function of $x - a$ and b in Figs. 2.11(b) and 2.12. These V and V_c are distinct from those in (2.91)–(2.95).

Let u be another scalar variable, and consider the function

$$L[x, u] = xu - V_c[x], \qquad (2.106)$$

which will be multivalued in cases like those just described, and whose negative is a simplification of $(2.75)_1$. First let us construct, as far as possible for this particular example, the chain of global functions indicated in Fig. 2.26. Since $L[x, u]$ is linear in u we have a case, like (2.80), in which only two of the other three functions are defined unambiguously. From

(2.106) with (2.105) we find $\partial L/\partial x = u - q$ and $L - x\partial L/\partial x = V$ so that

$$j\left[\frac{\partial L}{\partial x}, u\right] = V\left[u - \frac{\partial L}{\partial x}\right]. \tag{2.107}$$

That is, the function of the two variables on the left can be expressed entirely in terms of the potential energy function of the one variable on the right. A similar property appeared in $(2.75)_2$. Interchanging the active and passive roles of u and $\partial L/\partial x$ to construct a second Legendre transformation is then easy. Since $\partial L/\partial u = x$ and $h = ux - j$ we find

$$h\left[\frac{\partial L}{\partial x}, \frac{\partial L}{\partial u}\right] = \frac{\partial L}{\partial x}\frac{\partial L}{\partial u} + V_c\left[\frac{\partial L}{\partial u}\right]. \tag{2.108}$$

The function of two variables on the left is thus expressed in terms of the complementary energy function on the right, even when the latter is multi-valued as, for example, in the swallowtail case. The function $k[x, \partial L/\partial u]$ does not exist, as a function of two variables, because its domain degenerates to the line $\partial L/\partial u = x$. However, we can still say that $k = L - u\partial L/\partial u = -V_c[x]$ in value there, just as we identified the value $Z = -\frac{1}{2}x^2$ after (2.80).

The first and second gradients of (2.106) are

$$\frac{\partial L}{\partial x} = u - V_c'[x], \quad \frac{\partial L}{\partial u} = x,$$

$$L_{xx} = -V_c''[x], \quad L_{xu} = 1, \quad L_{uu} = 0. \tag{2.109}$$

Evidently $L[x, u]$ is a saddle function strictly concave in x and weakly

Fig. 2.27. Example of Legendre transforms (2.105).

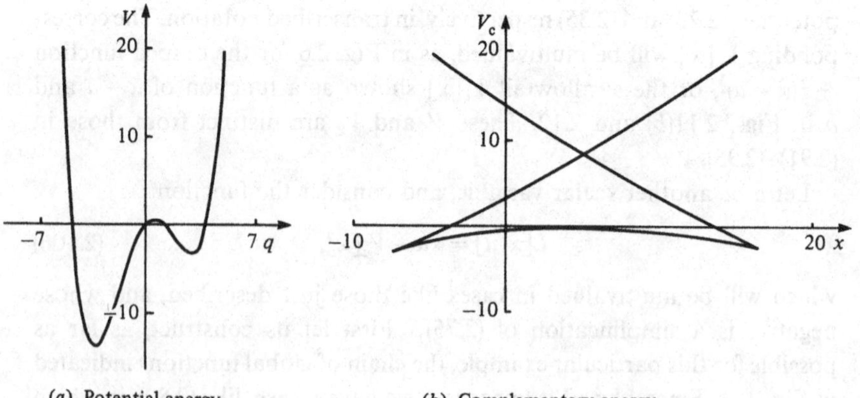

(*a*) Potential energy
$V[q] = \frac{1}{4}q^4 - 5q^2 + 5q$

(*b*) Complementary energy
Swallowtail function $V_c[x]$

convex in u if $V_c[x]$ is strictly convex. In the swallowtail example $V_c[x]$ of Fig. 2.11(*b*) this saddle property applies when $b > 0$ to the only branch of $L[x, u]$ which then exists, but as the control parameter b is decreased to $b < 0$, this single regular branch of $L[x, u]$ trifurcates into three regular branches, because $V_c[x]$ itself trifurcates into two convex branches and one concave branch, as shown in Fig. 2.27(*b*). Two of the new branches of $L[x, u]$ are saddle-shaped in the original sense, and one in the inverted sense. The other two well-defined functions $j[\partial L/\partial x, u]$ and $h[\partial L/\partial x, \partial L/\partial u]$ in the quartet of Fig. 2.26 experience a corresponding trifurcation, and the result for (2.107) is illustrated in Fig. 2.27(*a*). It is in such ways that a single quartet of Legendre dual functions may evolve by bifurcation or trifurcation into quartets of global functions, each having two or three or more branches.

In corresponding problems like (1.50)–(1.55) the multiplicity of the solutions will change as the control parameter changes, because these problems are defined in terms of the gradients of the evolving $L[x, u]$, and because one solution can be identified with each of the above mentioned branches. For example, on a branch where $V_c''[x] > 0$ such uniqueness is assured by Theorem 1.7(a) and the unique solution minimizes $V[q]$ among the solutions of the α subset. This *minimum potential energy principle* is a consequence of Theorem 1.10(b) and (2.116). The complementary energy is $-k = V_c[x]$, and so the prototype (2.106) is too simple to give a non-degenerate complementary energy principle. For in (1.53) and in (1.55) the β equation fixes $x_\beta = 0$ and there is no maximizing of $k_\beta = -V_c[0]$ to do. This degeneracy does not apply in the analogous compressible flow problem generated by (5.35) below with multivalued integrand.

The most familiar case is the stationary value problem (1.53), which by (2.109) is

$$u = V_c'[x] \quad (\alpha), \qquad x = 0 \quad (\beta).$$

The β equation is equivalent to

$$V'[q] = 0, \tag{2.110}$$

which is how the stationary energy property conventionally appears, and the α equation has merely defined u as q.

For the swallowtail $V_c[x]$, the solutions of (2.110) are the roots of the cubic

$$q^3 + bq + a = 0,$$

and $q^3 - 10q + 5 = 0$ identifies the three stationary points shown in Fig. 2.27(*a*). They are also the three gradients of $V_c[x]$ in Fig. 2.27(*b*) where its three branches pass through $x=0$. If a and b are imagined to vary, the cubic defines an 'equilibrium surface' of solutions in q, b, a space, like that shown in Fig. 2.7 or 2.8 after transcription of notation. If a is fixed and b is gradually reduced, the multiplicity of the real solution $q(b)$ changes from one to three as the bifurcation set $b^3=-27a^2/4$ is crossed (recall (2.22)). In the case $a=0$, when the quartic and swallowtail functions illustrated in Fig. 2.27 become symmetric, the solution $q(b)$ trifurcates according to the standard 'pitchfork' diagram, which here is $q(q^2+b)=0$, i.e. a parabola $q^2+b=0$ together with its axis $q=0$. It corresponds to the bottom of the two valleys and the centre line in Fig. 2.11(b). The pitchfork is also the intercept on the model shown in Fig. 2.8 made by a central vertical plane perpendicular to the axis of the photograph.

Among the other three cases is (1.54), which by (2.109) is

$$V'[q]\geqslant 0 \quad (\beta), \qquad q\geqslant 0 \quad (\alpha), \qquad qV'[q]=0. \tag{2.111}$$

A transcription of notation shows this to be the same as (1.30), which we have already treated in detail. For $V=\frac{1}{4}q^2+\frac{1}{2}bq^2$ only half of the pitchfork survives. However, if asymmetry is introduced by restoring the term aq to V, the solution $q=0$ will be additional to the stationary solutions of (2.111), as Fig. 2.27(b) indicates.

Bifurcation theory is a very large subject. The purpose of this subsection is to indicate that its connection with the saddle function viewpoint is represented by the appearance of a possibly multivalued complementary energy $V_c[x]$ in the $L[x,u]$, via (2.106); and also that the gradients $V_c'[0]$ of the different branches of $V_c[x]$ are the solutions of the stationary value problem $V'[q]=0$.

Mechanical equilibrium problems is one major area where the bifurcation of solutions to stationary value problems is illustrated. The books by Britvec (1973), Thompson and Hunt (1973, 1984), Zeeman (1977) and Poston and Stewart (1978) describe many examples, particularly from the finite dimensional viewpoint. Pioneering work in the infinite dimensional theory by Koiter (1945) in elasticity and Hill (1958, 1962) in plasticity began much earlier, and has been developed by Sewell (1966, 1972), Hutchinson (1974), Needleman and Tvergaard (1982), Ogden (1984) and Bushnell (1985). Even now the complementary energy aspect indicated in (2.106) is not well understood, perhaps because the associated multivaluedness has not been made explicit. We shall give another related example in §5.3 (see Fig. 5.3(b)).

Exercises 2.6

1. Fig. 2.28(a) shows a model for the simplest mode of dynamic snap-through

buckling experienced with shallow spherical caps under load as in beer-can lids, the contact switch under the buttons of hand calculators, and various domestic and industrial reversible 'snap-shut' fastenings. It can also be used to model the click mechanism of an insect in flight (e.g. see Sewell, 1978*c*, 1981). Three smooth pivots join two equal rods which remain straight but are compressible lengthwise, and loaded symmetrically, such that the potential energy in nondimensionalized variables is

$$V[q; p, a] = \tfrac{1}{2}(\sec a - \sec q)^2 + p \tan q,$$

where a is the value of the angle q when the load $p = 0$. Find the equilibrium surface and bifurcation set analytically, as defined in §2.2(viii), sketch them and show that they have the same qualitative shape as those for the general quartic $V[q]$. Prove that the bifurcation set is the projection onto the control plane of a *stability boundary* on the equilibrium surface where $V''[q] = 0$.

2. Fig. 2.28(*b*) shows a model for the fundamental mode of Euler strut buckling by quasistatic bifurcation. A pivoted rigid T-piece has a small flange of length ε simulating the imperfection induced by an off-set axial load p. The strut fibres which resist bending are modelled by two springs with fixed lines of action parallel to the load, so that the non-dimensionalized potential energy is

$$V[q; p, p\varepsilon] = \tfrac{1}{2}\tan^2 q - p(1 - \cos q + \varepsilon \sin q).$$

Repeat the investigation required in the previous question.

3. Reverse the spring attachments in Fig. 2.28(*b*) so that the lines of action are fixed to the model (instead of fixed in space) and therefore move together as the rotation q increases. Repeat the investigation again (cf. Sewell, 1966, 1978*c*) with the new potential energy

$$V[q; p, p\varepsilon] = \tfrac{1}{2}\sin^2 q - p(1 - \cos q + \varepsilon \sin q).$$

This is a prototype of imperfection sensitivity in the buckling of certain shells.

Fig. 2.28. Model structures. (*a*) Snap-buckling model. (*b*) Bifurcation buckling model.

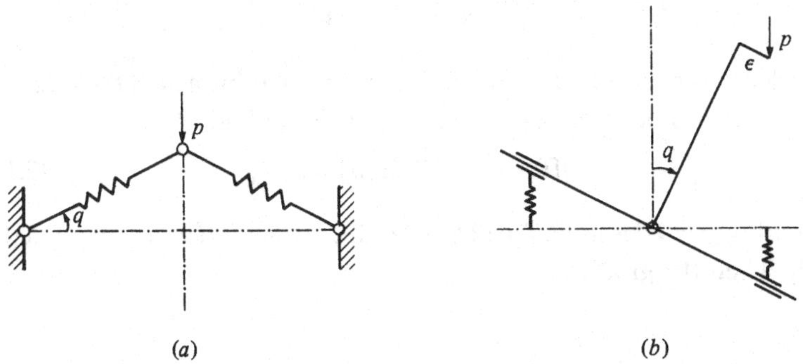

(iii) Generalized Hamiltonian and Lagrangian aspects

Returning to the general function $L[x,u]$ of $n+m$ scalar variables, in very many problems this can be represented in the form

$$L[x,u] = \langle u, Tx \rangle - X[x,u], \tag{2.112}$$

where neither T nor its transpose appear in the given function $X[x,u]$, which will be nonquadratic in nonlinear problems. In other words, T has a particular prominence because of its early appearance in the description of the problem, and this role is emphasized by separating out its associated bilinear term from any other terms $-X[x,u]$ which $L[x,u]$ may contain. One illustration is in (2.102), and another is the constraint matrix with elements $A_{\gamma i}$ in (1.147). Two more examples are the node-branch and loop-branch incidence matrices for networks, to be illustrated in (2.137) and (2.160) below.

Remembering that $\langle u, Tx \rangle = (x, T^*u)$, the gradients of (2.112) are

$$\frac{\partial L}{\partial x} = T^*u - \frac{\partial X}{\partial x}, \quad \frac{\partial L}{\partial u} = Tx - \frac{\partial X}{\partial u} \tag{2.113}$$

or, in the alternative matrix notation,

$$\frac{\partial L}{\partial \mathbf{x}} = T^{\mathrm{T}}\mathbf{u} - \frac{\partial X}{\partial \mathbf{x}}, \quad \frac{\partial L}{\partial \mathbf{u}} = T\mathbf{x} - \frac{\partial X}{\partial \mathbf{u}}.$$

The statements of the governing conditions in (1.50) and (1.53)–(1.55) can then be written in terms of $X[x,u]$, T and T^* instead of $L[x,u]$. It is noteworthy that the initial formulae for the *values* of the upper and lower bounds do not contain T or T^*, since

$$\begin{aligned} j &= L - \left(x, \frac{\partial L}{\partial x}\right) = \left(x, \frac{\partial X}{\partial x}\right) - X = Y, \\ k &= L - \left\langle u, \frac{\partial L}{\partial u}\right\rangle = \left\langle u, \frac{\partial X}{\partial u}\right\rangle - X = Z. \end{aligned} \tag{2.114}$$

Under the constraint $T^*u = \partial X/\partial x$, on a regular branch of the Legendre transformation between $X[x,u]$ and $Y[y,u]$, we find

$$j[0,u] = Y[T^*u, u] = J[u] \tag{2.115}$$

by writing $y = T^*u$ in (2.63), (2.65), $(2.114)_1$, where $J[u]$ is as defined in (1.61). Then the gradient

$$\frac{dJ}{du} = T\left[\frac{\partial Y}{\partial T^*u}\right] + \frac{\partial Y}{\partial u} \tag{2.116}$$

has a generalized Euler–Lagrange structure.

Under the constraint $Tx = \partial X/\partial u$, on a regular branch of the Legendre transformation between $X[x,u]$ and $Z[x,v]$, we find

$$k[x,0] = Z[x, -Tx] = K[x] \tag{2.117}$$

by writing $v = -Tx$ in Fig. 2.18, with (1.63). Then the gradient

$$\frac{dK}{dx} = \frac{\partial Z}{\partial x} + T^*\left[\frac{\partial Z}{\partial Tx}\right]. \tag{2.118}$$

In the hypotheses required for bounds in Theorem 1.6, any saddle hypothesis for $L[x,u]$ implies the same saddle hypothesis for $-X[x,u]$, because a bilinear term $(x^*, Tu) = \langle u, Tx\rangle$ contributes zero to the saddle quantity S in (1.57).

The stationary value problem (1.53) is, from (2.113),

$$T^*u = \frac{\partial X}{\partial x} \quad (\alpha), \qquad Tx = \frac{\partial X}{\partial u} \quad (\beta). \tag{2.119}$$

These equations have a structure which, when generalized to infinite dimensional problems as we shall do in (3.65), includes Hamilton's equations

$$\dot{q} = \frac{\partial \mathscr{H}}{\partial p}, \qquad -\dot{p} = \frac{\partial \mathscr{H}}{\partial q} \tag{2.120}$$

in classical mechanics, where $\mathscr{H}[p,q,t]$ is the Hamiltonian function of momentum p and position q of a particle, and of time t. The dot denotes d/dt. This is why we say that problems generated by (2.112), in cases (1.50), (1.54) and (1.55) as well as (1.53), represent the Hamiltonian aspect. In boundary value problems we shall see in Chapter 3 that T and T^* are composite operators called true adjoints which may contain differential operators and boundary terms, not merely formal adjoints like d/dt and $-d/dt$ in (2.120). Systematic identification of T and T^* early in the development of a particular infinite dimensional theory helps in the construction of a functional $X[x,u]$ suitable for the derivation of upper and lower bounds. This was first recognised by Noble (1964), who gave an infinite dimensional version of (2.119) with $X[x,u]$ as a functional in the form of an integral whose single valued integrand had second derivatives of opposite sign, and illustrated it by application to the Sturm-Liouville equation (see §3.7 here). In (3.65) we shall write $H[x,u]$ for a functional, in place of $X[x,u]$ which is a function in (2.119).

The Legendre transformation of classical mechanics is usually nonsingular, and relates the Hamiltonian $\mathscr{H}[p,q,t]$ to a Lagrangian $\mathscr{L}[\dot{q},q,t]=p\dot{q}-\mathscr{H}$ with $p=\partial\mathscr{L}/\partial\dot{q}$ and $\partial\mathscr{L}/\partial q=-\partial\mathscr{H}/\partial q$; then (2.120) converts to Lagrange's equation of motion $-\mathrm{d}(\partial\mathscr{L}/\partial\dot{q})/\mathrm{d}t+\partial\mathscr{L}/\partial q=0$. This corresponds to setting (2.116) zero. Evidently it is our $Y[T^*u,u]$ which is the immediate generalization of this classical Lagrangian, and not our $L[x,u]$. Further generalization leads back to $J[u]$, as in (2.115) and in §2.7(iv) below and, before the imposition of constraints, to our own original $L[x,u]$. It is in this widened sense that our use of L to represent the Lagrangian aspect is to be understood, in line with what has been called 'Lagrangian theory' by other authors (e.g. see Luenberger, 1969; Whittle, 1971; Sewell, 1973*a*, *b* 1982) and 'augmented Lagrangians' (e.g. Fortin and Glowinski, 1982).

(iv) Supplementary constraints

We consider here any general function $L[x,u]$, irrespective of whether it has the form (2.112), but for simplicity we now suppose that it is a single valued C^2 function. When $|L_{xx}|\neq 0$, the condition $\partial L/\partial x=0$ can be solved *a priori* and the function $J[u]$ defined in (1.61) exists with the property $J[u]=j[0,u]$. Then the remaining governing conditions in (1.54) can be rewritten

$$\frac{\mathrm{d}J}{\mathrm{d}u}\geqslant 0 \quad (\beta), \qquad u\geqslant 0 \quad (\alpha), \qquad \left\langle u,\frac{\mathrm{d}J}{\mathrm{d}u}\right\rangle=0 \tag{2.121}$$

(cf. the one dimensional problem (1.30) or (2.111)), while the stationary value problem (1.53) becomes simply

$$\frac{\mathrm{d}J}{\mathrm{d}u}=0 \quad (\beta). \tag{2.122}$$

Now suppose instead that $|L_{uu}|\neq 0$, allowing $\partial L/\partial u=0$ to be satisfied as a constraint *a priori*. Then the function $K[x]$ defined in (1.63) exists with the property $K[x]=k[x,0]$, and the remaining governing conditions in (1.55) can be rewritten

$$\frac{\mathrm{d}K}{\mathrm{d}x}\leqslant 0 \quad (\alpha), \qquad x\geqslant 0 \quad (\beta), \qquad \left(x,\frac{\mathrm{d}K}{\mathrm{d}x}\right)=0, \tag{2.123}$$

while the stationary value problem (1.53) reduces to

$$\frac{\mathrm{d}K}{\mathrm{d}x}=0 \quad (\alpha). \tag{2.124}$$

Next we introduce the extended definitions of $J[u]$ and $K[x]$ promised in §1.3(iv). Instead of $\partial L/\partial x = 0$, we seek the general solution of the trio

$$\frac{\partial L}{\partial x} \leqslant 0, \quad x \geqslant 0, \quad \left(x, \frac{\partial L}{\partial x}\right) = 0. \tag{2.125}$$

Suppose that every $r \times r$ principal minor of L_{xx} is nonsingular, for $1 \leqslant r \leqslant n$. For any r components of x which are strictly positive (with the remaining $n-r$ identically zero), the corresponding r components of $\partial L/\partial x$ are zero by (2.125). The latter can be solved parametrically as in §1.3(iv), just by replacing n by r. A corresponding part of a function $J[u]$ can be constructed satisfying (1.61) and (1.62), except that L_{xx} is replaced by its appropriate $r \times r$ minor. There are $\sum_{r=0}^{n}(n!/(n-r)!r!) = N$ (say) ways of selecting a set of positive components of x in (2.125), and so the general solution of (2.125) will be a composite piecewise C^1 function $x[u]$, not necessarily single valued, having N parts (see Exercises 1.5.7 and 2.5.5). There will be a corresponding composite function $J[u]$, not necessarily single valued (cf. (2.91) when $h < 0$), with N parts, and C^1 (but only piecewise C^2) because the formula $\mathrm{d}J/\mathrm{d}u = \partial L/\partial u$ in (1.61) still applies and $\partial L/\partial u$ is continuous by the C^2 hypothesis for $L[x, u]$. When $n = 1$, for example, $N = 2$ and

$$J[u] = \begin{cases} j[0, u] & \text{where} \quad x > 0, \\ L[0, u] & \text{where} \quad \partial L/\partial x < 0, \end{cases} \tag{2.126}$$

as in (2.91). The case $n = 2$ is illustrated by (2.96) (n and m are interchanged there). In terms of such composite $J[u]$, for general n, the governing equations in (1.50) can be written as (2.121), and the governing equations in (1.55) can be written as (2.122).

Similar arguments, when every principal minor of L_{uu} is nonsingular, allow the general solution $u = u[x]$ of the trio

$$\frac{\partial L}{\partial u} \geqslant 0, \quad u \geqslant 0, \quad \left\langle u, \frac{\partial L}{\partial u} \right\rangle = 0 \tag{2.127}$$

to be found and used to define a composite piecewise C^2 function $K[x]$ satisfying (1.64). In terms of this $K[x]$ the governing equations in (1.50) can be written as (2.123), and the governing equations in (1.54) can be written as (2.124). When $m = 1$, for example, we have

$$K[x] = \begin{cases} k[x, 0] & \text{where} \quad u > 0, \\ L[x, 0] & \text{where} \quad \partial L/\partial u > 0. \end{cases} \tag{2.128}$$

It is convenient to summarize the last three paragraphs as follows.

Theorem 2.16

Any one of the four sets (1.50) *and* (1.53)–(1.55), *when it can be partially solved* *to allow the definition of a function $J[u]$ or $K[x]$, may be reduced to one of the four sets* (2.121)–(2.124) *as indicated in the chart below.*

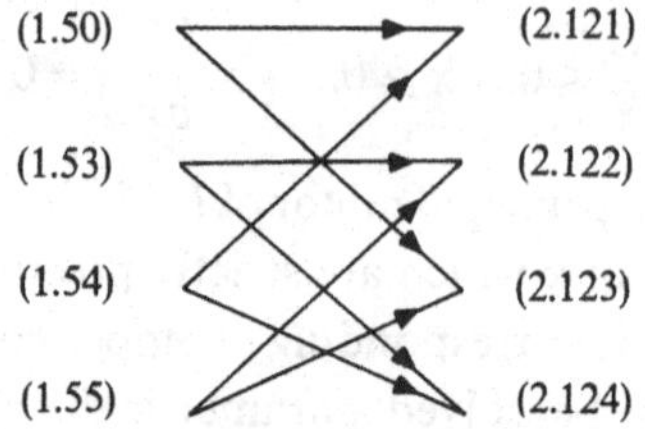

The facts that the nonstationary problems (1.55) and (1.54) can be reduced in this way to stationary problems (2.122) or (2.124), respectively, using a smaller number of variables, is important in principle for applications. Its importance is obviously tempered by how convenient it is to solve (2.125) or (2.127) explicitly at the outset to give the multivalued $J[u]$ or $K[x]$. This is easiest to do for a small number of single valued parts, especially two, as (2.126) and (2.128) illustrate, and is common in plasticity theory in terms of (2.91) or (2.93). One may say that the prior solution of (2.125) or (2.127) entails the exploitation of *supplementary constraints*, over and above those contained in the original α-subset, or in the original β-subset.

If it were practicable, in any one of the original four problems (1.50) or (1.53)–(1.55), to find *both* types of partial solution directly, i.e. the $x[u]$ which leads to $J[u]$ and the $u[x]$ which leads to $K[x]$, the intersections of these partial solutions would be the solutions of the whole problem. If these intersections could also be determined directly there would be no more work to do, and no need of extremum principles. We proceed on the basis that either it is not practicable to find both partial solutions or, if it is, then their intersections cannot be found directly.

We now impose stronger hypotheses on $L[x, u]$ then those so far adopted in this section, to see how the bounds and extremum principles stated in Theorems 1.6–1.13 can be expressed in terms of $J[u]$ and $K[x]$. Those theorems were proved for (1.50), and Theorem 2.16 indicates that by imposing supplementary constraints we can expect corresponding results for (2.121) or (2.123) (with the simpler versions for (1.53)–(1.55) implying corresponding results for (2.121)–(2.124)). Table 2.3 will show that an original upper bound $L_\alpha - (x_\alpha, \partial L/\partial x|_\alpha)$ can be decreased by supplementary constraints and a lower bound $L_\beta - \langle u_\beta, \partial L/\partial u|_\beta\rangle$ can be increased.

Note that, for example, although $|L_{xx}| \neq 0$ is sufficient to construct $j[\partial L/\partial x, u]$ and also to solve $\partial L/\partial x = 0$ uniquely locally for $x[u]$, it is not sufficient to solve the trio (2.126) uniquely or to construct an associated single valued $J[u]$. It would be sufficient for this if L_{xx} were (say) negative definite. It is necessary to prove that $J[u]$ is single valued before we can consider whether it is convex.

Theorem 2.17
Let $L[x,u]$ be a given single valued C^2 function over a rectangular domain.

(a) Let L_{xx} be negative definite. Then $J[u]$ is single valued, whether defined via the solution of $\partial L/\partial x = 0$, or of the trio (2.125).

(b) When the trio (2.121) *applies, any minimizer $\hat{u}_\alpha$ of $J[u_\alpha]$ with respect to all $u_\alpha \geqslant 0$ satisfies that whole trio. When* (2.122) *applies, any minimizer of $J[u_\alpha]$ with respect to arbitrary u_α satisfies* (2.122).

(c) Let L_{xx} be negative definite and J_{uu} be nonnegative definite. Then $J[u]$ is weakly convex. The simultaneous bounds (cf. Theorem 1.3*(b))*

$$J[u_\alpha] \geqslant L_0 \geqslant J[u_\beta] - \left\langle u_\beta, \left.\frac{\mathrm{d}J}{\mathrm{d}u}\right|_\beta \right\rangle \tag{2.129}$$

apply to (2.121) *(the lower bound is vacuous for* (2.122)*). Any solution of* (2.121) *or of* (2.122) *minimizes the corresponding $J[u_\alpha]$ so that, with* (b), *we have a minimum principle. The minimum is stationary for* (2.122).

(d) Let L_{xx} be negative definite, and J_{uu} be positive definite. Then $J[u]$ is strictly convex. There is at most one solution of (2.121), *and of* (2.122). *There is a maximum principle, that such a solution of* (2.121) *is equivalent to a maximizer of the lower bound in* (2.129).

Proof

(a) Negative definite L_{xx} is sufficient for the strict concavity of $L[x,u]$ with respect to x at each fixed u (by specializing Theorem 1.5), which implies (cf. (1.93))

$$-\left(x_+ - x_-, \left.\frac{\partial L}{\partial x}\right|_+ - \left.\frac{\partial L}{\partial x}\right|_- \right) > 0$$

for any distinct x_+ and x_-. This would be contradicted if x_+ and x_- could be two distinct solutions either of $\partial L/\partial x = 0$ or of (2.125). Therefore a parametric solution $x[u]$ for either case is unique, even when composite in the case of (2.125). Hence $J[u] = L[x[u], u]$ is single valued because $L[x,u]$ is so.

(b) This is a generalization of the *first* part of the proof of Theorem 1.3(a) and, as there, the convexity of $J[u]$ is not required. With negative definite L_{xx}, the proof can be constructed along the lines of Theorem 1.10(a). The upper bound $L_\alpha - (x_\alpha, \partial L/\partial x|_\alpha)$ now becomes $J[u_\alpha]$ (see Table 2.3), either because $\partial L/\partial x|_\alpha = 0$, or because the supplementary constraint $(2.125)_3$ is applied.

(c) The composite $J[u]$ is piecewise C^2 because J_{uu} on each of its N parts is given by a formula like that for j_{uu} in (2.101), but with the appropriate principal minor replacing L_{xx}. Nonnegative definite L_{uu} and negative definite L_{xx} are sufficient but not necessary for nonnegative definite J_{uu}. Weak convexity follows from the appropriate second mean value theorem (e.g. by rewriting Theorem 1.5 for $J[u]$ instead of $L[x,u]$). The bounds (2.129) follow by adapting the simple method of Theorem 1.4. The minimum principle is seen to be a direct generalization of Theorem 1.3(a), including its second part. It can be regarded as a variant of Theorem 1.10(b), with the hypothesis $S \geqslant 0$ there needed only under the imposition of supplementary constraints; for with the latter we have

$$J_+ - J_- - \left\langle u_+ - u_-, \frac{\mathrm{d}J}{\mathrm{d}u}\bigg|_-\right\rangle = S - \left(x_-, \frac{\partial L}{\partial x}\bigg|_+\right) \geqslant S$$

for any u_+, u_- and corresponding $x_+[u_+]$ and $x_-[u_-]$.

(d) Positive definite L_{uu} and negative definite L_{xx} are sufficient but not necessary for positive definite J_{uu}. Strict convexity of $J[u]$ also follows from the mean value theorem, and implies

$$\left\langle u_+ - u_-, \frac{\mathrm{d}J}{\mathrm{d}u}\bigg|_+ - \frac{\mathrm{d}J}{\mathrm{d}u}\bigg|_-\right\rangle > 0$$

for distinct u_+ and u_-. If these could both satisfy (2.121) (or (2.122)) there is a contradiction, which proves uniqueness. The maximum principle is seen to be a generalization of Theorem 1.3(b). It can also be approached via Theorem 1.11. □

The *stationary* minimum principle established here for (2.122) applies not only to (1.53), which itself was a stationary value problem, but also, by the imposition of supplementary constraints (Theorem 2.16) to (1.55) which was not originally a stationary value problem.

When L_{uu} is positive definite and K_{xx} is negative (semi-) definite, we leave the reader to formulate via Theorem 1.9 the analogue of Theorem 2.17 associated with bounds

$$K[x_\alpha] - \left(x_\alpha, \frac{\mathrm{d}K}{\mathrm{d}x}\bigg|_\alpha\right) \geqslant L_0 \geqslant K[x_\beta]. \tag{2.130}$$

The relations between the bounds given in (1.80), (2.129) and (2.130) are as set out in Table 2.3, by considering each case in turn, whenever the implied alternatives are available. In the table, u_α and x_β are arbitrary or (where designated) nonnegative and otherwise arbitrary; x_α and u_β are subject to the (residual) α and β inequalities respectively.

As an illustration, when L_{xx} is negative definite and L_{uu} is positive definite the minimum and maximum principles apply simultaneously, and in all three versions. The upper bounds in a given problem then all have the values of the same jointly convex function $j[\partial L/\partial x, u]$ (see Fig. 2.19), and the lower bounds all have the values of the same jointly concave function $k[x, \partial L/\partial u]$. The inequalities in Table 2.3 each express a restriction to a subdomain of such a function: for example, by applying $\partial k/\partial x = 0$ to the lower bound in (1.54) gives (for the same value of $\partial L/\partial u|_\beta$) that for (2.121); and by applying (2.125) rewritten as $\partial k/\partial x \leqslant 0$, $x \geqslant 0$, $(x, \partial k/\partial x) = 0$ to the lower bound in (1.50) gives that for (2.121), since

$$k_+ - k_- \leqslant (x_+ - x_-, \partial k/\partial x|_-) \leqslant 0$$

by the concavity in x, where the minus point satisfies the trio and $x_+ \geqslant 0$.

Table 2.3. *Related bounds in original and reduced problems*

Problem	Upper bound	Lower bound
(2.121)	$J[u_\alpha \geqslant 0]$	$J[u_\beta] - \langle u_\beta, \mathrm{d}J/\mathrm{d}u|_\beta\rangle$
	$=$	$\geqslant$
(1.50)	$L_\alpha - \left(x_\alpha, \left.\dfrac{\partial L}{\partial x}\right\|_\alpha\right) = j\left[\left.\dfrac{\partial L}{\partial x}\right\|_\alpha \leqslant 0, u_\alpha \geqslant 0\right]$	$L_\beta - \left\langle u_\beta, \left.\dfrac{\partial L}{\partial u}\right\|_\beta\right\rangle = k\left[x_\beta \geqslant 0, \left.\dfrac{\partial L}{\partial u}\right\|_\beta \geqslant 0\right]$
	$\geqslant$	$=$
(2.123)	$K[x_\alpha] - (x_\alpha, \mathrm{d}K/\mathrm{d}x\|_\alpha)$	$K[x_\beta \geqslant 0]$
(2.122)	$J[u_\alpha]$	vacuous
	$=$	
(1.53)	$L_\alpha = j[0, u_\alpha]$	$L_\beta = k[x_\beta, 0]$
		$=$
(2.124)	vacuous	$K[x_\beta]$
(2.121)	$J[u_\alpha \geqslant 0]$	$J[u_\beta] - \langle u_\beta, \mathrm{d}J/\mathrm{d}u\|_\beta\rangle$
	$=$	$\geqslant$
(1.54)	$L_\alpha = j[0, u_\alpha \geqslant 0]$	$L_\beta - \left\langle u_\beta, \left.\dfrac{\partial L}{\partial u}\right\|_\beta\right\rangle = k\left[x_\beta, \dfrac{\partial L}{\partial u_\beta} \geqslant 0\right]$
		$=$
(2.124)	vacuous	$K[x_\beta]$
(2.122)	$J[u_\alpha]$	vacuous
	$=$	
(1.55)	$L_\alpha - \left(x_\alpha, \left.\dfrac{\partial L}{\partial x}\right\|_\alpha\right) = j\left[\left.\dfrac{\partial L}{\partial x}\right\|_\alpha \leqslant 0, u_\alpha\right]$	$L_\beta = k[x_\beta \geqslant 0, 0]$
	$\geqslant$	$=$
(2.123)	$K[x_\alpha] - (x_\alpha, \mathrm{d}K/\mathrm{d}x\|_\alpha)$	$K[x_\beta \geqslant 0]$

Even if K_{xx} (rather than L_{xx}) is negative definite and L_{uu} is positive definite (so that the solution is still unique), the lower bound in (2.130) has the values of a function which is concave in x. The dependence of $k[x, \partial L/\partial u]$ on $\partial L/\partial u$ reappears in the way $K[x]$ is composed of several single valued parts. In some problems the upper bound is now vacuous.

Table 2.3 shows how upper and lower bounds are evaluated from functions which are the Legendre duals of each other. In particular, when $J[u]$ is defined by solving $\partial L/\partial x = 0$, $(2.100)_3$ shows its Legendre dual function to be

$$\left\langle u, \frac{\mathrm{d}J}{\mathrm{d}u} \right\rangle - J[u] = J_c[\bar{v}], \tag{2.131}$$

if we write

$$\bar{v} = \partial L/\partial u, \quad J_c[\bar{v}] = h[0, \bar{v}], \quad \mathrm{d}J_c/\mathrm{d}\bar{v} = u.$$

Similarly, when $K[x]$ is defined by solving $\partial L/\partial u = 0$,

$$\left(x, \frac{\mathrm{d}K}{\mathrm{d}x} \right) - K[x] = K_c[\bar{y}], \tag{2.132}$$

if we write

$$\bar{y} = \partial L/\partial x, \quad K_c[\bar{y}] = h[\bar{y}, 0], \quad \mathrm{d}K_c/\mathrm{d}\bar{y} = x.$$

When $J[u]$ is defined by solving the trio (2.125), and when $n = 1$, we find from (2.100) and (2.126) that the Legendre dual of $J[u]$ is

$$J_c[\bar{v}] = \begin{cases} h[0, \bar{v}] & \text{where } x > 0, \\ -k[0, \bar{v}] & \text{where } \partial L/\partial x < 0, \end{cases} \tag{2.133}$$

and, as in (2.95),

$$J + J_c = \langle u, \bar{v} \rangle, \quad \bar{v} = \mathrm{d}J/\mathrm{d}u, \quad u = \mathrm{d}J_c/\mathrm{d}\bar{v}.$$

The function $G[v]$ introduced in (1.33) was the forerunner of both of the present $J_c[\bar{v}]$.

Exercises 2.7

1. When $K[x]$ is defined by solving the trio (2.127), and, when $m = 1$, show that the Legendre dual of $K[x]$ is

$$K_c[\bar{y}] = \begin{cases} h[\bar{y}, 0] & \text{where } u > 0, \\ -j[\bar{y}, 0] & \text{where } \partial L/\partial u > 0, \end{cases}$$

and

$$K + K_c = (x, \bar{y}), \quad \bar{y} = \mathrm{d}K/\mathrm{d}x, \quad x = \mathrm{d}K_c/\mathrm{d}\bar{y}.$$

2. For the example (2.102) with general n and m, show that second gradients of

(2.103) and (2.104) are the operators

$$j_{uu} = I - \frac{TT^*}{\lambda}, \quad k_{xx} = \lambda I - T^*T,$$

and show that when their respective inverses exist

$$\begin{aligned} h[\bar{y}, \bar{v}] &= \tfrac{1}{2}\left\langle \bar{v} - \frac{T}{\lambda}(\bar{y} - a), j_{uu}^{-1}\left[\bar{v} - \frac{T}{\lambda}(\bar{y} - a) \right] \right\rangle + \frac{1}{2\lambda}(\bar{y} - a, \bar{y} - a) \\ &= \tfrac{1}{2}(\bar{y} - T^*\bar{v} - a, k_{xx}^{-1}[\bar{y} - T^*\bar{v} - a]) + \tfrac{1}{2}\langle \bar{v}, \bar{v} \rangle, \end{aligned}$$

where $\bar{y} = \partial L/\partial x$, $\bar{v} = \partial L/\partial u$.

3. In the example (2.102), when $\lambda > 0$ is less than the minimum eigenvalue of T^*T, verify the equivalence of the lower bounds in Table 2.3 for problems (1.55) and (2.123). Repeat for $\lambda < 0$, and in this case verify also the relations between the upper bounds in Table 2.3 for (1.55), (2.122) and (2.123). What can be said along these lines for the remainder of Table 2.3? (cf. §1.6(iii)).

2.8. Network theory

(i) Introduction

Here we continue the sequence of concrete illustrations of simultaneous extremum principles in finite dimensional problems which we began in §§1.6–1.9. Network theory involves a geometrical figure consisting of points connected by lines, upon which some physical phenomena are superimposed. Examples of the latter include (a) electrical current flow in the presence of voltage sources, resistors, diodes, etc. (e.g. see Iri, 1969; or Rockafellar, 1982), (b) distortion of mechanical frameworks of jointed elastic and plastic rods under the action of external forces and supports at the joints (e.g. see Pestel and Leckie, 1963, Britvec, 1973), and (c) flow of goods in transportation networks.

The underlying geometrical figure has interconnection properties which alone induce the early appearance, in any such theory, of a rectangular matrix corresponding to the T in (2.112), regardless of the subsequent physical considerations. This makes it natural to approach the construction of a generating function with values L via the Hamiltonian aspect of (2.119) as an intermediate step, rather than directly as we did in §§1.6–1.9. This section is built upon unpublished notes by Noble (1983, private communication).

Consider a set of n points, called *nodes*, in two or three dimensional space. Consider also a set of b finite line segments, called *branches*, each of which has a direction arbitrarily associated with it. Every branch begins and ends on two distinct nodes, and every node lies at the end of at least one branch. Such a geometrical figure is called a *directed graph.* A simple example is shown in Fig. 2.29, where $n = 4$ and $b = 6$, we have labelled nodes and branches with numbers, and the arrows specify the directions assigned to the branches.

(ii) A simple electrical network

Let i_j $(j = 1,\ldots,b)$ denote the electrical current in the jth branch, reckoned positive when it flows in the direction of the arrow. Then Kirchhoff's current law, that the algebraic sum of all currents entering and leaving a node is zero, gives

$$\begin{aligned} -i_1 - i_3 + i_4 &= 0, \\ -i_2 + i_3 + i_5 &= 0, \\ -i_4 - i_5 + i_6 &= 0, \\ i_1 + i_2 - i_6 &= 0, \end{aligned} \tag{2.134}$$

when applied to each node in turn in Fig. 2.29. Any three of these equations are independent, and their sum gives the fourth.

Let V_γ $(\gamma = 1,\ldots,n)$ denote the electrical potential at the γth node, relative to an arbitrary common datum, and let v_j $(j = 1,\ldots,b)$ denote the potential drop across the jth branch, reckoned positive when the potential falls in the direction of the arrow. Then, by definition, in Fig. 2.29 we have

$$\begin{aligned} v_1 = V_4 - V_1, \quad v_2 = V_4 - V_2, \quad v_3 = V_2 - V_1, \\ v_4 = V_1 - V_3, \quad v_5 = V_2 - V_3, \quad v_6 = V_3 - V_4. \end{aligned} \tag{2.135}$$

These equations satisfy Kirchhoff's voltage law, that the algebraic sum of the potential drops around any closed loop is zero (e.g. $v_1 - v_2 - v_3 = (V_4 - V_1) - (V_4 - V_2) - (V_2 - V_1) = 0$). We could arbitrarily choose one node to have reference potential zero (say $V_4 = 0$).

We may say that equations (2.134) and (2.135) are *balance equations,*

Fig. 2.29. Directed graph with four nodes and six branches.

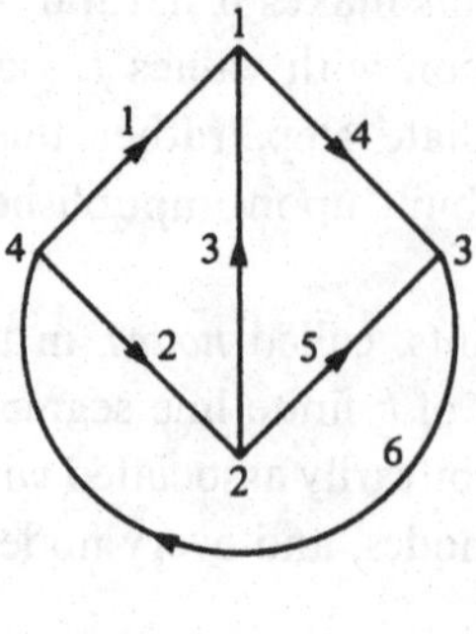

expressing balance of current through nodes and of voltage across branches. It is important to recognize that they hold irrespective of what resistors, diodes, voltage sources or other electrical elements may be present in the network. These elements have not yet been specified.

Typical members of equations (2.135) and (2.134) may be rewritten

$$T_{\gamma j}V_\gamma = v_j \quad \text{and} \quad T_{\gamma j}i_j = 0, \tag{2.136}$$

respectively, in terms of a matrix

$$T = \begin{bmatrix} -1 & 0 & -1 & +1 & 0 & 0 \\ 0 & -1 & +1 & 0 & +1 & 0 \\ 0 & 0 & 0 & -1 & -1 & +1 \\ +1 & +1 & 0 & 0 & 0 & -1 \end{bmatrix} \tag{2.137}$$

with typical element $T_{\gamma j}$. Assembling (2.136) for all branches and nodes gives

$$T^{\mathrm{T}}\mathbf{V} = \mathbf{v} \quad \text{and} \quad T\mathbf{i} = 0 \tag{2.138}$$

in the column matrix notation of (1.45).

(iii) Node-branch incidence matrix

For *any* directed graph, not necessarily that of Fig. 2.29 and without necessarily having the electrical connotation, we can define an $n \times b$ node-branch incidence matrix T as follows. Its typical element is

$$T_{\gamma j} = \begin{cases} +1 & \text{if the } \gamma\text{th node is at the beginning of the } j\text{th branch,} \\ -1 & \text{if the } \gamma\text{th node is at the end of the } j\text{th branch,} \\ 0 & \text{if the } \gamma\text{th node is not on the } j\text{th branch,} \end{cases}$$

where $\gamma = 1,\ldots,n$ and $j = 1,\ldots,b$. Plainly (2.137) is a particular illustration. The rank of T is $n-1$.

For any electrical network equations (2.138) will hold with this general T, where $\mathbf{V}$ is an $n \times 1$ column matrix, and $\mathbf{v}$ and $\mathbf{i}$ are $b \times 1$ column matrices. Subsequently we regard (2.138) as having this generality, thus defining an underdetermined system of $n+b$ scalar equations for $n+2b$ scalar variables. The branch current and voltage matrices will be orthogonal in any solution of this system, i.e.

$$\mathbf{v}^{\mathrm{T}}\mathbf{i} = \mathbf{V}^{\mathrm{T}}T\mathbf{i} = 0. \tag{2.139}$$

This result is called Tellegen's theorem. The antecedents and influence of it are indicated by Penfield, Spence and Duinker (1970). Being a property of the underdetermined system (2.138), it is true regardless of what branch characteristics are subsequently specified.

(iv) Electrical branch characteristics

The generalized equations (2.138) can become well-posed when b further relations between the elements of $\mathbf{V}$, $\mathbf{i}$ and $\mathbf{v}$ are specified. Any specified relation between the current i_j through and the voltage drop v_j across the jth branch is called a response characteristic, or branch characteristic for that branch. One of the simplest examples is Ohm's law $v_j = R_j i_j$ (no sum) for a linear resistor with given resistance $R_j > 0$. A voltage source is just $v_j = e_j$ for given e_j. A single diode has voltage/current characteristic like the step function in Fig. 1.6. Quite general branch characteristics arise from a combination of different electrical elements, and one such example of a voltage–current curve which is nondecreasing was sketched in Fig. 2.15(a) (replace x and y there by i_j and v_j respectively here). Another example is sketched in Fig. 2.30.

For simplicity we suppose that every characteristic is nondecreasing in both senses $v_j[i_j]$ and $i_j[v_j]$, and that it goes to infinity at both ends, but not necessarily parallel to a coordinate axis there. We find it desirable to distinguish between characteristics which go to infinity parallel to neither axis, as illustrated in Fig. 2.30, from those which go to infinity parallel to an axis at least at one end. Fig. 2.15(a) goes to infinity parallel to an axis at both ends.

Consider first the type of characteristic which goes to infinity parallel to neither axis. We have illustrated this in Fig. 2.30 by tilting the two end segments A and H in Fig. 2.15(a) to an acute angle with the axes. Any such characteristic has only finite values $v_j[i_j]$ and $i_j[v_j]$ for every finite value of its argument, and they can be expressed as

$$v_j = \frac{\mathrm{d}X_j}{\mathrm{d}i_j} \quad \text{or} \quad i_j = \frac{\mathrm{d}Y_j}{\mathrm{d}v_j} \tag{2.140}$$

Fig. 2.30. Nondecreasing branch characteristic defined for all v_j and i_j.

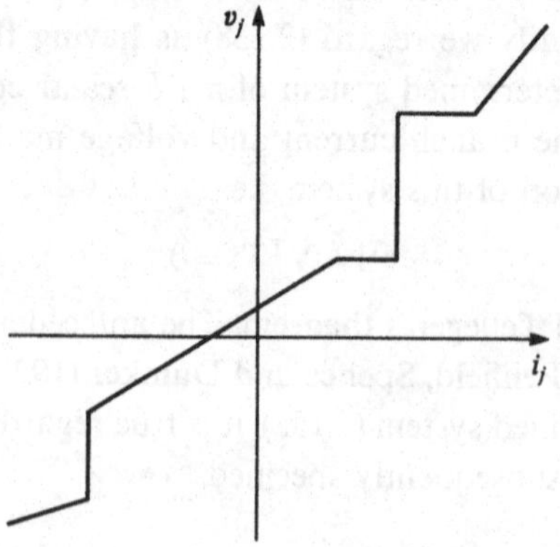

in terms of an electrical energy $X_j[i_j]$ and co-energy $Y_j[v_j]$. The latter are piecewise C^1 Legendre dual functions, each of one variable, whch satisfy

$$Y_j[v_j] + X_j[i_j] = v_j i_j \quad \text{(no sum)}, \tag{2.141}$$

and which are weakly convex because the branch characteristic is non-decreasing. Accumulated and distributed finite singularities of the Legendre transformation appear as vertices and finite straight segments on the functions. The $X_j[i_j]$ and $Y_j[v_j]$ for Fig. 2.30 will be the same as the $X[x]$ and $Y[y]$ in Fig. 2.15, except that their end segments will go to infinity as convex parabolas instead of straight lines.

When every branch characteristic in the network can be expressed as in (2.140), we can define the functions

$$\begin{aligned} X[i_1, i_2, \ldots, i_n] &= X_1[i_1] + X_2[i_2] + \cdots + X_n[i_n], \\ Y[v_1, v_2, \ldots, v_n] &= Y_1[v_1] + Y_2[v_2] + \cdots + Y_n[v_n], \end{aligned} \tag{2.142}$$

such that the branch characteristics (2.140) can be assembled in matrix form as

$$\mathbf{v} = \frac{dX}{d\mathbf{i}}, \quad \mathbf{i} = \frac{dY}{d\mathbf{v}}. \tag{2.143}$$

Then by adding all b equations of type (2.141),

$$Y[\mathbf{v}] + X[\mathbf{i}] = \mathbf{v}^T\mathbf{i}. \tag{2.144}$$

We defer until §2.8(ix) further consideration of the other type of characteristic, i.e. that for which either $v_j[i_j]$ or $i_j[v_j]$ does not have a finite value for every finite value of the argument.

(v) Saddle function and governing equations

By eliminating $\mathbf{v}$ from (2.138) and $(2.143)_1$ the governing equations for the whole system can be seen to be the particular example

$$T^T\mathbf{V} = \frac{dX}{d\mathbf{i}} \quad (\alpha), \qquad T\mathbf{i} = 0 \quad (\beta), \tag{2.145}$$

of the Hamiltonian structure (2.119), as pointed out by Noble (1967). The second variable $\mathbf{V}$ happens to be absent from the Hamiltonian function $X[\mathbf{i}]$. These equations (2.145) illustrate the point made in §2.8(i) that the Hamiltonian aspect emerges after early identification of T.

The system (2.145) can be rewritten as

$$\frac{\partial L}{\partial \mathbf{i}} = 0 \quad (\alpha), \qquad \frac{\partial L}{\partial \mathbf{V}} = 0 \quad (\beta), \tag{2.146}$$

in terms of the generating function

$$L[\mathbf{i}, \mathbf{V}] = \mathbf{V}^T T\mathbf{i} - X[\mathbf{i}]. \tag{2.147}$$

When each $X_j[i_j]$ is convex, $X[\mathbf{i}]$ is convex. Then $L[\mathbf{i}, \mathbf{V}]$ is a saddle function, concave in $\mathbf{i}$ and linear in $\mathbf{V}$. We can identify (2.146) as an illustration of (1.53), with $\mathbf{i}$ and $\mathbf{V}$ playing the role of $\mathbf{x}$ and $\mathbf{u}$ respectively in the general theory of Chapter 1. When a solution pair $\mathbf{i}, \mathbf{V}$ can be found, we have no reason to expect that such $\mathbf{V}$ will be unique, and such $\mathbf{i}$ may not be unique either unless the convexity of $X[\mathbf{i}]$ is strict.

(vi) Bounds and extremum principles

Theorem 2.18 (Upper bound and minimum principle)

(a) Any choice of $\mathbf{V}$ *will provide an upper bound* $Y[T^{\mathrm{T}}\mathbf{V}]$ *to the value which* Y *has for any solution of the system of equations* (2.145).

(b) If each $X_j[i_j]$ *is strictly convex, every minimizer of* $Y[T^{\mathrm{T}}\mathbf{V}]$ *with respect to arbitrary* $\mathbf{V}$ *will belong to a solution of* (2.145).

Proof

(a) The proof of Theorem 1.6(a) can be repeated for (1.53) in place of (1.50), and we require an extension of this for the generating function (2.147) which, instead of being C^1, contains a sum $(2.142)_1$ of piecewise C^1 functions of one variable. The saddle quantity becomes

$$S = \sum_{j=1}^{n} \left[X_{j-} - X_{j+} - (i_{j-} - i_{j+}) \left.\frac{\mathrm{d}X_j}{\mathrm{d}i_j}\right|_{+} \right] \geqslant 0$$

for every pair of distinct points in the domain of $L[\mathbf{i}, \mathbf{V}]$. The inequality is true even when there is a finite step in some or all the gradients $\mathrm{d}X_j/\mathrm{d}i_j$ at the plus point, by the weak convexity of the piecewise C^1 functions $X_j[i_j]$. It follows (cf. (1.38)) that an upper bound will be any value of

$$L - \mathbf{i}^{\mathrm{T}}\frac{\partial L}{\partial \mathbf{i}} = \mathbf{i}^{\mathrm{T}}\frac{\mathrm{d}X}{\mathrm{d}\mathbf{i}} - X = Y \tag{2.148}$$

which satisfies (2.145α) but not necessarily (2.145β). Any chosen $\mathbf{V}$ will lead, via $\mathbf{v} = T^{\mathrm{T}}\mathbf{V}$ in $(2.143)_2$, to an $\mathbf{i} = \mathrm{d}Y/\mathrm{d}\mathbf{v}$ which satisfies (2.145α) alone. An upper bound is therefore any value of the function

$$J[\mathbf{V}] = Y[T^{\mathrm{T}}\mathbf{V}] \tag{2.149}$$

for any $\mathbf{V}$ (cf. (2.115)).

(b) The converse minimum principle requires an application of Theorem 1.10(a) and (b), simplified because we are dealing with an example of (1.53) rather than (1.50), but extended to allow for the finite steps in each $\mathrm{d}X_j/\mathrm{d}i_j$. The details can be followed in the proof of Theorem 1.9(a) and (b), with an interchange of the roles of x and u afterwards. Thus for C^2 $L[x, u]$ the inverse mapping which leads to (1.96) converts (1.53) into

$$\frac{\partial G}{\partial x}=0 \quad (\alpha), \qquad v=0 \quad (\beta),$$

and converts the problem of maximizing $L-\langle u, \partial L/\partial u\rangle$ subject to $\partial L/\partial u=0$ into the problem of minimizing $G[x,0]$ without constraint on x. The latter is a simpler version of (1.99) with all $v_\beta=\hat{v}_\beta=0$, and it is then relatively easy to see that $\{\hat{x}_\beta\}\subseteq\{x_0\}$, where x_0 is any solution of $\partial G/\partial x=0$. With Theorem 1.6(b) we have $\{x_0\}\subseteq\{\hat{x}_\beta\}$, which establishes the maximum principle.

The particular version of Theorem 1.10(a) and (b) needed here calls for, at the stage corresponding to (1.94), the use of a new variable (say) $\mathbf{c}=\partial L/\partial \mathbf{i}=\mathbf{c}[\mathbf{i},\mathbf{V}]$ and inverse mapping $\mathbf{i}[\mathbf{c},\mathbf{V}]$ with Legendre transform

$$G[\mathbf{c},\mathbf{V}]=\mathbf{c}^{\mathrm{T}}\mathbf{i}[\mathbf{c},\mathbf{V}]-L[\mathbf{i}[\mathbf{c},\mathbf{V}],\mathbf{V}],$$

such that

$$\frac{\partial G}{\partial \mathbf{c}}=\mathbf{i} \quad \text{and} \quad \frac{\partial G}{\partial \mathbf{V}}=-\frac{\partial L}{\partial \mathbf{V}}.$$

The problem (2.146) is thereby rewritten

$$\mathbf{c}=0 \quad (\alpha), \qquad \frac{\partial G}{\partial \mathbf{V}}=0 \quad (\beta).$$

The extremum problem is that of minimizing

$$-G[0,\mathbf{V}]=L[\mathbf{i}[0,\mathbf{V}],\mathbf{V}]=Y[T^{\mathrm{T}}\mathbf{V}]$$

with respect to arbitrary $\mathbf{V}$, and the two problems are plainly equivalent. We are effectively using the Legendre transformation (2.143) and (2.144), which is not invalidated by corners on $X[\mathbf{i}]$. □

Theorem 2.19 (Lower bound and maximum principle)

(a) Any choice of $\mathbf{i}$ *which satisfies* $T\mathbf{i}=0$ *will provide* $-X[\mathbf{i}]$ *with a lower bound to the value which* $-X=Y$ *has for any solution of the system* (2.145).

(b) If each $X_j[i_j]$ *is weakly convex every maximizer of* $-X[\mathbf{i}]$ *with respect to all* $\mathbf{i}$ *satisfying* $T\mathbf{i}=0$ *will belong to a solution of* (2.145).

Proof

(a) The proof of Theorem 1.6(b) can be augmented, along similar lines to that of Theorem 2.18(a) above, to show that a lower bound will be any value of

$$L-\mathbf{V}^{\mathrm{T}}\frac{\partial L}{\partial \mathbf{V}}=-X[\mathbf{i}] \tag{2.150}$$

which satisfies (2.145β) but not necessarily (2.145α). Alternatively, Theorem 1.8(b) can be adapted, with a linear constraint there. The constraint $T\mathbf{i}=0$ is a genuine restriction, but its general solution can be expressed as $(2.157)_2$ below.

(b) The converse maximum principle can be approached by seeking to adapt the proof of Theorem 1.13(a) and (b), and using Farkas' lemma in Theorem 1.12. Specifically, let $\{\hat{\mathbf{i}}_\beta\}$ be the set of maximizers of $-X[\mathbf{i}_\beta]$ subject to the linear constraints $T\mathbf{i}_\beta=0$, and let $\{\mathbf{i}_0\}$ be the set of $\mathbf{i}$ which satisfy both of (2.145) for some $\mathbf{V}$. To show that $\{\hat{\mathbf{i}}_\beta\}\subseteq\{\mathbf{i}_0\}$, we note first that its maximum property $-X[\hat{\mathbf{i}}_\beta]\geqslant -X[\mathbf{i}_\beta]$ implies that

$$-(\hat{\mathbf{i}}_\beta-\mathbf{i}_\beta)^{\mathrm{T}}\left.\frac{\mathrm{d}X}{\mathrm{d}\mathbf{i}}\right|_{\bar{\mathbf{i}}}\geqslant 0$$

for some $\bar{\mathbf{i}}$ on the straight join of any $\hat{\mathbf{i}}_\beta$ to any $\mathbf{i}_\beta$, and for every subgradient if there is a gradient step at $\bar{\mathbf{i}}$. If $\mathbf{d}$ belongs to the cone of unit vectors pointing from $\hat{\mathbf{i}}_\beta$ to $\mathbf{i}_\beta$, then as in (1.112) we find

$$\mathbf{d}^{\mathrm{T}}\left.\frac{\mathrm{d}X}{\mathrm{d}\mathbf{i}}\right|_{\hat{\mathbf{i}}_\beta}\geqslant 0.$$

Consistency holds if this is true subject to (1.113) in the form

$$\mathbf{d}^{\mathrm{T}}T^{\mathrm{T}}=0,$$

and Farkas' lemma for such a case implies

$$\left.\frac{\mathrm{d}X}{\mathrm{d}\mathbf{i}}\right|_{\hat{\mathbf{i}}_\beta}=T^{\mathrm{T}}\mathbf{V}$$

for some $\mathbf{V}$, which is (2.145α), so that $\{\hat{\mathbf{i}}_\beta\}\subseteq\{\mathbf{i}_0\}$. The equivalence then follows by introducing the weak convexity of each $X_j[i_j]$. □

(vii) Equivalent underdetermined systems

Here we prove a theorem in matrix algebra, which applies to underdetermined equations of type (2.138) wherein the variables are $\mathbf{V}$, $\mathbf{v}$ and $\mathbf{i}$. The theorem does not use any specification of branch characteristics. To facilitate the proof we introduce temporary suffixes on the $b\times 1$ column matrices $\mathbf{v}$ and $\mathbf{i}$ for emphasis.

Thus let $\{\mathbf{v}_t\}$ and $\{\mathbf{i}_t\}$ denote the sets of such matrices which satisfy

$$T^{\mathrm{T}}\mathbf{V}=\mathbf{v}_t \quad\text{and}\quad T\mathbf{i}_t=0, \tag{2.151}$$

respectively, for some choice of $\mathbf{V}$. Recall that T is $n\times b$, and let R denote a nonzero $l\times n$ matrix such that the $n\times l$ matrix

$$TR^{\mathrm{T}}=0. \tag{2.152}$$

This notation for the matrix R is unrelated to any measure of resistance. Let $\{\mathbf{v}_r\}$ and $\{\mathbf{i}_r\}$ denote the sets of matrices $\mathbf{v}$ and $\mathbf{i}$ which satisfy

$$R\mathbf{v}_r = 0 \quad \text{and} \quad R^{\mathrm{T}}\mathbf{C} = \mathbf{i}_r, \tag{2.153}$$

respectively, for some choice of an $l \times 1$ matrix $\mathbf{C}$. This $\mathbf{C}$ has a similar status in (2.153) to that of $\mathbf{V}$ in (2.151), neither being preassigned (suffixes t and r are merely intended to be associated with T and R respectively).

Theorem 2.20
The sets $\{\mathbf{v}_t\}$ and $\{\mathbf{v}_r\}$ are equivalent. The sets $\{\mathbf{i}_t\}$ and $\{\mathbf{i}_r\}$ are also equivalent. The general solution of $(2.151)_2$ *is* $(2.153)_2$ *for arbitrary* $\mathbf{C}$, *and the general solution of* $(2.153)_1$ *is* $(2.151)_1$ *for arbitrary* $\mathbf{V}$.

Proof

$$R\mathbf{v}_t = RT^{\mathrm{T}}\mathbf{V} = 0 \text{ for any } \mathbf{V}, \text{ therefore } \{\mathbf{v}_t\} \subseteq \{\mathbf{v}_r\}.$$

The converse, that $\{\mathbf{v}_r\} \subseteq \{\mathbf{v}_t\}$, requires Theorem 1.12(a) to be rewritten in matrix notation and reinterpreted as follows. Choose a there to be $\mathbf{v}_r$. The set of admissible x there can be regarded as consisting of the set of rows of R here, by identifying the set of equations $Tx = 0$ there with successive columns of $TR^{\mathrm{T}} = 0$ here. Then each $(x, a) = 0$ is the corresponding row of $R\mathbf{v}_r = 0$ here. We must therefore have the representation $\mathbf{v}_r = T^{\mathrm{T}}\mathbf{V}$ here for some $\mathbf{V}$, by identifying with $a = T^*u$ there. Hence $\{\mathbf{v}_r\} \subseteq \{\mathbf{v}_t\}$ by $(2.151)_1$.

Hence $\{\mathbf{v}_r\} \equiv \{\mathbf{v}_t\}$, and similarly $\{\mathbf{i}_r\} \equiv \{\mathbf{i}_t\}$. □

In passing we notice that an analogue of this result for differential operators in three dimensions can be obtained by replacing (2.151)–(2.153) by

$$\nabla\phi = \mathbf{v}, \quad \nabla\cdot\mathbf{u} = 0, \tag{2.154}$$

$$\nabla\cdot(\nabla \wedge \boldsymbol{\psi}) = 0 = \nabla \wedge (\nabla\phi), \tag{2.155}$$

$$\nabla \wedge \mathbf{v} = 0, \quad \nabla \wedge \boldsymbol{\psi} = \mathbf{u}, \tag{2.156}$$

respectively. Here ∇ is the vector gradient operator, ϕ is a scalar, and $\mathbf{u}$, $\mathbf{v}$ and $\boldsymbol{\psi}$ are vectors. The general solution of $\nabla\cdot\mathbf{u} = 0$ is $\mathbf{u} = \nabla \wedge \boldsymbol{\psi}$ for any $\boldsymbol{\psi}$, and the general solution of $\nabla \wedge \mathbf{v} = 0$ is $\mathbf{v} = \nabla\phi$ for any ϕ. We shall describe in §3.3(iii) a sense in which the operators div and grad are formally adjoint to each other, and in which the curl is formally adjoint to itself.

Theorem 2.20 shows that when nonzero R can be found to satisfy $TR^{\mathrm{T}} = 0$, the underdetermined system

$$R\mathbf{v} = 0, \quad R^{\mathrm{T}}\mathbf{C} = \mathbf{i} \tag{2.157}$$

for the variables $\mathbf{C}$, $\mathbf{v}$ and $\mathbf{i}$ is equivalent to (2.138) to the extent that both are satisfied by the same set of **i** and **v**. An alternative derivation of (2.139) is from (2.157).

(viii) Loop-branch formulation

Matrices interpretable as this R and **C** appear at the outset when the network analysis is begun in the following alternative way. Instead of defining node potentials and writing down (2.134) and (2.135) to reveal the node-branch incidence matrix T, let the directed graph in Fig. 2.29 be re-designated to contain the three loops and six branches shown in Fig. 2.31. A *loop* is any set of distinct branches and the same number of distinct nodes which can be traversed continuously without repetition, starting and ending in the same place, with a direction arbitrarily assigned to it.

Let C_1, C_2, C_3 denote electrical currents supposed to be flowing in each loop, reckoned positive in the directions shown. Then, by definition, the branch currents must be

$$\begin{aligned} &i_1 = C_1, \quad i_2 = -C_1 + C_3, \quad i_3 = -C_1 + C_2, \\ &i_4 = C_2, \quad i_5 = -C_2 + C_3, \quad i_6 = C_3. \end{aligned} \tag{2.158}$$

These satisfy Kirchhoff's current law (2.134), for any loop currents. Kirchhoff's voltage law around the three loops requires

$$v_1 - v_2 - v_3 = 0, \quad v_3 + v_4 - v_5 = 0, \quad v_2 + v_5 + v_6 = 0. \tag{2.159}$$

Equations (2.159) and (2.158) illustrate (2.157) with

$$R = \begin{bmatrix} 1 & -1 & -1 & 0 & 0 & 0 \\ 0 & 0 & 1 & 1 & -1 & 0 \\ 0 & 1 & 0 & 0 & 1 & 1 \end{bmatrix}. \tag{2.160}$$

This is called the loop-branch incidence matrix.

For a general directed graph its typical element is $+1$ or -1 for a loop containing a branch with the same or opposite direction respectively, and 0 for a branch absent from a loop. It then follows that the general loop-

Fig. 2.31. A loop-branch designation of Fig. 2.29.

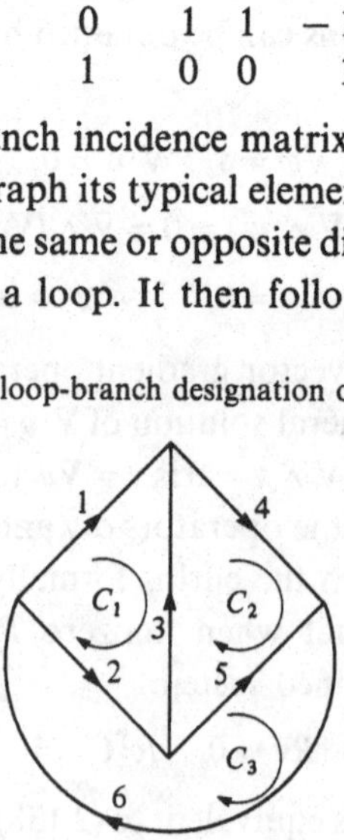

branch incidence matrix R has the property $TR^{\mathrm{T}} = 0$. The rank of R is $b - n + 1$. In the illustration (2.160) we have chosen R to have full rank by considering three independent loops, but (as with T above) the choice of larger R would not inhibit computation of upper and lower bounds.

If we had begun with (2.157) instead of (2.138), and if the branch characteristics could again be assembled as in (2.143), we could eliminate $\mathbf{i}$ to obtain

$$-R\mathbf{v} = 0 \quad (\alpha), \qquad -R^{\mathrm{T}}\mathbf{C} = -\frac{\mathrm{d}Y}{\mathrm{d}\mathbf{v}} \quad (\beta). \tag{2.161}$$

This is another system of Hamiltonian governing equations, which is an alternative to (2.145). Now we seek first to determine $\mathbf{v}$ and $\mathbf{C}$ rather than $\mathbf{i}$ and $\mathbf{V}$, and not necessarily uniquely for a similar reason to the previous one. Theorem 2.20 leads to the conclusion that the two systems are equivalent to the extent that the sets of $\mathbf{i}$ which satisfy (2.145) and of $\mathbf{v}$ which satisfy (2.161) are related by the branch characteristic (2.143).

We can express (2.161) as the zero gradients of the new generating saddle function

$$L[\mathbf{C}, \mathbf{v}] = -\mathbf{C}^{\mathrm{T}}R\mathbf{v} + Y[\mathbf{v}], \tag{2.162}$$

which is linear in $\mathbf{C}$ and convex in $\mathbf{v}$. There will be an upper bound having the values of

$$L - \mathbf{C}^{\mathrm{T}}\frac{\partial L}{\partial \mathbf{C}} = Y[\mathbf{v}] \tag{2.163}$$

for any solution of (2.161α), i.e. of $J[\mathbf{V}] = Y[T^{\mathrm{T}}\mathbf{V}]$ for any $\mathbf{V}$ exactly as in (2.149). There will be a lower bound having the values of

$$L - \mathbf{v}^{\mathrm{T}}\frac{\partial L}{\partial \mathbf{v}} = Y - \mathbf{v}^{\mathrm{T}}\frac{\mathrm{d}Y}{\mathrm{d}\mathbf{v}} = -X[\mathbf{i}] \tag{2.164}$$

for any solution of (2.161β), i.e. having the values of (2.150) in the form

$$K[\mathbf{C}] = -X[R^{\mathrm{T}}\mathbf{C}] \tag{2.165}$$

for any $\mathbf{C}$ (cf. (2.117)).

Thus the set of upper bounds $J[\mathbf{V}]$ is the same in either formulation (2.145) or (2.161), and the set of lower bounds $K[\mathbf{C}]$ is also the same. We leave the reader to write out the converse minimum and maximum principles along the lines of §2.8(vi).

These correspondences between the node-branch and loop-branch viewpoints have analogues in other problems, including infinite dimensional ones, as we summarise in Figs. 3.10 and 3.11, and illustrate in §§3.12(vi) and (vii). Other examples include the equilibrium and compatibility viewpoints in continuum mechanics.

(ix) Branch characteristics with inequalities

Here we illustrate briefly the type of branch characteristic which, as it goes to infinity, is parallel to a coordinate axis, at least at one end. This is the case postponed from §2.8(iv). We use Fig. 2.15(*a*) as an example, with x and y there replaced by i and v respectively here.

If *every* branch had this type of characteristic, we could assemble all the branch characteristics in matrix form as

$$\mathbf{v} \geqslant \frac{\mathrm{d}\phi}{\mathrm{d}\mathbf{i}}, \quad \mathbf{i} \leqslant \mathbf{i}_G, \quad (\mathbf{i}-\mathbf{i}_G)^{\mathrm{T}}\left(\mathbf{v}-\frac{\mathrm{d}\phi}{\mathrm{d}\mathbf{i}}\right)=0, \tag{2.166}$$

or equivalently

$$\mathbf{i} \leqslant \frac{\mathrm{d}\psi}{\mathrm{d}\mathbf{v}}, \quad \mathbf{v} \geqslant \mathbf{v}_B, \quad (\mathbf{v}-\mathbf{v}_B)^{\mathrm{T}}\left(\mathbf{i}-\frac{\mathrm{d}\psi}{\mathrm{d}\mathbf{v}}\right)=0, \tag{2.167}$$

where $\mathbf{i}_G$ and $\mathbf{v}_B$ are assigned, by rewriting the equations in §2.3(vii).

The system of governing conditions now consists of either (2.166) or (2.167) with either (2.138) or (2.157). We seek to write these as examples of (1.54) or (1.55) rather than (1.53). If we eliminate $\mathbf{v}$ at the outset from (2.138) and (2.166), we can write the result as an example of (1.55) of the form

$$\frac{\partial L}{\partial(\mathbf{i}_G-\mathbf{i})} \leqslant 0 \quad (\alpha), \qquad \mathbf{i}_G-\mathbf{i} \geqslant 0 \quad (\beta), \qquad (\mathbf{i}_G-\mathbf{i})^{\mathrm{T}}\frac{\partial L}{\partial(\mathbf{i}_G-\mathbf{i})}=0,$$
$$\frac{\partial L}{\partial \mathbf{V}}=0 \quad (\beta) \tag{2.168}$$

generated by the function

$$L[\mathbf{i}_G-\mathbf{i},\mathbf{V}]=[\mathbf{i}_G-(\mathbf{i}_G-\mathbf{i})]^{\mathrm{T}}T^{\mathrm{T}}\mathbf{V}-\phi[\mathbf{i}_G-(\mathbf{i}_G-\mathbf{i})]. \tag{2.169}$$

This is concave in $\mathbf{i}_G-\mathbf{i}$ and linear in $\mathbf{V}$, and so is still a saddle function of the type required for upper and lower bounds, but in adjusted variables. It has the same values as (2.147) where $\mathbf{i}<\mathbf{i}_G$.

An upper bound will be any value of

$$L-(\mathbf{i}_G-\mathbf{i})^{\mathrm{T}}\frac{\partial L}{\partial(\mathbf{i}_G-\mathbf{i})}=\mathbf{i}_G{}^{\mathrm{T}}\left(T^{\mathrm{T}}\mathbf{V}-\frac{\mathrm{d}\phi}{\mathrm{d}\mathbf{i}}\right)+\mathbf{i}^{\mathrm{T}}\frac{\mathrm{d}\phi}{\mathrm{d}\mathbf{i}}-\phi \tag{2.170}$$

which satisfies (2.168α), i.e. $T^{\mathrm{T}}\mathbf{V} \geqslant \mathrm{d}\phi/\mathrm{d}\mathbf{i}$. A lower bound will be any value of

$$L-\mathbf{V}^{\mathrm{T}}\frac{\partial L}{\partial \mathbf{V}}=-\phi \tag{2.171}$$

which satisfies (2.168β), i.e. $T\mathbf{i}=0$ and $\mathbf{i}\leqslant\mathbf{i}_G$.

If instead we eliminate **i** at the outset from (2.157) and (2.167), we can write the result as an example of (1.54) of the form

$$\frac{\partial L}{\partial \mathbf{C}} = 0 \quad (\alpha),$$
$$\frac{\partial L}{\partial (\mathbf{v}-\mathbf{v}_B)} \leqslant 0 \quad (\beta), \qquad \mathbf{v}-\mathbf{v}_B \geqslant 0 \quad (\alpha), \qquad (\mathbf{v}-\mathbf{v}_B)^{\mathrm{T}} \frac{\partial L}{\partial (\mathbf{v}-\mathbf{v}_B)} = 0, \tag{2.172}$$

generated by the function

$$L[\mathbf{C}, \mathbf{v}-\mathbf{v}_B] = -[\mathbf{v}-\mathbf{v}_B+\mathbf{v}_B]^{\mathrm{T}} R^{\mathrm{T}} \mathbf{C} + \psi[\mathbf{v}-\mathbf{v}_B+\mathbf{v}_B]. \tag{2.173}$$

This is linear in **C** and convex in $\mathbf{v}-\mathbf{v}_B$, and so it is also the requisite type of saddle function. It has the same values as (2.162) where $\mathbf{v} > \mathbf{v}_B$.

An upper bound will be any value of

$$L - \mathbf{C}^{\mathrm{T}} \frac{\partial L}{\partial \mathbf{C}} = \psi \tag{2.174}$$

which satisfies (2.172α), i.e. $R\mathbf{v} = 0$ and $\mathbf{v} \geqslant \mathbf{v}_B$. A lower bound will be any value of

$$L - (\mathbf{v}-\mathbf{v}_B)^{\mathrm{T}} \frac{\partial L}{\partial (\mathbf{v}-\mathbf{v}_B)} = -\mathbf{v}_B{}^{\mathrm{T}}\left(R^{\mathrm{T}}\mathbf{C} - \frac{\mathrm{d}\psi}{\mathrm{d}\mathbf{v}}\right) - \mathbf{v}^{\mathrm{T}}\frac{\mathrm{d}\psi}{\mathrm{d}\mathbf{v}} + \psi \tag{2.175}$$

which satisfies (2.172β), i.e. $R^{\mathrm{T}}\mathbf{C} \leqslant \mathrm{d}\psi/\mathrm{d}\mathbf{v}$. It can be shown directly that the two sets of upper bounds (2.170) and (2.174) are the same, and also that the two sets of lower bounds (2.171) and (2.175) are the same.

More complicated considerations are required if every branch cannot be regarded as having the same type of characteristic. We refer to Iri (1969) and Rockafellar (1982).

3

Upper and lower bounds via saddle functionals

3.1. Introduction

This is the core of the book. We show how to construct upper and lower bounds in a number of representative infinite dimensional problems.

The approach is to generalize Theorem 1.6, which was proved for the finite dimensional case. The basic result is Theorem 3.4. We set the problems in a pair of infinite dimensional inner product spaces. We construct some simple tools which allow this setting to mimic the finite dimensional development in Chapters 1 and 2.

These tools include a straightforward but unfamiliar type of inner product space element (§3.2(ii)), a pair of linear operators T and T^* which are adjoint to each other in the true sense, not merely the formal sense (§3.3(ii) and (iii)), and the gradient of a functional whose domain is an inner product space (§3.4(ii) and (iii)).

Application to differential equations with one type of mixed boundary conditions is explained by the detailed working of an example (§3.7) and the extension of similar methods to other cases is indicated in §§3.8–3.10. A connection with variational inequalities is mentioned in §3.11. We conclude (§3.12) with an account of equations containing a nonnegative operator, typically decomposable as T^*T. These admit not only simultaneous bounds but also, in some cases, alternative pairs of such bounds not unlike those already indicated for networks in §2.8. This is one situation in which integral equations become prominent, as we illustrate for a wave scattering problem in fluid mechanics.

Except for an illustration of the fundamental lemma of the calculus of variations (Theorem 3.8) we do not discuss the converse theorems which would convert, for example, an upper bound statement into a minimum principle. We proved similar results, and so indicated their significance, in §1.5. In the infinite dimensional case here it appears that the details of such proofs would vary from case to case. Such aspects, and the further technicalities which are required to treat them, are discussed from some viewpoints elsewhere in the literature. We prefer to give space in this book to other matters not so well appreciated, such as the simplicity of the theoretical structure for bounds which we summarize in §3.9, and also the features underlying that structure which Chapter 2 is designed to convey, especially for nonlinear problems.

3.2. Inner product spaces

(i) Linear spaces

In §1.3(i) we introduced inner product space notation in the context of a pair of finite dimensional spaces. In this section we indicate some ideas needed to carry this notation over to infinite dimensional spaces, where it is especially useful for proving upper and lower bounds.

We recall first that a real *linear space U* (sometimes alternatively called a vector space) is a set of elements $u, v, \ldots$ for which the operations of addition and of multiplication by real numbers $\lambda, \mu, \ldots$ are defined with the following property. For any $u \in U$ and $v \in U$, the element $\lambda u + \mu v$ also belongs to U with any λ, μ.

As a simple example, the set of real scalar functions $u(t)$ that are defined and continuous on a fixed interval such as $0 \leqslant t \leqslant 1$, with unrestricted end values, comprise a linear space. This follows because the sum of any two such functions is another one. For the same reason the set of piecewise C^1 functions $u(t)$ which each satisfy the homogeneous boundary conditions $u(0) = u'(1) = 0$ form another linear space.

These two illustrations belong to a hierarchy of similar examples obtained by progressive strengthening of smoothness hypotheses on $u(t)$ over $0 \leqslant t \leqslant 1$. Thus, square integrable functions form a linear space, regardless of whether they are continuous or smooth, because $(u+v)^2 \leqslant 2(u^2 + v^2)$ shows that if u and v are square integrable then so is $u + v$. Linear spaces are also formed, for example, by functions which are: piecewise continuous with unrestricted end values; or continuous with zero end values; or piecewise C^1 with unrestricted end values; or C^1 with $u(0) = u'(0) = 0$; or piecewise C^2 with homogeneous end values; or C^2 with no end restrictions; and so on.

By contrast, for example, the set of C^1 functions subject to nonhomogeneous end restrictions such as $u(0) - 1 = u'(1) = 0$ do not form a linear space, because the sum of any two of them does not satisfy $u(0) = 1$. Nonhomogeneous assigned boundary values such as $a = u(0)$, $b = u'(1)$ can often be subtracted out by working in the linear space of $w(t) = u(t) - a - bt$.

Another type of linear space, of which we shall make much use, is made up of composite elements $\tilde{u}$ of the type defined in $0 \leqslant t \leqslant 1$ by

$$\tilde{u} = \begin{bmatrix} u(t) \\ u(0) \\ u(1) \end{bmatrix}. \tag{3.1}$$

Here the function $u(t)$, of specified smoothness and no end restrictions, is used to construct a matrix $\tilde{u}$ whose first entry is the function itself in $0 \leqslant t \leqslant 1$, and whose second and third entries are the end values of the function. We shall use $\tilde{u}$ rather than u to designate space elements of this and similar composite type, to distinguish in context the matrix $\tilde{u}$ from the value u of the function used to construct $\tilde{u}$. If no end point entries are needed we omit the tilde, then writing u as an abbreviation for $u(t)$, as is common. The sum of two such matrices is another one, so that the matrices themselves form a linear space. In general theory, which applies whether the space elements are of composite type or not, we omit the tilde.

A *normed space* U is a linear space equipped with a definition of length or *norm* $\|u\|$, which is a real number associated with any $u \in U$ such that

(a) $\|u\| \geqslant 0$, with $\|u\| = 0$ if and only if $u = 0$,

(b) $\|\lambda u\| = |\lambda| \, \|u\|$ for any real λ,

(c) $\|u + v\| \leqslant \|u\| + \|v\|$ for any $v \in U$ (triangle inequality).

(ii) Inner product spaces

A real *inner product space* U is a linear space equipped with an *inner product*, say $\langle u, v \rangle$, which is a real number associated with every $u \in U$ and $v \in U$ and having the properties of

(a) linearity: $\langle u, \lambda v + \mu w \rangle = \lambda \langle u, v \rangle + \mu \langle u, w \rangle$
for any $w \in U$ and all real λ, μ;

(b) symmetry: $\langle u, v \rangle = \langle v, u \rangle$;

(c) positivity: $\langle u, u \rangle \geqslant 0$, with $\langle u, u \rangle = 0$ if and only if $u = 0$.

An inner product space has a norm defined by $\|u\| = \langle u, u \rangle^{\frac{1}{2}}$. By considering $\langle u + \lambda v, u + \lambda v \rangle \geqslant 0$ for all real λ, we deduce the Schwarz inequality

$$|\langle u, v \rangle| \leqslant \|u\| \, \|v\|, \tag{3.2}$$

from which the triangle inequality follows.

We shall frequently consider a second inner product space X, consisting of elements $x, y, \ldots$, and having an inner product (x, y) whose definition may be different from that of $\langle u, v \rangle$, although it must also have the three properties (a), (b) and (c) just listed. Finite dimensional illustrations of two different inner products were given in (1.46). The elements of X and U there were sets of n and of m real numbers respectively or, from another viewpoint but with the same effect, $n \times 1$ and $m \times 1$ matrices. The context will obviate any confusion between the use of X and U to label spaces, and uses of the same letters to denote the values of certain functions, as in §2.6(i) and (ii), for example.

Occasionally we shall need a third inner product space, whose inner product will be denoted by an overbar $\overline{\ ,\ }$. We never use [,] as an inner product; such square brackets will enclose the arguments of a functional. We shall sometimes use in a subsidiary way the dot product notation $\mathbf{u}\cdot\mathbf{v}$ for the scalar product of two physical vectors $\mathbf{u}$ and $\mathbf{v}$; of course this is a particular illustration of a finite dimensional inner product.

Any one of the linear spaces in the hierarchy described for the function $u(t)$ in §(i) above can be endowed with a norm (and no inner product) or an inner product (and therefore a norm) in various ways, depending on the subsequent purpose. For example, the space of square integrable functions could be given the simple inner product

$$\langle u, v\rangle = \int_0^1 \rho(t)u(t)v(t)\,\mathrm{d}t, \tag{3.3}$$

where $\rho(t) > 0$ is a suitably given weighting function, for example $\rho(t) \equiv 1$. The space of C^1 functions which have square integrable derivatives could be given the inner product

$$\int_0^1 (u(t)v(t) + u'(t)v'(t))\,\mathrm{d}t.$$

We shall find it particularly helpful, when considering linear spaces whose elements are of the composite type (3.1), to define an inner product of the type

$$\langle \tilde{u}, \tilde{v}\rangle = \int_0^1 \rho(t)u(t)v(t)\mathrm{d}t + au(0)v(0) + bu(1)v(1), \tag{3.4}$$

where $a \geqslant 0$ and $b \geqslant 0$ are given numbers, and $\rho(t)$ is a suitably given function such that $\rho(t) > 0$ everywhere in $0 \leqslant t \leqslant 1$. Positivity of $\langle \tilde{u}, \tilde{v}\rangle$ is assured for all square integrable functions. For some applications it will prove possible and desirable to use a weighting function $\rho(t)$ which is positive 'almost everywhere', but we defer this until §3.7(xvi).

Inner products of the type (3.4) were used systematically by Noble and Sewell (1972). As remarked above the functions $u(t)$, $v(t)$, ... have specified smoothness (at least square integrable, but perhaps continuous or, say, piecewise C^1), but with no end restrictions at this stage.

Problems in physics and mechanics of solids and fluids are frequently posed over a two or three dimensional spatial region V with boundary

Σ. Even the simplest problems entail the interaction of scalar valued variables, say $\phi, \psi, \ldots$, for example potentials, with vector valued variables, say $\mathbf{u}$, $\mathbf{v}, \ldots$, for example forces or displacements. Such variables may be sought as functions of position over V and Σ, with continuity and smoothness properties specified within a hierarchy like that already described for functions of one variable. Two different inner product spaces now suggest themselves, as follows.

The first space is the space of composite or compound elements

$$\tilde{\phi} = \begin{bmatrix} \phi(V) \\ \phi(\Sigma_1) \\ \phi(\Sigma_2) \end{bmatrix}, \tag{3.5}$$

defined as a column matrix like (3.1), and given the inner product

$$(\tilde{\phi}, \tilde{\psi}) = \int \phi\psi \, \mathrm{d}V + \int \phi\psi \, \mathrm{d}\Sigma_1 + \int \phi\psi \, \mathrm{d}\Sigma_2. \tag{3.6}$$

Here $\Sigma = \Sigma_1 + \Sigma_2$, the surface being regarded as having two contiguous parts Σ_1 and Σ_2 in anticipation of the *later* imposition of different kinds of boundary conditions on the two parts. The entries in the column matrix (3.5) are scalar functions of position over the indicated regions. For simplicity we use V in the first entry of (3.5) to denote the closed region, and likewise in (3.7) and (3.9) and all later examples. Such $\tilde{\phi}$ is itself defined over the closed region also, thus generalizing (3.1) to higher dimensions.

The second space is that of composite elements

$$\tilde{\mathbf{u}} = \begin{bmatrix} \mathbf{u}(V) \\ \mathbf{u}(\Sigma_1) \\ \mathbf{u}(\Sigma_2) \end{bmatrix}, \tag{3.7}$$

also defined as a column matrix like (3.1), but this time with entries which are vector functions of position, and given the inner product

$$\langle \tilde{\mathbf{u}}, \tilde{\mathbf{v}} \rangle = \int \mathbf{u}\cdot\mathbf{v} \, \mathrm{d}V + \int \mathbf{u}\cdot\mathbf{v} \, \mathrm{d}\Sigma_1 + \int \mathbf{u}\cdot\mathbf{v} \, \mathrm{d}\Sigma_2. \tag{3.8}$$

The two definitions of inner products in (3.6) and (3.8) are plainly different, and hence it is desirable to use different symbols for them, such as the round and pointed brackets. Suitable weighting functions could be incorporated in (3.6) and (3.8), as indicated in (3.4).

Tensors of order k include scalars ($k = 0$) and physical vectors ($k = 1$). Second order tensors with cartesian components σ_{ij} ($i, j = 1, 2, 3$) can also be used to construct a linear space with composite elements defined as the matrix

$$\tilde{\sigma} = \begin{bmatrix} \sigma_{ij}(V) \\ \sigma_{ij}(\Sigma_1) \\ \sigma_{ij}(\Sigma_2) \end{bmatrix} . \tag{3.9}$$

Each of the three entries in $\tilde{\sigma}$ is itself a matrix, consisting of the nine cartesian components of the tensor, evaluated over the closed interior and the boundary parts according to the conventions laid down above for $\tilde{\phi}$. Such elements will first be used in (3.27) below, and in §5.5 we give a fuller discussion of such matters.

As alternatives we could use three components in each entry of (3.7), and invariant dyads or 'arrows' of the tensor in (3.9).

3.3. Linear operators and adjointness

(i) Operators

We give some brief terminology for operators, which are sometimes also called mappings, transformations, functions or functionals, and which need not be linear until so specified in §(ii) below.

An operator M is a specification for converting each element of a set $\mathscr{D}(M)$, called the *domain* of M, into some element of another set $\mathscr{R}(M)$, called the *range* of M. We write M: $\mathscr{D}(M) \to \mathscr{R}(M)$, and alternatively $u = M(x)$ or $u = M[x]$ for every element $x \in \mathscr{D}(M)$, so that $\mathscr{R}(M)$ is the set of all elements u which can be so obtained from elements of $\mathscr{D}(M)$. The elements of domain and range are generally of different types.

We shall be particularly concerned with situations where the set $\mathscr{D}(M)$ is actually a subspace of an inner product space X, and where the set $\mathscr{R}(M)$ is a subspace (or the whole) of a second inner product space U. The elements of a linear *subspace* themselves comprise a linear space, which is contained within the original space. For example, the space of C^1 functions is a subspace of the space of continuous functions.

It is a common convention in functional analysis to reserve the names operator, mapping, etc. for the single valued case in which a single $u \in \mathscr{R}(M)$ is induced by each $x \in \mathscr{D}(M)$, as indicated in Fig. 3.1(*a*). We were able to follow this viewpoint for functions in Chapter 1, but not in Chapter 2, where we needed to speak of multivalued functions having several single-valued *parts*, usually contiguous. For example, the parabola $u = \pm x^{\frac{1}{2}}$ is described by a double valued function having two contiguous single valued parts with a common domain $0 \leqslant x < \infty$ but different ranges $-\infty < u \leqslant 0$

Fig. 3.1. Pictorial representation of mappings. (*a*) A single valued mapping M maps $\mathscr{D}(M) \subset X$ *into* U and *onto* $\mathscr{R}(M) \subset U$. Note that one element $u \in \mathscr{R}(M)$ may correspond to more than one x in $\mathscr{D}(M)$. (*b*) A double valued mapping. (*c*) If M is both onto and one-to-one, then M^{-1} exists which is also both onto and one-to-one.

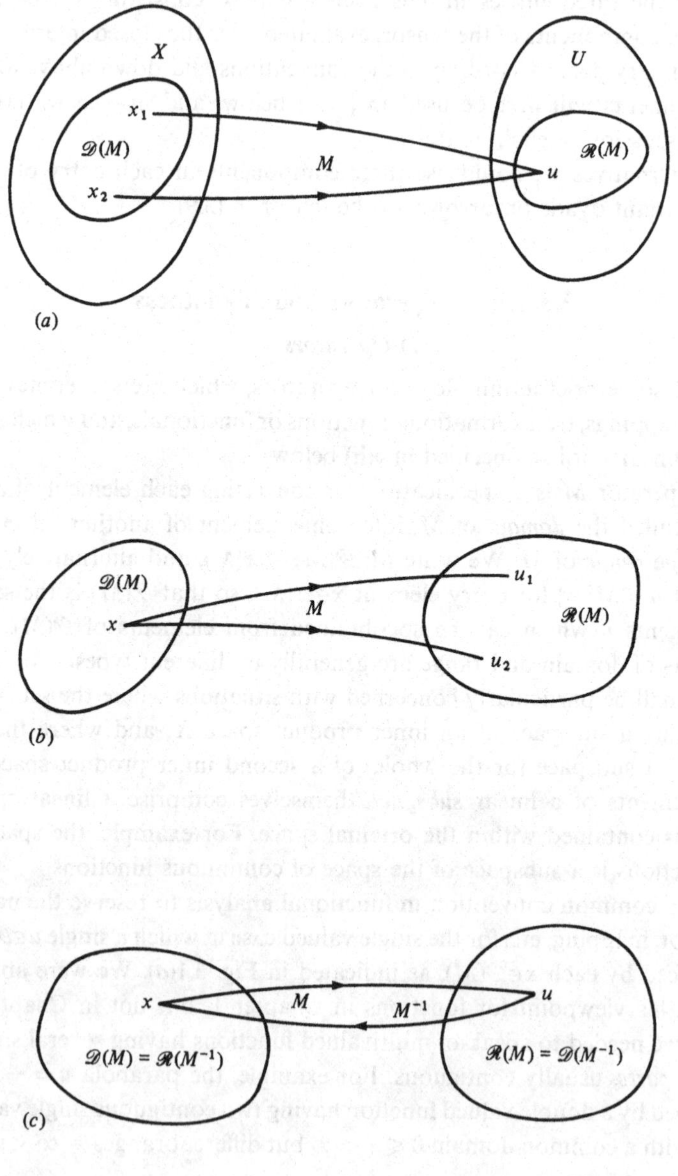

and $0 \leqslant u < +\infty$. We shall always regard a part as single valued, by definition, and we shall allow the context to decide whether an operator needs the adjective single valued for emphasis or not. Fig. 3.1(*b*) represents a double valued operator. A graphic illustration of a multivalued mapping of a plane onto a plane is provided by flattening an arbitrarily folded plastic sheet (the range) onto the working plate (the domain) of an overhead projector (see Fig. 2.25).

So far an element $u \in \mathscr{R}(M)$ may correspond to more than one x in $\mathscr{D}(M)$ (see Fig. 3.1(*a*)), as in the parabola $u = x^2$. If M is single valued and if for every $u \in \mathscr{R}(M)$ there corresponds exactly one $x \in \mathscr{D}(M)$ such that $u = M(x)$, then M is said to be *one-to-one*. The situation when a mapping is both onto *and* one-to-one is illustrated in Fig. 3.1(*c*). There $u = M(x)$ associates a unique $u \in \mathscr{R}(M)$ with each $x \in \mathscr{D}(M)$, and there exists an *inverse* operator M^{-1} with $\mathscr{D}(M^{-1}) = \mathscr{R}(M)$ and $\mathscr{R}(M^{-1}) = \mathscr{D}(M)$ that associates a unique $x = M^{-1}(u) \in \mathscr{R}(M^{-1})$ with each $u \in \mathscr{D}(M^{-1})$. The inverse operator maps $\mathscr{R}(M)$ onto $\mathscr{D}(M)$.

(ii) Adjointness of linear operators

An operator T is called *linear* if

$$T(\lambda x + \mu y) = \lambda T x + \mu T y \quad \forall x, y \in \mathscr{D}(T)$$

and for any real numbers λ, μ. We follow the convention that when the element acted upon by a linear operator T is denoted by a single symbol, such as x, we write the result as Tx rather than $T(x)$. The latter usage, such as $M(x)$ above, indicates a mapping which in general will be nonlinear. We also assume that $T0 = 0$, whence $0 = Tx - Tx$, so that T is single valued.

Consider any pair of inner product spaces, X with inner product $(.,.)$ and elements $x, y, \ldots,$ and U with inner product $\langle .,. \rangle$ and elements u, $v, \ldots$. Suppose that, corresponding to a linear operator T whose domain $\mathscr{D}(T) \subseteq X$ and whose range $\mathscr{R}(T) \subseteq U$, there exists a second linear operator T^* whose domain $\mathscr{D}(T^*) \subseteq U$ and whose range $\mathscr{R}(T^*) \subseteq X$, with the property that

$$(x, T^*u) = \langle u, Tx \rangle \tag{3.10}$$

for all $x \in \mathscr{D}(T)$ and for all $u \in \mathscr{D}(T^*)$. We then say that T^* is the *adjoint* of T, and that T is the adjoint of T^*. Note that since $Tx \in U$ and $T^*u \in X$, equation (3.10) is consistent with the facts that $u \in U$ and $x \in X$.

A finite dimensional example of (3.10) was given in (1.75). The adjoint linear operators T and T^* there represent rectangular matrices which are the transposes of each other, such that $\mathscr{D}(T)=\mathscr{R}(T^*)=X=\mathbb{R}^n$ and $\mathscr{D}(T^*)=\mathscr{R}(T)=U=\mathbb{R}^m$.

In infinite dimensional cases where T and T^* contain differential operators it is common but not universal to find that $\mathscr{D}(T)\subset\mathscr{R}(T^*)\subseteq X$ and $\mathscr{D}(T^*)\subset\mathscr{R}(T)\subseteq U$, as Fig. 3.2 indicates. The examples in §(iii) below illustrate this. On the other hand, when T and T^* are integral operators we may have $\mathscr{R}(T^*)\subseteq\mathscr{D}(T)\subseteq X$ and $\mathscr{R}(T)\subseteq\mathscr{D}(T^*)\subseteq U$ instead. In either case $\mathscr{D}(T)$ is often a linear subspace (rather than merely a subset) of X, and $\mathscr{D}(T^*)$ is often a subspace of U.

(iii) Examples of adjoint operators

In this chapter so far we have been assembling some ingredients of an infinite dimensional theory. It has been logical to introduce a pair of inner product spaces first, and then to explain how a pair of adjoint operators acts between them.

When faced with a specific context however, especially if it is a boundary value problem, we shall find that it is a natural practice to write down a statement of adjointness first, and then to infer from that the implied definitions of adjoint operators and of inner product spaces.

Accordingly we now give a list of some of the more common statements of adjointness in their familiar notation. Custom has assigned a different *name* to almost every one of these different statements of adjointness, and we give these names to help the reader feel more comfortable. However, this custom has tended to obscure the emphasis which is due to adjointness as the pivotal idea in the structure of each theory at this stage.

For each entry in the list we show how to construct the adjoint operators and inner product spaces.

Fig. 3.2. Representation of some adjoint operators including differentiation.

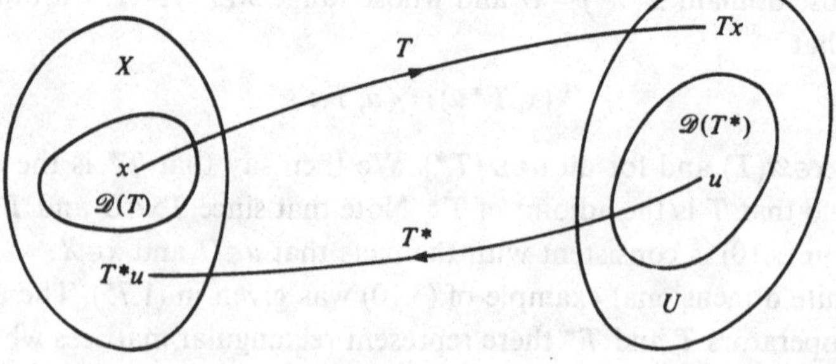

1. Transposition of a matrix

$$\mathbf{x}^{\mathrm{T}}T^{\mathrm{T}}\mathbf{u}=\mathbf{u}^{\mathrm{T}}T\mathbf{x} \quad \forall \mathbf{x} \text{ and } \mathbf{u}. \tag{1.67 bis}$$

This is the finite dimensional case already referred to, and previously rewritten as (1.75) in anticipation of (3.10). Here $\mathbf{x}$ and $\mathbf{u}$ are $n \times 1$ and $m \times 1$ column matrices respectively, T is $m \times n$ and T^{T} is its $n \times m$ transpose. Thus T^* is T^{T}.

2. Integration by parts

$$-\int_0^1 x(t)u'(t)\,\mathrm{d}t - x(0)u(0) = \int_0^1 u(t)x'(t)\,\mathrm{d}t - u(1)x(1). \tag{3.11}$$

Here $x(t)$ and $u(t)$ are any two real scalar piecewise C^1 functions of t on the same fixed interval of the real line, chosen here to be $0 \leqslant t \leqslant 1$ for simplicity. Any other single interval is treated similarly. The isolated finite jumps permitted in the first derivatives of $x(t)$ and $u(t)$ do not contribute any extra term to (3.11).

We have arbitrarily chosen one particular way of distributing terms on the two sides of (3.11), among the several possible ways. This choice suggests that we use linear spaces having composite elements of type (3.1) in terms of both $u(t)$ and $x(t)$, but assign to them the different inner products

$$(\tilde{x}, \tilde{y}) = \int_0^1 x(t)y(t)\,\mathrm{d}t + x(0)y(0), \tag{3.12}$$

$$\langle \tilde{u}, \tilde{v} \rangle = \int_0^1 u(t)v(t)\,\mathrm{d}t + u(1)v(1), \tag{3.13}$$

in conjunction with the definitions

$$\tilde{x} = \begin{bmatrix} x(t) \\ x(0) \end{bmatrix}, \quad T^*\tilde{u} = \begin{bmatrix} -u'(t) \\ -u(0) \end{bmatrix}, \tag{3.14}$$

$$\tilde{u} = \begin{bmatrix} u(t) \\ u(1) \end{bmatrix}, \quad T\tilde{x} = \begin{bmatrix} x'(t) \\ -x(1) \end{bmatrix}. \tag{3.15}$$

The first row in each of these four elements is associated with $0 \leqslant t \leqslant 1$, the second rows of $\tilde{x}$ and $T^*\tilde{u}$ are associated with $t = 0$, and the second rows of $\tilde{u}$ and $T\tilde{x}$ with $t = 1$.

The effect of the operator T on the composite $\tilde{x}$ is thus to differentiate the first entry, and replace the second entry $x(0)$ by $-x(1)$; the effect of T^* is similar. The linearity of these operators is clear. No end restrictions on $x(t)$ and $u(t)$ are imposed at this stage; they will enter later when some ordinary differential equation problem is posed. We are able to use smaller matrices than (3.1) in this particular case.

We can now see that $\mathscr{D}(T)$ is the inner product space of composite elements $(3.14)_1$ which are defined in terms of piecewise C^1 functions $x(t)$, and has inner product (3.12); and that $\mathscr{R}(T)$ belongs to the inner product space of composite elements $(3.15)_1$ which are defined in terms of piecewise continuous functions and have inner product (3.13). Similarly, $\mathscr{D}(T^*)$ is the space of composite elements $\tilde{u}$ defined in terms of piecewise C^1 functions $u(t)$ and having inner product (3.13), while $\mathscr{R}(T^*)$ belongs to the space of composite elements $\tilde{x}$ defined in terms of piecewise continuous functions and having inner product (3.12). Thus $\mathscr{D}(T) \subset \mathscr{R}(T^*)$ and $\mathscr{D}(T^*) \subset \mathscr{R}(T)$ as foreshadowed above, and (3.11) says that

$$(\tilde{x}, T^*\tilde{u}) = \langle \tilde{u}, T\tilde{x} \rangle \quad \forall \tilde{x} \in \mathscr{D}(T) \quad \text{and} \quad \forall \tilde{u} \in \mathscr{D}(T^*).$$

The terms in (3.11) may be distributed differently, for example as

$$-\int_0^1 xu' \, \mathrm{d}t - \mu x(0)u(0) + (1-\lambda)x(1)u(1)$$
$$= \int_0^1 ux' \, \mathrm{d}t + (1-\mu)u(0)x(0) - \lambda u(1)x(1) \tag{3.16}$$

with any fixed λ and μ. Other choices of spaces and operators then become appropriate (see Exercises 3.1.1, 3.1.2). When given positive coefficients different from unity also appear in an integration by parts formula, these can be used as weighting functions in appropriate inner products (Exercise 3.1.3). In any of these alternative choices of operators, the first entries in $T\tilde{x}$ and $T^*\tilde{u}$ (i.e. leaving aside the specifically end point terms) contain what may be called *formal adjoints*, which we shall denote by D and D^* respectively. For example $Dx = x'(t)$ and $D^*u = -u'(t)$ in (3.14) and (3.15) illustrate that the differential operators

$$D = \frac{\mathrm{d}}{\mathrm{d}t} \quad \text{and} \quad D^* = -\frac{\mathrm{d}}{\mathrm{d}t} \text{ are formal adjoints} \tag{3.17}$$

of each other, as illustrated by Hamilton's equations in mechanics (see §2.7(iii)). In general such formal adjoints are distinct from the composite true adjoints as we have defined them, and then we write $D \subset T$ and $D^* \subset T^*$.

When homogeneous end restrictions can be anticipated, spaces may be defined so that formal adjoints do coincide with true adjoints. For example (cf. Exercise 3.1.3), if $\rho(t) > 0$ is a given C^1 function, and if $u(t)$ is any piecewise C^1 function such that $u(0) = u(1) = 0$,

$$\int_0^1 x(t)(\rho(t)u(t))' \, \mathrm{d}t = -\int_0^1 \rho(t)u(t)x'(t) \, \mathrm{d}t, \tag{3.18}$$

where $x(t)$ is any piecewise C^1 function not subject to any end restrictions at this stage. Then (3.18) can be written $(x, T^*u) = \langle u, Tx \rangle$ with

$$T^*u = (\rho(t)u(t))', \quad Tx = -x'(t),$$
$$(x, y) = \int_0^1 x(t)y(t)\mathrm{d}t, \quad \langle u, v \rangle = \int_0^1 \rho(t)u(t)v(t)\mathrm{d}t. \tag{3.19}$$

Here, then, $\mathscr{D}(T^*)$ is simply the space of piecewise C^1 functions $u(t)$ satisfying $u(0) = u(1) = 0$ with the weighted $\langle .,. \rangle$, and $\mathscr{D}(T)$ is the space of piecewise C^1 functions $x(t)$ unrestricted at the ends and with the stated $(.,.)$. We do not require composite elements like those in (3.14) and (3.15). Again $\mathscr{D}(T) \subset \mathscr{R}(T^*)$ and $\mathscr{D}(T^*) \subset \mathscr{R}(T)$, because differentiation reduces smoothness. When $\rho(t) \equiv 1$ we also have $\mathscr{D}(T^*) \subset \mathscr{D}(T)$, because fewer functions satisfy the end conditions than do not.

When isolated discontinuities are present not only in derivatives, but also in the functions $x(t)$ and $u(t)$ themselves, the integration by parts formula (3.11) is augmented by extra terms which can be entered in enlarged composite adjoint operators T and T^* (see Exercise 3.1.4).

3. Green's/Gauss'/divergence theorem

$$-\int \phi \nabla \cdot \mathbf{u} \, \mathrm{d}V + \int \phi \mathbf{n} \cdot \mathbf{u} \, \mathrm{d}\Sigma_2 = \int \mathbf{u} \cdot \nabla \phi \, \mathrm{d}V - \int \mathbf{u} \cdot \mathbf{n} \phi \, \mathrm{d}\Sigma_1. \tag{3.20}$$

Here the scalar ϕ and vector $\mathbf{u}$ are piecewise C^1 functions, in the following sense, of position over a fixed two or three dimensional region V with surface $\Sigma = \Sigma_1 + \Sigma_2$ whose unit outward normal is $\mathbf{n}$. The vector gradient operator is denoted by ∇. We allow isolated lines or surfaces within V, when V is two or three dimensional respectively, across which there may be discontinuities in the gradients of ϕ and/or $\mathbf{u}$. Such jumps will not contribute extra terms to (3.20) (see Exercise 3.1.6).

This formula suggests two different spaces of composite elements (3.5) and (3.7), with inner products (3.6) and (3.8) such that (3.20) can be written $(\tilde{\phi}, T^*\tilde{\mathbf{u}}) = \langle \tilde{\mathbf{u}}, T\tilde{\phi} \rangle$ in terms of composite adjoint operators such that

$$T^*\tilde{\mathbf{u}} = \begin{bmatrix} -\nabla \cdot \mathbf{u} \\ 0 \\ \mathbf{n} \cdot \mathbf{u} \end{bmatrix}, \quad T\tilde{\phi} = \begin{bmatrix} \nabla \phi \\ -\mathbf{n}\phi \\ 0 \end{bmatrix} \text{ over } \begin{bmatrix} V \\ \Sigma_1 \\ \Sigma_2 \end{bmatrix}. \tag{3.21}$$

Here, then, $\mathscr{D}(T^*)$ is the space of composite elements (3.7) constructed from piecewise C^1 vector fields and with inner product (3.8); and $\mathscr{D}(T)$ is the space of composite elements (3.5) constructed from piecewise C^1 scalar fields and with inner product (3.6). From (3.21) we see that $\mathscr{D}(T) \subset \mathscr{R}(T^*)$ and $\mathscr{D}(T^*) \subset \mathscr{R}(T)$, the ranges being larger than the domains because they are constructed from fields which need only be piecewise continuous rather than piecewise C^1. The zeros can be deleted from (3.21) (cf. (3.14) and (3.15)) if we also delete the Σ_1 terms from (3.5) and (3.6), and the Σ_2 terms from (3.7) and (3.8). Such abbreviation is often a matter of convenience only, but for some purposes it can become important to delete the zeros, as §4.5(ii) shows.

All the variants detailed above for integration by parts can be repeated here. The reader may find it helpful, for example, to specify explicitly the effects of rearranging (3.20), of the adequacy of briefer inner products, of anticipating homogeneous boundary conditions, and of discontinuities in the scalar and vector fields themselves.

For (3.21), and for most of its variants, we can say that

$$D = \text{grad} \quad \text{and} \quad D^* = -\text{div are formal adjoints} \qquad (3.22)$$

only, but not true adjoints in general. Exceptionally they may be true adjoints when homogeneous boundary conditions justify it in the way indicated after (3.18) above. Other authors do not make our distinction between formal and true adjoints, because they do not use the composite space elements deployed here, and so the reader needs to be aware that their terminology is different from ours. For example, Arthurs (1980) calls collective boundary terms such as those in (3.11) and (3.20) the 'conjunct' of the two variables, and he treats such terms separately from the differential operators themselves.

4. A vector identity

$$\int \mathbf{v}\cdot\nabla\wedge\mathbf{u}\,dV - \int \mathbf{v}\cdot(\mathbf{n}\wedge\mathbf{u})\,d\Sigma = \int \mathbf{u}\cdot\nabla\wedge\mathbf{v}\,dV. \qquad (3.23)$$

Here the vectors $\mathbf{u}$ and $\mathbf{v}$ are piecewise C^1 functions of position, and we have chosen to write all the boundary terms on one side of the equation (cf. Exercise 3.1.1). The formula (3.23) may be written $(\tilde{\mathbf{v}}, T^*\mathbf{u}) = \langle \mathbf{u}, T\tilde{\mathbf{v}}\rangle$ in terms of adjoint operators such that

$$T^*\mathbf{u} = \begin{bmatrix} \nabla\wedge\mathbf{u} \\ -\mathbf{n}\wedge\mathbf{u} \end{bmatrix} \quad \text{over} \quad \begin{bmatrix} V \\ \Sigma \end{bmatrix}, \quad T\tilde{\mathbf{v}} = \nabla\wedge\mathbf{v}. \qquad (3.24)$$

These induce a choice of two different inner product spaces with elements $\mathbf{u}$ and $\tilde{\mathbf{v}}$ which the reader may find it helpful to construct following the ideas described above. We may add to (3.17) and (3.22) the statement that

$$D = \text{curl} = D^* \text{ is formally self-adjoint.} \qquad (3.25)$$

5. Virtual work transformation

$$\int \sigma_{ij}\frac{\partial u_j}{\partial x_i}dV - \int \sigma_{ij}n_i u_j\,d\Sigma_2 = -\int u_j\frac{\partial \sigma_{ij}}{\partial x_i}dV + \int u_j n_i \sigma_{ij}\,d\Sigma_1. \qquad (3.26)$$

The suffixes here refer to cartesian coordinates x_i $(i = 1, 2, 3)$, and the summation convention is used on repeated suffixes. The cartesian components σ_{ij} of any piecewise C^1 second order tensor field, not necessarily symmetric, and u_j of any piecewise C^1 first order tensor (i.e. vector) field appear in (3.26), and n_i is the ith component of the unit outward normal vector on $\Sigma = \Sigma_1 + \Sigma_2$.

We may write (3.26) as $(\tilde{\sigma}, T^*\tilde{\mathbf{u}}) = \langle \tilde{\mathbf{u}}, T\tilde{\sigma} \rangle$ in terms of composite adjoint linear operators such that

$$T^*\tilde{\mathbf{u}} = \begin{bmatrix} \dfrac{\partial u_j}{\partial x_i} \\ 0 \\ -n_i u_j \end{bmatrix}, \quad T\tilde{\sigma} = \begin{bmatrix} -\dfrac{\partial \sigma_{ij}}{\partial x_i} \\ n_i \sigma_{ij} \\ 0 \end{bmatrix} \text{ over } \begin{bmatrix} V \\ \Sigma_1 \\ \Sigma_2 \end{bmatrix} \qquad (3.27)$$

and appropriate inner product spaces, whose elements are indicated in §3.2(ii). We shall exploit these and associated variants fully in §5.5. The zeros in (3.27) may help book keeping, but they can be suppressed as we indicated for (3.21).

No particular interpretations for the tensor σ_{ij} and the vector u_j are essential at this stage, but a common example in continuum mechanics has σ_{ij} as a stress tensor and u_j as a displacement or as a velocity vector. The common name 'virtual work transformation' is of course prompted by this example, but (3.26) is not the virtual work principle itself. The latter is an assertion (for example (5.70)) which implies, for example, local equations of equilibrium; whereas the adjointness statement (3.26) is really no more than an elaborated version of integration by parts.

A generalization of the result in Exercise 3.1.5, for a tensor of order k in n dimensional space with any k and n, is stated by Ericksen (1960, p. 816) and ascribed to Poincaré. From this follows not only Green's theorem, but also the so-called Stokes' theorem (really due to Kelvin). Equation (3.27) illustrates, for $k = 2$, the property that the negative divergence of a tensor of order k and the gradient of a tensor of order $k - 1$ are formal adjoints of each other.

6. Integral operators

Interchange of the order of integration

$$\int_a^b x(t) \int_a^b k[t, s] u(s) \, ds \, dt = \int_a^b u(s) \int_a^b k[t, s] x(t) \, dt \, ds \qquad (3.28)$$

is valid when the double integral of the product of the three functions $k[t, s], x(t), u(s)$ exists over the given square $a \leqslant s \leqslant b, a \leqslant t \leqslant b$. The simplest case is when the product is a continuous function, and the result is sometimes called Fubini's theorem. A proof is given by Apostol (1974, Theorem 14.6). A generalization sufficient for our purpose allows the product to be a step function (Apostol, *op. cit.*, Theorem 15.2). We regard (3.28) as an example of a less common type of adjointness $\langle x, T^*u\rangle = \langle u, Tx\rangle$ in which a single inner product space is adequate. Hence we induce definitions of integral operators T and T^* such that

$$Tx = \int_a^b k[t,s]x(t)\mathrm{d}t, \quad T^*u = \int_a^b k[t,s]u(s)\mathrm{d}s. \tag{3.29}$$

Here $\mathscr{D}(T) = \mathscr{D}(T^*)$ is deemed to be a single space of functions $x(t)$ and $u(s)$ with inner product

$$\langle x, u\rangle = \int_a^b x(t)u(t)\mathrm{d}t. \tag{3.30}$$

Another example of adjointness, different from (3.28), is the formula

$$\int_a^b x(t)\int_a^t k[t,s]u(s)\mathrm{d}s\,\mathrm{d}t = \int_a^b u(s)\int_s^b k[t,s]x(t)\mathrm{d}t\,\mathrm{d}s, \tag{3.31}$$

in which integration is carried out over only that half of the square lying on one side of the diagonal joining opposite corners a, a to b, b. Then

$$Tx = \int_s^b k[t,s]x(t)\mathrm{d}t, \quad T^*u = \int_a^t k[t,s]u(s)\mathrm{d}s. \tag{3.32}$$

Again $\mathscr{D}(T) = \mathscr{D}(T^*)$ using (3.30).

In such examples $\mathscr{R}(T) = \mathscr{R}(T^*) \subset \mathscr{D}(T) = \mathscr{D}(T^*)$ in general, because of the smoothing effect of integration. In §4.2(v)(b), however, we give an illustration of (3.31) in which Dirac delta functions are admitted in $\mathscr{D}(T)$ or $\mathscr{D}(T^*)$ but not both, so that Tx or T^*u may be Heaviside step functions. In other cases $\mathscr{R}(T) = \mathscr{R}(T^*) = \mathscr{D}(T) = \mathscr{D}(T^*)$, for example when $x(t)$, $u(s)$ and $k[t,s]$ are all required to be square integrable, by the Schwarz inequality.

7. *Sequences and integrable functions*

$$\int_a^b x(t)\sum_{i=1}^N u_i y_i(t)\mathrm{d}t = \sum_{i=1}^N u_i \int_a^b y_i(t)x(t)\mathrm{d}t. \tag{3.33}$$

Here the u_i are N real scalars, where N may be finite or infinite, and integrable $x(t)$ and the continuous $y_i(t)$ are $1+N$ given functions on the given interval $a \leqslant t \leqslant b$. If N is infinite we require $\sum_{i=1}^{\infty} u_i y_i(t)$ to be uniformly convergent in order that the interchange of integration and summation in (3.33) be valid. We treat (3.33) as an illustration of adjointness $(x, T^*u) = \langle u, Tx\rangle$ in terms of integrable functions $x = x(t)$ and sequences

$u = \{u_1, u_2, \ldots, u_N\}$ with inner products

$$(x, y) = \int_a^b x(t)y(t)\mathrm{d}t, \quad \langle u, v \rangle = \sum_{i=1}^{N} u_i v_i. \tag{3.34}$$

This viewpoint induces the operators T and T^* such that

$$Tx = \{(Tx)_1, (Tx)_2, \ldots, (Tx)_N\} \quad \text{where} \quad (Tx)_i = \int_a^b y_i(t)x(t)\mathrm{d}t,$$

$$T^*u = \sum_{i=1}^{N} y_i(t)u_i. \tag{3.35}$$

Here $\mathcal{R}(T^*) = \mathcal{D}(T)$ is the space of integrable functions with the inner product $(.,.)$ of (3.34), and $\mathcal{R}(T) = \mathcal{D}(T^*)$ is the space of sequences with $\langle .,. \rangle$ (and with the convergence property when N is infinite). Thus T maps an integrable function into a whole sequence of scalars, and T^* maps a sequence into an integrable function.

A simple illustration is provided when N is infinite and the $y_i(t)$ form a complete orthonormal set of functions, for example $(2/\pi)^{\frac{1}{2}} \sin(it)$ on $0 \leqslant t \leqslant \pi$. Then T^* maps a sequence of generalized Fourier coefficients onto the corresponding function, and T maps a function onto its complete sequence of Fourier coefficients.

(iv) The operator T^*T

Operators which have this structure occur in integral equations, and second (and higher even) order differential operators often have the decomposition D^*D in terms of formal adjoints. We make some brief remarks about T^*T here at the level of generality in §(ii), for true adjoints T and T^* and a pair of inner product spaces. We also list some illustrations which we shall need in §3.12.

For the operator T^*T to exist it is necessary that $\mathcal{R}(T) \subseteq \mathcal{D}(T^*)$. The smoothing effect of integral operators may enforce this. However, §(iii) above indicates that there are many instances where $\mathcal{D}(T^*) \subset \mathcal{R}(T)$, broadly because when differential operators are present more smoothness is required in the domain than in the range, and also because of the presence of boundary terms in the composite elements of the spaces. These contrasting inequalities suggest that $\mathcal{R}(T) = \mathcal{D}(T^*)$ may be sufficient to facilitate a worthwhile class of illustrations of T^*T. It is also sometimes possible to redefine $\mathcal{D}(T)$ more narrowly, for example by imposing extra smoothness, as in Examples 4 and 5 below, so that we do achieve $\mathcal{R}(T) \subseteq \mathcal{D}(T^*)$. In either case it does not follow without parallel requirements that TT^* is also available, and when both do exist they need not be the same.

Consider the following illustrations.

1. Any rectangular matrix T and its transpose T^{T} allow the square matrices $T^{\mathrm{T}}T$ and TT^{T} to exist and be different. We illustrated the emergence of $T^{\mathrm{T}}T$ in (1.135), with $\mathscr{R}(T)=\mathscr{D}(T^{\mathrm{T}})$ as the space of $m\times 1$ column matrices.
2. The integral operators (3.29) permit both T^*T and TT^* to be defined and rearranged in the form

$$\begin{aligned} T^*Tx &= \int_a^b p[t,\tau]x(\tau)\mathrm{d}\tau, \quad p[t,\tau]=\int_a^b k[t,s]k[\tau,s]\mathrm{d}s,\\ TT^*u &= \int_a^b q[t,\tau]u(\tau)\mathrm{d}\tau, \quad q[t,\tau]=\int_a^b k[s,t]k[s,\tau]\mathrm{d}s, \end{aligned} \tag{3.36}$$

supposing that the order of integration over the square can be reversed. If $k[t,s]=k[s,t]$, then T^*T and TT^* are the same.

The operators T and T^* in (3.32) are each associated with Volterra integral equations. Both T^*T and TT^* can be formed from them, as we indicate in Exercise 3.1.8, and each is associated with a Fredholm integral equation.

3. The adjoint operators (3.35) permit the two definitions

$$\begin{aligned} T^*Tx &= \sum_{i=1}^{N} y_i(t)\int_a^b y_i(\tau)x(\tau)\mathrm{d}\tau,\\ TT^*u &= \{w_1,w_2,\ldots,w_N\}, \quad w_i=\int_a^b y_i(t)\sum_{j=1}^{N} u_j y_j(t)\mathrm{d}t. \end{aligned} \tag{3.37}$$

4. We saw that $\mathscr{D}(T^*)\subset\mathscr{R}(T)$ and $\mathscr{D}(T)\subset\mathscr{R}(T^*)$ for the operators in (3.14) and (3.15), so that neither T^*T nor TT^* can exist for those definitions as they stand. However, these latter operators can be constructed by assuming extra smoothness provided also, in some cases, certain homogeneous end conditions are adopted. For example, let $\mathscr{D}(T^*)$ be as specified after (3.15), but redefine $\mathscr{D}(T)$ there to be the space of $\tilde{x}$ constructed from piecewise C^2 functions $x(t)$ such that $x'(1)=-x(1)$. Then adjointness is satisfied *a fortiori*, and in addition $\mathscr{R}(T)=\mathscr{D}(T^*)$. Hence

$$T^*T\tilde{x}=\begin{bmatrix}-x''(t)\\ -x'(0)\end{bmatrix}. \tag{3.38}$$

We still have $\mathscr{D}(T)\subset\mathscr{R}(T^*)$ with the $\mathscr{D}(T)$ just specified, because $\mathscr{R}(T^*)$ is constructed from continuous functions without end restrictions. Therefore TT^* does not exist with this choice of operators, even though T^*T does.

Another example of $\mathscr{R}(T) = \mathscr{D}(T^*)$ is provided by (3.19) if we choose $\mathscr{D}(T)$ there to be the space of piecewise C^2 functions $x(t)$ such that $x'(0) = x'(1) = 0$ with the stated (.,.). Then,

$$T^*Tx = -(\rho(t)x'(t))'$$

in terms of true adjoints, not merely formal adjoints.

5. The operators (3.21) have the property $\mathscr{D}(T^*) \subset \mathscr{R}(T)$ as they stand, but they too can be amended to achieve $\mathscr{R}(T) \subseteq \mathscr{D}(T^*)$ and hence to allow the construction of T^*T. When the formal adjoints are given by (3.22), we can expect T^*T to contain the Laplacian $-\nabla^2 = D^*D$, but the other details will vary from problem to problem, depending on the boundary conditions. For example, Green's theorem (3.20) applies *a fortiori* in the form

$$-\int \phi \nabla \cdot \mathbf{u} \, \mathrm{d}V + \int \phi \mathbf{n} \cdot \mathbf{u} \, \mathrm{d}\Sigma_2 = \int \mathbf{u} \cdot \nabla \phi \, \mathrm{d}V \tag{3.39}$$

for any C^2 scalar function ϕ such that $\phi = 0$ on Σ_1, and for any C^1 vector function $\mathbf{u}$. This can be expressed as $(\tilde{\phi}, T^*\mathbf{u}) = \langle \mathbf{u}, T\tilde{\phi} \rangle \; \forall \tilde{\phi} \in \mathscr{D}(T)$ and $\forall \; u \in \mathscr{D}(T^*)$ where $\tilde{\phi} = [\phi(V) \; \phi(\Sigma_2)]^{\mathrm{T}}$, $(\phi, \psi) = \int \phi\psi \, \mathrm{d}V + \int \phi\psi \, \mathrm{d}\Sigma_2$, $\langle \mathbf{u}, \mathbf{v} \rangle = \int \mathbf{u} \cdot \mathbf{v} \, \mathrm{d}V$,

$$T^*\mathbf{u} = \begin{bmatrix} -\nabla \cdot \mathbf{u} \\ \mathbf{n} \cdot \mathbf{u} \end{bmatrix} \text{ over } \begin{bmatrix} V \\ \Sigma_2 \end{bmatrix}, \quad T\tilde{\phi} = \nabla \phi, \tag{3.40}$$

and $\mathscr{D}(T)$ and $\mathscr{D}(T^*)$ are constructed from the functions ϕ and $\mathbf{u}$ just specified. With this choice we see that $\mathscr{R}(T)$ is the space of C^1 vector functions which are gradients of a scalar having zero values on Σ_1, so that $\mathscr{R}(T) \subset \mathscr{D}(T^*)$. Then

$$T^*T\tilde{\phi} = \begin{bmatrix} -\nabla^2 \phi \\ \mathbf{n} \cdot \nabla \phi \end{bmatrix} \text{ over } \begin{bmatrix} V \\ \Sigma_2 \end{bmatrix}. \tag{3.41}$$

This is an example in which a homogeneous boundary condition ($\phi = 0$ on Σ_1) is built into the definition of one of the spaces ($\mathscr{D}(T)$).

We conclude this section with a theorem for a general pair of inner product spaces, which requires two preliminary definitions. A linear operator A will be called *self-adjoint* with respect to (.,.) if

$$(x, Ay) = (y, Ax) \quad \forall x, y \in \mathscr{D}(A). \tag{3.42}$$

This is not the same as formal self-adjointness, which we introduced via an illustration in (3.25).

A linear operator P will be called *nonnegative* with respect to (.,.) if

$$(x, Px) \geqslant 0 \quad \forall x \in \mathscr{D}(P). \tag{3.43}$$

There may be $x \neq 0$ for which $(x, Px) = 0$. P is called *positive* if $(x, Px) > 0$ for all $x \neq 0$.

For the case of matrices in Chapters 1 and 2 we used the conventional extra adjective 'definite', but this is really superfluous and we omit it in the case of infinite dimensional operators. Thus nonnegative means the same as positive semidefinite.

Theorem 3.1
When $\mathscr{R}(T) \subseteq \mathscr{D}(T^*)$ the operator T^*T is self-adjoint and nonnegative.

Proof
For any x and y in $\mathscr{D}(T)$, Tx and Ty belong to $\mathscr{D}(T^*)$, and (3.10) then implies

$$(x, T^*Ty) = \langle Tx, Ty \rangle = \langle Ty, Tx \rangle = (y, T^*Tx).$$

Here we have also used the symmetry of the inner product $\langle .,. \rangle$. It follows from (3.42) that T^*T is self-adjoint with respect to $(.,.)$.

Choosing $y = x$ we have $(x, T^*Tx) = \langle Tx, Tx \rangle \geqslant 0$ by the positivity of $\langle .,. \rangle$, so that T^*T is nonnegative with respect to $\langle .,. \rangle$ by (3.43). If $Tx = 0$ it does not follow that $x = 0$ without other hypotheses, so we do not conclude that T^*T is positive. □

It is easy to verify that (3.38) and (3.41) illustrate this Theorem. For example, since $\phi = \psi = 0$ on Σ_1, (3.40) gives

$$(\tilde{\psi}, T^*T\tilde{\phi}) = -\int \psi \nabla^2 \phi \, \mathrm{d}V + \int \psi \mathbf{n} \cdot \nabla \phi \, \mathrm{d}\Sigma_2 = \int \nabla \psi \cdot \nabla \phi \, \mathrm{d}V$$
$$= (\tilde{\phi}, T^*T\tilde{\psi}).$$

The question of a converse to Theorem 3.1 arises: under what conditions does a given nonnegative operator P admit the decomposition

$$P = T^*T$$

in terms of a pair of true adjoints T and T^*? Even when the decomposition exists it might not be unique, and we gave illustrations for a square matrix P in Exercise 1.8.1. We also illustrated in (1.139) how, for the purpose of calculating upper and lower bounds, it may be enough to know that T and T^* exist for a given P, without needing to know what they actually are. There is a substantial literature on T^*T operators in infinite dimensional spaces, going back at least to von Neumann in 1932, and we shall later cite a notable paper by Fujita (1955) on linear problems associated with such operators. We return to the topic in §3.12.

Exercises 3.1

1. When the integration by parts formula (3.11) is rewritten as

$$-\int_0^1 x(t)u'(t)\,dt + x(1)u(1) - x(0)u(0) = \int_0^1 u(t)x'(t)\,dt,$$

give a precise description of the two inner product spaces needed to represent it as an example of the adjointness formula (3.10) with the operators

$$T^*u = \begin{bmatrix} -u'(t) \\ -u(0) \\ u(1) \end{bmatrix}, \quad T\tilde{x} = x'(t).$$

Specify the domains and ranges of these operators in terms of the two spaces.

2. Repeat Exercise 1 for the alternative

$$-\int_0^1 x(t)u'(t)\,dt + x(1)u(1) = \int_0^1 u(t)x'(t)\,dt + u(0)x(0)$$

with operators

$$T^*\tilde{u} = \begin{bmatrix} -u'(t) \\ u(1) \end{bmatrix}, \quad T\tilde{x} = \begin{bmatrix} x'(t) \\ x(0) \end{bmatrix}.$$

3. When $\rho(t) > 0$ is a given C^1 function on $0 \leqslant t \leqslant 1$ and integration by parts is written as

$$-\int_0^1 x(t)(\rho(t)u(t))'\,dt + \rho(1)x(1)u(1)$$
$$= \int_0^1 \rho(t)u(t)x'(t)\,dt + \rho(0)u(0)x(0),$$

repeat Exercise 1 in the case when the operators are chosen as

$$T^*\tilde{u} = \begin{bmatrix} -(\rho(t)u(t))' \\ u(1) \end{bmatrix}, \quad T\tilde{x} = \begin{bmatrix} x'(t) \\ x(0) \end{bmatrix}.$$

Verify in particular that, for this choice, the appropriate inner products are (see §3.7(iii) for an application)

$$(\tilde{x}, \tilde{y}) = \int_0^1 x(t)y(t)\,dt + \rho(1)x(1)y(1),$$

$$\langle \tilde{u}, \tilde{v} \rangle = \int_0^1 \rho(t)u(t)v(t)\,dt + \rho(0)u(0)v(0).$$

Give another choice of operators and spaces in which $\rho(t)$ is disposed differently.

4. If there is an isolated location $t = a$ within $0 < t < 1$ where the otherwise piecewise C^1 $x(t)$ has a discontinuity $[\![x]\!]_a$ (= right value minus left value), and a different isolated location $t = b$ where the otherwise piecewise C^1 $u(t)$ has a

discontinuity $[\![u]\!]_b$, show that (3.11) may be replaced by

$$-\int_0^1 x(t)u'(t)\,\mathrm{d}t - x(0)u(0) - x(b)[\![u]\!]_b$$
$$= \int_0^1 u(t)x'(t)\,\mathrm{d}t - u(1)x(1) + u(a)[\![x]\!]_a.$$

Hence define inner product spaces which allow adjointness to be written as $(\tilde{x}, T^*\tilde{u}) = \langle \tilde{u}, T\tilde{x}\rangle$ in terms of composite adjoint operators

$$T\tilde{x} = \begin{bmatrix} x'(t) \\ -x(1) \\ [\![x]\!]_a \end{bmatrix}, \quad T^*\tilde{u} = \begin{bmatrix} -u'(t) \\ -u(0) \\ -[\![u]\!]_b \end{bmatrix}.$$

5. Prove that if ψ is a C^1 scalar function of position in a two or three dimensional region V with boundary Σ, then

$$\int \frac{\partial \psi}{\partial x_i}\mathrm{d}V = \int n_i \psi\,\mathrm{d}\Sigma,$$

where x_i is the ith cartesian coordinate and n_i the corresponding component of the outward unit normal vector on Σ. This is the basic formula upon which versions of Green's theorem are built, and proofs are found in advanced calculus books (e.g. Friedman 1971, pp. 339–40). Replace ψ by the product ϕu_j, where ϕ is a scalar and u_j is the jth component of a vector, and write out the result. Hence obtain (3.20) in the case of C^1 functions ϕ and $\mathbf{u}$.

6. Show that if an internal surface S with unit normal $\mathbf{m}$ is imagined to divide V into two parts, it will contribute to (3.20) an integral over S of

$$\phi_1 \mathbf{m}\cdot\mathbf{u}_1 - \phi_2 \mathbf{m}\cdot\mathbf{u}_2,$$

where ϕ_1, ϕ_2 and $\mathbf{u}_1, \mathbf{u}_2$ are the values of ϕ and $\mathbf{u}$ on the two sides of Σ. Hence show that, for this contribution to be zero, it is sufficient that ϕ and $\mathbf{m}\cdot\mathbf{u}$ each be continuous across S, and that this allows not only jumps in the gradients of ϕ and $\mathbf{u}$, but also in the tangential component of $\mathbf{u}$ itself.

7. Specify precise operators and inner product spaces which allow

$$\int_a^b x(t)\int_0^\infty q[t,s]u(s)\,\mathrm{d}s\,\mathrm{d}t = \int_0^\infty u(s)\int_a^b q[t,s]x(t)\,\mathrm{d}t\,\mathrm{d}s$$

to be written as an example of $(x, T^*u) = \langle u, Tx\rangle$ for all $x \in \mathscr{D}(T)$ and all $u \in \mathscr{D}(T^*)$.

Show that T^*T and TT^* exist and are the counterparts of (3.37).

8. Examples of the operators in (3.29) are

$$Tx = \frac{2}{\pi^{\frac{1}{2}}}\int_0^1 \frac{\sin ts}{(1-t^2)^{\frac{1}{4}}}x(t)\,\mathrm{d}t, \quad T^*u = \frac{2}{\pi^{\frac{1}{2}}}\int_0^1 \frac{\sin st}{(1-t^2)^{\frac{1}{4}}}u(s)\,\mathrm{d}s.$$

Show that the Bessel function, $J_0(t) = (2/\pi)\int_0^1((\cos t\sigma)/(1-\sigma^2)^{\frac{1}{2}})\,\mathrm{d}\sigma$, of the first kind of order zero, has the property

$$\int_0^1 [J_0(t-\tau)-J_0(t+\tau)]x(\tau)\,d\tau = T^*Tx.$$

9. Prove that the operators in (3.32) lead to

$$T^*Tx = \int_a^b \hat{p}[t,\tau]x(\tau)\,d\tau, \quad \hat{p}[t,\tau] = \int_a^{\min[t,\tau]} k[t,s]k[\tau,s]\,ds,$$

$$TT^*u = \int_a^b \hat{q}[t,\tau]u(\tau)\,d\tau, \quad \hat{q}[t,\tau] = \int_{\max[t,\tau]}^b k[s,t]k[s,\tau]\,ds,$$

by integrating in each case over a trapezium which can be regarded as a rectangle joined to a 45° triangle.

Prove directly that these T^*T and TT^* are each self-adjoint and positive operators in the sense of Theorem 3.1.

Write out (3.32) and illustrate the above results for

$$k[t,s] = (t-s)^{-\frac{1}{2}} \quad \text{with} \quad a=0, \quad b=1$$

and show that, for $t \neq \tau$,

$$\hat{p}[t,\tau] = \ln\left|\frac{t^{\frac{1}{2}}+\tau^{\frac{1}{2}}}{t^{\frac{1}{2}}-\tau^{\frac{1}{2}}}\right|, \quad \hat{q}[t,\tau] = \hat{p}[1-t,1-\tau].$$

10. With the inner products required in Exercise 1, show that if $\mathscr{D}(T)$ is the space of $\tilde{x} = [x(t)\ x(0)\ x(1)]^{\mathrm{T}}$ constructed from piecewise C^2 functions $x(t)$, then $\mathscr{R}(T) = \mathscr{D}(T^*)$ and

$$T^*T\tilde{x} = \begin{bmatrix} -x''(t) \\ -x'(0) \\ x'(1) \end{bmatrix}.$$

Hence $\mathscr{D}(T^*)$ is the space of piecewise C^1 functions $u(t)$.

3.4. Gradients of functionals

(i) Derivatives of general operators

Our main concern from §(ii) below will be with the gradient of a functional. This is a particular illustration of the derivative of a general operator. We indicate the latter concept briefly here, in order to set the scene. Further details can be found in Luenberger (1969), Liusternik and Sobolev (1961), or Kantorovich and Akilov (1964).

Let $M(x)$ be an operator with domain $\mathscr{D}(M)$ (cf. §3.3(i)). Let μ be a real scalar. If the limit

$$\lim_{\mu\to 0} \frac{1}{\mu}[M(x+\mu h) - M(x)] \tag{3.44}$$

exists at some fixed $x \in \mathscr{D}(M)$ for every $h \in \mathscr{D}(M)$, it may be called the *Gateaux differential* of M at x. It is plainly necessary that $x + \mu h \in \mathscr{D}(M)$ for all sufficiently small μ. Whenever $\mathscr{D}(M)$ is a linear space we are assured that $x + \mu h \in \mathscr{D}(M)$ for any μ. The Gateaux differential depends on both x and h. It is found frequently to depend *linearly* on h at each x, and this linearity is illustrated by the following three examples.

1. A linear operator Tx has Gateaux differential

$$\lim_{\mu \to 0} \frac{1}{\mu}[T(x + \mu h) - Tx] = Th, \tag{3.45}$$

which is thus the same for every x.

2. A C^1 function $\phi[x]:\mathbb{R}^n \to \mathbb{R}$ mapping the n-tuples x of n real scalars x_i (cf. (1.44)) onto the real line has

$$\lim_{\mu \to 0} \frac{1}{\mu}[\phi[x + \mu h] - \phi[x]] = \frac{\partial \phi}{\partial x_i} h_i. \tag{3.46}$$

3. Let $f[t, x]:\mathbb{R}^2 \to \mathbb{R}$ be a function of two scalars such that $\partial f/\partial x$ is continuous with respect to both t and x (it would be sufficient that $f[t, x]$ be a C^2 function). Let $x(t)$ and $h(t)$ belong to the space of continuous scalar functions on $0 \leqslant t \leqslant 1$. Then the functional

$$F[x] = \int_0^1 f[t, x(t)]\,\mathrm{d}t$$

has

$$\lim_{\mu \to 0} \frac{1}{\mu}[F[x + \mu h] - F[x]] = \int_0^1 \frac{\partial f}{\partial x} h\,\mathrm{d}t \tag{3.47}$$

since, under the stated assumptions, we can interchange the order of integration with respect to t and of differentiation with respect to x.

The linearity in h which appears in such examples allows us to identify a *linear operator*, with domain $\mathscr{D}(M)$, which we call the *Gateaux derivative* of $M(x)$ and write as $\mathrm{D}M/\mathrm{d}x$. It depends on x but not on h. Thus when it exists we can write

$$\lim_{\mu \to 0} \frac{1}{\mu}[M(x + \mu h) - M(x)] = \frac{\mathrm{D}M}{\mathrm{d}x} h. \tag{3.48}$$

If $M[x, u]$ is an operator acting on two different spaces, the notation $\mathrm{D}M/\partial x$ and $\mathrm{D}M/\partial u$ denotes partial Gateaux derivatives, holding u and then x fixed respectively. We use D to distinguish the Gateaux derivative of a general operator from the gradient of a functional defined next, where D is replaced by a prime or by d or ∂. The meaning of this D is quite distinct from the same letter used for differential operators in (3.17).

(ii) Gradients of functionals

A functional is an operator whose range is part or all of the real line. Our concern here is with single valued functionals whose domain belongs to an inner product space.

Let $E[x]$ denote such a functional whose domain $\mathscr{D}(E)$ belongs to a linear space X with inner product $(.,.)$. When the Gateaux derivative $\mathrm{D}E/\mathrm{d}x$ exists it is a linear operator whose domain is $\mathscr{D}(E) \subseteq X$ by (3.48), and it will often be isomorphic to an element $E'[x]$ (say) of X such that

$$\frac{\mathrm{D}E}{\mathrm{d}x} h = (E'[x], h) \tag{3.49}$$

for every $h \in \mathscr{D}(E)$. We call any such $E'[x]$ the gradient of $E[x]$.

We shall meet many examples (e.g. (3.55)) in which $\mathscr{D}(E)$ is a subspace of X, contained strictly within X so that $\mathscr{D}(E) \subset X$. The use of adjointness (3.10) in such examples leads to a clear choice of gradient $E'[x]$ which belongs to X but not to $\mathscr{D}(E)$. For this reason results like the Riesz representation theorem, which gives a unique representation theorem for a bounded linear functional on a Hilbert space (e.g. see §5.3 of Luenberger, 1969), are too strong for our purpose.

The *gradient of a functional* $E[x]$ whose domain $\mathscr{D}(E)$ belongs to an inner product space X may be defined directly as an element $E'[x]$ of X which is such that

$$\lim_{\mu \to 0} \frac{1}{\mu}[E[x + \mu h] - E[x]] = (E'[x], h) \tag{3.50}$$

for every $h \in \mathscr{D}(E)$.

In concrete cases it is often convenient to determine gradients by using the classical notation of the calculus of variations. That is, as alternative notation we write

$$\mu h = \delta x, \quad E[x + \delta x] - E[x] - \mathrm{o}(\mu) = \delta E$$

for sufficiently small μ, and identify $E'[x]$ from the coefficient of δx in

$$\delta E = (E'[x], \delta x) \tag{3.51}$$

whose limiting form after division by μ is (3.50).

The following examples illustrate the gradient of a functional, the first two being rephrased versions of examples 2 and 3 respectively in §(i) above.

1. $\delta\phi = (\phi'[x], \delta x)$ with $(x, y) = x_i y_i$ so that the gradient of $\phi[x]$ is the n-tuple of partial derivatives

$$\phi'[x] = \left\{\frac{\partial\phi}{\partial x_1}, \frac{\partial\phi}{\partial x_2}, \ldots, \frac{\partial\phi}{\partial x_n}\right\}.$$

2. $\delta F = (F'[x], \delta x)$ with $(x, y) = \int_0^1 x(t)y(t)\,dt$ so that
$$F'[x] = \frac{\partial f}{\partial x}[t, x(t)] \tag{3.52}$$
from (3.47). This belongs to the stated inner product space of functions continuous in t because $x(t)$ does so, and because $\partial f/\partial x$ is continuous with respect to both t and x as we assumed above.
3. Let real matrices be given as follows: A is $n \times n$, T is $m \times n$, $\mathbf{a}$ is $n \times 1$ and $\mathbf{u}$ is $m \times 1$. Using the simple inner product $\mathbf{x}^{\mathrm{T}}\mathbf{y}$ for the space of $n \times 1$ matrices $\mathbf{x}, \mathbf{y}, \ldots$, the function
$$E[\mathbf{x}] = -\tfrac{1}{2}\mathbf{x}^{\mathrm{T}}A\mathbf{x} + \mathbf{u}^{\mathrm{T}}T\mathbf{x} + \mathbf{a}^{\mathrm{T}}\mathbf{x}$$
is easily seen to have gradient (cf. (1.66))
$$E'[\mathbf{x}] = -\tfrac{1}{2}(A + A^{\mathrm{T}})\mathbf{x} + T^{\mathrm{T}}\mathbf{u} + \mathbf{a}.$$
Special cases can be illustrated by taking any two of A, T and $\mathbf{a}$ to be zero in turn. We could have chosen A to be symmetric, without loss of generality in $E[\mathbf{x}]$, so that $A^{\mathrm{T}} = A$.
4. Let a be any given element of a space X with inner product $(.,.)$. The linear functional $E[x] = (a, x)$ has domain $\mathscr{D}(E) = X$ and Gateaux differential (a, h) for every $h \in X$, by (3.45) or from the left side of (3.50). 'Comparing coefficients' with the desired representation on the right side of (3.50), we choose the gradient
$$E'[x] = a \quad \text{if} \quad E[x] = (a, h). \tag{3.53}$$
5. Let X and U be any two inner product spaces for which the definition of adjointness in (3.10) can be constructed. Let x be an arbitrary member of $\mathscr{D}(T)$ and let b be a fixed member of $\mathscr{D}(T^*)$. The linear functional $E[x] = \langle b, Tx \rangle$ has $\mathscr{D}(E) = \mathscr{D}(T)$ and Gateaux differential $\langle b, Th \rangle = (h, T^*b)$ for every $h \in \mathscr{D}(T)$. Comparing with the right side of (3.50) we are led to choose the gradient
$$E'[x] = T^*b \quad \text{if} \quad E[x] = \langle b, Tx \rangle. \tag{3.54}$$
Notice, however, that $T^*b \in \mathscr{R}(T^*) \subseteq X$, and recall how frequently we have seen that $\mathscr{D}(T) \subset \mathscr{R}(T^*)$, in which case the chosen gradient need not belong to $\mathscr{D}(T)$.
6. Let $x(t)$ and $u(t)$ be piecewise C^1 functions on $0 \leqslant t \leqslant 1$, with no end restrictions but with $u(t)$ supposed given. The functional
$$F[x(t)] = \int_0^1 u(t)x'(t)\,dt$$
then depends only on $x(t)$ and has Gateaux differential

$$\int_0^1 u(t)h'(t)\,\mathrm{d}t = -\int_0^1 u'(t)h(t)\,\mathrm{d}t + u(1)h(1) - u(0)h(0)$$

by (3.11) in the form of Exercise 3.1.1. With the T and T^* given there, and

$$\langle u, v\rangle = \int_0^1 u(t)v(t)\,\mathrm{d}t, \quad (\tilde{x}, \tilde{y}) = \int_0^1 x(t)y(t) + x(0)y(0) + x(1)y(1)$$

chosen as inner products, we have $F[\tilde{x}] = \langle u, T\tilde{x}\rangle$. Hence $\mathscr{D}(F) = \mathscr{D}(T)$ is the space of matrices $\tilde{x} = [x(t)\ x(0)\ x(1)]^{\mathrm{T}}$ constructed from piecewise C^1 functions $x(t)$, and having the stated $(\tilde{x}, \tilde{y})$ as inner product. The Gateaux differential can also be written $\langle u, T\tilde{h}\rangle = (\tilde{h}, T^*u)$. This has the representation $(F'[\tilde{x}], \tilde{h})$ if, following (3.54), we choose as gradient of $F[\tilde{x}]$ the composite element

$$F'[\tilde{x}] = \begin{bmatrix} -u'(t) \\ -u(0) \\ u(1) \end{bmatrix} = T^*u. \tag{3.55}$$

This is independent of $\tilde{x}$, and belongs to $\mathscr{R}(T^*)$, which is the space of matrices $\tilde{x}$ constructed from piecewise continuous functions, so that $\mathscr{D}(T) \subset \mathscr{R}(T^*)$. This illustrates the point made in the previous example that the gradient of $F[\tilde{x}]$ need not belong to the domain of $F[\tilde{x}]$.

7. With the generality of example 5 adjointness shows that

$$E'[x] = T^*Tx \quad \text{if} \quad E[x] = \tfrac{1}{2}\langle Tx, Tx\rangle.$$

We can choose

$$E'[x] = x/\|x\| \quad \text{if} \quad E[x] = \|x\|$$

for $x \neq 0$, because

$$\frac{\mathrm{d}}{\mathrm{d}\mu}(x + \mu h, x + \mu h)^{\frac{1}{2}}\Big|_{\mu=0} = (x, h)(x, x)^{-\frac{1}{2}}.$$

(iii) Partial gradients of functionals

Let $L[x, u]$ be a single valued functional whose domain is the product of two domains $\mathscr{D}_x(L) \subseteq X$ and $\mathscr{D}_u(L) \subseteq U$, where X and U are spaces with inner products $(.,.)$ and $\langle .,. \rangle$ respectively in the generality of §3.3(ii). The *partial gradients* of $L[x, u]$ at x, u are defined as the elements $\partial L/\partial x$ and $\partial L/\partial u$ of X and U, respectively, which are such that

$$\begin{aligned} \lim_{\mu \to 0} \frac{1}{\mu}[L[x + \mu h, u] - L[x, u]] &= \left(\frac{\partial L}{\partial x}, h\right), \\ \lim_{\lambda \to 0} \frac{1}{\lambda}[L[x, u + \lambda k] - L[x, u]] &= \left\langle \frac{\partial L}{\partial u}, k\right\rangle, \end{aligned} \tag{3.56}$$

for every $h \in \mathscr{D}_x(L)$ and for every $k \in \mathscr{D}_u(L)$, where μ and λ are real scalars (cf. Fig. 1.5).

If we use the alternative notation $\mu h = \delta x$, $\lambda k = \delta u$ and

$$L[x+\delta x, u+\delta u] - L[x,u] - R_2[\mu,\lambda] = \delta L$$

for sufficiently small μ and λ, where $R_2[\mu,\lambda]$ is the remainder after two terms in Maclaurin's expansion in μ and λ, we can identify $\partial L/\partial x$ and $\partial L/\partial u$ from the coefficients of δx and δu in

$$\delta L = \left(\frac{\partial L}{\partial x}, \delta x\right) + \left\langle \frac{\partial L}{\partial u}, \delta u \right\rangle. \tag{3.57}$$

We shall say that $L[x,u]$ is a C^1 *functional* at any point of its domain where the partial gradients $\partial L/\partial x$ and $\partial L/\partial u$ defined in (3.56) can be chosen unambiguously.

For the next theorem let T and T^* be given linear operators which are adjoint in the sense of (3.10), and let A and B be given linear operators which are self-adjoint in the sense of (3.42) and of

$$\langle u, Bv \rangle = \langle v, Bu \rangle \quad \forall u, v \in \mathscr{D}(B) \tag{3.58}$$

respectively (cf. the finite dimensional precursors in (1.74)). Suppose $a \in X$ and $b \in U$ are given elements.

Theorem 3.2

The quadratic functional

$$L[x,u] = -\tfrac{1}{2}(x, Ax) + \langle u, Tx \rangle + \tfrac{1}{2}\langle u, Bu \rangle + (a,x) + \langle b, u \rangle \tag{3.59}$$

has partial gradients

$$\begin{aligned} \frac{\partial L}{\partial x} &= -Ax + T^*u + a, \\ \frac{\partial L}{\partial u} &= Tx + Bu + b. \end{aligned} \tag{3.60}$$

Proof

The Gateaux differential on the left of $(3.56)_1$ is

$$\begin{aligned} &-\tfrac{1}{2}(h, Ax) - \tfrac{1}{2}(x, Ah) + \langle u, Th \rangle + (a,h) \\ &= (-Ax + T^*u + a, h) \end{aligned}$$

for every $h \in \mathscr{D}_x(L) = \mathscr{D}(T) \cap \mathscr{D}(A)$, after using (3.42) and (3.10), so that $(3.60)_1$ follows from the representation on the right of $(3.56)_1$.

Using the alternative δ notation with $\delta x = 0$ and $\delta u = \lambda k$ in place of $(3.56)_2$, we have, with fixed x,

$$\begin{aligned} \delta L &= \langle \delta u, Tx \rangle + \tfrac{1}{2}\langle \delta u, Bu \rangle + \tfrac{1}{2}\langle u, B\delta u \rangle + \langle b, \delta u \rangle \\ &= \langle Tx + Bu + b, \delta u \rangle \end{aligned}$$

for every $k \in \mathscr{D}_u(L) = \mathscr{D}(T^*) \cap \mathscr{D}(B)$, after using (3.58) and (3.10), so that $(3.60)_2$ follows from (3.57). □

The important bilinear particular case of Theorem 3.2 is that

$$L[x,u] = \langle u, Tx \rangle = (x, T^*u) \Rightarrow \frac{\partial L}{\partial x} = T^*u, \quad \frac{\partial L}{\partial u} = Tx. \tag{3.61}$$

Theorem 3.3

Let $H[x,u]$ be any single valued C^1 functional such that

$$L[x,u] = (x, T^*u) - H[x,u]. \tag{3.62}$$

Then

$$\frac{\partial L}{\partial x} = T^*u - \frac{\partial H}{\partial x}, \quad \frac{\partial L}{\partial u} = Tx - \frac{\partial H}{\partial u}. \tag{3.63}$$

The proof is an immediate consequence of, for example, (3.57) with (3.10).

This theorem is a generalization of (2.113) to the infinite dimensional case. The functional $H[x,u]$ need not be quadratic. It is so in the particular case (3.59), however, for then

$$H[x,u] = \tfrac{1}{2}(x, Ax) - \tfrac{1}{2}\langle u, Bu \rangle - (a, x) - \langle b, u \rangle. \tag{3.64}$$

In many of our applications to boundary value problems $H[x,u]$ will be a sum of volume and surface integrals whose integrands do not contain derivatives of the field variables, such derivatives appearing only in T and T^*.

The notation (3.62) is chosen to stimulate a comparison of the structure of (3.63) with Hamilton's equations in mechanics. Their generalization here is

$$T^*u = \frac{\partial H}{\partial x} \quad (\alpha), \qquad Tx = \frac{\partial H}{\partial u} \quad (\beta). \tag{3.65}$$

In the finite dimensional version (2.119) it was more convenient to replace the present functional $H[x,u]$ by the function $X[x,u]$, because of the symmetric notation required in Fig. 2.18, which we shall not seek to generalize fully to infinite dimensions here. The comparison of (3.65) with Hamilton's equations is only partly appropriate, however, as we indicated in §2.7(iii); that is, if $H[x,u]$ is thought of as a Hamiltonian functional, the analogue of the classical Lagrangian is not $L[x,u]$ as it stands in (3.62), but instead is the functional of u only obtained by solving the equation $T^*u = \partial H/\partial x$ to express x in terms of T^*u and u, and inserting this expression for x into $L[x,u]$. We shall not need to press for a full comparison along these lines, but instead we shall handle examples of type (3.62) on their own merits. Equations (3.65) have the structure introduced by Noble (1964), as we remarked before.

3.5. Saddle functional

(i) Saddle quantity

Let $L[x,u]$ be a single valued functional which is here assumed to be C^1 at all points of its domain. As in §3.4(iii), this domain belongs to the product of a general pair of infinite dimensional inner product spaces X and U, so that $x \in X$ and $u \in U$, with respective inner products $(.,.)$ and $\langle .,. \rangle$.

Guided by the developments in Chapter 1, especially Theorem 1.2(a), and by the remarks about saddle functions of $n+m$ variables in §1.3(iii), we generalize (1.57) to define the saddle quantity

$$S = L_+ - L_- - \left(x_+ - x_-, \frac{\partial L}{\partial x}\bigg|_+ \right) - \left\langle u_+ - u_-, \frac{\partial L}{\partial u}\bigg|_- \right\rangle \tag{3.66}$$

for any pair of points x_+, u_+ and x_-, u_- in the infinite dimensional domain of $L[x,u]$. The requisite shape of this domain is discussed in §(ii)(a) below.

The single valuedness of $L[x,u]$ is required in order to be sure that expressions like S do have an unambiguous value for a chosen pair of points in the domain. Nevertheless, as was foreshadowed in Chapter 2, it will be possible to discuss certain questions even when $L[x,u]$ is a multi-valued functional, provided we can identify it as consisting of single valued parts which can be handled in turn. It is desirable to be aware of this possibility, which occurs in (2.106) and (5.35) for example, so that one can identify when it becomes a live issue. For simplicity, however, a first reading of what follows can be based on the initial presumption of a single valued $L[x,u]$.

(ii) Definition of a saddle functional

We define $L[x,u]$ to be a strict or weak *saddle functional concave in x and convex in u* over some particular subdomain if

$$S > 0 \quad \text{or} \quad S \geqslant 0, \tag{3.67}$$

respectively, for every pair of distinct points in that subdomain.

Our choice of definition raises the following questions.

(a) Can the definition be justified in terms of equivalence to other possible definitions? Theorem 1.2(a) can be extended not only to the $n+m$ dimensional case, as we indicated in §1.3(iii), but also to the present infinite dimensional case. This requires the definition of a rectangular domain, as the product of a convex domain in X and a convex domain in U. These are defined exactly as before, and the extension of Theorem 1.2(a) is then immediate. Therefore, over a rectangular domain, (3.67) is equivalent to a saddle functional definition which has the structure of (1.22) or of Exercise 1.5.5, and which does not use gradients.

Over nonrectangular domains we shall still express our hypotheses in terms of (3.66), recognizing that in some such cases it may still be appropriate to call $L[x,u]$ a saddle functional and in other cases not (see the remarks about wedge-shaped and line subdomains after (f) in §1.3(iii)).

(b) Can such hypotheses be readily verified? It would be possible at this point to introduce more analytical machinery defining, for example, the second order gradients of functionals. Under natural restrictions the generalizations of $-\partial^2 L/\partial x^2$ and $\partial^2 L/\partial u^2$ in Chapter 1 would emerge as self-adjoint linear operators, of which A and B in (3.42), (3.58) and (3.59) are examples. Then we may expect to generalize Theorem 1.2(b) and Theorem 1.5, at least over rectangular domains, to give sufficient conditions for (3.67) to hold in terms of the positivity of such operators.

We prefer, however, to emphasize that there are many circumstances when (3.67), or weaker hypotheses on S like (1.58b–f), can be verified *directly*. Notice particularly that the bilinear term $\langle u, Tx\rangle = (x, T^*u)$ contributes zero to S, and is thus always a saddle functional, trivially concave in x and convex in u, like (1.1) when referred to the asymptotes of the hyperbolic sections. In the case (3.62) it then follows that $L[x,u]$ is concave in x and convex in u if and only if $H[x,u]$ is convex in x and concave in u. We remarked that $H[x,u]$ is often a sum of integrals whose integrands do not contain derivatives. Such integrands are functions which are often easily seen to be convex in x and concave in u. We give a tangible example of this with quadratic integrand in (3.83), where A and B in (3.87) do not contain derivatives. Later we shall give examples of nonquadratic integrands which can also be shown to be convex–concave without difficulty.

(c) Can the subdomain mentioned in the definition (3.67) be described explicitly? This is so if it is taken to be rectangular, in which case we have the generalization of hypothesis (a) in the list after (1.58). But this hypothesis is too strong for some purposes, and instead we shall often require the generalization of one of (b)–(f) after (1.58). Then, in the case of (c) there for example, the question is whether the α-constraints can be solved to describe $\{x_\alpha, u_\alpha\}$ explicitly. This issue presents genuine difficulties in applications. Such constraints may consist of a system of inequalities or equations or both. One may need to explore, for example, the extent to which an infinite dimensional Legendre transformation is available, or whether Lagrange multiplier techniques are applicable.

In the next section we do not underestimate these various difficulties, but we separate them from the problem we address there of proving general upper and lower bounds under what may be minimal hypotheses.

3.6. Upper and lower bounds

(i) Introduction

Let $L[x,u]$ again be a given C^1 functional on the two general infinite dimensional spaces with inner products $(.,.)$ and $\langle .,. \rangle$. In terms of the partial gradients $\partial L/\partial x$ and $\partial L/\partial u$ of this $L[x,u]$, we consider the infinite dimensional versions of the four sets of governing conditions (1.50) and (1.53)–(1.55). We repeat these because it is convenient to have them before us in this chapter. For example, (1.50) is

$$\begin{aligned} &\frac{\partial L}{\partial x} \leqslant 0, \qquad && u \geqslant 0, \quad (\alpha) \\ &x \geqslant 0, && \frac{\partial L}{\partial u} \geqslant 0, \quad (\beta) \\ &\left(x, \frac{\partial L}{\partial x}\right) = 0, && \left\langle u, \frac{\partial L}{\partial u}\right\rangle = 0, \end{aligned} \qquad (1.50\ bis)$$

and (1.53) is

$$\frac{\partial L}{\partial x} = 0 \quad (\alpha), \qquad \frac{\partial L}{\partial u} = 0 \quad (\beta). \qquad (1.53\ bis)$$

Also (1.54) is (1.53α) with the second trio of (1.50), and (1.55) is (1.53β) with the first trio of (1.50). We also have in view the basic equivalence problem posed for problems I–IV in §1.3(ii). The notation is unchanged in going from the finite dimensional case there to the infinite dimensional case here, as it was designed to be. Almost all of §1.3(ii) can be repeated verbatim, except that the C^1 function $L[x,u]$ there is replaced by the C^1 functional $L[x,u]$, now that we have defined the meaning of this in §3.4(iii).

In order that inequalities such as $\partial L/\partial x \leqslant 0$ and $u \geqslant 0$ be meaningful in the infinite dimensional case, it must be supposed that the elements of the inner product spaces are built up from real functions and real numbers in some composite way, for example as in the matrix (3.1), so that the respective inequalities can be applied to each of these functions and numbers.

The infinite dimensional version of (1.53) will still be called a stationary value problem, because any solution of $\partial L/\partial x = \partial L/\partial u = 0$ will imply $\delta L = 0$ in (3.57). From the viewpoint of the classical calculus of variations, however, the converse result is only justified when the so-called fundamental lemma of the calculus of variations, illustrated in Theorem 3.8, is available. This lemma will allow us to conclude that when $\delta L = 0$ for all sufficiently small δx and δu, it is implied that $\partial L/\partial x = \partial L/\partial u = 0$ (cf. (1.14) and the remarks there). We do not need to invoke this fundamental lemma for the upper and lower bounds established next.

In what follows it will be understood without further comment that, when (1.50) and (1.53)–(1.55) are mentioned, it is their infinite dimensional generalizations which are under consideration. Problems of this type and generality, in which we do not need to assume that the generating $L[x,u]$ has the particular form (3.62), were considered by Sewell (1973*a*, *b*). These papers also worked out the comprehensive examples of (3.62) from solid mechanics described in Chapter 5.

(ii) A central theorem in infinite dimensions

We can now give one of our main results, albeit in unconventional form which exploits the preparation which we have developed.

Theorem 3.4

The whole of Theorems 1.6 *and* 1.7, *and their proofs, can be repeated verbatim using the definition of S in* (3.66).

This is an immediate consequence of the way in which we have structured the notation, first of all for matrices rewritten as in §1.4, and then for the infinite dimensional inner product spaces under consideration now.

In particular the expressions for the bounds in (1.78)–(1.80) can be repeated verbatim, and are

$$L_\alpha - \left(x_\alpha, \frac{\partial L}{\partial x}\bigg|_\alpha\right) \geqslant L_0 \quad \text{and} \quad L_0 \geqslant L_\beta - \left\langle u_\beta, \frac{\partial L}{\partial u}\bigg|_\beta\right\rangle. \tag{3.68}$$

These are valid before any attempt is made to determine the sets of points $\{x_\alpha, u_\alpha\}$ and $\{x_\beta, u_\beta\}$ which actually solve the α and β constraints respectively.

Theorem 3.5

If $L[x,u] = (x, T^*u) - H[x,u]$ *the expressions for the bounds may be written in terms of H via*

$$\begin{aligned} L - \left(x, \frac{\partial L}{\partial x}\right) &= \left(x, \frac{\partial H}{\partial x}\right) - H, \quad (\alpha) \\ L - \left\langle u, \frac{\partial L}{\partial u}\right\rangle &= \left\langle u, \frac{\partial H}{\partial u}\right\rangle - H. \quad (\beta) \end{aligned} \tag{3.69}$$

The proof is immediate using adjointness. Notice that these expressions do not contain T and T^*. In boundary value problems this means that, since derivatives are often absent from $H[x,u]$, they are then also absent from these initial expressions for the bounds, until the solutions of the α and β conditions are inserted into (3.69α) and (3.69β) respectively.

The two bounds evidently have the same *values* as the two infinite dimensional Legendre transforms of the functional $H[x,u]$, and of $-L[x,u]$ also, as we can infer from §2.7, when such transforms exist.

(iii) Quadratic generating functional

Theorem 3.6

The quadratic functional (3.59) *generates the following formulae, whichever one of the four sets of governing conditions applies.*

$$L-\left(x,\frac{\partial L}{\partial x}\right)=\tfrac{1}{2}(x,Ax)+\tfrac{1}{2}\langle u,Bu\rangle+\langle b,u\rangle. \tag{3.70}$$

$$L-\left\langle u,\frac{\partial L}{\partial u}\right\rangle=-\tfrac{1}{2}(x,Ax)-\tfrac{1}{2}\langle u,Bu\rangle+(a,x). \tag{3.71}$$

$$L_0=\tfrac{1}{2}(a,x_0)+\tfrac{1}{2}\langle b,u_0\rangle. \tag{3.72}$$

$$S=\tfrac{1}{2}(x_+-x_-,A\overline{x_+-x_-})+\tfrac{1}{2}\langle u_+-u_-,B\overline{u_+-u_-}\rangle. \tag{3.73}$$

L_0 is the quantity bounded by upper and/or lower bounds which have the values of (3.70) *and* (3.71) *respectively. Upper and lower bounds apply simultaneously if A and B are both nonnegative operators.*

If both parts x_0 and u_0 of a solution of the governing conditions exist, they are unique if A and B are positive operators.

Proof

Formulae (3.70), (3.71) and (3.73) are immediate, from (3.59) and (3.66), without using any governing conditions. We can also rearrange (3.59) in the form

$$2L=\left(x,\frac{\partial L}{\partial x}\right)+\left\langle u,\frac{\partial L}{\partial u}\right\rangle+(a,x)+\langle b,u\rangle. \tag{3.74}$$

Then (3.72) follows using the orthogonality conditions alone, and regardless of whether the solution x_0, u_0 of all the governing conditions is unique or not. The result applies whichever one of the four sets of governing conditions (1.50) and (1.53)–(1.55) is under consideration.

The fact that L_0 is bounded by (3.70) and/or (3.71) follows from (3.68) via Theorems 1.6 and 3.4. If A and B are both nonnegative, then by (3.73) $S\geqslant 0$ for every pair of points in the domain of $L[x,u]$, so the hypothesis of Theorem 1.6(c) holds *a fortiori*.

If A and B are both positive, then $S>0$ and uniqueness of x_0 and u_0 follows from Theorems 1.7 and 3.4. □

As soon as a quadratic generating functional $L[x,u]$ has been identified, the quantity L_0 bounded by either or both of the bounds is a linear functional of the solution(s) which can be read off from $L[x, u]$ simply by taking half of the linear terms therein, as (3.72) shows.

Further progress in evaluating the bounds requires that we address the problem of solving the α and/or β constraints to determine $\{x_\alpha, u_\alpha\}$ and/or $\{x_\beta, u_\beta\}$, and insert such results into the expressions for the bounds in (3.68).

With a general functional $L[x, u]$, perhaps nonquadratic, we may seek to mimic the developments given for the finite dimensional case in §2.7(i), (iii) and (iv) to obtain expressions for the bounds like those summarized in Table 2.3. In the quadratic case this will lead to specializations of (3.70) and (3.71). Then in (1.53) or (1.54), for example, we need to solve

$$Ax_\alpha = T^*u_\alpha + a \tag{3.75}$$

for x_α, subject to $u_\alpha \geqslant 0$ in (1.54), in order to express $\{x_\alpha, u_\alpha\}$ explicitly. And in (1.53) or (1.55) we need to solve

$$Bu_\beta = -(Tx_\beta + b) \tag{3.76}$$

for u_β, subject to $x_\beta \geqslant 0$ in (1.55), in order to express $\{x_\beta, u_\beta\}$ explicitly. If inverse operators A^{-1} and B^{-1} can be defined, it can be seen that the bounds in (3.68) with (3.70) and (3.71) can be sought as

$$L_\alpha = J[u_\alpha] \quad \text{and} \quad L_\beta = K[x_\beta], \tag{3.77}$$

respectively, in terms of a functional $J[u_\alpha]$ of u_α alone, and of another functional $K[x_\beta]$ of x_β alone.

For a problem to possess a solution, it is often necessary that consistency conditions on the data in it be satisfied. This question arises when we try to solve the α or the β constraints separately. For example, when we try to find $\{x_\alpha, u_\alpha\}$ from (3.75) or from $T^*u_\alpha = \partial H/\partial x_\alpha$, we are effectively treating u_α as a part of the data and seeking to express x_α in terms of u_α by constructing a Legendre transformation between functionals $H[x, u]$ and $Y[T^*u, u]$, generalizing (2.115). Certain *consistency conditions* on this data u_α emerge, as necessary conditions for solutions $x_\alpha(u_\alpha)$ of such α constraints to exist. Similar remarks apply to the data x_β involved in solving $Tx_\beta = \partial H/\partial u_\beta$ as $u_\beta(x_\beta)$. We give some general discussion of such consistency in the linear case in §3.8, but before that we explore a concrete example in §3.7.

3.7. An ordinary differential equation

(i) Introduction

The main object here is to give our first detailed illustration of how an infinite dimensional problem, given *ab initio*, may be converted into a form generated from a saddle functional, from which formulae for upper and lower bounds characterizing the problem can be derived.

Consider the problem of finding a real scalar C^2 function $x(t)$ of the real scalar t on $0 \leqslant t \leqslant 1$ which satisfies

$$
\begin{gathered}
-\frac{\mathrm{d}}{\mathrm{d}t}\left(\rho(t)\frac{\mathrm{d}x}{\mathrm{d}t}\right) + r(t)x = s(t), \\
a_1 x + a_2 \frac{\mathrm{d}x}{\mathrm{d}t} = a_3 \quad \text{at} \quad t = 0, \\
b_1 x + b_2 \frac{\mathrm{d}x}{\mathrm{d}t} = b_3 \quad \text{at} \quad t = 1.
\end{gathered}
\tag{3.78}
$$

Here $a_1, a_2, a_3, b_1, b_2, b_3$ are given constants, $\rho(t)$ is a given C^1 function, and $r(t)$ and $s(t)$ are given continuous functions. We also suppose that $\rho(t)$ is positive everywhere in $0 \leqslant t \leqslant 1$, until §(xvi) where we explain how $\rho(t)$ can be zero at isolated locations.

(ii) Intermediate variable

The first step is to convert the second order equation into a pair of first order equations, by introducing a new function $u(t)$ as intermediate variable defined by $u(t) = -\mathrm{d}x/\mathrm{d}t$. The problem becomes

$$
\begin{gathered}
\frac{\mathrm{d}x}{\mathrm{d}t} = -u, \quad -\frac{\mathrm{d}}{\mathrm{d}t}(\rho u) = rx - s, \\
a_1 x(0) - a_2 u(0) = a_3, \quad b_1 x(1) - b_2 u(1) = b_3.
\end{gathered}
\tag{3.79}
$$

From this we are to determine a C^1 $u(t)$ and a C^2 $x(t)$.

(iii) Inner product spaces and adjoint operators

We now write down the integration by parts formula (3.11) and its possible variants, such as those in (3.16) and Exercises 3.1.1–3, to see if they suggest appropriate inner products and true adjoint operators T and T^*.

This choice will depend on which one of various possible cases occur in the boundary conditions. Let us fix attention on the case $a_1 b_2 \neq 0$. Then we can divide by a_1 to isolate $x(0)$ and by b_2 to isolate $u(1)$. In this case the appropriate integration by parts formula is that of Exercise 3.1.3, namely

$$
-\int_0^1 x(t)(\rho(t)u(t))'\,\mathrm{d}t + \rho(1)x(1)u(1) = \int_0^1 \rho(t)u(t)x'(t)\mathrm{d}t + \rho(0)u(0)x(0).
$$

For this to have the form $(\tilde{x}, T^*\tilde{u}) = \langle \tilde{u}, T\tilde{x} \rangle$ of (3.10) we choose the inner products to be

$$\begin{aligned}(\tilde{x}, \tilde{y}) &= \int_0^1 x(t)y(t)\mathrm{d}t + \rho(1)x(1)y(1),\\ \langle \tilde{u}, \tilde{v} \rangle &= \int_0^1 \rho(t)u(t)v(t)\mathrm{d}t + \rho(0)u(0)v(0).\end{aligned} \tag{3.80}$$

This requires that X and U be the inner product spaces whose elements are column matrices

$$\tilde{x} = \begin{bmatrix} x(t) \\ x(1) \end{bmatrix}, \quad \tilde{u} = \begin{bmatrix} u(t) \\ u(0) \end{bmatrix} \tag{3.81}$$

constructed from square integrable functions $x(t)$ and $u(t)$, and using the inner products (3.80). Then the above integration by parts is satisfied if we choose the operators T and T^* defined in Exercise 3.1.3, and their domains $\mathscr{D}(T) \subset X$ and $\mathscr{D}(T^*) \subset U$ to be the subspaces of X and U constructed by using piecewise C^1 functions $x(t)$ and $u(t)$. Less smoothness is not admissible unless the inner products are augmented with terms reflecting discontinuities (cf. Exercise 3.1.4), and there is no substantial advantage in imposing more smoothness at this stage, even though an eventual solution $x(t)$ will be smoother.

Except for recognizing their type, the actual boundary conditions themselves have not yet been used in the choice of the inner products (3.80), space elements (3.81), or operators T and T^* (repeated in (3.82)). To simplify the algebra we now choose $a_1 = b_2 = 1$, and rewrite (3.79) in matrix style as

$$\begin{aligned} T^*\tilde{u} &= \begin{bmatrix} -(\rho(t)u(t))' \\ u(1) \end{bmatrix} = \begin{bmatrix} r(t)x(t) - s(t) \\ b_1 x(1) - b_3 \end{bmatrix}, \quad (\alpha) \\ T\tilde{x} &= \begin{bmatrix} x'(t) \\ x(0) \end{bmatrix} = \begin{bmatrix} -u(t) \\ a_3 + a_2 u(0) \end{bmatrix}. \quad (\beta) \end{aligned} \tag{3.82}$$

The two equations on the left here are definitions of the operators T and T^*.

The alternative change of variable $u(t) = -\rho(t)\mathrm{d}x/\mathrm{d}t$ is prompted by the formal adjointness of $\mathrm{d}/\mathrm{d}t$ and $-\mathrm{d}/\mathrm{d}t$ established in (3.17). The reader may examine what subsequent changes are necessitated by this second choice.

(iv) Hamiltonian functional

Next we ask whether the matrices on the right in (3.82) can be written as the gradients $\partial H/\partial \tilde{x}$ and $\partial H/\partial \tilde{u}$ of a generalized Hamiltonian functional $H[\tilde{x}, \tilde{u}]$. It is necessary for this that the first rows on the right in (3.82α) and (3.82β) be the derivatives of an integrand with respect to $x(t)$ and $u(t)$

respectively, and that the second rows be the partial derivatives of a function of $x(1)$ and $u(0)$. Such partial integration works rather easily because of the absence of t-derivatives from the process, and gives

$$H[\tilde{x},\tilde{u}] = \int_0^1 [\tfrac{1}{2}rx^2 - sx - \tfrac{1}{2}\rho u^2]dt + (\tfrac{1}{2}b_1x(1) - b_3)x(1)\rho(1)$$
$$+ (a_3 + \tfrac{1}{2}a_2u(0))u(0)\rho(0). \qquad (3.83)$$

This is the procedure originally advocated by Noble (1964) and illustrated in terms of Sturm–Liouville equations of the type considered here. This partial integration also works for nonlinear problems (3.65), where $H[\tilde{x},\tilde{u}]$ does not turn out to be quadratic as it is here and in (3.64).

We have omitted the argument t from the functions in the integrand in (3.83), since in this context there should be no danger from using the same symbol x for both the function $x(t)$ and its value, and similarly for u.

(v) Generating functional

We can now subtract (3.83) from either side $(\tilde{x}, T^*\tilde{u}) = \langle\tilde{u}, T\tilde{x}\rangle$ of the integration by parts formula in §(iii), and so write down the generating functional

$$L[\tilde{x},\tilde{u}] = (\tilde{x}, T^*\tilde{u}) - H[\tilde{x},\tilde{u}]$$
$$= \int_0^1 [\rho ux' - \tfrac{1}{2}rx^2 + sx + \tfrac{1}{2}\rho u^2]dt + \rho(0)u(0)x(0)$$
$$-(\tfrac{1}{2}b_1x(1) - b_3)x(1)\rho(1) - (a_3 + \tfrac{1}{2}a_2u(0))u(0)\rho(0). \qquad (3.84)$$

The domain $\mathscr{D}(L)$ of this functional consists of those pairs of elements $\tilde{x}$ and $\tilde{u}$ defined by (3.81) and constructed from piecewise C^1 functions $x(t)$ and $u(t)$ which are otherwise unrestricted, since its definition only requires a valid specification of integration by parts.

The partial gradients of (3.84) may be found either from (3.57), using integration by parts, or by subtracting the two sides of (3.82), to be

$$\frac{\partial L}{\partial \tilde{x}} = \begin{bmatrix} -(\rho u)' - rx + s \\ u(1) - b_1x(1) + b_3 \end{bmatrix}, \quad \frac{\partial L}{\partial \tilde{u}} = \begin{bmatrix} x' + u \\ x(0) - a_3 - a_2u(0) \end{bmatrix}. \qquad (3.85)$$

The functions of t in the first rows of these partial gradients need not be piecewise C^1, but only piecewise continuous, so $\partial L/\partial\tilde{x}$ and $\partial L/\partial\tilde{u}$ need *not* belong to $\mathscr{D}_x(L) = \mathscr{D}(T)$ and $\mathscr{D}_u(L) = \mathscr{D}(T^*)$ respectively.

(vi) Identification of A and B

Since the original problem (3.78) is linear, we can examine the rewritten version (3.82) to see if it can be expressed as

$$T^*\tilde{u} = A\tilde{x} - \tilde{a} \quad (\alpha), \qquad T\tilde{x} = -B\tilde{u} - \tilde{b} \quad (\beta), \tag{3.86}$$

i.e. as an example of (1.53) using the gradients (3.60). Comparison of the right sides of (3.82) and (3.86) allows us to identify the self-adjoint operators A and B and the elements $\tilde{a}$ and $\tilde{b}$ to be such that

$$\begin{aligned} A\tilde{x} &= \begin{bmatrix} r(t)x(t) \\ b_1 x(1) \end{bmatrix}, \quad \tilde{a} = \begin{bmatrix} s(t) \\ b_3 \end{bmatrix}, \\ B\tilde{u} &= \begin{bmatrix} u(t) \\ -a_2 u(0) \end{bmatrix}, \quad \tilde{b} = \begin{bmatrix} 0 \\ -a_3 \end{bmatrix}. \end{aligned} \tag{3.87}$$

No smoothness is required in the domain or range of A and B, when considered by themselves, so we can choose $\mathscr{D}(A) = \mathscr{R}(A) = X \supset \mathscr{D}(T)$ and $\mathscr{D}(B) = \mathscr{R}(B) = U \supset \mathscr{D}(T^*)$. The first rows of $\tilde{a}$ and $\tilde{b}$ refer to $0 \leqslant t \leqslant 1$, and their second rows to $t = 1$ and $t = 0$ respectively. In (3.87) we have determined the coefficients in the quadratic (3.59), where the inner products are here (3.80). This is another way of arriving at the generating functional (3.84), alternative to that of §(iv) but restricted to linear problems.

(vii) Stationary principle

The original problem (3.78) with $a_1 = b_2 = 1$ has been shown to be equivalent to (3.82), and hence to (1.53), i.e. to $\partial L/\partial \tilde{x} = \partial L/\partial \tilde{u} = 0$ using (3.85). Therefore $\delta L = ((\partial L/\partial \tilde{x}), \delta \tilde{x}) + \langle(\partial L/\partial \tilde{u}), \delta \tilde{u}\rangle$ shows that every solution of the original problem makes $\delta L = 0$, and we may call this result a stationary property.

The converse result would be that $\delta L = 0$ *only* if $\partial L/\partial \tilde{x} = \partial L/\partial \tilde{u} = 0$, i.e. only if the original problem is satisfied. In this the $\delta \tilde{x}$ and $\delta \tilde{u}$ are expressed via (3.81) in terms of first variations of piecewise C^1 functions $x(t)$ and $u(t)$ which are otherwise unrestricted. This converse requires the fundamental lemma of the calculus of variations. The stationary property *with* its converse is called a stationary principle, or a free variational principle. We defer the proof of these results until §§(xiv) and (xv).

(viii) Saddle functional $L[\tilde{x}, \tilde{u}]$

It follows from (3.87) that for the piecewise C^1 functions $x(t)$ and $u(t)$ which make up the domain of L, B is positive and A is nonnegative in the sense of (3.43) if, in addition to our original assumption that $\rho(t) > 0$ everywhere, the given data is subject to the extra restrictions

$$r(t) \geqslant 0, \quad b_1 \geqslant 0, \quad a_2 \leqslant 0. \tag{3.88}$$

Then $L[\tilde{x}, \tilde{u}]$ is a weak saddle functional concave in $\tilde{x}$ and convex in $\tilde{u}$.

There are various alternative ways of seeing this. It follows by inspection of (3.59) since

$$(\tilde{x}, A\tilde{x}) = \int_0^1 rx^2\,dt + b_1\rho(1)(x(1))^2,$$

$$\langle \tilde{u}, B\tilde{u} \rangle = \int_0^1 \rho u^2\,dt - a_2\rho(0)(u(0))^2,$$

or because $S \geqslant 0$ from (3.73), for every pair of points in the domain of $L[\tilde{x}, \tilde{u}]$, or because $H[\tilde{x}, \tilde{u}]$ in (3.83) is convex in $\tilde{x}$ and concave in $\tilde{u}$ when (3.88) holds (cf. §3.5(ii)(b)).

(ix) Upper and lower bounds

With (3.88) we are assured by Theorems 1.6(c) and 3.4 that simultaneous upper and lower bounds are available for the problem recast as (3.82). These bounds will have the values of the expressions (3.70) and (3.71). From (3.84), (3.85) and (3.80) they become

$$L - \left(\tilde{x}, \frac{\partial L}{\partial \tilde{x}} \right) = \frac{1}{2}\int_0^1 [rx^2 + \rho u^2]\,dt - a_3 u(0)\rho(0) + \tfrac{1}{2}b_1\rho(1)(x(1))^2 - \tfrac{1}{2}a_2(u(0))^2\rho(0), \tag{3.89}$$

$$L - \left\langle \tilde{u}, \frac{\partial L}{\partial \tilde{u}} \right\rangle = -\frac{1}{2}\int_0^1 [rx^2 + \rho u^2 - 2sx]\,dt + b_3\rho(1)x(1) - \tfrac{1}{2}b_1\rho(1)(x(1))^2 + \tfrac{1}{2}a_2(u(0))^2\rho(0). \tag{3.90}$$

In writing these expressions we have not yet used the constraints. Bounds for (3.82) are obtained by inserting any solution of (3.82α) alone into (3.89), and any solution of (3.82β) alone into (3.90). We carry this process further in §§(xi) and (xii) below.

(x) Associated inequality problems

At this point it is appropriate to notice that the structure of our theory yields exactly the same expressions (3.89) and (3.90) as upper and lower bounds for three other problems, corresponding to (1.50), (1.54) and (1.55), using the same $L[\tilde{x}, \tilde{u}]$ in (3.84) which we have deduced by regarding (3.82) as an example of (1.53).

For example, the infinite dimensional form of (1.55) with (3.84) is equivalent to the problem

$$\begin{aligned} &0 \leqslant (\rho(t)u(t))' + r(t)x(t) - s(t), \quad u(1) \leqslant b_1 x(1) - b_3, \quad (\alpha) \\ &x'(t) = -u(t), \quad x(t) \geqslant 0, \quad x(0) = a_3 + a_2 u(0), \quad (\beta) \\ &\int_0^1 x(t)[-(\rho(t)u(t))' - r(t)x(t) + s(t)]\,dt + \rho(1)x(1)[u(1) - b_1 x(1) + b_3] = 0. \end{aligned} \tag{3.91}$$

An upper bound is obtained with any solution of (3.91α) alone substituted into (3.89). A lower bound is obtained with any solution of (3.91β) alone substituted into (3.90).

It is worth noting that any solution of the whole system (3.91) must in fact have the pointwise orthogonality properties

$$x(t)[-(\rho(t)u(t))' - r(t)x(t) + s(t)] = 0,$$
$$\rho(1)x(1)[u(1) - b_1x(1) + b_3] = 0.$$

The problem (3.91) may have arisen *ab initio* either as it stands, and we give physical examples of this in Exercise 3.2.4 and §3.10(ii) below, or expressed in terms of $x(t)$ alone. For example, if $r = 0$, $\rho = constant$ and $s = \rho w''(t)$, then the distance $x(t) = y(t) - w(t)$ of an unknown $y(t)$ from the given curve $w(t)$ is required by (3.91) to satisfy

$$y'' \leqslant 0, \quad y \geqslant w, \quad (y - w)y'' = 0$$

together with end conditions. Hence $y(t)$ can represent the shape of a string drawn taut over an obstacle of given profile $w(t)$. Contact can take place where the obstacle is concave, and the string is straight where there is separation. The bounds which we obtain in §3.10(ii) for such *obstacle problems* are closely connected with the superficially different method of variational inequalities, as we suggest in §3.11.

(xi) Solution of individual constraints

We return to the problem (3.82) to give a detailed illustration of the evaluation of bounds. We now solve (3.82α) alone and use L_α to denote any corresponding value of (3.89); and we solve (3.82β) alone and use L_β to denote any corresponding value of (3.90). Thus we give expressions for the bounds in the form (3.77).

We carry out these solutions of the individual constraints under less restrictive assumptions than (3.88). Even when the latter are not available to guarantee that L_α and L_β are simultaneous bounds, we can expect from Theorems 1.6(a), (b) and 3.4 that L_α or L_β will provide a single upper or lower bound, respectively, under other assumptions (to be specified in §3.8(iii) and (iv)).

In this and in other concrete examples of the general theory we shall often attach the subscripts α, β and 0 to the variables themselves also, as in the previous general theory where we used $\{x_\alpha, u_\alpha\}$, $\{x_\beta, u_\beta\}$ and $\{x_0, u_0\}$.

Since the domain of (3.84) is built from piecewise C^1 functions $x(t)$ and $u(t)$, the solutions of (3.82α) and (3.82β), considered separately, may be sought as follows.

Consider first (3.82α), taken by itself. Existence of piecewise C^1 solutions $x_\alpha(t)$ and $u_\alpha(t)$ of the differential equation there insists that $u_\alpha(t)$ be actually

C^1, for consistency, since $r(t)$ and $s(t)$ are continuous. If $r(t) \neq 0$ at every point in $0 \leqslant t \leqslant 1$, and if $b_1 \neq 0$, we can rearrange (3.82α) as

$$\begin{bmatrix} x_\alpha(t) \\ x_\alpha(1) \end{bmatrix} = \begin{bmatrix} [s(t) - (\rho(t)u_\alpha(t))']/r(t) \\ (b_3 + u_\alpha(1))/b_1 \end{bmatrix}. \tag{3.92}$$

The continuity of $x_\alpha(t)$ at $t = 1$ then imposes a consistency condition on $u_\alpha(1)$, namely

$$\frac{s(1) - \rho'(1)u_\alpha(1) - \rho(1)u_\alpha'(1)}{r(1)} = x_\alpha(1) = \frac{b_3 + u_\alpha(1)}{b_1}, \tag{3.93}$$

which it will always be possible to satisfy, since $\rho(1) \neq 0$ by hypothesis. This condition mimics the given boundary condition $(3.78)_3$ at $t = 1$, but we are not obliged to assume that $u_\alpha(1) = -x_\alpha'(1)$. If $r(t) = 0$ at any point in $0 \leqslant t \leqslant 1$ there is an intrinsic constraint

$$(\rho(t)u_\alpha(t))' = s(t) \tag{3.94}$$

to be satisfied there, and (3.92) does not give $x_\alpha(t)$ but such $x_\alpha(t)$ disappears from (3.89). If $b_1 = 0$ there is an intrinsic constraint

$$u_\alpha(1) = -b_3 \tag{3.95}$$

replacing (3.93), and $x_\alpha(1)$ disappears from (3.89).

Consider next (3.82β), taken by itself. Existence of piecewise C^1 solutions $x_\beta(t)$ and $u_\beta(t)$ of $u_\beta(t) = -x_\beta'(t)$ insists that $x_\beta(t)$ be actually piecewise C^2. If $a_2 \neq 0$ we can rearrange (3.82β) as

$$\begin{bmatrix} u_\beta(t) \\ u_\beta(0) \end{bmatrix} = \begin{bmatrix} -x_\beta'(t) \\ (x_\beta(0) - a_3)/a_2 \end{bmatrix}. \tag{3.96}$$

The continuity of $u_\beta(t)$ at $t = 0$ then imposes a consistency condition on $x_\beta(0)$, namely

$$-x_\beta'(0) = u_\beta(0) = (x_\beta(0) - a_3)/a_2. \tag{3.97}$$

If $a_2 = 0$ $u_\beta(0)$ disappears from (3.90), and (3.97) is replaced by

$$x_\beta(0) = a_3, \tag{3.98}$$

so that the original boundary conditions $(3.78)_2$ at $t = 0$ is recovered in the form $x_\beta(0) + a_2 x_\beta'(0) = a_3$ for any a_2.

In arriving at the inversions (3.92) and (3.96) we have, in effect, applied an infinite dimensional regular Legendre transformation to the functional $H[\tilde{x}, \tilde{u}]$, with $\tilde{x}$ and then $\tilde{u}$ active in turn (cf. Fig. 2.18 with H here taking the place of X there). Consistency conditions (3.93) and (3.97) are necessary for existence of the inverses described by such transformations. Singularities of the distributed/accumulated type correspond to the cases $r(t) \equiv 0$, $b_1 = 0$ and $a_2 = 0$, as the then linear parts of (3.83) indicate.

(xii) Evaluation of simultaneous bounds

We can now use the foregoing properties (3.92)–(3.95) of $\{\tilde{x}_\alpha, \tilde{u}_\alpha\}$, and (3.96)–(3.98) of $\{\tilde{x}_\beta, \tilde{u}_\beta\}$, to get the following result.

Theorem 3.7
The problem (3.78), in which the data also satisfies (3.88) and $a_1 = b_2 = 1$, can be characterized by the simultaneous upper and lower bounds

$$L_\alpha \geqslant L_0 \geqslant L_\beta$$

whose explicit form is as follows.

The bounded quantity is the value which any C^2 solution $x_0(t)$ of (3.78) gives to the functional

$$L_0 = \frac{1}{2}\int_0^1 sx_0\,dt + \tfrac{1}{2}b_3\rho(1)x_0(1) + \tfrac{1}{2}a_3\rho(0)x_0'(0). \tag{3.99}$$

The upper bound is $L_\alpha = J[u_\alpha(t)]$ where

$$J[u_\alpha(t)] = \frac{1}{2}\int_0^1\left[\frac{(s-(\rho u_\alpha)')^2}{r} + \rho u_\alpha^2\right]dt - a_3\rho(0)u_\alpha(0)$$
$$+\frac{1}{2}\frac{\rho(1)}{b_1}(b_3+u_\alpha(1))^2 - \tfrac{1}{2}a_2\rho(0)(u_\alpha(0))^2 \tag{3.100}$$

for any C^1 function $u_\alpha(t)$ which satisfies
(a) $(\rho(t)u_\alpha(t))' = s(t)$ whenever $r(t) = 0$, and
(b) $b_1[s(1) - \rho'(1)u_\alpha(1) - \rho(1)u_\alpha{}'(1)]$ (3.101)
$= r(1)[b_3 + u_\alpha(1)]$ if $r(1) > 0$.

It is to be understood that, if r or b_1 are zero, $J[u_\alpha(t)]$ is defined with the corresponding terms omitted from (3.100).

The lower bound is $L_\beta = K[x_\beta(t)]$ where

$$K[x_\beta(t)] = -\frac{1}{2}\int_0^1 [rx_\beta^2 + \rho x_\beta'^2 - 2sx_\beta]\,dt + b_3\rho(1)x_\beta(1)$$
$$-\tfrac{1}{2}b_1\rho(1)(x_\beta(1))^2 + \tfrac{1}{2}a_2\rho(0)(x_\beta'(0))^2 \tag{3.102}$$

for any piecewise C^2 function $x_\beta(t)$ which satisfies

$$x_\beta(0) + a_2x_\beta'(0) = a_3. \qquad (3.78)_2\ bis$$

If (3.88) is strengthened to require $r(t) > 0$ everywhere, instead of $r(t) \geqslant 0$, then the original problem has at most one solution.

Proof
We know from (3.72) that the bounded quantity is the value which any solution of the original problem gives to half of the linear terms in the

generating functional (3.84), and this is (3.99) after we have used the property $u_0(0) = -x_0'(0)$ of any such solution.

Simultaneous upper and lower bounds follow by Theorems 1.6(c) and 3.4, because (3.88) is sufficient to ensure that $S \geqslant 0$ for every pair of points in the domain of $L[\tilde{x}, \tilde{u}]$. The expressions (3.100) and (3.102) for the bounds are obtained by inserting the foregoing solutions for the individual constraints (3.82α) and (3.82β) into (3.89) and (3.90) respectively.

Uniqueness follows from Theorem 3.7 because, if $r(t) > 0$ everywhere in (3.88), then A is positive definite as well as B. □

(xiii) *Specific example*

If both ρ and r are positive constants and s is a nonzero constant, we can write $\rho = s = 1$ and $r = m^2$ without loss of generality in (3.78), where $m \neq 0$ is given. With $a_1 = b_2 = 1$ already, let us examine the particular case $a_2 = b_1 = 0$, $a_3 = 1$, $b_3 = b$ (say). The *ab initio* problem is then

$$-x'' + m^2 x = 1 \quad \text{with} \quad x(0) = 1, \quad x'(1) = b.$$

This has the unique solution

$$x_0(t) = \frac{(m^2 - 1)\cosh m(t-1) + mb \sinh mt}{m^2 \cosh m} + \frac{1}{m^2}$$

and it gives to the generating saddle functional (3.84) the value

$$L_0 = \frac{1}{2}\int_0^1 x_0(t)\,dt + \tfrac{1}{2}bx_0(1) + \tfrac{1}{2}x_0'(0)$$

$$= \frac{[b^2m^2 - (m^2-1)^2]\sinh m + 2mb(m^2-1)}{2m^3 \cosh m} + \frac{2b+1}{2m^2}.$$

The availability of this result permits a direct comparison of the upper and lower bounds in Theorem 3.7 with the quantity they are designed to estimate. A value of the upper bound (3.100) is now provided by any trial function $u_\alpha(t)$ which is C^1 and which satisfies $u_\alpha(1) = -b$. The simplest such choice is the constant function $u_\alpha(t) = -b$, giving an upper bound

$$L_\alpha = J[-b] = \frac{1}{2}\left(\frac{1}{m^2} + b^2\right) + b.$$

A value of the lower bound (3.102) is provided by a trial function $x_\beta(t)$ which may be any piecewise C^2 function which satisfies $x_\beta(0) = 1$, and the simplest such choice is the constant function $x_\beta(t) = 1$, giving a lower bound

$$L_\beta = K[1] = -\tfrac{1}{2}m^2 + b + 1.$$

As a simple check we notice that the difference between these bounds is

$$L_\alpha - L_\beta = \frac{1}{2}\left(\frac{1}{m^2} + m^2\right) - 1 + \tfrac{1}{2}b^2 \geqslant 0,$$

and this is zero when $m^2 = 1$ and $b = 0$.

In Fig. 3.3 we show how L_α, L_0 and L_β depend on m for $b = 0$ and $b = 1$. This not only confirms the validity of the bounds, but also emphasizes that if b is not small and/or if m^2 is not near 1, more elaborate trial functions are needed than the constant ones chosen above.

(xiv) *The fundamental lemma of the calculus of variations*

We explain this in the context of the current example, because it allows us to see how the lemma may be true for a more general functional notwithstanding the difficulty mentioned after (3.49) that the gradient frequently does not belong to the domain of the functional.

The current example of the first variation (3.57) is, from (3.80) and (3.85),

$$\delta L = \int_0^1 [-(\rho u)' - rx + s]\delta x\, dt + \rho(1)[u(1) - b_1 x(1) + b_3]\delta x(1)$$

$$+ \int_0^1 \rho(x' + u)\delta u\, dt + \rho(0)[x(0) - a_2 u(0) - a_3]\delta u(0), \qquad (3.103)$$

where $\delta x(t) = \mu h(t)$ and $\delta u(t) = \lambda k(t)$ for any piecewise C^1 functions $h(t)$ and $k(t)$, and μ and λ are sufficiently small scalars.

Fig. 3.3. Bounds for $-x'' + m^2 x = 1$ with $x(0) = 1$, $x'(1) = b$.

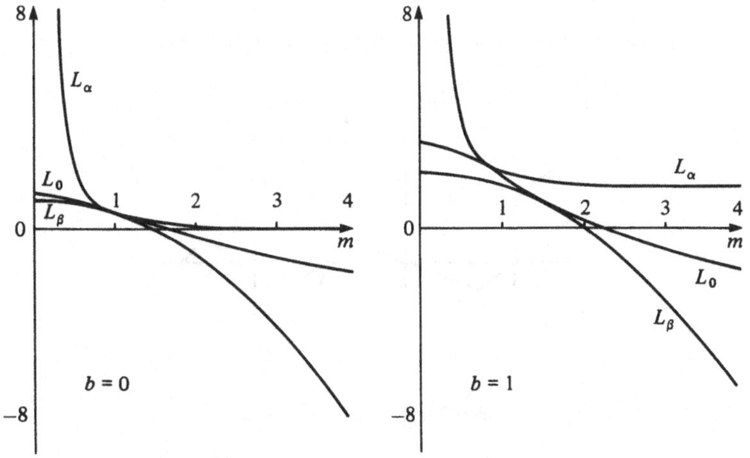

Theorem 3.8 (Fundamental lemma)
If $\delta L = 0$ for every such choice of $\delta x(t)$ and $\delta u(t)$, then the integrands and the boundary coefficients in (3.103) *must be zero, i.e.* $\partial L/\partial \tilde{x} = \partial L/\partial \tilde{u} = 0$.

Proof
First select $\delta u(t) \equiv 0$ in $0 \leqslant t < 1$ and $\delta x(1) = 0$, still otherwise allowing any piecewise C^1 $\delta x(t)$. Its coefficient $-(\rho u)' - rx + s$ $(=g(t)$, say) may be piecewise continuous. Suppose such $g(t)$ is nonzero somewhere, perhaps also changing sign. We can always choose $\delta x(t)$ to have the same sign locally as $g(t)$ in $0 \leqslant t < 1$, as illustrated in Fig. 3.4. Then $\int_0^1 g\delta x\,\mathrm{d}t > 0$ if $g \not\equiv 0$, which contradicts the hypothesis that $\delta L = 0$. Hence $g(t) \equiv 0$ for all t in $0 \leqslant t \leqslant 1$. We can then choose $g(1) = 0$ since u' is not defined from the right at $t = 1$.

Next again select $\delta u(t) \equiv 0$ but not $\delta x(1) = 0$. Instead use the conclusion $g(t) \equiv 0$ and choose any $\delta x(t)$ having $\delta x(1) \neq 0$. Then $\delta L = 0$ for all such $\delta x(t)$ implies that the coefficient of $\delta x(1)$ is zero. Hence $\partial L/\partial \tilde{x} = 0$.

Repeat with the roles of δx and δu reversed. Hence $\partial L/\partial \tilde{u} = 0$. □

The proof evidently works because each inner product (3.80) can be written as a sum of integrals over subintervals of $0 \leqslant t \leqslant 1$, such that the contribution to the gradient from within any one subinterval *does* belong to the same space (of continuous functions with that subinner product) from which the variations are chosen – so that completeness becomes available there, in the sense that orthogonality of a pair of elements with one arbitrary implies the other one zero.

A tacit assumption of this kind is widely adopted in applied mathematics, and we shall adopt it in certain boundary value problems without further discussion of detail.

Fig. 3.4. Fundamental lemma with piecewise continuous $g(t) \not\equiv 0$.

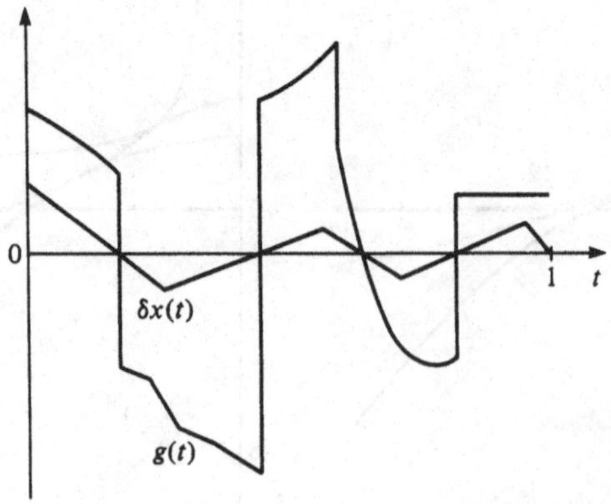

(xv) Complementary stationary principles

A *stationary principle* is a stationary property together with its converse (see §§1.1(iv) and 3.7(vii)), e.g. for (1.53). It is synonymous with what used to be commonly called a *variational principle.* The latter name, however, has also become attached to 'pseudo-variational' principles, in which no functional takes a stationary property, although there may be an orthogonality statement. The name 'stationary principle' is an informative literal description, and we use it to give contrast with our parallel discussions of inequality governing conditions (1.50) or (1.54) or (1.55), whose solutions may be equivalent to those of a so-called 'variational inequality' (see §3.11).

Theorem 3.9 (Free stationary principle)
The generating functional (3.84) *is stationary with respect to unconstrained first variations of $\tilde{x}$ and $\tilde{u}$, built as in* (3.81) *from piecewise C^1 functions $x(t)$ and $u(t)$, if and only if the original problem* (3.82) *is satisfied.*

The proof is a combination of the fundamental lemma with the remarks of §(vii) above. The adjective 'free' is attached to the theorem to emphasize the absence of any constraints on the variations.

The formula (3.103) is an example of the general first variation

$$\delta L = \left(\frac{\partial L}{\partial x}, \delta x\right) + \left\langle \frac{\partial L}{\partial u}, \delta u\right\rangle \qquad (3.57\ bis)$$

for any C^1 functional $L[x, u]$. The corresponding free stationary principle would be that $\delta L = 0$ for unconstrained δx and δu if and only if $\partial L/\partial x = 0$ and $\partial L/\partial u = 0$.

Suppose next that δx and δu are chosen from among x and u which, either by design or by accident, satisfy $\partial L/\partial x = 0$. Then it is self-evident that $\delta L = 0$ if such x and u also satisfy $\partial L/\partial u = 0$. When it can also be shown, as is often the case in practice, that $\delta L = 0$ *only if* such constrained x and u also satisfy $\partial L/\partial u = 0$, we have a constrained stationary principle: that $L[x, u]$ is stationary subject to the constraint $\partial L/\partial x = 0$ if and only if the 'natural condition' $\partial L/\partial u = 0$ is satisfied (cf. Courant and Hilbert, 1953, p. 231). It may also be possible to prove that $L[x, u]$ is stationary subject to $\partial L/\partial u = 0$ if and only if $\partial L/\partial x = 0$ holds. The roles of constraints and natural conditions are then reversible, and in this way we obtain a pair of *complementary stationary principles.*

These remarks apply in particular when the subsets (3.82α) and (3.82β) are selected to define the complementary constraints and natural conditions. We can use the constraints to express $L[\tilde{x}, \tilde{u}]$ more explicitly as (3.100) and (3.102), respectively. These complementary stationary principles are valid without needing to assume the inequalities (3.88) used to establish upper and lower bounds.

The qualifications which affect the 'only if' part of the proof of natural conditions are illustrated by the discussion after Theorem 1.14. By regarding the first variation

$$\delta\phi = \frac{\partial\phi}{\partial x_1}\delta x_1 + \frac{\partial\phi}{\partial x_2}\delta x_2$$

of the function $\phi[x_1, x_2]$ as a precursor of (3.57), we need to examine whether possibilities of the type represented by (1.125) are excluded, so that the sets C and U of constrained and unconstrained stationary points are the same. Evidently this will be the case when the solution value of an associated Lagrange multiplier is necessarily zero (as in (1.124)) or can be deemed zero without loss of generality (as in (1.127)). For the functional $L[\tilde{x}, \tilde{u}]$ of (3.84), these two circumstances arise according as $r(t)$ or b_1 or a_2 are nonzero or zero, respectively.

For general $L[x, u]$ subject to $\partial L/\partial x = 0$, the Lagrange multiplier method entails the introduction of a multiplier λ in the same space as x, and the setting to zero of the first variation of an augmented functional

$$\bar{L}[x, u, \lambda] = L[x, u] + \left(\lambda, \frac{\partial L}{\partial x}\right)$$

with respect to free variations $\delta x, \delta u, \delta\lambda$. If $L[x, u]$ is linear in x, then $x + \lambda$ appears only linearly in $\partial\bar{L}/\partial u$ and is absent from $\partial\bar{L}/\partial(x + \lambda)$, as in (1.127). We give illustrations in Exercise 3.3.3, and in Theorems 5.3, 5.12, 5.19.

(xvi) Weighting function with isolated zeros

We required $\rho(t)$ to be positive everywhere in the class of inner products (3.4) proposed for a space of square integrable functions, and we illustrated the use of this in (3.80).

We now consider the possibility that $\rho(t)$ be *positive almost everywhere* in the sense that $\rho(t) > 0$ in $0 \leqslant t \leqslant 1$, except at a finite number of isolated locations where $\rho(t) = 0$. This would not guarantee positivity of (3.4) taken by itself, for the $u(t)$ and $v(t)$ considered there would be integrable if they were zero almost everywhere, with finite values at a finite number of isolated locations. It could then happen that $\langle\tilde{u}, \tilde{u}\rangle = 0$ with $\tilde{u} \not\equiv 0$.

In a specific application like (3.78), however, the generating functional (3.84) is found to be built from piecewise C^1 functions $x(t)$ and $u(t)$. There is then sufficient smoothness in the functions actually used in the proof of the upper and lower bounds, as we can see from Theorem 3.7, that difficulties of the above type cannot occur. There is, however, another difficulty if $\rho(0)=0$ or if $\rho(1)=0$, which is that the imposed end conditions cannot then be accommodated unambiguously in the definitions of the gradients, as we can clearly see from (3.103). Some further progress in this direction can be made if such end conditions are deleted from the outset, as indicated in Exercise 3.2.5. Alternatively, it may be possible to retain such end conditions by building them into the definition of $\mathscr{D}(T)$ or of $\mathscr{D}(T^*)$ instead of into the gradients, as we illustrate in §3.10(iv). This widens the scope of our theory.

Notice that the alternative change of variable $u=-\rho x'$ is precluded when $\rho(t)$ can have an isolated zero because ρ appears in denominators.

Exercises 3.2

1. When $a_1b_2=0$ in (3.78) the subsequent details in the development of upper and lower bounds have to be modified. Assuming $a_2b_2\neq 0$, show how the operators given in Exercise 3.1.1 can be used to carry this modification through §§(ii)–(xiii) of §3.7 when $\rho(t)>0$ everywhere and $u=-\rho x'$ is used as the intermediate variable.
2. Provide the modification required to repeat the previous exercise when $a_1b_1\neq 0$, and obtain a theorem stating the upper and lower bounds.
3. As a specific example of Exercise 2 consider the *ab initio* problem of finding a C^2 function $x(t)$ on $0\leqslant t\leqslant 1$ such that

$$-x''+m^2x=-1 \quad \text{with} \quad x(0)=a, \quad x(1)=b.$$

Show that this is equivalent to setting the gradients of

$$L[\tilde{x},\tilde{u}]=\int_0^1 [\tfrac{1}{2}u^2-x(u'+1)-\tfrac{1}{2}m^2x^2]\,dt-au(0)+bu(1)$$

to zero, where $L[\tilde{x},\tilde{u}]$ is defined for any pair of piecewise C^1 functions $x(t)$ and $u(t)$.

Establish upper and lower bounds $J[u_\alpha(t)]\geqslant L_0\geqslant K[x_\beta(t)]$ for the solution value L_0 of $-\frac{1}{2}\int_0^1 x(t)\,dt-\frac{1}{2}au(0)+\frac{1}{2}bu(1)$, where

$$J[u_\alpha(t)]=\frac{1}{2}\int_0^1\left[u_\alpha^2+\frac{1}{m^2}(u_\alpha'+1)^2\right]dt-au_\alpha(0)+bu_\alpha(1)$$

in which $u_\alpha(t)$ can be any C^1 function whatever, and

$$K[x_\beta(t)]=-\frac{1}{2}\int_0^1 [x_\beta'^2+m^2x_\beta^2+2x_\beta]\,dt$$

in which $x_\beta(t)$ can be any piecewise C^2 function such that $x_\beta(0)=a$, $x_\beta(1)=b$. Show that the family of trial functions $u_\alpha(t)=p(1-2t)$ with constant p leads to a family of upper bounds whose least value is

$$\frac{1}{2m^2}-\frac{[m^2(a+b)+2]^2}{m^2(\frac{2}{3}m^2+8)}\left(=\frac{1}{24+2m^2}\quad \text{if} \quad a+b=0\right).$$

Show that, if $a = b = 0$, the family of trial functions $x_\beta(t) = qt(1-t)$ with constant q leads to a family of lower bounds whose greatest value is $1/(24 + 2.4m^2)$ (Noble 1964). Compare with the exact solution.

4. This exercise includes a description of Moreau's (1967) model of incipient cavitation in incompressible inviscid fluid, which we mentioned in §1.7. Flow takes place in a rigid curved pipe thin enough to permit a one dimensional approximation with distance s along the pipe as the only spatial coordinate, and $0 \leqslant s \leqslant 1$. The height $h(s)$ of the pipe above the end $s = 0$ is a given function, and given constants ρ and g represent fluid density and the acceleration due to gravity. Pressure $p(s)$ and acceleration $a(s)$ of a fluid particle are the primary unknown functions, to be determined at a given instant of unsteady flow when the velocity distribution is supposed known.

   Initial motion without cavitation is governed by the equations of momentum $\rho a = -p' - \rho g h'$ and (differentiated) continuity $a' = 0$, with end conditions $p(0) = c$ and $a(1) = k$, where c and k are given constants. Show that these equations are an example of (3.79), write out a generating saddle functional and the two bounds, and show that the quantity enclosed by simultaneous bounds is effectively the solution value of $ca(0) + kp(1)$, i.e. the 'work' supplied by the forcing agents at the two ends. Show that the exact solution $p(s) = c - \rho ks - \rho gh(s)$ is in fact easily found.

   If $q(s)$ is an assigned *vaporization pressure* function, then where $p > q$ cavitation cannot occur. Elsewhere, cavitation might be possible but, to find out, the problem must be redefined by replacing $a' = 0$ and $a(1) = k$ by the following conditions. Both continuity and the inception of a cavity within the fluid are accommodated by the properties

$$a' \geqslant 0, \quad p \geqslant q, \quad (p - q)a' = 0,$$

   and the inception of a cavity on the 'piston' at $s = 1$ will entail

$$a(1) \leqslant k, \quad p(1) \geqslant q(1), \quad (p(1) - q(1))(a(1) - k) = 0.$$

   Breakdown of continuity is expressed as a positive time derivative of velocity divergence, possible if and only if $p = q$. With the other conditions as before, show that the redefined problem is an example of (3.91), not necessarily with the same identifications of x and u, as before. Write out the saddle functional and the redefined theorem of upper and lower bounds.

   Sketch a graphical solution of the cavitation problem in terms of the following obstacle problem (see §3.10(ii) below). The given function $w(s) = q(s) + \rho gh(s)$ represents an obstacle in the s, y plane such that a solution for the unknown function $y(s) = p(s) + \rho gh(s)$, if it exists, has the shape of a taut string drawn from $s = 0$, $y = c$ so as to pass over the obstacle and arrive at the ordinate $s = 1$ with slope not less $-\rho k$ if $y(1) = w(1)$, or with slope $-\rho k$ if $y(1) > w(1)$. Incipient cavitation can occur where the string is in contact with the obstacle, either where the obstacle is concave or at $s = 1$ or both. Discuss when $q(s) \equiv 0$ and the pipe is wrapped n times round a horizontal circular cylinder of radius $1/(2\pi n)$ showing that the obstacle is n cycles of a sine wave.

5. (i) Show that any solution of Legendre's equation

$$(1-t^2)x'' - 2tx' + \nu(\nu+1)x = -s(t)$$

which is C^2 on $0 \leqslant t \leqslant 1$ and subject to the single end condition $x(0) + a_2 x'(0) = a_3$ can be characterized by a free stationary principle for any given ν, and by simultaneous upper and lower bounds if $-1 \leqslant \nu \leqslant 0$ and $a_2 \leqslant 0$.
(ii) Show that any solution of an amended Bessel's equation

$$tx'' + x' + \left[(t+\lambda) - \frac{\nu^2}{t+\lambda}\right]x = -s(t)$$

which is C^2 on $0 \leqslant t \leqslant 1$ and subject to the single end condition $b_1 x(1) + x'(1) = b_3$, with given $\lambda > 0$ and any given ν, can be characterized by a free stationary principle generated by

$$L = \int_0^1 [tux' - \tfrac{1}{2}rx^2 + sx + \tfrac{1}{2}tu^2]\,dt - (\tfrac{1}{2}b(1)x(1) - b_3)x(1),$$

whose domain is the set of piecewise C^1 functions $x(t)$ and $u(t)$, with $r(t) = \nu^2/(t+\lambda) - (t+\lambda)$. Show further that if $\nu - 1 \geqslant \lambda > 0$ and $b_1 \geqslant 0$, then upper and lower bounds apply. State them, and apply them with simple trial functions.
Note that $\rho(t) = 0$ at an end point in both these problems, and associated with this fact are well known singular solutions which are not C^2 on the closed interval $0 \leqslant t \leqslant 1$, and which we cannot expect to cover by the foregoing principles.

6. Show that the end value $x(1)$ and end slope $x'(0)$ of the exact solution of Airy's equation $-x'' + tx = 0$ subject to any of the end conditions in (3.78) with $b_1 \geqslant 0$ and $a_2 \leqslant 0$ satisfy the simultaneous bounds

$$1.14{b_3}^2 \geqslant b_3 x(1) + a_3 x'(0) \geqslant -(b_1 + \tfrac{1}{3}){a_3}^2 + 2b_3 a_3.$$

Verify that the trial functions $u(t) = -b_3 t^2(4t - 3t^2)$ and $x(t) = a_3$ used here, respectively, satisfy the conditions required by Theorem 3.7, and explore other choices. Compare the result with the known series solution.

3.8. Solution of linear constraints

(i) Consistency conditions

We return to the general problem postponed from §3.6(iii) of finding more explicit forms for $\{x_\alpha, u_\alpha\}$ and for $\{x_\beta, u_\beta\}$ when these sets must satisfy the linear equations (3.75) and (3.76) respectively. We treat the two cases separately, as we did in the particular example discussed in §3.7(xi), which we use for illustration.

Consider first

$$Ax_\alpha = T^* u_\alpha + a. \qquad (3.75\ bis)$$

Because this is $\partial L/\partial x = 0$ we must have

$$x_\alpha \in \mathscr{D}_x(L) = \mathscr{D}(T) \cap \mathscr{D}(A) \quad \text{and} \quad u_\alpha \in \mathscr{D}_u(L) = \mathscr{D}(T^*) \cap \mathscr{D}(B).$$

In addition the data demands that $T^*u_\alpha + a$ belongs to that part of $\mathscr{R}(A)$ which maps from $\mathscr{D}_x(L)$. An example of this in §3.7(xi) is the deduction, just before (3.92), that $u_\alpha(t)$ there be actually C^1.

Furthermore, if an inverse operator A^{-1} exists with (cf. §3.3(i)) $\mathscr{D}(A^{-1}) = \mathscr{R}(A)$ and $\mathscr{R}(A^{-1}) = \mathscr{D}(A)$, then

$$x_\alpha = A^{-1}(T^* u_\alpha + a) \qquad (3.104)$$

for any of the above u_α which also satisfies

$$A^{-1}(T^* u_\alpha + a) \in \mathscr{D}(T) \cap \mathscr{D}(A) \qquad (3.105)$$

for consistency. In fact we never need to apply A^{-1} to any elements other than $T^*u_\alpha + a$, and an example of (3.104) is (3.92). We illustrate (3.105) by $(3.108)_1$ and the remarks thereafter.

Consider next

$$Bu_\beta = -(Tx_\beta + b). \qquad (3.76\ bis)$$

Because this is $\partial L/\partial u = 0$ we must have

$$x_\beta \in \mathscr{D}(T) \cap \mathscr{D}(A) \quad \text{and} \quad u_\beta \in \mathscr{D}(T^*) \cap \mathscr{D}(B).$$

In addition $-(Tx_\beta + b)$ must belong to that part of $\mathscr{R}(B)$ which maps from $\mathscr{D}(T^*) \cap \mathscr{D}(B)$. An example of this in §3.7(xi) is the deduction, just before (3.96), that $x_\beta(t)$ there be actually piecewise C^2.

Furthermore, if an inverse operator B^{-1} exists with $\mathscr{D}(B^{-1}) = \mathscr{R}(B)$ and $\mathscr{R}(B^{-1}) = \mathscr{D}(B)$, then

$$u_\beta = -B^{-1}(Tx_\beta + b) \qquad (3.106)$$

for any of the above x_β which also satisfies

$$-B^{-1}(Tx_\beta + b) \in \mathscr{D}(T^*) \cap \mathscr{D}(B) \qquad (3.107)$$

for consistency. We never need to apply B^{-1} to any elements other than $Tx_\beta + b$, and an example of (3.106) is (3.96).

In §3.7(iii)–(vi) we saw that $\mathscr{D}(T) \subset \mathscr{D}(A) = X$ and $\mathscr{D}(T^*) \subset \mathscr{D}(B) = U$, so that $\mathscr{D}_x(L) = \mathscr{D}(T)$ and $\mathscr{D}_u(L) = \mathscr{D}(T^*)$. Then (3.105) and (3.107) simplify to

$$A^{-1}(T^* u_\alpha + a) \in \mathscr{D}(T) \quad \text{and} \quad -B^{-1}(Tx_\beta + b) \in \mathscr{D}(T^*). \qquad (3.108)$$

Consequences of these are illustrated, respectively, by (3.93) and by (3.97). In such cases the consistency conditions in (3.108) each impose, via the continuity properties required of $\mathscr{D}(T)$ and $\mathscr{D}(T^*)$, an equation which either is, or mimics, a given boundary condition.

We may expect that, in other boundary value problems too, similar equations will be implied by the consistency conditions (3.105) and (3.107), since u_α and then x_β respectively are acting as extra data in the sense explained at the end of §3.6(iii). This qualifies the facts that (3.104) and (3.106) represent the general solution of (3.75) and (3.76) respectively.

Although individual boundary conditions, illustrated by (3.93) and (3.97), are present in the form just given, and therefore have to be satisfied during the process of solving individual constraints, it will often be the case that they can in fact be satisfied, i.e. that trial functions can be chosen to satisfy them. Another illustration is contained within Exercise 3.3.2.

By contrast, there may be choices of data for which no solution x_0, u_0 exists for the whole given problem, i.e. of all constraints taken together. In such a case an implicit consistency condition cannot be satisfied, as illustrated at the end of §(iv) below. Put otherwise, the possibility is left open that trial functions might exist allowing bounds to be calculated, even when no solution x_0, u_0 exists.

(ii) General formulae for $J[u_\alpha]$ and $K[x_\beta]$

The object here is to describe what general formulae (3.77) can be given, of which (3.100) and (3.102) are particular illustrations.

Theorem 3.10

(a) In either of the problems (1.53) *and* (1.54) *generated by* (3.59)

$$L_\alpha = J[u_\alpha] = \tfrac{1}{2}\langle u_\alpha, Bu_\alpha + TA^{-1}(T^*u_\alpha + a)\rangle + \langle b, u_\alpha\rangle + \tfrac{1}{2}(a, A^{-1}(T^*u_\alpha + a)) \qquad (3.109)$$

for any u_α which satisfies the consistency conditions associated with (3.75), (3.104) *and* (3.105), *and $u_\alpha \geqslant 0$ in the case of* (1.54).

(b) In either of the problems (1.53) *or* (1.55) *generated by* (3.59)

$$L_\beta = K[x_\beta] = -\tfrac{1}{2}(x_\beta, Ax_\beta + T^*B^{-1}(Tx_\beta + b)) + (a, x_\beta) - \tfrac{1}{2}\langle b, B^{-1}(Tx_\beta + b)\rangle \qquad (3.110)$$

for any x_β which satisfies the consistency conditions associated with (3.76), (3.106) *and* (3.107), *and $x_\beta \geqslant 0$ in the case of* (1.55).

The proof is immediate by substituting (3.104) and (3.106) into (3.70) and (3.71) respectively, and using adjointness.

Note particularly, however, that although $A^{-1}(T^*u_\alpha + a)$ belongs to $\mathscr{D}(T)$, it is not in general true that $A^{-1}T^*u_\alpha$ and $A^{-1}a$ each separately belong to $\mathscr{D}(T)$, and so $TA^{-1}T^*u_\alpha$ and $TA^{-1}a$ are not defined (unless $a=0$). In other words

$$TA^{-1}(T^*u_\alpha + a) \neq TA^{-1}T^*u_\alpha + TA^{-1}a \text{ unless } a=0. \quad (3.111)$$

For a similar reason

$$T^*B^{-1}(Tx_\beta + b) \neq T^*B^{-1}Tx_\beta + T^*B^{-1}b \text{ unless } b=0. \quad (3.112)$$

This means that we cannot rearrange (3.109) or (3.110) in a way which collects together terms of the same degree.

These points are illustrated by (3.92) and by (3.96). For example, comparing the latter with (3.106) we infer

$$B^{-1}T\tilde{x}_\beta = \begin{bmatrix} x'_\beta(t) \\ -x_\beta(0)/a_2 \end{bmatrix} \quad B^{-1}\tilde{b} = \begin{bmatrix} 0 \\ a_3/a_2 \end{bmatrix}. \quad (3.113)$$

The sum of these two belongs to the domain of T^*, by the continuity of $x'(t)$ and by the consistency condition $a_3 - x(0) = a_2 x'(0)$, but individually neither belongs to the domain of T^* unless $a_3 = 0$. Therefore although $\langle Tx_\beta, B^{-1}(Tx_\beta + b)\rangle = (x_\beta, T^*B^{-1}(Tx_\beta + b))$ may be used to get (3.110), we must be aware that $\langle Tx_\beta, B^{-1}Tx_\beta\rangle \neq (x_\beta, T^*B^{-1}Tx_\beta)$ and $\langle Tx_\beta, B^{-1}b\rangle \neq (x_\beta, T^*B^{-1}b)$. Similar remarks apply to (3.92).

(iii) Separate upper and lower bounds, not simultaneous

We now describe conditions under which an upper bound is available when a lower bound is not, for the problem (1.53) or the problem (1.54) generated by the general quadratic functional (3.59). Each of these problems includes the equation $Ax = T^*u + a$. As an auxiliary device we shall need to refer also to the homogeneous equation $Ax = T^*u$. When A^{-1} exists there will be consistency conditions qualifying the general solution $x = A^{-1}T^*u$ of this auxiliary homogeneous equation. We can obtain them by putting $a=0$ in §(i) above, and in a boundary value problem they may imply a homogeneous boundary condition.

Theorem 3.11

If $B + TA^{-1}T^*$ is nonnegative in the sense that

$$\langle u, \overline{B + TA^{-1}T^*u}\rangle \geqslant 0 \quad (3.114)$$

for all u which satisfy the consistency conditions associated with the auxiliary homogeneous equation $Ax = T^*u$, then the original problem (1.53) *or* (1.54) *possesses the upper bound $L_\alpha \geqslant L_0$, where $L_\alpha = J[u_\alpha]$ is given by* (3.109).

If $B + TA^{-1}T^*$ is positive in the sense that the strict inequality applies in (3.114) *for all $u \neq 0$ specified there, then when a solution x_0, u_0 of the original problem exists it is unique.*

Proof
Consider the saddle quantity (3.73). Any distinct plus and minus points which both belong to $\{x_\alpha, u_\alpha\}$ will each satisfy $Ax_\alpha = T^*u_\alpha + a$ and have differences Δx, Δu (say) which satisfy $A\Delta x = T^*\Delta u$. The general solution of this auxiliary homogeneous equation will be described by (3.104) and (3.105) with $a = 0$ there, and the corresponding saddle quantity is

$$S = \tfrac{1}{2}\langle \Delta u, \overline{B + TA^{-1}T^*\,\Delta u}\rangle \geqslant 0$$

by our hypothesis. The upper bound then follows by invoking Theorems 1.6(a) and 3.4. The inequality is established for (1.53) and then applies *a fortiori* to (1.54).

Positivity of $B + TA^{-1}T^*$ implies $S > 0$, whence Theorems 1.7(b) and 3.4 establish unique u_0, and unique x_0 then follows from $x_\alpha = A^{-1}(T^*u_\alpha + a)$. □

Conditions under which a lower bound is available when an upper bound is not, for (1.53) or for (1.55) generated by the general quadratic functional (3.59), require a different auxiliary homogeneous equation with solution $u = -B^{-1}Tx$. Arguments parallel to those above prove the following result.

Theorem 3.12
If $A + T^*B^{-1}T$ is nonnegative in the sense that

$$(x, \overline{A + T^*B^{-1}Tx}) \geqslant 0 \tag{3.115}$$

for all x which satisfy the consistency conditions associated with the auxiliary homogeneous equation $Bu = -Tx$, then the original problem (1.53) *or* (1.55) *possesses the lower bound $L_\beta \leqslant L_0$, where $L_\beta = K[x_\beta]$ is given by* (3.110).

If $A + T^*B^{-1}T$ is positive in the sense that the strict inequality applies in (3.115) *for all $x \neq 0$ specified there, then when a solution of the original problem exists it is unique.*

Either of the functionals in (3.114) or (3.115) may be called an *exclusion functional* in the sense that, when positive, they ensure uniqueness of solution and hence exclude bifurcation. The name was used by Hill (1978) in plasticity theory (see Theorem 5.26). If A and B are not both nonnegative, so that simultaneous bounds are not assured (Theorem 3.6), it may not be trivial to verify (3.114) or (3.115). For example, emphasis may then shift from the lower bound itself to the problem of ensuring that an eigenvalue problem

$$(A + T^*B^{-1}T)x = \lambda Cx, \tag{3.116}$$

for any suitable positive operator C, gives no negative eigenvalues to the scalar λ. In this way hypothesis (3.115) can be verified, and the limit of its availability may herald bifurcation in an associated nonlinear problem. The stationary minimization of an associated Rayleigh quotient is closely linked to the lower bound statement. Similar remarks apply to $B + TA^{-1}T^*$ and the upper bound.

(iv) Example of a single bound

Included within (3.78) is the problem

$$-x'' - n^2 x = s, \quad x(0) = a, \quad x(1) = b, \tag{3.117}$$

with fixed $n > 0$, a and b, and with a given continuous function $s(t)$. A variant of the procedure in §3.7(ii)–(vii) leads (cf. Exercise 3.2.2) via $x = x(t)$, $\tilde{u} = [u(t)\ u(0)\ u(1)]^{\mathrm{T}}$ and

$$\begin{aligned} T^*\tilde{u} &= u'(t) = -n^2 x(t) - s(t) = \frac{\partial H}{\partial x}, \quad (\alpha) \\ Tx &= \begin{bmatrix} x'(t) \\ x(0) \\ -x(1) \end{bmatrix} = \begin{bmatrix} -u(t) \\ a \\ -b \end{bmatrix} = \frac{\partial H}{\partial \tilde{u}}, \quad (\beta) \end{aligned} \tag{3.118}$$

to the quadratic generating functional

$$L[x, \tilde{u}] = \int_0^1 [-xu' + \tfrac{1}{2}n^2x^2 + sx + \tfrac{1}{2}u^2]\,dt - au(0) + bu(1) \tag{3.119}$$

of piecewise C^1 functions $x(t)$ and of $\tilde{u}$ built from piecewise C^1 functions $u(t)$. This leads to a free stationary principle, but $L[x, \tilde{u}]$ cannot be an unqualified saddle functional concave in x and convex in $\tilde{u}$ because the coefficients of x^2 and u^2 both have positive sign.

To look for a single bound we first identify

$$Ax = -n^2 x, \quad B\tilde{u} = [u\ 0\ 0]^{\mathrm{T}}$$

so that, from (3.73), the saddle quantity is

$$S = -\tfrac{1}{2}n^2 \int_0^1 (\Delta x)^2\,dt + \tfrac{1}{2}\int_0^1 (\Delta u)^2\,dt.$$

The α equation can be solved as $x_\alpha = (u_\alpha' - s)/n^2$ for any given C^1 function $u_\alpha(t)$, whence $\Delta x_\alpha = \Delta u_\alpha'/n^2$ and

$$S = \frac{1}{2}\int_0^1 [(\Delta u_\alpha)^2 - (\Delta u_\alpha')^2/n^2]\,dt$$

as in the proof of Theorem 3.11. Hence $S \geqslant 0$ for every pair of distinct points which satisfy the α equation if

$$n^2 \geq \frac{\int_0^1 u_\alpha'^2\, dt}{\int_0^1 u_\alpha^2\, dt}$$

for every C^1 function $u_\alpha(t)$; but this is impossible with given n since the quotient is unbounded. For example, the inequality is violated with $u_\alpha(t) = \sin m(\pi/2)t$ for any $m(\pi/2) > n$. Therefore the hypothesis of Theorems 1.6(a) with 3.4 cannot hold, and no upper bound L_α can be established.

The β equation can be solved as $u_\beta = -x_\beta'$ for any given piecewise C^2 function $x_\beta(t)$ which satisfies $x_\beta(0) = a, x_\beta(1) = b$, whence

$$S = \frac{1}{2}\int_0^1 [(\Delta x_\beta')^2 - n^2(\Delta x_\beta)^2]\, dt$$

with $\Delta x_\beta(0) = \Delta x_\beta(1) = 0$. Hence $S \geq 0$ for every pair of distinct points which satisfy the β equation if

$$n^2 \leq \frac{\int_0^1 x_\beta'^2\, dt}{\int_0^1 x_\beta^2\, dt}$$

for every piecewise C^2 function $x_\beta(t)$ satisfying $x_\beta(0) = x_\beta(1) = 0$. The minimum of the Rayleigh quotient on the right is π^2, provided by $x_\beta(t) = \sin \pi t$. Therefore if $n \leq \pi$ the hypothesis of Theorems 1.6(b) with 3.4 is satisfied, and there is a lower bound $L_\beta \leq L_0$ where

$$L_\beta = \frac{1}{2}\int_0^1 [n^2 x_\beta - x_\beta'^2 + 2s x_\beta]\, dt \tag{3.120}$$

may be evaluated with any piecewise $C^2 x_\beta(t)$ satisfying $x_\beta(0) = a$, $x_\beta(1) = b$. This L_β bounds the quantity

$$L_0 = \frac{1}{2}\left[\int_0^1 s x_0\, dt + a x_0'(0) - b x_0'(1)\right],$$

evaluated for any solution $x_0(t)$ of the given problem. This is an example of the first part of Theorem 3.12, in which here $(A + T^*B^{-1}T)x = -n^2 x - x''$.

If $n < \pi$ uniqueness is assured in the given problem. The hypothesis $S > 0$ required for this is the same as the strict lower bound inequality for the associated homogeneous case $a = b = s = 0$, and this is the same as the statement that the exclusion functional $\int_0^1 [x'^2 - n^2x^2]\,dt > 0$ for all piecewise $C^2\,x(t)$ satisfying $x(0) = x(1) = 0$. Minimization of the above Rayleigh quotient entails solving the homogeneous version of the given problem, or in other words, applying the homogeneous version of the lower bound statement.

The given problem in this subsection in fact has solution

$$x_0(t) = a\cos nt + c\sin nt - \frac{1}{n}\int_0^t \sin n(t-\tau)s(\tau)\,d\tau, \tag{3.121}$$

where

$$b = a\cos n + c\sin n - \frac{1}{n}\int_0^1 \sin n(1-\tau)s(\tau)\,dt,$$

so that a unique solution exists unless $n = m\pi$, where m is an integer. In the latter case there are infinitely many solutions provided

$$b = a\cos m\pi - \frac{1}{m\pi}\int_0^1 \sin m\pi(1-\tau)s(\tau)\,d\tau \tag{3.122}$$

and otherwise none. The last equation therefore illustrates a consistency condition on the data $a, b, m, s(\tau)$, which must be satisfied in order that a solution for the given problem shall exist; but the condition is obtained after the solution in this example, not before it.

3.9. A procedure for the derivation of bounds

(i) Introduction

It seems helpful to pause for a résumé at this stage.

We summarize our procedure for the construction of upper and/or lower bounds in §(ii) below. The procedure applies in particular to problems which are posed in terms of ordinary or partial differential equations or inequalities, with associated boundary conditions. We gave a detailed illustration in §3.7.

In §3.10 we give a brief catalogue of examples illustrating some of the steps in the procedure. The aim of this catalogue is to indicate how the procedure may be applied in some diverse situations without offering every detail such as we did in §3.7. The reader may amplify each example in the catalogue by supplying the full procedure and the omitted detail. The central point is that the bounds themselves can always be validated by Theorem 3.4. Any application is on firm ground as soon as the validity of the bounds is established, and so is improvement of either bound within their respective families.

Given the bounds, the procedure also lists steps of more theoretical importance, such as the development of converse theorems which are required to confirm minimum and maximum principles, and stationary principles. The 'fundamental lemma' type of argument will often be available to prove these, but in general we leave them to be implied, not supplying the details which may vary from case to case. Finite dimensional guidelines were given in §1.5, but it is beyond the present scope to relate these to what is said in some parts of functional analysis.

Detailed numerical application of the bounds is also a large specialist field now, represented by many texts, for example that of Fortin and Glowinski (1982).

(ii) Steps in the procedure

We define our procedure as a list of steps, often in the form of questions.

1. Is the problem decomposable by introducing some intermediate dependent variable? For example, if it contains a differential operator of even order, can it be split into a problem twice the size but half the order? Practical clues to this include the recognition of a formally adjoint pair D and D^* of differential operators in the combination D^*Dx (cf. §3.3(iv)), when $u = Dx$ may be the intermediate variable; or the recognition of a gradient structure $D^*(\partial f/\partial Dx) + \partial f/\partial x$ (cf. (2.118)) when we could choose $u = \partial f/\partial Dx$ and invert by a Legendre transformation. This first question is closely linked to whether the problem is self-adjoint. If not, or if the problem is of odd order, it may be possible to embed it usefully in a doubled system of the given order, by introducing the adjoint system (see §4.3).
2. What are the possible statements of adjointness (there may be many) associated with each such choice of intermediate variable, in the first instance without regard to the given boundary values?
3. What pair of inner product spaces with composite elements and inner products (x, y) and $\langle u, v\rangle$ is suggested by each statement of adjointness?
4. What pair of true adjoint operators T and T^* with

$$(x, T^*u) = \langle u, Tx\rangle \quad \forall x \in \mathscr{D}(T) \quad \text{and} \quad \forall u \in \mathscr{D}(T^*) \qquad (3.10\ bis)$$

are then implied?

5. Can the rearranged problem, with its second variable and its boundary conditions now incorporated in it, be written as

$$T^*u = \frac{\partial H}{\partial x} \quad (\alpha), \qquad Tx = \frac{\partial H}{\partial u} \quad (\beta) \qquad (3.65\ bis)$$

in terms of the gradients of some (not necessarily quadratic) generalized

Hamiltonian functional $H[x,u]$? Or are inequalities required in trios, such as

$$T^*u \leqslant \frac{\partial H}{\partial x} \quad (\alpha), \qquad x \geqslant 0 \quad (\beta), \qquad \left(x, T^*u - \frac{\partial H}{\partial x}\right) = 0$$

in place of $T^*u = \partial H/\partial x$?

6. Write down the generating functional $L[x,u]$ as

$$L[x,u] = (x, T^*u) - H[x,u]. \tag{3.62 bis}$$

This is a pivotal step. The domain of this functional will often be the product of $\mathscr{D}(T)$ and $\mathscr{D}(T^*)$, and this fact is another reason for the importance of (3.10).

7. Can the problem be written as an example of (1.53)?

$$\frac{\partial L}{\partial x} = 0 \quad (\alpha), \qquad \frac{\partial L}{\partial u} = 0 \quad (\beta). \tag{1.53 bis}$$

8. Can a 'fundamental lemma' be proved, so validating a free stationary principle stating that $\delta L = 0$ if and only if (1.53) is satisfied? Here δL is given by (3.57). This is a sidestep, albeit an important one, on the way to establishing bounds. It may be a more important step when bounds do not exist. Write out any complementary stationary principles which are of interest.
9. What are the other three problems (1.50), (1.54) and (1.55) which are generated by the same $L[x,u]$ now supposedly established for (1.53)? Do any or all of them have applications in their own right?
10. Write out the expressions

$$L - \left(x, \frac{\partial L}{\partial x}\right) \quad \text{and} \quad L - \left\langle u, \frac{\partial L}{\partial u}\right\rangle$$

whose values will be those of upper and lower bounds, respectively, in all four problems (1.50) and (1.53)–(1.55), when such bounds can be established below.

11. Does $L[x,u]$ have the properties required to make it a saddle functional in a strong enough sense $S \geqslant 0$ to validate simultaneous bounds via Theorems 1.6(c) and 3.4?
12. If not, does $L[x,u]$ satisfy the lighter assumptions required to validate an upper bound *or* a lower bound?
13. What can be said about uniqueness of solution?
14. What is the precise definition of a solution of the α conditions alone? And similarly of the β conditions alone? Each may entail more smoothness than is required of a member of the domain of $L[x,u]$ (see 6

above), but less smoothness than may be required of a solution of the original problem.

15. Find a family of solutions of the α conditions, calculate the corresponding upper bounds, and minimize them within that family; similarly maximize lower bounds within a family of solutions of the β conditions. Repeat with any of the other three problems which may have applications in their own right. Of course it may be a difficult and specialist matter to find such families of constraint solutions and optimize within them. Our Chapter 2 is a contributions towards this question for nonlinear constraints.
16. What can be said about Legendre transformations, in either finite or infinite dimensions? Are any of the functions in Fig. 2.18, or the functionals in the infinite dimensional version of Fig. 2.26, multivalued?
17. What can be said about bifurcation of solution?
18. Pursue the question of converse theorems in context.

(iii) Hierarchy of smoothness in admissible fields

It is remarked in step 6 of §(ii) above that the domain of $L[x, u]$ in (3.62) will often be the product of $\mathscr{D}(T)$ and $\mathscr{D}(T^*)$. In particular, for problems posed in terms of differential equations, this means that L is built from functions validating integration by parts, in order that its gradients can be calculated by using adjointness in that appropriate form. Typically such $L = L[\tilde{x}, \tilde{u}]$, where $\tilde{x}$ and $\tilde{u}$ are composite inner product space elements built from piecewise C^1 functions $x(V)$ and $u(V)$ of position in the region V.

Existence of the gradients $\partial L/\partial \tilde{x}$ and $\partial L/\partial \tilde{u}$ in such cases also presupposes that differentiation can be carried through the integral sign. Standard sufficient conditions for this, like that stated after (3.47), are not also necessary in certain important cases. This is illustrated by $(\tilde{x}, T^*\tilde{u}) = \langle \tilde{u}, T\tilde{x} \rangle$, whose gradients are $T^*\tilde{u}$ and $T\tilde{x}$ by (3.61), even when the formal adjoints D^*u and Dx within them are piecewise continuous because $x(V)$ and $u(V)$ are piecewise C^1. The case (3.11) exemplifies this. Further, it will frequently happen, as in §3.7(vi), that when $L[\tilde{x}, \tilde{u}] = (\tilde{x}, T^*\tilde{u}) - H[\tilde{x}, \tilde{u}]$ applies with T and T^* absent from $H[\tilde{x}, \tilde{u}]$, existence of the gradients of $H[\tilde{x}, \tilde{u}]$ does not require more smoothness than does adjointness itself.

More elaborate cases will appear in Exercises 3.3.3, 5.2.3(a) and 5.3.5, in §4.5(ii), and in (5.79), (5.145), etc.; in these the generating functional can be of the type $L[\tilde{x}, \tilde{u}, \tilde{\lambda}]$ whose domain includes a third inner product space, with elements $\tilde{\lambda}$ built from functions $\lambda(V)$ which do not have to participate in integration by parts. Such $\lambda(V)$ may therefore need only to be integrable for the purpose of deriving the gradient $\partial L/\partial\tilde{\lambda}$ provided they appear in a function having suitably smooth coefficients. Examples are e in the terms $-\sigma e + U[e]$ in the L of Exercise 3.3.3, the λ_γ in (5.111), and the F_{ji} in (5.145). As a working *general assumption* we shall suppose that the domain of $L[\tilde{x}, \tilde{u}, \tilde{\lambda}]$ is built from piecewise C^1 functions $x(V)$ and $u(V)$ and integrable $\lambda(V)$; and also that this is sufficient to validate the existence of any gradients which we need, while being aware that exceptions are possible.

More smoothness is imposed in any case when we come to satisfying the constraints, as we have seen in §§3.7(xi) and 3.8(ii). For example, gradients must not only exist, but some of them may need to have zero value. Therefore, theorems establishing upper or lower bounds will usually entail such enhanced smoothness in their α or β admissible fields respectively (and similarly for complementary stationary principles). In statements of bounds we shall often keep track of this enhanced smoothness, but in some nonlinear situations it may be easier and adequate to simplify smoothness hypotheses.

An actual solution, since it must satisfy *all* the constraints (i.e. both α and β subsets and any orthogonality conditions), is likely to have even more smoothness than need be possessed by either an α or a β class of admissible fields.

This last remark leads us to expect, for example, that since a minimum principle must show that the minimizers are equivalent to the actual solutions, any such minimizer will be smoother than most of the admissible fields in the associated upper bound theorem; and similarly for maxima. In fact, it is known in the calculus of variations that classes of variations can have less smoothness than is required by an actual solution (cf. Hilbert's theorem on page 21 of Akhiezer (1962)).

In practice one will make choices of admissible fields which do mimic, as far as possible, the expected properties of the actual solution. One does have the option of seeking to prove extremum principles within admissible fields having the same smoothness as that expected of an actual solution, rather than within the wider class allowed for above by individual constraints.

Summarizing, our statements of bounds will often specify the smoothness which the α or β constraints add to the adjointness requirement. In what follows, and in Chapter 5 in particular, we shall not give explicit proof of the converse theorems needed to justify extremum principles *per se*, apart from some conventional appeals to the fundamental lemma. Nevertheless, it is a reasonable assumption that such extremum principles can often be validated (cf. §1.5).

3.10. A catalogue of examples

(i) Introduction

The purpose of this section is to focus the procedure of §3.9(ii) by selecting a few examples of some diversity, and performing some of the steps in a much briefer style than the illustration detailed in §3.7. We omit equation numbering. We aim to deal quickly with each step in turn, without providing all of the amplification and cross-referencing which has previously seemed desirable. There are many other examples which could be similarly treated.

(ii) Obstacle problem

Given problem: find $x(t)$ satisfying

$$-x'' \geqslant w'', \quad x \geqslant 0, \quad x(-x''+w'')=0 \quad \text{in} \quad 0 \leqslant t \leqslant 1,$$
$$x(0)=a, \quad x(1)=b$$

with fixed nonnegative a and b, where $w(t)$ is a given function such that $w(0)=w(1)=0$.

Physical interpretation: $x(t)=y(t)-w(t)$ is the distance, from an obstacle with given profile $w(t)$, of a string whose shape is $y(t)$ drawn taut between given end points $y(0)=a$, $y(1)=b$. The string is in contact with the obstacle where the latter obstructs a straight join of the end points. The string and obstacle profile are in a single plane (recall §3.7(x) and Exercise 3.2.4).

Associated problem: $-x''(t)=w''(t)$, $x(0)=a$, $x(1)=b$.

Adjointness: integration by parts (3.16)

$$-\int_0^1 xu'\,dt-\mu x(0)u(0)+(1-\lambda)x(1)u(1)$$
$$=\int_0^1 ux'\,dt+(1-\mu)u(0)x(0)-\lambda u(1)x(1)$$

applies with any fixed λ and μ and any piecewise $C^1\,x(t)$ and $u(t)$.
New variable: the boundary conditions indicate $\mu=0$ and $\lambda=1$, and the decomposable positive operator (cf. (3.38)) associated with $-x''$ prompts the choice of new variable $u=-x'$.

Spaces and inner products: select $x=x(t)$, $\tilde{u}=[u(t)\ u(0)\ u(1)]^{\mathrm{T}}$,

$$(x,y)=\int_0^1 xy\,dt, \quad \langle \tilde{u},\tilde{v}\rangle=\int_0^1 uv\,dt+u(0)v(0)+u(1)v(1).$$

True adjoints: $(x, T^*\tilde{u}) = \langle \tilde{u}, Tx \rangle$ is to be true for all piecewise $C^1 x(t)$ and $u(t)$, and indicates

$$T^*\tilde{u} = -u'(t), \quad Tx = [x'(t) \quad x(0) \quad -x(1)]^{\mathrm{T}}.$$

Hamiltonian: the associated problem $-u' = -w''$, $x' = -u$ entails

$$\partial H/\partial x = -w'', \quad \partial H/\partial \tilde{u} = [-u \quad a \quad -b]^{\mathrm{T}}$$

and so therefore does the given problem. Partial integration gives

$$H[x, \tilde{u}] = \int_0^1 (-xw'' - \tfrac{1}{2}u^2)\,\mathrm{d}t + au(0) - bu(1).$$

Generating functional: $(x, T^*\tilde{u}) = -\int_0^1 xu'\,\mathrm{d}t$, less H, gives

$$L[x, \tilde{u}] = \int_0^1 (-xu' + xw'' + \tfrac{1}{2}u^2)\,\mathrm{d}t - au(0) + bu(1).$$

This is a saddle functional weakly concave in x and strictly convex in $\tilde{u}$, over a domain of such elements constructed from piecewise C^1 functions $x(t)$ and $u(t)$. These are otherwise unrelated and unrestricted for the purpose of defining the domain. First variations show the gradients to be

$$\partial L/\partial x = -u' + w'', \quad \partial L/\partial \tilde{u} = [x' + u \quad x(0) - a \quad b - x(1)]^{\mathrm{T}}.$$

Governing conditions: the given problem illustrates (1.55),

$$-u' + w'' \leqslant 0 \quad (\alpha), \qquad x \geqslant 0 \quad (\beta), \qquad x(-u' + w'') = 0,$$
$$u = -x', \qquad x(0) = a, \qquad x(1) = b \quad (\beta).$$

Upper and lower bounds: these apply simultaneously, with values

$$L - \left(x, \frac{\partial L}{\partial x}\right) = \int_0^1 \tfrac{1}{2}u^2\,\mathrm{d}t - au(0) + bu(1) \qquad \text{subject to } (\alpha),$$

$$L - \left\langle \tilde{u}, \frac{\partial L}{\partial \tilde{u}} \right\rangle = \int_0^1 (xw'' - \tfrac{1}{2}u^2)\,\mathrm{d}t \qquad \text{subject to } (\beta).$$

Evaluation: a solution of the single α inequality is $u_\alpha(t) = w'(t) + u_\alpha(0) - w'(0)$, giving an upper bound

$$L_\alpha = \frac{1}{2}\int_0^1 [w'(t) + u_\alpha(0) - w'(0)]^2\,\mathrm{d}t$$
$$- (a - b)u_\alpha(0) + b(w'(1) - w'(0))$$

which is a quadratic function of the arbitrary $u_\alpha(0)$. A solution of the β

conditions is $x_\beta(t) = a + t(b-a)$, $u_\beta(t) \equiv a - b$, giving a fixed lower bound

$$L_\beta = -\tfrac{1}{2}(b-a)^2 + bw'(1) - aw'(0).$$

For $w(t) = -\tfrac{1}{2}t(t-1)$ the minimizing $u_\alpha(0)$ gives $L_\alpha - L_\beta = \tfrac{1}{24}$. (The same bounds would apply to the associated equation, but we know its solution $x(t) = -w(t) + t(b-a) + a$ anyway. The latter could be used to give a particular upper bound for the given problem, but not a lower bound because it will violate $x_\beta(t) \geqslant 0$ where the obstacle rises above a straight line joining the ends of the string.)

Bounded quantity: this is the exact solution value of

$$\frac{1}{2}\int_0^1 xw''\,\mathrm{d}t + \tfrac{1}{2}ax'(0) - \tfrac{1}{2}bx'(1).$$

Generalization: similar ideas apply to higher dimensions, for example when a stretched membrane may be intercepted by an obstacle rising from a plane. The points where the membrane or string leaves contact with the obstacle form a so-called 'free boundary', which is unspecified in advance and is to be found as one of the unknowns in the problem. Another such problem is discussed in §(vi) below.

(iii) Euler equation and Hamilton's principle

Given problem: find $q(t)$ in $0 \leqslant t \leqslant 1$ satisfying

$$\frac{\partial \mathscr{L}}{\partial q} - \frac{\mathrm{d}}{\mathrm{d}t}\left(\frac{\partial \mathscr{L}}{\partial \dot{q}}\right) = 0, \quad q(0) = a, \quad \dot{q}(1) = b,$$

with fixed a and b, where $\mathscr{L}[\dot{q}, q, t]$ is a given function and $\dot{q} = \mathrm{d}q/\mathrm{d}t$.

Physical interpretation: $q(t)$ can represent the position of a particle in classical dynamics, as a solution of Lagrange's equation of motion in a unit time interval, with assigned position at the beginning and assigned velocity at the end, and $\mathscr{L}$ as the Lagrangian function. More generally the equation is the Euler equation which one meets when beginning calculus of variations, and many interpretations are possible. Other end conditions are also important.

New variable: Let $p = \partial\mathscr{L}/\partial\dot{q}$. Invert this as

$$\dot{q} = \frac{\partial \mathscr{H}}{\partial p} \quad \text{where} \quad \mathscr{H}[p, q, t] = p\dot{q} - \mathscr{L}$$

represents the classical Hamiltonian function for the particle, and a Legendre transformation describes the inversion, so that $\partial\mathscr{L}/\partial q = -\partial\mathscr{H}/\partial q$.

Hamilton's equations: we have now to find $q(t)$ and $p(t)$ such that

$$\dot{q} = \partial\mathscr{H}/\partial p, \quad -\dot{p} = \partial\mathscr{H}/\partial q,$$
$$q(0) = a, \quad p(1) = \mathrm{d}Q/\mathrm{d}q(1).$$

Here a given function $Q[q(1)]$ of the unknown terminal position $q(1)$ is defined via the function $\partial\mathscr{L}/\partial\dot{q} = p[\dot{q}, q, t]$ evaluated at $t = 1$, by the formula $Q[q(1)] = \int_c^{q(1)} p[b, s, 1]\,\mathrm{d}s$ (Q is *not* the classical generalized force) with any fixed c.

Adjointness: if $q(t)$ and $p(t)$ are any unrelated piecewise C^1 functions,

$$\int_0^1 p\dot{q}\,\mathrm{d}t + p(0)q(0) = -\int_0^1 q\dot{p}\,\mathrm{d}t + q(1)p(1).$$

Inner product space: define elements and inner products

$$\tilde{p} = [p(t) \quad p(0)]^{\mathrm{T}}, \quad \tilde{q} = [q(t) \quad q(1)]^{\mathrm{T}},$$

$$(\tilde{p}_1, \tilde{p}_2) = \int_0^1 p_1 p_2\,\mathrm{d}t + p_1(0)p_2(0),$$

$$\langle \tilde{q}_1, \tilde{q}_2 \rangle = \int_0^1 q_1 q_2\,\mathrm{d}t + q_1(1)q_2(1).$$

True adjoint operators: $(\tilde{p}, T^*\tilde{q}) = \langle \tilde{q}, T\tilde{p} \rangle$ leads to

$$T^*\tilde{q} = \begin{bmatrix} \dot{q} \\ q(0) \end{bmatrix} = \begin{bmatrix} \partial\mathscr{H}/\partial p \\ a \end{bmatrix} = \frac{\partial H}{\partial \tilde{p}}, \quad (\alpha)$$

$$T\tilde{p} = \begin{bmatrix} -\dot{p} \\ p(1) \end{bmatrix} = \begin{bmatrix} \partial\mathscr{H}/\partial q \\ \mathrm{d}Q/\mathrm{d}q(1) \end{bmatrix} = \frac{\partial H}{\partial \tilde{q}}, \quad (\beta)$$

where $H[\tilde{p}, \tilde{q}] = \int_0^1 \mathscr{H}[p(t), q(t), t]\,\mathrm{d}t + ap(0) + Q[q(1)]$.

Generating functional: $L = (\tilde{p}, T^*\tilde{q}) - H$ gives

$$L[\tilde{p}, \tilde{q}] = \int_0^1 (p\dot{q} - \mathscr{H})\,\mathrm{d}t + (q(0) - a)p(0) - Q[q(1)].$$

The domain of this functional is all $\tilde{p}$ and $\tilde{q}$ constructed from piecewise C^1 functions $p(t)$ and $q(t)$, which are otherwise unrelated and unrestricted for the purpose of defining the domain. L has gradients

$$\frac{\partial L}{\partial \tilde{p}} = \left[\dot{q} - \frac{\partial \mathscr{H}}{\partial p} \quad q(0) - a \right]^{\mathrm{T}}, \quad \frac{\partial L}{\partial \tilde{q}} = \left[-\dot{p} - \frac{\partial \mathscr{H}}{\partial q} \quad p(1) - \frac{\mathrm{d}Q}{\mathrm{d}q(1)} \right]^{\mathrm{T}}.$$

Free stationary principle: this states that L is stationary with respect to any piecewise C^1 small variations $\delta p(t)$ and $\delta q(t)$ if and only if the given problem is satisfied.

Hamilton's principle: this states that L is stationary, with respect to any small variations of piecewise C^1 functions $p(t)$ and $q(t)$ which satisfy the α conditions, if and only if the β conditions, and therefore the given problem, is satisfied. In other words, the β conditions are natural in the sense of §3.7(xv). The result is more commonly stated in the case when $q(1)$ rather than $\dot{q}(1)$ is assigned (e.g. Whitaker, 1937, §99), and such a terminal condition belongs to the α instead of the β conditions.

Singularities: there is an implicit assumption so far that $\partial^2 \mathscr{L}/\partial \dot{q}^2 \neq 0$ except possibly at isolated locations. At each given q and t this permits isolated but not distributed singularities in the transformation from $\mathscr{L}(\dot{q})$ to $\mathscr{H}(p)$, for example the inflexion and cusp respectively in Fig. 2.6 are allowed. In particular, therefore, $\mathscr{H}(p)$ could be multivalued so far, like $X(x)$ there.

Saddle functional: $L[\tilde{p}, \tilde{q}]$ is concave in $\tilde{p}$ and convex in $\tilde{q}$ if $\mathscr{H}[p, q, t]$ is convex in p and concave in q, and if $Q[q(1)]$ is at least weakly concave.

Simultaneous bounds: these then have the values of

$$L - \left(\tilde{p}, \frac{\partial L}{\partial \tilde{p}} \right) = \int_0^1 \left(p\frac{\partial \mathscr{H}}{\partial p} - \mathscr{H} \right) \mathrm{d}t - Q \qquad \text{subject to } \alpha,$$

$$L - \left\langle \tilde{q}, \frac{\partial L}{\partial \tilde{q}} \right\rangle = \int_0^1 \left(q\frac{\partial \mathscr{H}}{\partial q} - \mathscr{H} \right) \mathrm{d}t - ap(0) + q(1)\frac{\mathrm{d}Q}{\mathrm{d}q(1)} - Q \quad \text{subject to } \beta.$$

Until an upper bound is assured it can be a misnomer to describe solutions of Hamilton's equations as 'extremals' of Hamilton's principle.

Evaluation: p_α can be eliminated from the upper bound via the inverted α equation $p_\alpha = \partial \mathscr{L}/\partial \dot{q}_\alpha$, giving the functional

$$L_\alpha[q_\alpha(t)] = \int_0^1 \mathscr{L}[\dot{q}_\alpha(t), q_\alpha(t), t]\,\mathrm{d}t - Q[q_\alpha(1)]$$

of any $q_\alpha(t)$ satisfying $q_\alpha(0) = a$ and with smoothness (often piecewise C^2) consistent with both $p_\alpha(t)$ and $q_\alpha(t)$ being piecewise C^1 in $\dot{q}_\alpha = \partial\mathcal{H}/\partial p_\alpha$. The statement that $L_\alpha[q_\alpha(t)]$ is minimized if and only if the minimizing $\hat{q}_\alpha(t)$ is a solution of the given problem would be a strengthened version of Hamilton's principle, which itself is only a stationary principle. The principle of least action, in its original form (Whitaker, 1937, §100), is not over a fixed time interval but over one implicitly defined by the constraint that $\mathcal{H}$ has a fixed value. This is consistent with solutions of Hamilton's equations if $\mathcal{L}[\dot{q}, q]$ does not contain t explicitly.

q_β can be eliminated from the lower bound if the β equation $-\dot{p}_\beta = \partial\mathcal{H}/\partial q_\beta$ can be written as

$$q_\beta = \partial g/\partial \dot{p}_\beta \quad \text{for} \quad g[p, \dot{p}, t] = \dot{p}q + \mathcal{H}$$

via another Legendre transformation. Then the lower bound becomes the functional

$$L_\beta[p_\beta(t)] = -\int_0^1 g[p_\beta(t), \dot{p}_\beta(t), t]\,\mathrm{d}t - ap_\beta(0) + P[p_\beta(1)],$$

where $P[p(1)]$ is the Legendre transform of $Q[q(1)]$ with $q(1) = \mathrm{d}P/\mathrm{d}p(1)$ and $P + Q = p(1)q(1)$; L_β is a functional of any $p_\beta(t)$ with smoothness (often piecewise C^2) consistent with both $p_\beta(t)$ and $q_\beta(t)$ being piecewise C^1 in $-\dot{p}_\beta = \partial\mathcal{H}/\partial q_\beta$. Also, $p_\beta(1)$ is assigned when $\partial\mathcal{L}/\partial\dot{q}$ does not contain q, so that $Q[q(1)]$ is linear and $P = 0$ (an accumulated singularity).

Legendre transformations: parts of Fig. 2.18 can be recognized in both the interior and boundary terms of this problem as in Fig. 3.5.

Fig. 3.5. Legendre transformations.

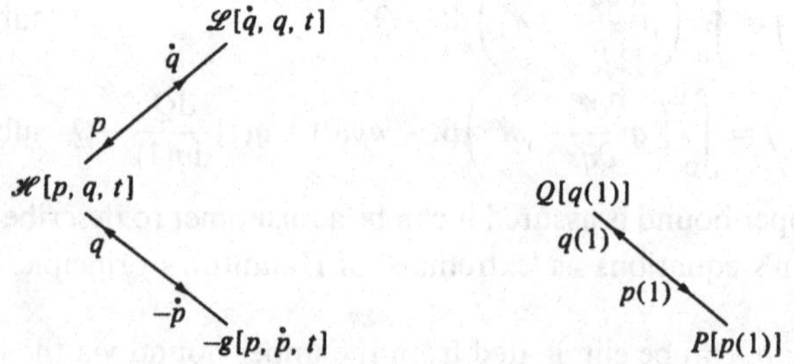

Uniqueness: recalling Fig. 2.19, sufficient conditions for $L[\tilde{p}, \tilde{q}]$ to be a saddle functional are that $\mathcal{L}[\dot{q}, q, t]$ is a jointly strictly convex function of $\dot{q}$ and q (in particular that $\partial^2\mathcal{L}/\partial\dot{q}^2 > 0$, $\partial^2\mathcal{L}/\partial q^2 > 0$, $\partial^2\mathcal{L}/\partial\dot{q}\partial q = 0$) with weakly concave $Q[q(1)]$. Then there is at most one solution $q(t)$ of the given problem.

Example: A particle with fixed mass $m>0$ and given concave potential energy $V(q)$ (e.g. $V=-\mu/q$ for the inverse square law of force in $q>0$, with fixed $\mu>0$) has $\mathscr{L}[\dot{q},q]=\frac{1}{2}m\dot{q}^2-V(q)$. Then the bounds are

$$L_\alpha[q_\alpha(t)]=\int_0^1[\tfrac{1}{2}m\dot{q}_\alpha{}^2-V(q_\alpha)]\mathrm{d}t-mbq_\alpha(1)$$

for any piecewise C^2 $q_\alpha(t)$ such that $q_\alpha(0)=a$; and

$$L_\beta[p_\beta(t)]=-\int_0^1 g(\dot{p}_\beta)\mathrm{d}t-ap_\beta(0)$$

for any piecewise C^2 $p_\beta(t)$ such that $p_\beta(1)=mb$, where $g(\dot{p})$ is now the Legendre dual of suitably smooth $-V(q)$ via $-\dot{p}=\partial V/\partial q$.

Generalization: the same structure applies when the scalar $q(t)$ is replaced by a matrix $\mathbf{q}(t)$ of n unknown scalar functions. Then it will be markedly easier to satisfy the constraints required to calculate the bounds $L_\alpha[\mathbf{q}_\alpha(t)]$ and $L_\beta[\mathbf{p}_\beta(t)]$ than it will be to find the exact value of L which is bracketed by them. A terminal condition is less welcome in mechanics than another initial condition would be, but it may be quite appropriate in other problems of the calculus of variations.

(iv) Föppl-Hencky equation

Given problem: find $x(t)$ in $0\leqslant t\leqslant 1$ satisfying

$$x''+\frac{3}{t}x'=-\frac{k}{x^2},\quad x'(0)=0,\quad x'(1)+bx(1)=0,$$

where k and b are positive constants. This is one of a number of nonlinear problems for which bounds are described by Arthurs (1980, §6.3), and we discuss it here so that the reader may compare the two approaches. The variable x relates to the stress in a simply supported circular elastic membrane under given normal pressure, and $1-b$ is Poisson's ratio.

New variable: if $u=-x'$ then $u'+3ut=k/x^2$.

Adjointness: with some choice of formally adjoint differential operators D and D^* a pair of differential equations

$$Dx=f[x,u,t],\quad D^*u=g[x,u,t]$$

have Hamiltonian form if and only if the given functions on the right satisfy the integrability condition $\partial f/\partial x=\partial g/\partial u$. In the present case this cannot be satisfied with the simplest adjoints $Dx=x'$ and $D^*u=-u'$. However, the integrating factor t^3 leads to $-(t^3u)'=-kt^3/x^2$. Then $Dx=x'$, $D^*u=$

$-(t^3u)'$ gives

$$Dx = \frac{\partial h}{\partial u}, \quad D^*u = \frac{\partial h}{\partial x} \quad \text{with} \quad h[x, u, t] = \frac{kt^3}{x} - \tfrac{1}{2}u^2.$$

Such D and D^* are formal adjoints in the sense of

$$-\int_0^1 x(t^3u)'\,dt + x(1)u(1) = \int_0^1 t^3ux'\,dt.$$

This is valid for any unrelated piecewise C^1 $x(t)$ and $u(t)$.

Inner products: we can now recognise the integrating factor t^3 as a weighting function in a choice of spaces with elements $\tilde{x} = [x(t)\ x(1)]^T$, $u = u(t)$ and inner products $(\tilde{x}, \tilde{y}) = \int_0^1 xy\,dt + x(1)y(1)$ and $\langle u, v\rangle = \int_0^1 t^3uv\,dt$.

Hamiltonian version: the given problem may now be rewritten in terms of true adjoints T and T^* such that $(\tilde{x}, T^*u) = \langle u, T\tilde{x}\rangle$ as

$$T^*u = \begin{bmatrix} -(t^3u)' \\ u(1) \end{bmatrix} = \begin{bmatrix} -\dfrac{kt^3}{x^2} \\ bx(1) \end{bmatrix} = \frac{\partial H}{\partial \tilde{x}}, \quad (\alpha)$$

$$T\tilde{x} = x' = -u = \frac{\partial H}{\partial u} \quad (\beta)$$

where

$$H[\tilde{x}, u] = \int_0^1 t^3\left(\frac{k}{x} - \tfrac{1}{2}u^2\right)dt + \tfrac{1}{2}bx(1)^2.$$

Here $\mathscr{D}(T)$ is the space of $\tilde{x}$ constructed from piecewise C^1 functions $x(t)$, and $\mathscr{D}(T^*)$ is the space of piecewise C^1 functions $u(t)$ which satisfy $u(0) = 0$. We are obliged to build this homogeneous condition at $t = 0$ into the definition of $\mathscr{D}(T^*)$ (and therefore into the α conditions) because the weighting function t^3 vanishes at $t = 0$, and so prevents $u(0)$ appearing in integration by parts $(\tilde{x}, T^*u) = \langle u, T\tilde{x}\rangle$.

Generating functional: $\langle u, T\tilde{x}\rangle = \int_0^1 t^3ux'\,dt$, therefore

$$L[\tilde{x}, u] = \int_0^1 t^3\left[ux' - \frac{k}{x} + \tfrac{1}{2}u^2\right]dt - \tfrac{1}{2}bx(1)^2.$$

The domain of this is all piecewise C^1 functions $u(t)$ satisfying $u(0) = 0$, with $\tilde{x} = [x(t)\ x(1)]^T$ constructed from piecewise C^1 functions $x(t)$ of one sign. Its gradients are

$$\frac{\partial L}{\partial \tilde{x}} = \left[-(t^3u)' + \frac{kt^3}{x^2} \quad u(1) - bx(1)\right]^T, \quad \frac{\partial L}{\partial u} = x' + u.$$

Bounds: under the extra restriction $x(t) > 0$ $L[\tilde{x}, u]$ is a strict saddle functional concave in $\tilde{x}$ and convex in u. Then there are simultaneous bounds whose values are

$$L - \left(\tilde{x}, \frac{\partial L}{\partial \tilde{x}}\right) = \int_0^1 t^3\left[-\frac{2k}{x} + \tfrac{1}{2}u^2\right]\mathrm{d}t + \tfrac{1}{2}bx(1)^2 \quad \text{subject to } \alpha,$$

$$L - \left\langle u, \frac{\partial L}{\partial u}\right\rangle = -\int_0^1 t^3\left[\frac{k}{x} + \tfrac{1}{2}u^2\right]\mathrm{d}t - \tfrac{1}{2}bx(1)^2 \quad \text{subject to } \beta.$$

Evaluation: the α conditions are satisfied by $1/x_\alpha = +[(t^3u_\alpha)'/kt^3]^{\frac{1}{2}}$ for any piecewise C^2 $u_\alpha(t)$ such that

$$u_\alpha(1)[u_\alpha'(1) + 3u_\alpha(1)]^{\frac{1}{2}} = bk^{\frac{1}{2}} \quad \text{and} \quad u_\alpha(0) = 0.$$

As a functional of such $u_\alpha(t)$, the upper bound is

$$L_\alpha[u_\alpha(t)] = \int_0^1 t^3\left[\tfrac{1}{2}u_\alpha{}^2 - 2\left(\frac{k(t^3u_\alpha)'}{t^3}\right)^{\frac{1}{2}}\right]\mathrm{d}t + \tfrac{1}{2}\frac{u_\alpha(1)^2}{b}.$$

Perhaps the simplest trial function is $u_\alpha(t) = (\tfrac{1}{2}bk^{\frac{1}{2}})^{\frac{2}{3}}t$ giving $L_\alpha = (\tfrac{1}{2}bk^2)^{\frac{1}{3}}((b-18))/24$.

The β conditions are satisfied by $u_\beta = -x_\beta{}'$ with any piecewise C^2 positive function $x_\beta(t)$, giving the lower bound

$$L_\beta[x_\beta(t)] = -\int_0^1 t^3\left[\frac{k}{x_\beta} + \tfrac{1}{2}x_\beta'^2\right]\mathrm{d}t - \tfrac{1}{2}bx_\beta(1)^2.$$

The simplest such trial function is a positive constant, and the greatest lower bound so obtained is $L_\beta = -\tfrac{3}{2}b^{\frac{1}{3}}(k/4)^{\frac{2}{3}}$.

The gap between even these crude bounds is $L_\alpha - L_\beta = bL_\alpha/(18-b)$, implying a maximum error of 5.6% for any realistic Poisson ratio, corresponding to $b = 1$. Arthurs refers to methods of improving bounds. Here we have preferred to indicate how simple bounds can be obtained quickly.

(v) A partial differential equation

Given problem: in a fixed two or three dimensional spatial region V with boundary $\Sigma = \Sigma_1 + \Sigma_2$, find a scalar function ϕ of position such that

$$-\nabla\cdot(\rho\nabla\phi) + r\phi = s \quad \text{in } V,$$

$$\phi = a \quad \text{on } \Sigma_1, \quad \mathbf{n}\cdot\nabla\phi = b \quad \text{on } \Sigma_2,$$

where $\mathbf{n}$ is the unit outward normal to Σ. Here ρ, r, s, a and b are given functions of position in the indicated regions. The smoothness assumed for them, and for Σ, will affect the smoothness which can be expected of a solution ϕ. In formal respects the problem is a direct generalization to higher dimensions of one of the cases of (3.78) which we treated in detail in §3.7. We assume that ρ is a C^1 function and positive almost everywhere.

Physical significance: this equation includes all those classical problems, for example in electromagnetic theory and fluid mechanics, which are associated with the equations of Poisson ($\rho = 1, r = 0, s \not\equiv 0$) and Laplace ($\rho = 1, r = s = 0$), including Dirichlet ($\Sigma_2 = 0$) and Neumann ($\Sigma_1 = 0$) boundary conditions. It also includes certain diffusion problems, as mentioned below.

Adjointness: this will be a variant of (3.20), namely

$$-\int \phi\nabla\cdot(\rho\mathbf{u})\,\mathrm{d}V + \int \phi\mathbf{n}\cdot\mathbf{u}\rho\,\mathrm{d}\Sigma_2 = \int \rho\mathbf{u}\cdot\nabla\phi\,\mathrm{d}V - \int \rho\mathbf{u}\cdot\mathbf{n}\phi\,\mathrm{d}\Sigma_1,$$

valid for any scalar ϕ and any vector $\mathbf{u}$ which are each piecewise C^1.

Inner product spaces: recalling (3.5) and (3.7) choose

$$\tilde{\phi} = [\phi(V) \quad \phi(\Sigma_2)]^{\mathrm{T}}, \quad \tilde{\mathbf{u}} = [\mathbf{u}(V) \quad \mathbf{u}(\Sigma_1)]^{\mathrm{T}},$$

constructed from piecewise C^1 functions with values ϕ and $\mathbf{u}$, and

$$(\tilde{\phi}, \tilde{\psi}) = \int \phi\psi\,\mathrm{d}V + \int \rho\phi\psi\,\mathrm{d}\Sigma_2, \quad \langle\tilde{\mathbf{u}}, \tilde{\mathbf{v}}\rangle = \int \rho\mathbf{u}\cdot\mathbf{v}\,\mathrm{d}V + \int \rho\mathbf{u}\cdot\mathbf{v}\,\mathrm{d}\Sigma_1.$$

Hamiltonian version: adjointness $(\tilde{\phi}, T^*\tilde{\mathbf{u}}) = \langle\tilde{\mathbf{u}}, T\tilde{\phi}\rangle$ implies

$$\begin{matrix} V \\ \Sigma_2 \end{matrix} \quad T^*\tilde{\mathbf{u}} = \begin{bmatrix} -\nabla\cdot(\rho\mathbf{u}) \\ \mathbf{n}\cdot\mathbf{u} \end{bmatrix} = \begin{bmatrix} r\phi - s \\ -b \end{bmatrix} = \frac{\partial H}{\partial\tilde{\phi}}, \quad (\alpha)$$

$$\begin{matrix} V \\ \Sigma_1 \end{matrix} \quad T\tilde{\phi} = \begin{bmatrix} \nabla\phi \\ -\mathbf{n}\phi \end{bmatrix} = \begin{bmatrix} -\mathbf{u} \\ -\mathbf{n}a \end{bmatrix} = \frac{\partial H}{\partial\tilde{\mathbf{u}}}, \quad (\beta)$$

with

$$H[\tilde{\phi}, \tilde{\mathbf{u}}] = \int [\tfrac{1}{2}r\phi^2 - s\phi - \tfrac{1}{2}\rho\mathbf{u}^2]\,\mathrm{d}V - \int \rho b\phi\,\mathrm{d}\Sigma_2 - \int \rho a\mathbf{n}\cdot\mathbf{u}\,\mathrm{d}\Sigma_1.$$

Generating functional:

$$L[\tilde{\phi}, \tilde{\mathbf{u}}] = \int [\rho(\mathbf{u}\cdot\nabla\phi + \tfrac{1}{2}\mathbf{u}^2) - \tfrac{1}{2}r\phi^2 + s\phi]\,\mathrm{d}V$$
$$+ \int \rho\mathbf{n}\cdot\mathbf{u}(a - \phi)\,\mathrm{d}\Sigma_1 + \int \rho b\phi\,\mathrm{d}\Sigma_2.$$

The domain of this consists of all $\tilde{\phi}$ and $\tilde{\mathbf{u}}$ constructed from piecewise C^1 functions of position ϕ and $\mathbf{u}$, and the gradients of L are

$$\begin{matrix} V \\ \Sigma_2 \end{matrix} \quad \frac{\partial L}{\partial \tilde{\phi}} = \begin{bmatrix} -\nabla\cdot(\rho\mathbf{u}) - r\phi + s \\ \mathbf{n}\cdot\mathbf{u} + b \end{bmatrix},$$

$$\begin{matrix} V \\ \Sigma_1 \end{matrix} \quad \frac{\partial L}{\partial \tilde{\mathbf{u}}} = \begin{bmatrix} \nabla\phi + \mathbf{u} \\ \mathbf{n}(a-\phi) \end{bmatrix}.$$

Saddle functional: with $\rho > 0$ almost everywhere already, L is weakly concave in $\tilde{\phi}$ and strictly convex in $\mathbf{u}$ if $r \geqslant 0$ everywhere.

Simultaneous bounds: these then have the values of

$$L - \left(\tilde{\phi}, \frac{\partial L}{\partial \tilde{\phi}}\right) = \frac{1}{2}\int [r\phi^2 + \rho\mathbf{u}^2]\,\mathrm{d}V + \int \rho a\mathbf{n}\cdot\mathbf{u}\,\mathrm{d}\Sigma_1 \qquad \text{subject to } \alpha,$$

$$L - \left\langle \tilde{\mathbf{u}}, \frac{\partial L}{\partial \tilde{\mathbf{u}}} \right\rangle = -\frac{1}{2}\int [r\phi^2 + \rho u^2 - 2s\phi]\,\mathrm{d}V + \int \rho b\phi\,\mathrm{d}\Sigma_2 \qquad \text{subject to } \beta.$$

Inequalities problem: if the current problem $\partial L/\partial\tilde{\phi} = \partial L/\partial\tilde{\mathbf{u}} = 0$ is replaced by any of the three variants (1.50), (1.54) or (1.55) generated by the same $L[\tilde{\phi}, \tilde{\mathbf{u}}]$, exactly the same expressions for the bounds apply, under the same saddle hypothesis. With (1.55), for example, the problem becomes

$$-\nabla\cdot(\rho\mathbf{u}) - r\phi + s \leqslant 0 \quad \text{in} \quad V, \qquad \mathbf{n}\cdot\mathbf{u} + b \leqslant 0 \quad \text{on} \quad \Sigma_2, \quad (\alpha)$$

$$\phi \geqslant 0, \quad \nabla\phi + \mathbf{u} = 0 \quad \text{in} \quad V, \qquad \phi = a \quad \text{on} \quad \Sigma_1, \quad (\beta)$$

$$\int \phi[-\nabla\cdot(\rho\mathbf{u}) - r\phi + s]\,\mathrm{d}V + \int \rho\phi[\mathbf{n}\cdot\mathbf{u} + b]\,\mathrm{d}\Sigma_2 = 0.$$

If there is a known scalar function w such that $s = \nabla\cdot(\rho\nabla w) - rw$, we can rewrite this problem in terms of the unknown shape $y = \phi + w$ of a generalized membrane stretched over an obstacle of given shape w (cf. §§3.7(x), 3.10(i) and Exercise 3.2.4).

Evaluation of bounds: the α conditions can be satisfied, in both the given problem and the inequalities problem, by choosing a vector field $\mathbf{u}_\alpha$ such that $\mathbf{n}\cdot\mathbf{u}_\alpha = -b$ on Σ_2, and $-\nabla\cdot(\rho\mathbf{u}_\alpha) + s = 0$ where $r = 0$ or, where $r > 0$, $\phi_\alpha = [s - \nabla\cdot(\rho\mathbf{u}_\alpha)]/r$. The upper bound is (cf. (3.100))

$$L_\alpha[\mathbf{u}_\alpha] = \frac{1}{2}\int\left[\frac{(s - \nabla\cdot(\rho\mathbf{u}_\alpha))^2}{r} + \rho\mathbf{u}_\alpha{}^2\right]\mathrm{d}V + \int \rho a\mathbf{n}\cdot\mathbf{u}_\alpha\,\mathrm{d}\Sigma_1,$$

with the r term omitted where $r = 0$. This reduces to Thomson's principle for Laplace's equation (cf. §5.3(v)). The bound can be minimized with respect to a scale factor in $\mathbf{u}_\alpha$ if $s=0$ where $r=0$ and if $b=0$ or $\Sigma_2=0$.

The lower bound is

$$L_\beta[\phi_\beta] = -\frac{1}{2}\int[r\phi_\beta{}^2 + \rho(\nabla\phi_\beta)^2 - 2s\phi_\beta]\,\mathrm{d}V + \int\rho b\phi_\beta\,\mathrm{d}\Sigma_2$$

for a scalar field ϕ_β such that $\phi_\beta = a$ on Σ_1, (and $\phi_\beta \geqslant 0$ everywhere for the inequalities problem). This bound can be maximized with respect to a scale factor in ϕ_β if $a=0$ or $\Sigma_1=0$.

A case treated by Arthurs (1980, §4.4) is $-\nabla^2\phi + m^2\phi = s$ with fixed nonzero m and s, and $\phi=0$ on Σ (i.e. $\Sigma_2=0$). The bounded quantity is then the value of $\frac{1}{2}s\int\phi\,\mathrm{d}V$ in a solution. It is proportional to a quantity called the absorption probability of a diffusion process. Arthurs gives simple numerical bounds when V is a sphere.

(vi) A free boundary problem

Given problem: diffusion through a porous dam. We use this example to represent a variety of free and moving boundary problems. Elliott and Ockendon (1982) describe how a number of such problems, including this one, each lead to a variational inequality. Numerical solutions of such problems for dams are detailed by Oden and Kikuchi (1980). Here we demonstrate how the dam problem may be regarded as an example of (1.55), and we derive a generating saddle functional and expressions for upper and lower bounds. Fig. 3.6 shows a side view of a straight-walled dam occupying $0 \leqslant x \leqslant l$. Water diffuses slowly through, according to D'Arcy's law (4.140) relating velocity to pressure gradient, from the reservoir in $x \leqslant 0$ to the river in $x \geqslant l$, thus defining a surface $y=h(x)$ which is called a *free boundary* of the fluid because the function $h(x)$ is unknown in advance. The

Fig. 3.6. Porous dam.

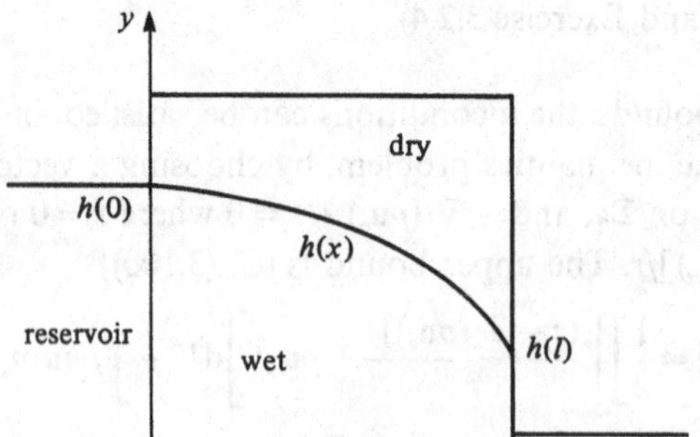

ab initio governing equations in the *wet region* $0 \leqslant y \leqslant h(x)$ can be expressed in terms of a second unknown function, the pressure $p[x, y] \geqslant 0$, as follows.

$$\nabla^2 p = 0 \quad \text{in} \quad 0 \leqslant x \leqslant l, \qquad 0 \leqslant y \leqslant h(x),$$

$$p = \rho g(h(0) - y) \quad \text{on} \quad x = 0, \qquad p = \rho g(h(l) - y) \quad \text{or} \quad 0 \quad \text{on} \quad x = l,$$

$$\frac{\partial p}{\partial y} + \rho g = 0 \quad \text{on} \quad y = 0,$$

$$p = 0 \quad \text{and} \quad \frac{\partial p}{\partial y} - \frac{\mathrm{d}h}{\mathrm{d}x}\frac{\partial p}{\partial x} + \rho g = 0 \quad \text{on} \quad y = h(x).$$

Thus hydrostatic conditions are applied on the side walls by reservoir and river, and $p = 0$ on the 'seepage face' between the outlet height $h(l)$ and the river surface. Density ρ and acceleration g due to gravity are constants.

First new variable: introduce a new dependent variable

$$\theta[x, y] = \int_y^{h(x)} p[x, s]\,\mathrm{d}s.$$

This equation is often called a 'Baiocchi transformation' (1972), but was earlier used by Schatz (1969). Using $p[x, h(x)] = 0$ gives

$$\frac{\partial \theta}{\partial x} = \int_y^{h(x)} \frac{\partial p}{\partial x}[x, s]\,\mathrm{d}s, \qquad \frac{\partial \theta}{\partial y} = -p.$$

Note that these derivatives are zero on the free boundary, unlike the partial derivatives of $p[x, y]$. Using $\nabla^2 p = 0$, and then the second (mass conservation) free surface condition gives, at any point in the wet region,

$$\frac{\partial^2 \theta}{\partial x^2} = \int_y^{h(x)} \frac{\partial^2 p}{\partial x^2}[x, s]\,\mathrm{d}s + \frac{\mathrm{d}h}{\mathrm{d}x}\frac{\partial p}{\partial x}[x, h(x)] = \frac{\partial p}{\partial y} + \rho g,$$

$$\frac{\partial^2 \theta}{\partial y^2} = -\frac{\partial p}{\partial y}.$$

Hence $\nabla^2 \theta = \rho g$ in the wet region. On the free boundary we note that $\theta = 0$ from the definition, and $\partial\theta/\partial n = 0$ because $p = 0$.

Second new variable: we define the *dry region* to be the region of the dam which is above the free boundary. The horizontal top of the dry region is regarded as fixed, above the reservoir level. Let V denote the *fixed* spatial region consisting of wet and dry regions taken together, and define in it the new function

$$\phi[x, y] = \begin{cases} \theta[x, y] & \text{in the wet region,} \\ 0 & \text{in the dry region.} \end{cases}$$

This has the properties

$$\nabla^2\phi - \rho g \leqslant 0, \quad \phi \geqslant 0, \quad \phi(\nabla^2\phi - \rho g) = 0$$

since $\phi > 0$ with $\nabla^2\phi = \rho g$ within the wet region, and $\nabla^2\phi - \rho g = -\rho g < 0$ with $\phi \equiv 0$ in the dry region. Everywhere on the fixed rectangular boundary Σ (say) of V we have the Dirichlet condition $\phi = a$ where a denotes a given piecewise continuous function of position. Thus $a = 0$ on the sides and top of the dry region and on the seepage face; the given pressure conditions show that a is a quadratic function of y on the wet side walls, and a linear function of x on the bottom since $\partial^2\theta/\partial x^2 = 0$ there. Any $\phi[x, y]$ which solves the problem will be piecewise C^2.

Third new variable: putting $\mathbf{u} = -\nabla\phi$ gives the inequalities variant of a simplified version of the previous subsection, with $\Sigma_2 = 0$ and $\Sigma_1 = \Sigma$ there, as follows. Spaces of elements $\phi(V)$ and $\tilde{\mathbf{u}} = [\mathbf{u}(V) \;\; \mathbf{u}(\Sigma)]^{\mathrm{T}}$, constructed from piecewise C^1 functions $\phi(V)$ and $\mathbf{u}(V)$, are equipped with inner products

$$(\phi, \psi) = \int \phi\psi \, \mathrm{d}V, \quad \langle \tilde{\mathbf{u}}, \tilde{\mathbf{v}} \rangle = \int \mathbf{u}\cdot\mathbf{v} \, \mathrm{d}V + \int \mathbf{u}\cdot\mathbf{v} \, \mathrm{d}\Sigma,$$

so that

$$(\phi, T^*\tilde{\mathbf{u}}) = -\int \phi \nabla\cdot\mathbf{u} \, \mathrm{d}V = \int \mathbf{u}\cdot\nabla\phi \, \mathrm{d}V - \int \mathbf{u}\cdot\mathbf{n}\phi \, \mathrm{d}\Sigma = \langle \tilde{\mathbf{u}}, T\phi \rangle$$

identifies $T^*\tilde{\mathbf{u}} = -\nabla\cdot\mathbf{u}$, $T\phi = [\nabla\phi \;\; -\mathbf{n}\phi]^{\mathrm{T}}$.

Generating functional: temporarily considering the associated problem

$$-\nabla\cdot\mathbf{u} = \rho g, \qquad \nabla\phi = -\mathbf{u} \quad \text{in} \quad V, \qquad \phi = a \quad \text{on} \quad \Sigma$$

induces $H[\phi, \tilde{\mathbf{u}}]$ by partial integration, and hence

$$L[\phi, \tilde{\mathbf{u}}] = \int [\mathbf{u}\cdot\nabla\phi + \tfrac{1}{2}\mathbf{u}^2 - \rho g\phi] \, \mathrm{d}V + \int \mathbf{n}\cdot\mathbf{u}(a - \phi) \, \mathrm{d}\Sigma.$$

Governing conditions: the gradients

$$\partial L/\partial\phi = -\nabla\cdot\mathbf{u} - \rho g, \quad \partial L/\partial\tilde{\mathbf{u}} = [\nabla\phi + \mathbf{u} \quad \mathbf{n}(a - \phi)]^{\mathrm{T}}$$

allow the original dam problem to be expressed as (1.55) in the form

$$\frac{\partial L}{\partial\phi} \leqslant 0 \quad (\alpha), \qquad \phi \geqslant 0 \quad (\beta), \qquad \frac{\partial L}{\partial\tilde{\mathbf{u}}} = 0 \quad (\beta), \qquad \left(\phi, \frac{\partial L}{\partial\phi}\right) = 0.$$

The orthogonality condition is $\int \phi(\nabla\cdot\mathbf{u} + \rho g) \, \mathrm{d}V = 0$, but with the other conditions it implies $\phi(\nabla^2\phi - \rho g) = 0$ at every point of V, as required. A solution for $\phi[x, y]$ serves to define the free boundary, where $\phi \equiv 0$ ceases to apply, and hence $\theta[x, y]$ in the wet region, and hence $p[x, y] = -\partial\theta/\partial y$ there.

Simultaneous bounds: these apply since $L[\phi, \tilde{\mathbf{u}}]$ is a saddle functional weakly concave in ϕ and strictly convex in $\tilde{\mathbf{u}}$.

$$\text{Upper bound} = L - \left(\phi, \frac{\partial L}{\partial \phi}\right) = \frac{1}{2}\int \mathbf{u}_\alpha{}^2\,\mathrm{d}V + \int a\mathbf{n}\cdot\mathbf{u}_\alpha\,\mathrm{d}\Sigma$$

for any $\mathbf{u}_\alpha$ such that $\nabla\cdot\mathbf{u}_\alpha \geqslant -\rho g$.

$$\text{Lower bound} = L - \left\langle \tilde{\mathbf{u}}, \frac{\partial L}{\partial \tilde{\mathbf{u}}}\right\rangle = -\frac{1}{2}\int [\mathbf{u}_\beta{}^2 + 2\rho g\phi_\beta]\,\mathrm{d}V$$

$$= -\int [\tfrac{1}{2}(\nabla\phi_\beta)^2 + \rho g\phi_\beta]\,\mathrm{d}V$$

for any $\phi_\beta \geqslant 0$ such that $\phi_\beta = a$ on Σ.

Exercises 3.3

1. Brachistochrone. Consider the problem of finding the shape of a smooth wire in a vertical plane, down which a bead will slide between two given ordinates in the least time. This version was posed by Jakob Bernoulli in 1697. Troutman (1983) is one of many texts which discusses the classical difficulties. Show by elementary kinematics that the problem involves minimizing the functional

$$\int_0^1 \left[\frac{1 + q'^2}{q}\right]^{\frac{1}{2}} \mathrm{d}x$$

among all C^1 functions $q(x) \geqslant 0$ such that $q(0) = 0$, where x and q are horizontal and vertically downward coordinates, and $q' = \mathrm{d}q/\mathrm{d}x$.

Verify that the functions

$$\mathscr{L}[q', q] = \left[\frac{1 + q'^2}{q}\right]^{\frac{1}{2}} \quad \text{and} \quad \mathscr{H}[p, q] = -\left[\frac{1}{q} - p^2\right]^{\frac{1}{2}}$$

are Legendre dual functions with active $p = \partial\mathscr{L}/\partial q'$ and $q' = \partial\mathscr{H}/\partial p$ and passive q, and sketch their level contours. By elementary methods find the Euler–Lagrange differential equation whose C^1 solutions $q(x) \geqslant 0$ make $\int_0^1 \mathscr{L}[q', q]\,\mathrm{d}x$ stationary under suitable terminal conditions. Show that $\mathscr{H}$ = constant (say h) in any such solution, and that the latter belongs to the family of cycloids

$$q = \frac{1}{2h^2}(1 - \cos\theta), \quad x = \frac{1}{2h^2}(\theta - \sin\theta),$$

where θ is a parameter in $0 \leqslant \theta < 2\pi$. If q is not assigned at $x = 1$, show that the natural terminal condition assigns $q' = 0$ there and hence picks out the particular cycloid with $h = -(\pi/2)^{\frac{1}{2}}$.

Use the inner products and spaces of §3.10(iii), with x here replacing t there, to express the governing equations of the brachistochrone with assigned $q'[q(1+q'^2)]^{-\frac{1}{2}}=c$ (say) at $x=1$ as

$$\frac{\partial L}{\partial \tilde{p}}=\begin{bmatrix} q'-\dfrac{\partial \mathscr{H}}{\partial p} \\ q(0) \end{bmatrix}=0\ (\alpha),\quad \frac{\partial L}{\partial \tilde{q}}=\begin{bmatrix} -p'-\dfrac{\partial \mathscr{H}}{\partial q} \\ p(1)-c \end{bmatrix}=0\ (\beta),$$

in terms of the gradients of the functional

$$L[\tilde{p},\tilde{q}]=\int_0^1 [pq'-\mathscr{H}]\,\mathrm{d}x+q(0)p(0)+cq(1).$$

Elucidate the free stationary principle, and the complementary stationary principles, associated with

$$L-\left(\tilde{p},\frac{\partial L}{\partial \tilde{p}}\right)=\int_0^1 \mathscr{L}\,\mathrm{d}x-cq(1),\quad L-\left\langle \tilde{q},\frac{\partial L}{\partial \tilde{q}}\right\rangle=\int_0^1\left(q\frac{\partial \mathscr{H}}{\partial q}-\mathscr{H}\right)\mathrm{d}x.$$

Show that

$$\frac{\partial^2 \mathscr{H}}{\partial p^2}=-\frac{1}{q\mathscr{H}^3}>0,\quad \frac{\partial^2 \mathscr{H}}{\partial q^2}=\frac{3-4p^2q}{4q^4\mathscr{H}^3},$$

and hence that $\mathscr{H}[p,q]$ is convex in p, but concave in q only in the part $0\leqslant p^2q\leqslant\frac{3}{4}$ of its domain $0\leqslant p^2q\leqslant 1$. Show that it is saddle-shaped at a point in the sense of (1.9) in $0\leqslant p^2q\leqslant\frac{4}{5}$. Show that the saddle quantity

$$S=L_+-L_--\left(\tilde{p}_+-\tilde{p}_-,\left.\frac{\partial L}{\partial \tilde{p}}\right|_+\right)-\left\langle \tilde{q}_+-\tilde{q}_-,\left.\frac{\partial L}{\partial \tilde{q}}\right|_-\right\rangle$$

$$=\int_0^1\left[\mathscr{H}_--\mathscr{H}_+-(p_--p_+)\left.\frac{\partial \mathscr{H}}{\partial p}\right|_+-(q_--q_+)\left.\frac{\partial \mathscr{H}}{\partial q}\right|_-\right]\mathrm{d}x.$$

Hence show that the above cycloid with $c=0$ can be characterized by an upper bound property for the transit time, based on $S\geqslant 0$ for all C^1 $q(x)$ satisfying the α conditions (i.e. $p=q'[q(1+q'^2)]^{-\frac{1}{2}}$ with $q(0)=0$, the former being no real restriction because the Legendre transformation is non-singular); but that the *local* saddle shape of $\mathscr{H}[p,q]$ is insufficient to verify the hypothesis, because the solution has the property $p^2q=1-q\pi/2$, and when q is small enough this contour $\mathscr{H}=-(\pi/2)^{\frac{1}{2}}$ always passes outside the region where $\mathscr{H}[p,q]$ is saddle-shaped. Thus a more sophisticated verification of the hypothesis $S\geqslant 0$ is required, in terms of the integral and not just its integrand (see Troutman, 1983, p. 286).

Express the last paragraph alternatively in terms of the joint convexity of $\mathscr{L}[q',q]$ (cf. Theorem 2.11), which holds only if $q'^2\leqslant 3$. Investigate the multiplicity of the dual function obtainable from $\mathscr{H}[p,q]$ by treating q as active and p as passive, in a Legendre transformation which has an isolated singularity where $p^2q=\frac{3}{4}$.

2. Nonlinear diffusion with a mixed boundary condition. Consider the problem of finding a scalar function ϕ of position in a two or three dimensional region V with boundary Σ, such that

$$-\nabla^2\phi + W'[\phi] = 0 \quad \text{in} \quad V, \qquad a\phi + b\mathbf{n}\cdot\nabla\phi = c \quad \text{in} \quad \Sigma.$$

Here a, b and c are given functions of position on Σ, and $W[\phi]$ is a given function of ϕ in V, with $W' = \mathrm{d}W/\mathrm{d}\phi$, and $\mathbf{n}$ is the unit outward normal to Σ. Deduce that the governing equations are described by the zero gradients of either of the following two alternative generating functionals. If $a \neq 0$, then with $\phi = \phi(V)$ and $\tilde{\mathbf{u}} = [\mathbf{u}(V)\ \mathbf{u}(\Sigma)]^{\mathrm{T}}$,

$$L_1[\phi, \tilde{\mathbf{u}}] = \int [-\phi\nabla\cdot\mathbf{u} - W[\phi] + \tfrac{1}{2}\mathbf{u}^2]\,\mathrm{d}V + \iint\left[\frac{b}{2a}(\mathbf{n}\cdot\mathbf{u})^2 + \frac{c}{a}\mathbf{n}\cdot\mathbf{u}\right]\mathrm{d}\Sigma.$$

If $b \neq 0$, then with $\tilde{\phi} = [\phi(V)\ \phi(\Sigma)]^{\mathrm{T}}$ and $\mathbf{u} = \mathbf{u}(V)$,

$$L_2[\tilde{\phi}, \mathbf{u}] = \int [\mathbf{u}\cdot\nabla\phi - W[\phi] + \tfrac{1}{2}\mathbf{u}^2]\,\mathrm{d}V - \iint\left[\frac{a}{2b}\phi^2 - \frac{c}{b}\phi\right]\mathrm{d}\Sigma.$$

Develop the results corresponding to those of §§3.7(vii)–(xii) and 3.10(v), including simultaneous upper and lower bounds when $W[\phi]$ is convex and $ab \geqslant 0$, for both L_1 and L_2.

Verify in particular that if $a \neq 0$ and $b \neq 0$, then for any piecewise C^1 functions ϕ and $\mathbf{u}$ which satisfy $a\phi - b\mathbf{n}\cdot\mathbf{u} = c$ on Σ we have

$$L_2[\tilde{\phi}, \mathbf{u}] - L_1[\phi, \tilde{\mathbf{u}}] = \int \frac{c^2}{2ab}\,\mathrm{d}\Sigma.$$

Hence show that, when $ab > 0$, the solution values L_{20} and L_{10} (say) bounded by $J_1 \geqslant L_{10} \geqslant K_1$ and $J_2 \geqslant L_{20} \geqslant K_2$ only differ by a constant. Show also that

$$J_1 = \int [W_c[-\nabla\cdot\mathbf{u}_\alpha] + \tfrac{1}{2}\mathbf{u}_\alpha{}^2]\,\mathrm{d}V + \iint\left[\frac{b}{2a}(\mathbf{n}\cdot\mathbf{u}_\alpha)^2 + \frac{c}{a}\mathbf{n}\cdot\mathbf{u}_\alpha\right]\mathrm{d}\Sigma$$

for any suitably smooth $\mathbf{u}_\alpha$ and $\phi_\alpha = W_c'[-\nabla\cdot\mathbf{u}_\alpha]$, where $W_c = \phi W' - W$ is the Legendre dual of W; and that

$$K_2 = \int [-W[\phi_\beta] - \tfrac{1}{2}(\nabla\phi_\beta)^2]\,\mathrm{d}V - \iint\left[\frac{a\phi_\beta{}^2}{2b} - \frac{c}{b}\phi_\beta\right]\mathrm{d}\Sigma$$

for any piecewise $C^2\,\phi_\beta$. Notice therefore that the bounds in

$$J_1 + \int \frac{c^2}{2ab}\,\mathrm{d}\Sigma \geqslant L_{20} \geqslant K_2$$

can be calculated without requiring boundary or consistency conditions to be satisfied; but that in

$$J_2 \geqslant L_{20} \geqslant K_1 + \int \frac{c^2}{2ab}\,\mathrm{d}\Sigma$$

J_2 entails a consistency condition $aW_c'[-\nabla\cdot\mathbf{u}_\alpha] = b\mathbf{n}\cdot\mathbf{u}_\alpha + c$ on $\mathbf{u}_\alpha$, and K_1

entails the boundary condition on ϕ_β. There remains the question of which is the strongest pair among the four alternative pairs of bounds.

The special case $W[\phi]=\gamma|\phi|^3+\frac{1}{2}r\phi^2-s\phi$ for given $\gamma>0$, $r>0$, s and $a=c=0$ was used by Talbot and Willis (1985) to illustrate certain 'comparison theorems'. A general review of bounds in diffusion and reaction problems has been given by Strieder and Aris (1973).

3. A one-dimensional model for nonlinear elasticity. Let s denote the initial spatial coordinate of a typical material particle within a given interval $0 \leqslant s \leqslant 1$. Let the subsequent stress, displacement and strain for the particle be the values of functions $\sigma(s)$, $u(s)$ and $e(s)$ respectively. These are to be found from the following governing equations, in which a and c are given constants and $g(s)$ and $U[e]$ are given functions, the latter representing stored energy density. The dot denotes differentiation with respect to s, and $U'=\mathrm{d}U/\mathrm{d}e$.

Kinematics: $\dot{u}(s)=e(s), \quad u(0)=c, \quad (\alpha)$

Equilibrium: $\dot{\sigma}(s)+g(s)=0, \quad \sigma(1)=a, \quad (\beta)$

Constitutive: $\sigma(s)=U'[e(s)], \quad (\beta)$

We have labelled the two kinematical conditions α, and the three force-like conditions β. If $g(s)$ is continuous and $U[e]$ is smooth enough, an actual solution will be a set of functions $\sigma(s), u(s), e(s)$ which are C^1, C^2, C^1 respectively. This model is so simple that it is statically determinate, i.e. the equilibrium conditions can be solved alone. If g is constant $\sigma(s)=(1-s)g+a$.

Use elementary methods of the calculus of variations to calculate the first variation of the functional

$$L=\int_0^1 [-u\dot{\sigma}-\sigma e+U[e]-gu]\,\mathrm{d}s-c\sigma(0)+u(1)[\sigma(1)-a].$$

Hence show that an actual solution of the five governing equations makes L stationary with respect to arbitrary independent small piecewise C^1 variations $\delta\sigma(s)$, $\delta u(s)$ and $\delta e(s)$. Establish the converse, and hence a free stationary principle, by constructing a suitable version of the fundamental lemma (Theorem 3.8).

Verify that the *three* spaces with elements, operators and inner products

$$\tilde{\sigma}=\begin{bmatrix}\sigma(s)\\ \sigma(0)\end{bmatrix}, \quad T^*\tilde{u}=\begin{bmatrix}\dot{u}(s)\\ u(0)\end{bmatrix}, \quad (\tilde{\sigma},\tilde{\tau})=\int_0^1 \sigma\tau\,\mathrm{d}s+\sigma(0)\tau(0),$$

$$\tilde{u}=\begin{bmatrix}u(s)\\ u(1)\end{bmatrix}, \quad T\tilde{\sigma}=\begin{bmatrix}-\dot{\sigma}(s)\\ \sigma(1)\end{bmatrix}, \quad \langle\tilde{u},\tilde{v}\rangle=\int_0^1 uv\,\mathrm{d}s+u(1)v(1),$$

$$e=e(t), \qquad \overline{e,f}=\int_0^1 ef\,\mathrm{d}s$$

built from piecewise C^1 functions $\sigma(s)$, $u(s)$ and integrable $e(s)$ allow L to be regarded as a functional $L[\tilde{\sigma}, \tilde{u}, e]$ having gradients

$$\frac{\partial L}{\partial \tilde{\sigma}} = \begin{bmatrix} \dot{u} - e \\ u(0) - c \end{bmatrix}, \quad \frac{\partial L}{\partial \tilde{u}} = \begin{bmatrix} -\dot{\sigma} - g \\ \sigma(1) - a \end{bmatrix}, \quad \frac{\partial L}{\partial e} = U' - \sigma.$$

Hence show that

$$L - \left(\tilde{\sigma}, \frac{\partial L}{\partial \tilde{\sigma}}\right) = \int_0^1 [U[e] - gu]\,\mathrm{d}s - au(1),$$

$$L - \left\langle \tilde{u}, \frac{\partial L}{\partial \tilde{u}} \right\rangle - \overline{e, \frac{\partial L}{\partial e}} = \int_0^1 [U[e] - eU'[e]]\,\mathrm{d}s - c\sigma(0).$$

When $U[e]$ is convex, show that $L[\tilde{\sigma}, \tilde{u}, e]$ is weakly concave in $\tilde{\sigma}$ and convex in $\tilde{u}$ and e taken together, so that upper and lower bounds apply whose values are those of the above expressions, under α and β constraints respectively.

Regardless of whether $U[e]$ is convex or not, show that solutions of $\partial L/\partial e = 0$ considered alone may be expressed in terms of a complementary energy density function $U_c[\sigma]$ via a Legendre transformation such that $U_c' = e$ and $U + U_c = \sigma e$, where $U_c' = dU_c/d\sigma$. If $U''[e] \not\equiv 0$ the Legendre transformation may still have isolated singularities, and $U_c[\sigma]$ may be multivalued. Explore the cases

$$\begin{aligned} &U[e] = \tfrac{1}{2}e^2, && \sigma = e, && U_c[\sigma] = \tfrac{1}{2}\sigma^2, \\ &U[e] = \tfrac{1}{3}e^3 - pe, && \sigma = e^2 - p, && U_c[\sigma] = \pm\tfrac{2}{3}(\sigma + p)^{\frac{3}{2}}, \text{cusp}, \\ &U[e] = \tfrac{1}{4}e^4 + \tfrac{1}{2}be^2 - pe, && \sigma = e^3 + be - p, && U_c[\sigma] = \text{swallowtail}, \end{aligned}$$

for different given p and b. The stress–strain curve for rubber often has cubic appearance in suitably chosen variables. This model thus contains 'material nonlinearities', but not 'geometric nonlinearities' which are nonlinearities in the displacement-strain equation.

For constant g, show that every actual displacement solution of all the governing equations is included in the formula

$$u(s) = c + \frac{1}{g}[U_c[a + g] - U_c[a + (1 - s)g]].$$

In particular, use this formula in the case of the swallowtail $U_c[\sigma]$ to explore the bifurcations from one solution to three which can take place when the parameter b is allowed to vary (cf. Fig. 2.11).

Now putting to one side the fact that the solutions of the governing equations have been found directly, use the formula

$$\delta L = \left(\delta\tilde{\sigma}, \frac{\partial L}{\partial \tilde{\sigma}}\right) + \left\langle \delta\tilde{u}, \frac{\partial L}{\partial \tilde{u}} \right\rangle + \overline{\delta e, \frac{\partial L}{\partial e}}$$

to give explicit examples of complementary stationary principles (§2.7(xv)) and the Lagrange multiplier method (cf. Theorem 1.14). For example, impose the α condition $\partial L/\partial \tilde{\sigma} = 0$ on the free stationary principle. By solving this directly as

$e = \dot{u}$ subject to $u(0) = c$, show that the *total potential energy* functional

$$J[\tilde{u}] = \int_0^1 [U[\dot{u}] - gu]\,ds - au(1)$$

is stationary if and only if the β conditions taken together are satisfied. Instead of solving $\partial L/\partial\tilde{\sigma} = 0$ directly, introduce a Lagrange multiplier $\tilde{\tau}$ and an augmented functional

$$L_1 = L_1[\tilde{\sigma}, \tilde{u}, e, \tilde{\tau}] = L[\tilde{\sigma}, \tilde{u}, e] + \left(\tilde{\tau}, \frac{\partial L}{\partial\tilde{\sigma}}\right)$$

$$= L + \int_0^1 \tau(\dot{u} - e)\,ds + \tau(0)(u(0) - c),$$

as explained in §3.7(xv), and set $\delta L_1 = 0$ for arbitrary $\delta\tilde{\sigma}, \delta\tilde{u}, \delta e, \delta\tilde{\tau}$. Show that the original governing conditions emerge, but with $\sigma + \tau$ in place of σ because L is linear in $\tilde{\sigma}$, and finally deem $\tilde{\tau} = 0$ as explained after (1.127) (with $\tilde{\tau}$ here replacing u there).

Next assume $U''[e] \not\equiv 0$, and impose $\partial L/\partial e = 0$ instead of $\partial L/\partial\tilde{\sigma} = 0$. Show that the direct solution $e = U_c'[\sigma]$ of $\partial L/\partial e = 0$ allows the definition of a new functional

$$L_2[\tilde{\sigma}, \tilde{u}] = \int_0^1 [-u\dot{\sigma} - U_c[\sigma] - gu]\,ds - c\sigma(0) + u(1)[\sigma(1) - a],$$

for which $\partial L_2/\partial\tilde{\sigma} = \partial L_2/\partial\tilde{u} = 0$ have the same set of solutions as the original problem. If, instead of solving $\partial L/\partial e = 0$ directly, we introduce a Lagrange multiplier function $\lambda(s)$ and an augmented functional

$$L_3[\tilde{\sigma}, \tilde{u}, e, \lambda] = L[\tilde{\sigma}, \tilde{u}, e] + \overline{\lambda, \frac{\partial L}{\partial e}} = L + \int_0^1 \lambda(U'[e] - \sigma)\,ds,$$

demonstrate the equivalence of the original problem to $\partial L_3/\partial\tilde{\sigma} = \partial L_3/\partial\tilde{u} = \partial L_3/\partial e = \partial L_3/\partial\lambda = 0$. Show that these equations are an infinite dimensional example of (1.118), in which only the case corresponding to (1.124) applies since $U''[e] \not\equiv 0$, so that $\lambda = 0$ is finally implied, and not just deemed to hold. Show that $\partial L/\partial e = 0$ implies

$$L - \left\langle \tilde{u}, \frac{\partial L}{\partial\tilde{u}} \right\rangle = -\int_0^1 U_c[\sigma]\,ds - c\sigma(0).$$

This is the negative of the *total complementary energy* in this statically determinate problem.

4. A moving boundary problem for electro-chemical machining. A plane annular region (Fig. 3.7(a)) containing electrolyte has a fixed cathode Σ as external boundary and a contracting anode Σ_t as internal boundary. As time t increases from zero the anode dissolves from its initial position Σ_0. This process shapes a long cylindrical metal part perpendicular to the plane.

Let $\mathbf{r}$ be the planar position vector within Σ, and suppose that the shape of Σ_t

can be expressed as $t = l(\mathbf{r})$ for all $t \geq 0$, where $l(\mathbf{r})$ is a function to be found. Suppose that the space origin is always within the anode until the latter disappears, after time $l(0) = \tau$ (say), which is unknown. For example, if $l(\mathbf{r}) = \tau - |\mathbf{r}|$ the anode would be circular of radius $\tau - t$. Denote the space–time domains shown schematically in Fig. 3.7(*b*) as follows,

$$A\text{: within } \Sigma_0\text{, and for } 0 \leq t \leq l(\mathbf{r}),$$
$$B\text{: within } \Sigma_0\text{, and for } l(\mathbf{r}) \leq t,$$
$$C\text{: between } \Sigma \text{ and } \Sigma_0 \text{ for } 0 \leq t.$$

We could impose $t \leq \tau$ on $B \cup C$ without loss of generality. Let the given function $v(t) > 0$ be the potential difference across the electrodes, and $b > 0$ be a given constant representing rate of removal of anode material. The *ab initio* problem is to find the potential field $\phi[\mathbf{r}, t]$ and the function $l(\mathbf{r})$ satisfying

$$\nabla^2\phi = 0 \text{ in the liquid, i.e.}\quad \text{on}\quad B \cup C,$$
$$\phi = 0 \text{ on } \Sigma \text{ and } \phi = v(t) \quad \text{on} \quad \Sigma_t, \quad \forall t \geq 0,$$
$$\nabla\phi \cdot \nabla l = b \quad \text{on} \quad \Sigma_t \quad \forall t \geq 0.$$

By introducing the new variable

$$\theta[\mathbf{r}, t] = \begin{cases} \displaystyle\int_{l(\mathbf{r})}^{t} [v(s) - \phi[\mathbf{r}, s]]\,\mathrm{d}s & \text{in } B \\ \displaystyle\int_{0}^{t} [v(s) - \phi[\mathbf{r}, s]]\,\mathrm{d}s & \text{in } C \end{cases}$$

show, by the 'maximum principle' of potential theory, that $\theta \geq 0$ in $B \cup C$. Show also that

$$\nabla\theta = -\int_{l(\mathbf{r})}^{t} \nabla\phi[\mathbf{r}, s]\,\mathrm{d}s \text{ in } B, \quad \frac{\partial\theta}{\partial t} = v(t) - \phi[\mathbf{r}, t] \text{ in } B \cup C,$$

$$\nabla^2\theta = \begin{cases} b \text{ in } B, \\ 0 \text{ in } C. \end{cases}$$

Fig. 3.7. Electro-chemical machining. (*a*) Fixed cathode Σ and contracting anode Σ_t. (*b*) Schematic time–space region.

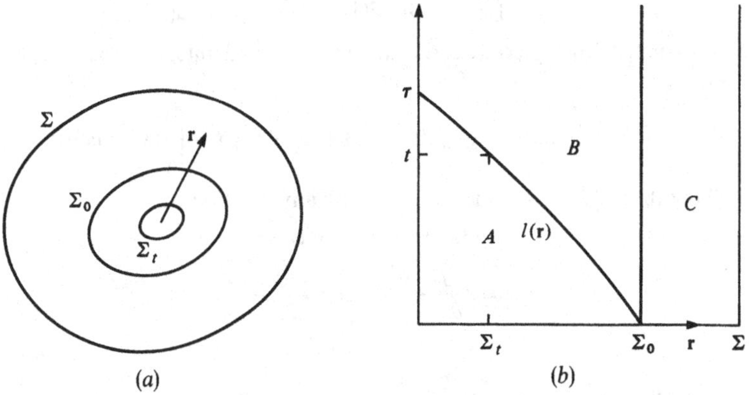

Next, by regarding A as the analogue of the dry region in the dam problem, introduce the new function

$$\xi[\mathbf{r},t]=\begin{cases}\theta[\mathbf{r},t] & \text{in } B\cup C,\\ 0 & \text{in } A,\end{cases}$$

and show that if

$$g=\begin{cases}b \text{ in } A\cup B,\\ 0 \text{ in } C,\end{cases}$$

then in the fixed space–time domain $A\cup B\cup C$

$$\nabla^2\xi-g\leqslant 0,\quad \xi\geqslant 0,\quad \xi(\nabla^2\xi-g)=0.$$

Show also that, $\forall t$, $\xi=\int_0^t v(s)\mathrm{d}s=a(t)$ (say) on Σ and $\xi=\partial\xi/\partial n=0$ on Σ_t. The results thus far were given in essence by Elliott (1980), who then expressed the problem as a variational inequality and solved it numerically.

We see that, for each given t, $\xi[\mathbf{r},t]$ must satisfy a linear complementarity problem (cf. (1.30) and Fig. 1.6) within the entire interior V of Σ, including the region occupied by the anode, and with a Dirichlet boundary condition on Σ. By introducing another new variable $\mathbf{u}=-\nabla\xi$ as in the dam problem, show how to represent the problem as an example

$$\frac{\partial L}{\partial\xi}\leqslant 0\ \ (\alpha),\qquad \xi\geqslant 0\ \ (\beta),\qquad \frac{\partial L}{\partial\tilde{\mathbf{u}}}=0\ \ (\beta),\qquad \left(\xi,\frac{\partial L}{\partial\xi}\right)=0.$$

of (1.55) generated by

$$L[\xi,\tilde{\mathbf{u}}]=\int[\mathbf{u}\cdot\nabla\xi+\tfrac{1}{2}\mathbf{u}^2-g\xi]\mathrm{d}V+\int\mathbf{n}\cdot\mathbf{u}(a-\xi)\mathrm{d}\Sigma$$

and develop expressions for the upper and lower bounds. Mackie (1986) gives a number of results along these lines, for this and similar physical problems.

5. Embedding of a simple control problem. The functional

$$L[\bar{\lambda},\bar{\xi}]=\int_0^1(F[x,u]-\lambda f[x,u]-\dot{\lambda}x)\mathrm{d}t-a\lambda(0)+b\lambda(1)$$

contains given C^1 functions $F[x,u]$, $f[x,u]$ and given scalars a, b. Its domain consists of inner product space elements

$$\bar{\lambda}=[\lambda(t)\quad\lambda(0)\quad\lambda(1)]^{\mathrm{T}},\quad \bar{\xi}=[x(t)\quad u(t)]^{\mathrm{T}},$$

constructed from piecewise C^1 $x(t)$ and $\lambda(t)$, and integrable $u(t)$, with inner products

$$(\bar{\lambda},\bar{\mu})=\int_0^1\lambda\mu\,\mathrm{d}t+\lambda(0)\mu(0)+\lambda(1)\mu(1),\quad \langle\bar{\xi},\bar{\eta}\rangle=\int_0^1(xy+uv)\,\mathrm{d}t.$$

Show that $L[\bar{\lambda},\bar{\xi}]$ is made stationary by any solution of

$$\dot{x}=f,\quad x(0)=a,\quad x(1)=b,\quad(\alpha)$$

$$-\dot{\lambda}=\lambda\frac{\partial f}{\partial x}-\frac{\partial F}{\partial x},\quad 0=\lambda\frac{\partial f}{\partial u}-\frac{\partial F}{\partial u}\quad(\beta).$$

Show that $L-(\bar{\lambda}, \partial L/\partial\bar{\lambda}) = \int_0^1 F[x,u]\,dt$. Show also that an upper bound to this integral is provided by any solution of the α conditions alone if $F[x,u]$ is jointly convex and if $f[x,u]$ is linear. In the control problem the α conditions are given as 'necessary', and the objective is to find a 'control' function $u(t)$ which minimizes $\int_0^1 F\,dt$.

3.11. Variational inequalities

(i) Introduction

We have already shown how the theoretical framework developed in this book delivers upper and lower bounds characterizing the solutions of the obstacle problem in §§3.7(x) and 3.10(ii), and the dam problem in §3.10(vi). These are problems which are governed by inequalities of the type (1.55), and each is expressed in terms of the gradients of a saddle functional. The latter are displayed explicitly, and a similar procedure applies, for example, to the problems of incipient cavitation in a pipe (Exercise 3.2.4) and of electro-chemical machining of an anode (Exercise 3.3.4), and to other problems having certain linear or nonlinear 'complementarity' features, or a free or moving boundary, or both aspects (as in plasticity, where the boundary between elastic and plastic regions is one of the unknowns).

The procedure which we have described is straightforward, in the sense that it requires only the identification of an underlying saddle functional, and its differentiation. The problems referred to above are often attacked in the literature by the method of 'variational inequalities', which in its standard form is superficially quite different from our procedure. There must, however, be a close connection between the two approaches, since their outcomes have features in common, in particular their scope as a basis for approximation.

It is important to note this connection, because a large amount of expertise in the theory and numerical solution of variational inequalities is available. Many examples of variational inequalities can be found in the books of Cottle *et al.* (1980), Elliott and Ockendon (1982), Friedman (1982), Kinderlehrer and Stampacchia (1980), and Lions (1971).

(ii) A general definition

The purpose in this section is to indicate briefly a connection which standard variational inequalities have with the problems (1.50)–(1.55) generated by a given functional $L[x,u]$, in the generality of §3.6(i).

We offer a fresh view here, in which we shall be guided by the following fact. Any solution x_0, u_0 of the infinite dimensional version of (1.50), (1.53),

(1.54) or (1.55) satisfies

$$-\left(x_0, \frac{\partial L}{\partial x}\bigg|_\alpha\right) \geqslant 0, \quad \left\langle u_\alpha, \frac{\partial L}{\partial u}\bigg|_0\right\rangle \geqslant 0, \quad \left\langle u_0, \frac{\partial L}{\partial u}\bigg|_0\right\rangle = 0, \qquad (3.123)$$

$$-\left(x_\beta, \frac{\partial L}{\partial x}\bigg|_0\right) \geqslant 0, \quad \left\langle u_0, \frac{\partial L}{\partial u}\bigg|_\beta\right\rangle \geqslant 0, \quad \left(x_0, \frac{\partial L}{\partial x}\bigg|_0\right) = 0. \qquad (3.124)$$

This is true regardless of whether $L[x,u]$ has any saddle or convexity properties.

Choose nonnegative constants a, b, c, d. Define an *α variational inequality* to be the problem of finding the set $\{x_1, u_1\}$ (say) of points which satisfy

$$-a\left(x_1, \frac{\partial L}{\partial x}\bigg|_\alpha\right) + b\left\langle u_\alpha - u_1, \frac{\partial L}{\partial u}\bigg|_1\right\rangle \geqslant 0 \qquad (3.125)$$

for every point x_α, u_α which satisfies (1.50α). Define a *β variational inequality* to be the problem of finding the set $\{x_2, u_2\}$ (say) of points which satisfy

$$c\left(x_2 - x_\beta, \frac{\partial L}{\partial x}\bigg|_2\right) + d\left\langle u_2, \frac{\partial L}{\partial u}\bigg|_\beta\right\rangle \geqslant 0 \qquad (3.126)$$

for every point x_β, u_β which satisfies (1.50β).

Such definitions will have most interest if it can be shown that $\{x_1, u_1\} \equiv \{x_0, u_0\}$ and/or $\{x_2, u_2\} \equiv \{x_0, u_0\}$. They are offered here for two reasons. First, we can see at once that $\{x_0, u_0\} \subseteq \{x_1, u_1\}$ and $\{x_0, u_0\} \subseteq \{x_2, u_2\}$, by comparing (3.123) with (3.125), and (3.124) with (3.126). Second, they imply familiar versions of standard variational inequalities in the cases of certain examples displayed in §(iii) below.

The two properties just mentioned may be possessed by other suggestions in the spirit of (3.125) and (3.126). A full comparative study relating such inequalities to the standard viewpoint referenced above is not available. In particular it will be necessary to answer the converse questions of what additional hypotheses lead to $\{x_1, u_1\} \subseteq \{x_0, u_0\}$ and $\{x_2, u_2\} \subseteq \{x_0, u_0\}$, and hence equivalence. We may expect, however, that $L[x, u]$ would need to satisfy $S > 0$ or $S \geqslant 0$ over some domain, in terms of the saddle quantity S defined in (3.66). In the literature it is often shown that solutions of variational inequalities are equivalent to the solutions of complementarity problems, or of saddle point problems (as in minimax theory – see Ekeland and Temam, 1976), or to certain minimizers. In the latter connection we recall the way in which (3.123) and (3.124) enter the proofs of the upper and lower bounds (3.68) in Theorem 3.4, by direct generalization of the steps following (1.38) and (1.39). Under hypothesis (1.58c), that $S \geqslant 0$ for all pairs of points satisfying the α conditions, we deduced that

$$L_\alpha - \left(x_\alpha, \frac{\partial L}{\partial x}\bigg|_\alpha \right) - L_0 \geqslant -\left(x_0, \frac{\partial L}{\partial x}\bigg|_\alpha \right) + \left\langle u_\alpha, \frac{\partial L}{\partial u}\bigg|_0 \right\rangle \geqslant 0. \quad (3.127)$$

Under hypothesis (1.58e) that $S \geqslant 0$ for all pairs of points satisfying the β conditions, we deduced that

$$L_0 - L_\beta - \left\langle u_\beta, \frac{\partial L}{\partial u}\bigg|_\beta \right\rangle \geqslant -\left(x_\beta, \frac{\partial L}{\partial x}\bigg|_0 \right) + \left\langle u_0, \frac{\partial L}{\partial u}\bigg|_\beta \right\rangle \geqslant 0. \quad (3.128)$$

(iii) Examples

1. If x is entirely absent so that $L[x, u] = J[u]$, a given functional of u alone, then (1.54) reduces to

$$J'[u] \geqslant 0 \quad (\beta), \qquad u \geqslant 0 \quad (\alpha), \qquad \langle u, J'[u] \rangle = 0, \qquad (3.129)$$

which is the generalization of (1.30). Then (3.125) with $b > 0$ becomes

$$\langle u_\alpha - u_1, J'[u_1] \rangle \geqslant 0 \qquad (3.130)$$

as posed in Exercise 1.3.5; and (3.126) with $d = 0$ becomes vacuous.

2. The obstacle problem in §3.10(ii) was shown to be an example of (1.55), in whch $\partial L/\partial u|_\beta = 0$. Choosing $c > 0$ in (3.126) and transcribing, the β variational inequality defines the problem of finding piecewise C^1 functions $x_2(t)$ and $u_2(t)$ such that

$$\int_0^1 (x_2 - x_\beta)(-u_2' + w'')\mathrm{d}t \geqslant 0 \qquad (3.131)$$

for all piecewise C^2 $x_\beta(t) \geqslant 0$ such that $x_\beta(0) = a$, $x_\beta(1) = b$. The fact that any solution $x_0(t), u_0(t)$ of the obstacle problem satisfies (3.131) allows the inequality to be rearranged as follows. Writing $x_2 = x_0$ and $u_2 = u_0 = -x_0'$ leads to

$$\int_0^1 x_0'(x_\beta' - x_0')\mathrm{d}t \geqslant \int_0^1 w''(x_\beta - x_0)\mathrm{d}t, \qquad (3.132)$$

which has the familiar form ensuing from a standard variational inequality.

3. The dam problem in §3.10(vi) was also shown to be an example of (1.55). Again choosing $c > 0$ in (3.126) and transcribing leads to the problem of finding piecewise C^1 scalars ϕ_2 and vectors $\mathbf{u}_2$ such that

$$\int_0^1 (\phi_2 - \phi_\beta)(-\nabla \cdot \mathbf{u}_2 - \rho g)\mathrm{d}V \geqslant 0 \qquad (3.133)$$

for all piecewise C^2 $\phi_\beta \geqslant 0$ in V, such that $\phi_\beta = a$ on Σ. Since any solution

ϕ_0, $\mathbf{u}_0$ of the dam problem satisfies (3.133), putting $\phi_2 = \phi_0$ and $\mathbf{u}_2 = \mathbf{u}_0 = -\nabla\phi_0$ leads to

$$\int \nabla\phi_0 \cdot \nabla(\phi_\beta - \phi_0)\mathrm{d}V \geqslant -\int \rho g(\phi_\beta - \phi_0)\mathrm{d}V. \qquad (3.134)$$

Thus we recover again the familiar version of a variational inequality, as we see from Elliott and Ockendon (1982, equation (4.34)), for example.

3.12. Nonnegative operator equations

(i) Introduction

We now return to the generality of any inner product space with elements x, y,... and inner product (x, y), for which we defined a nonnegative operator P in (3.43). Let $E[x]$ be a given functional having a gradient $E'[x]$ in the sense of (3.50). Consider the nonnegative operator equation

$$Px + E'[x] = 0. \qquad (3.135)$$

This can be nonlinear if $E[x]$ is not quadratic, for example if

$$E[x] = \gamma \| x \|^3 + \tfrac{1}{2}(x, Ax) - (a, x)$$

with given a, self-adjoint A (see (3.42)) and scalar γ, and $\| x \| = (x, x)^{\frac{1}{2}}$.

Equations of type (3.135) occur in particular when $P = T^*T$ (see Theorem 3.1). They may also occur, for example, if $P = A + T^*B^{-1}T$ in the sense of (3.115), where $Px = \lambda Cx$ gives no negative eigenvalues to λ with suitable positive C. Here we will confine attention to $P = T^*T$, where T and T^* are true adjoints in the sense of (3.10), not merely formal adjoints. Then

$$T^*Tx + E'[x] = 0 \qquad (3.136)$$

conveys the whole problem. This could be an integral equation, for example. Our use of true adjoints means that it could also be the whole composite description of a boundary value problem in terms of a differential equation together with boundary conditions. The latter do not have to be appended as auxiliary statements. The differential operator in such cases will be decomposable as D^*D into formal adjoints $D \subseteq T$ and $D^* \subseteq T^*$.

Linear equations

$$T^*Tx = a - Ax \qquad (3.137)$$

are included in (3.136) with $E[x] = \frac{1}{2}(x, Ax) - (a, x)$. If (3.137) is an integral equation it is said to be of the second kind if $A \neq 0$ and constant; if $A = 0$ it is of the first kind.

(ii) Examples

We gave a list of examples of the operator T^*T in §3.3(iv) which have the property $\mathscr{R}(T)=\mathscr{D}(T^*)$ or $\mathscr{R}(T)\subset\mathscr{D}(T^*)$. Associated nonnegative operator equations (3.136) include the following.

1. Find an integrable scalar function $x(t)$ on $a\leqslant t\leqslant b$ which satisfies the integral equation

$$\int_a^b p[t,\tau]x(\tau)\mathrm{d}\tau + e[x(t),t]=0, \tag{3.138}$$

where $e[x,t]$ is an assigned scalar function, $p[t,\tau]$ is given by $(3.36)_1$, and T and T^* are defined by (3.29). The inner product (3.30) implies $E[x(t),t]=\int_a^b\int_c^{x(t)}e[s,t]\,\mathrm{d}s\,\mathrm{d}t$ for any fixed c in order to achieve $E'[x]=e[x]$ at each fixed t. Equation (3.138) might originally appear in terms of the inverse function $x[e,t]$ of $e[x,t]$ as a Hammerstein equation

$$\int_a^b p[t,\tau]x[e(\tau),\tau]\mathrm{d}\tau + e(t)=0$$

for the determination of $e(t)=e[x(t),t]$.

2. Find an integrable function $x(t)$ on $0\leqslant t\leqslant d$ satisfying

$$\sum_{i=1}^{\infty} y_i(t)\int_0^d y_i(\tau)x(\tau)\mathrm{d}\tau = a(t), \tag{3.139}$$

where $a(t)$ and every $y_i(t)$ are given functions of t, and d is given. This is a linear equation (3.137) with $A=0$ and with T^*T as specified in (3.37). The inner products and T and T^* are those defined in (3.34) and (3.35) with that $a=0$ and $b=d$, but with $N=\infty$ there so that uniform convergence is required to justify the adjointness of (3.33).

The integral equation (3.139) arises, for example, in the problem of wave diffraction at a submerged weir in incompressible fluid (Fig. 3.12), where d is the depth of the top of the weir and $x(t)$ is related to the horizontal fluid speed at depth t below the surface in the gap directly above the weir.

3. Find scalar functions $x(t)$ in $0\leqslant t\leqslant 1$ which are at least piecewise C^2 and satisfy the following problems, where $e[x,t]$ and $f[x]$ are assigned functions;

$$\begin{aligned}
&\text{(a) } x''(t)=e[x(t),t],\quad x'(0)=f[x(0)], x'(1)=-x(1),\\
&\text{(b) } -(\rho(t)x'(t))'+e[x(t),t]=0,\quad x'(0)=x'(1)=0,\\
&\text{(c) } -x''(t)+e[x(t),t]=0,\quad x'(0)=2, x'(1)=-3.
\end{aligned} \tag{3.140}$$

Here (a) and (b) are illustrations of (3.136) by the reasoning associated with (3.38). In (c) we require the spaces of Exercise 3.1.10, and

$$E[x(t),t] = \int_0^1 \int_c^{x(t)} e[s,t]\,\mathrm{d}s\,\mathrm{d}t + 2x(0) + 3x(1)$$

with any fixed c to achieve $E'[x] = e[x]$ at each fixed t. In (b) $\rho(t) > 0$ is a given piecewise C^1 function.

4. Find a scalar C^2 function ϕ of position, in a two or three dimensional region V with surface $\Sigma = \Sigma_1 + \Sigma_2$, which satisfies

$$-\nabla^2\phi + e[\phi] = 0 \text{ in } V, \quad \phi = 0 \text{ on } \Sigma_1, \quad \frac{\partial\phi}{\partial n} + f[\phi] = 0 \text{ on } \Sigma_2, \tag{3.141}$$

where $e[\phi]$ and $f[\phi]$ are given scalar functions. The remarks leading to (3.41) show how (3.141) can be regarded as an example of (3.136). That is, the problem can be decomposed as

$$\begin{aligned} T^*\mathbf{u} &= \begin{bmatrix} -\nabla\cdot\mathbf{u} \\ \mathbf{n}\cdot\mathbf{u} \end{bmatrix} = \begin{bmatrix} e[\phi] \\ f[\phi] \end{bmatrix} \text{ over } \begin{bmatrix} V \\ \Sigma_2 \end{bmatrix}, \quad (\alpha) \\ T\tilde{\phi} &= \nabla\phi = -\mathbf{u} \quad \text{in} \quad V, \quad (\beta) \end{aligned} \tag{3.142}$$

in which, to be sure that T^*T exists, we require that $\mathscr{D}(T)$ consists of $\tilde{\phi} = [\phi(V)\ \phi(\Sigma_2)]^{\mathrm{T}}$ constructed from C^2 functions $\phi(V)$ such that $\phi(\Sigma_1) = 0$, and that $\mathscr{D}(T^*)$ is the space of C^1 vectors $\mathbf{u}$, each with the inner products specified after (3.39). Then $\mathscr{R}(T) \subset \mathscr{D}(T^*)$ and $\mathbf{u}$ can be eliminated from (3.142) to recover (3.141), and so we have an example of (3.136) with (3.41), i.e. $T^*T\tilde{\phi} + E'[\tilde{\phi}] = 0$. Here

$$E[\tilde{\phi}] = \iint_{c_1}^{\phi} e[s]\,\mathrm{d}s\,\mathrm{d}V + \iint_{c_2}^{\phi} f[s]\,\mathrm{d}s\,\mathrm{d}\Sigma_2,$$

where c_1 and c_2 are arbitrary constants. The functional generating (3.142) is

$$L[\tilde{\phi},\mathbf{u}] = \int [\mathbf{u}\cdot\nabla\phi + \tfrac{1}{2}\mathbf{u}^2]\,\mathrm{d}V - E[\tilde{\phi}].$$

The approach here, of building the homogeneous boundary condition $\phi = 0$ on Σ_1 into the definition of $\mathscr{D}(T)$, is different from that of §3.10(v), where such a condition would only enter the β condition, and not also the α condition as it does here (via $e[\phi]$ and $f[\phi]$). The approaches effectively coincide in certain special cases, for example if ϕ is absent from e and f, as in

$$\nabla^2\phi = 0 \text{ in } V, \quad \phi = 0 \text{ on } \Sigma,$$

and if Σ_1 is absent, as in

$$-\nabla^2\phi = a \text{ in } V, \quad \frac{\partial\phi}{\partial n} + k^2\phi = c \text{ on } \Sigma \tag{3.143}$$

with fixed scalars a, k and c. These last two cases are included in a list of illustrations of $T^*Tx = a$ given by Fujita (1955) in the course of his paper about the construction of pointwise bounds on a solution of the unknown x, which we discuss in §4.2.

(iii) General results

We can rewrite (3.136) as

$$T^*u = E'[x] \quad (\alpha), \qquad Tx = -u \quad (\beta), \tag{3.144}$$

which leads via the Hamiltonian $E[x] - \frac{1}{2}\langle u, u\rangle$ to the generating functional

$$L[x,u] = (x, T^*u) - E[x] + \tfrac{1}{2}\langle u, u\rangle. \tag{3.145}$$

If $\mathscr{D}(T) \subseteq \mathscr{D}(E)$, this $L[x,u]$ is defined over all $x \in \mathscr{D}(T)$ and all $u \in \mathscr{D}(T^*)$.

Theorem 3.13

If $E[x]$ is at least weakly convex, then simultaneous bounds $L_\alpha \geqslant L_0 \geqslant L_\beta$ apply to (3.136). *The bounds are*

$$L_\alpha = (x_\alpha, E'[x_\alpha]) - E[x_\alpha] + \tfrac{1}{2}\langle u_\alpha, u_\alpha\rangle \tag{3.146}$$

with any x_α, u_α such that $T^*u_\alpha = E'[x_\alpha]$; and

$$L_\beta = -E[x_\beta] - \tfrac{1}{2}(x_\beta, Px_\beta) \tag{3.147}$$

with any $x_\beta \in \mathscr{D}(T)$, where $P = T^*T$.

If $E[x]$ is strictly convex there is at most one solution x_0 of (3.136). *The bounded quantity is*

$$L_0 = \tfrac{1}{2}(x_0, E'[x_0]) - E[x_0] = -\tfrac{1}{2}\langle Tx_0, Tx_0\rangle - E[x_0]$$

whether x_0 is unique or not.

Suppose in addition that $E'[0] = 0$. Then if $E[x]$ is strictly convex the unique $x_0 = 0$; but if $E[x]$ is weakly convex then any solution x_0 must satisfy $Tx_0 = 0$.

Proof

Convex $E[x]$ implies that $L[x,u]$ is a saddle functional concave in x and convex in u, so that Theorem 3.4 applies. The bounds follow from Theorem 1.6(c) and uniqueness from Theorem 1.7(a).

If $E'[0] = 0$, then $x_0 = 0$ is one solution. If some $x_0 \neq 0$ is another solution, (3.136) and the weak convexity imply

$$\langle Tx_0, Tx_0\rangle = -(x_0, E'[x_0]) \leqslant 0$$

so that $Tx_0 = 0$ to avoid a contradiction. □

It is to be noted particularly that the lower bound L_β does not require explicit knowledge of the operators T and T^* individually, but only that some such decomposition $P = T^*T$ exists. L_β contains P, which is manifest in the given equation. Theorem 3.13 was given by Noble and Sewell (1972).

Suppose next that the gradient equation $e = E'[x]$ has an inverse $x = f[e]$. With suitable hypotheses (e.g. see W. D. Collins, 1982, §3) this can be expressed as the gradient $F'[e]$ of another functional $F[e]$ via a transformation of Legendre type, so that

$$e = E'[x], \quad x = F'[e], \quad E[x] + F[e] = (x, e).$$

We must also be aware that the inversion may entail certain consistency conditions of the type which we met for linear equations in §3.8(i). Then we can solve the α equation as $x_\alpha = f[T^*u_\alpha]$ for any $u_\alpha \in \mathscr{D}(T^*)$ and subject to the consistency conditions. Since $\mathscr{R}(T) \subseteq \mathscr{D}(T^*)$ is necessary for T^*T to exist, we are entitled to make the particular choice of $u_\alpha = -Tx_\beta$ for any $x_\beta \in \mathscr{D}(T)$ and which satisfies the consistency conditions. The α equation is then solved by

$$x_\alpha = f[-Px_\beta]. \tag{3.148}$$

The corresponding upper bound is

$$L_\alpha = -(Px_\beta, f[-Px_\beta]) - E[f[-Px_\beta]] + \tfrac{1}{2}(x_\beta, Px_\beta). \tag{3.149}$$

Like L_β, this L_α depends only on x_β and P, and does not require knowledge of T and T^* individually. The difference between the bounds (3.149) and (3.147) is then

$$L_\alpha - L_\beta = E[x_\beta] - E[x_\alpha] - (x_\beta - x_\alpha, E'[x_\alpha]) \tag{3.150}$$

for the x_α given by (3.148). In other words, when we choose $u_\alpha = u_\beta$ for any x_β in this way, the error in estimating L_0 cannot exceed the height at x_β of the convex functional $E[x_\beta]$ above its tangent plane at $f[-Px_\beta]$.

In the linear case (3.137) the α equation is solved exactly as in (3.104), with the same qualifications about consistency and the distributive property which attach to A^{-1} in §3.8(i)–(ii). Then (3.148), (3.149) and (3.150) respectively are

$$x_\alpha = -A^{-1}(Px_\beta - a),$$

$$L_\alpha = \tfrac{1}{2}(Px_\beta - a, A^{-1}(Px_\beta - a)) + \tfrac{1}{2}(x_\beta, Px_\beta),$$

$$L_\alpha - L_\beta = \tfrac{1}{2}(x_\alpha - x_\beta, A(x_\alpha - x_\beta)) \tag{3.151}$$

$$= \tfrac{1}{2}(Px_\beta + Ax_\beta - a, A^{-1}[Px_\beta + Ax_\beta - a]). \tag{3.152}$$

The infinite dimensional version of (1.137) with $u_\alpha = u_\beta$ also gives (3.151). When A^{-1} is positive definite we can define a norm $\|\cdot\| = (.,A^{-1}.)^{\frac{1}{2}}$, and (3.152) is half the square of the residual error $\|Px_\beta + Ax_\beta - a\|$ made in treating x_β as an approximate solution of $Px + Ax - a = 0$. If $A = -\lambda I$ with the unit operator I and given scalar $\lambda < 0$, we can define the alternative norm $\|\cdot\| = (.,.)^{\frac{1}{2}}$ and write (3.152) as $(-1/2\lambda)\|Px_\beta - \lambda x_\beta - a\|^2$, which is a generalization of (1.139).

(iv) Laplacian problems

If $e[\phi]$ and $f[\phi]$ in (3.141) are monotonically nondecreasing functions of ϕ, $E[\tilde{\phi}]$ there is weakly convex, and we have a class of Laplacian problems which are decomposable as (3.142) and generated by a functional $L[\tilde{\phi}, \mathbf{u}]$ weakly concave in $\tilde{\phi}$ and strictly convex in $\mathbf{u}$, $\forall\ \tilde{\phi} \in \mathscr{D}(T)$ and $\forall\ \mathbf{u} \in \mathscr{D}(T^*)$.

Theorem 3.13 applies. Recalling (3.41) and the inner products there, the lower bound (3.147) is

$$L_\beta = -E[\tilde{\phi}_\beta] + \frac{1}{2}\int \phi_\beta \nabla^2 \phi_\beta \,\mathrm{d}V - \frac{1}{2}\int \phi_\beta \mathbf{n}\cdot\nabla\phi_\beta \,\mathrm{d}\Sigma_2,$$

where we have written $\tilde{\phi}_\beta = [\phi_\beta(V)\ \phi_\beta(\Sigma_2)]^{\mathrm{T}}$ for any C^2 function $\phi_\beta(V)$ such that $\phi_\beta(\Sigma_1) = 0$.

To calculate an upper bound (3.146) we need to choose some $\tilde{\phi}_\alpha \in \mathscr{D}(T)$ and some $\mathbf{u}_\alpha \in \mathscr{D}(T^*)$ which satisfy (3.142α). To give an illustration of (3.148) and (3.152), suppose the original problem (3.141) is linear with

$$e[\phi] = m^2\phi, \quad f[\phi] = k^2\phi - c,$$

where the scalars $m \neq 0$ and k are given constants, and $c \not\equiv 0$ is a given function on Σ_2.

Then (3.142α) may be written

$$-\nabla\cdot\mathbf{u}_\alpha = m^2\phi_\alpha \quad \text{in} \quad V, \qquad \mathbf{n}\cdot\mathbf{u}_\alpha = k^2\phi_\alpha - c \quad \text{on} \quad \Sigma_2,$$

to be satisfied for a C^2 function $\phi_\alpha(V)$ such that $\phi_\alpha(\Sigma_1) = 0$, and for a C^1 function $\mathbf{u}_\alpha(V)$. In the notation of §3.8(i) this is an illustration of $T^*u_\alpha = Ax_\alpha - a$, which can be solved here as $x_\alpha = A^{-1}(T^*u_\alpha + a)$ provided certain consistency conditions are added. First $\nabla\cdot\mathbf{u}_\alpha$ must be C^2, for which it is sufficient that $\mathbf{u}_\alpha(V)$ be a C^3 function, rather than merely C^1 which was required for existence of T^*T. Then $\tilde{\phi}_\alpha = [\phi_\alpha(V)\ \phi_\alpha(\Sigma_2)]^{\mathrm{T}}$ is built from

$$\phi_\alpha = -\frac{1}{m^2}\nabla\cdot\mathbf{u}_\alpha$$

for any C^3 $\mathbf{u}_\alpha(V)$ which satisfies the consistency conditions

$$\nabla\cdot\mathbf{u}_\alpha = 0 \quad \text{on} \quad \Sigma_1, \qquad \mathbf{n}\cdot\mathbf{u}_\alpha = -\frac{k^2}{m^2}\nabla\cdot\mathbf{u}_\alpha - c \quad \text{on} \quad \Sigma_2.$$

The upper bound (3.146) becomes

$$L_\alpha = \frac{1}{2m^2}\int(\nabla\cdot\mathbf{u}_\alpha)^2\,\mathrm{d}V + \frac{1}{2}\int\mathbf{u}_\alpha{}^2\,\mathrm{d}V + \frac{1}{2k^2}\int(\mathbf{n}\cdot\mathbf{u}_\alpha + c)^2\,\mathrm{d}\Sigma_2.$$

Now define $\mathbf{u}_\beta = -\nabla\phi_\beta$ for those previous trial functions $\phi_\beta(V)$ satisfying $\phi_\beta(\Sigma_1) = 0$ which happen to be C^4 rather than merely C^2. For this subclass of such $\phi_\beta(V)$ we may choose

$$\mathbf{u}_\alpha = \mathbf{u}_\beta \quad \text{and therefore} \quad \phi_\alpha = \frac{1}{m^2}\nabla^2\phi_\beta, \tag{3.153}$$

provided *such* ϕ_β satisfy the conditions

$$\phi_\beta = 0 \quad \text{and} \quad \nabla^2\phi_\beta = 0 \quad \text{on} \quad \Sigma_1, \quad \mathbf{n}\cdot\nabla\phi_\beta = -\frac{k^2}{m^2}\nabla^2\phi_\beta + c \quad \text{on} \quad \Sigma_2. \tag{3.154}$$

These illustrate the restrictions which must be satisfied if $L_\alpha - L_\beta$ is to be expressed as the square of the error (3.152) (see Exercise 3.3.1). Similar remarks apply to the problems in (3.140).

(v) A comparison of equivalent differential and integral equations

Consider the specific example of §3.7(xiii) for which we calculated the bounds displayed in Fig. 3.3. That problem is an example of (3.82), namely

$$T^*\tilde{u} = \begin{bmatrix} -u'(t) \\ u(1) \end{bmatrix} = \begin{bmatrix} m^2x(t) - 1 \\ -b \end{bmatrix}, \quad T\tilde{x} = \begin{bmatrix} x'(t) \\ x(0) \end{bmatrix} = \begin{bmatrix} -u(t) \\ 1 \end{bmatrix}.$$

The operator T^*T does not exist in this formulation because, although we could restrict $\mathscr{D}(T)$ to be built from piecewise C^2 functions $x(t)$, to achieve $\mathscr{R}(T) \subseteq \mathscr{D}(T^*)$ we should also need to insist that $x'(0) = x(0)$, which is too restrictive in general because we have already imposed $x(0) = 1$ and $x'(1) = b$.

However, a positive operator can be introduced by converting the problem into an integral equation as follows. Integrating $x'' = m^2x - 1$ once and using $x'(1) = b$ gives

$$x'(t) = -m^2\int_t^1 x(\tau)\,\mathrm{d}\tau - t + b + 1.$$

Integrating again and using $x(0) = 1$ gives

$$x(t) = -m^2\int_0^t\int_s^1 x(\tau)\,\mathrm{d}\tau\,\mathrm{d}s - \tfrac{1}{2}t^2 + (b+1)t + 1. \tag{3.155}$$

This integral equation is equivalent to the original problem, as may be confirmed by differentiating twice, so recovering not only the differential equation but also the end conditions in the process. We now introduce the adjoint operators

$$Tx = \int_s^1 x(\tau)\,d\tau, \quad T^*u = \int_0^t u(s)\,ds,$$

which are particular illustrations of (3.32) with (3.31), and which allow the integral equation (3.155) to be written as

$$T^*Tx = a(t) - \frac{x(t)}{m^2}, \quad \text{where} \quad m^2 a(t) = -\tfrac{1}{2}t^2 + (b+1)t + 1.$$

This illustrates (3.137) with positive definite $A = I/m^2$, and

$$E[x(t)] = \frac{1}{2m^2}\int_0^1 x^2\,dt - \int_0^1 ax\,dt.$$

The generating functional (3.145) is quadratic, and is defined for all integrable functions $x(t)$ and $u(t)$ in $0 \leqslant t \leqslant 1$ with common inner product $\int_0^1 x(t)u(t)\,dt$ by (3.30). It is also strictly concave in $x(t)$ and strictly convex in $u(t)$, so that Theorem 3.13 applies and confirms, in particular, that the solution given in §3.7(xiii) is the unique solution of the integral equation too. There are simultaneous bounds, and the bracketed quantity is the solution value of $\frac{1}{2}\int_0^1 ax\,dt$, i.e.

$$L_0 = \frac{1}{2m^2}\left(1 - \frac{1}{m^2}\right)\int_0^1 x\,dt - \frac{b}{2m^4}x(1) + \frac{1}{2m^4}\left(b^2 + 2b + \tfrac{4}{3}\right)$$

after using $x = (1 + x'')/m^2$, $x(0) = 1$ and $x'(1) = b$. In this formulation we are therefore bounding a different functional of the exact solution from that which we bounded in Fig. 3.3, and so the two approaches give different information. This possibility is noteworthy, since in some problems it may be fruitful to exploit both approaches. Here, of course, we happen to know the unique solution and it gives to the above L_0 the value

$$L_0 = \frac{1}{2m^3\cosh m}\left[\left\{\left(1 - \frac{1}{m^2}\right)^2 - \frac{b^2}{m^2}\right\}\sinh m - \frac{2b}{m}\left(1 - \frac{1}{m^2}\right)\right]$$
$$+ \frac{1}{2m^4}\left[b^2 + b\left(3 - \frac{2}{m^2}\right) + \frac{7}{3} - \frac{1}{m^2}\right]. \tag{3.156}$$

The decomposition (3.144) of (3.155) is

$$\int_0^t u(s)\,ds = \frac{x(t)}{m^2} - a(t) \quad (\alpha), \qquad \int_s^1 x(\tau)\,d\tau = -u(s) \quad (\beta).$$

The lower bound (3.147) is

$$L_\beta = \int_0^1 ax_\beta\,dt - \frac{1}{2m^2}\int_0^1 {x_\beta}^2\,dt - \frac{1}{2}\int_0^1 x_\beta(t)\int_0^t\int_s^1 x(\tau)\,d\tau\,ds\,dt,$$

evaluated with any integrable function $x_\beta(t)$. Thus even a saw-tooth or a finitely stepped function are not excluded as trial functions, but with the same $x_\beta(t) \equiv 1$ which we used in §3.7(xiii) we now get

$$L_\beta = \frac{1}{m^2}(\tfrac{5}{6} + \tfrac{1}{2}b) - \tfrac{1}{6}. \tag{3.157}$$

The upper bound (3.146) can be evaluated by inverting the α equation as $x_\alpha(t) = m^2[a(t) + \int_0^t u_\alpha(s)\,ds]$ with any integrable $u_\alpha(s)$, without the need of additional consistency conditions which occur for differential equations. If we then choose $u_\alpha(s) = -\int_s^1 x_\beta(\tau)\,d\tau$ and hence

$$x_\alpha(t) = m^2\left[a(t) - \int_0^t \int_s^1 x_\beta(\tau)\,d\tau\,ds\right]$$

we have a linear example of (3.148). Hence, from (3.152),

$$L_\alpha = L_\beta + \frac{m^2}{2}\int_0^1 \left[\left[\int_0^t \int_s^1 x_\beta(\tau)\,d\tau\,ds + \frac{x_\beta(t)}{m^2} - a(t)\right]^2 dt.$$

The choice $x_\beta(t) \equiv 1$ gives

$$L_\alpha = L_\beta + \frac{m^2}{15}\left(1 - \frac{1}{m^2}\right)^2 - \frac{5b}{24}\left(1 - \frac{1}{m^2}\right) + \frac{b^2}{6m^2}. \tag{3.158}$$

In Fig. 3.8(a) we plot the bounds (3.157) and (3.158) and the exact value (3.156) as functions of m for $b = 1$. These functions for $b = 0$ are very similar, but displaced a little to the left and upward. The graphs are to be contrasted with Fig. 3.3. The three curves in Fig. 3.8(*a*) touch at the same point, and to resolve their divergence more clearly at lower values of positive m we have plotted $L_\alpha - L_0$ and $L_\beta - L_0$ in Fig. 3.8(*b*). A different choice of trial function $x_\beta(t)$ will give different bounds, and we leave the reader to explore the question of optimizing the bounds with respect to $x_\beta(t)$, for example

Fig. 3.8. Integral equation bounds with $b = 1$ and $x_\beta(t) \equiv 1$. (*a*) Bounds for $m > 1$. (*b*) Differences for $0 < m < 3$.

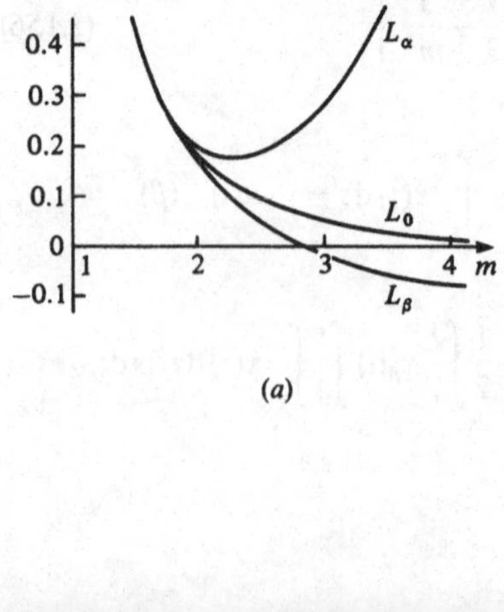

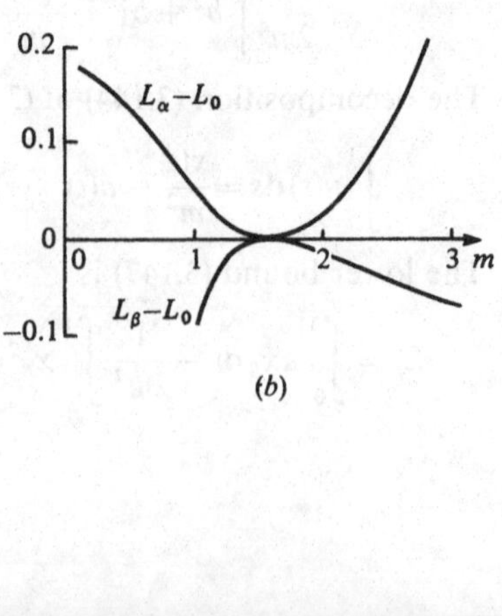

among other constant values. Our purpose here has been to give a quick comparison of the differential and integral equation approaches, with a very simple trial function.

(vi) Alternative bounds

There is a class of problems for which a pair of simultaneous bounds can be constructed in two distinct ways, and such that each pair brackets the same functional of the solution. We have already given an example in network theory (§§2.8(vii) and (viii)), where the node-branch and loop-branch formulations were distinct approaches to the same problem, and another example in Exercise 3.3.2. If one approach leads to a bound under constraints which can be solved explicitly in terms of the second approach then there is an obvious advantage in having both approaches available. The common difficulty that one bound of a pair is harder to find than the other may then be avoided.

Here we indicate a general infinite dimensional framework for such circumstances, and we give another application of it in the next subsection.

We introduce a third inner product space with elements $\xi, \eta, \ldots$ and inner product denoted by $\overline{\xi, \eta}$. In general such elements and inner products may have different definitions from the (x, y) and $\langle u, v \rangle$ already available. Let linear operators R and R^* be defined such that they are adjoint to each other in the sense (cf. (3.10)) that

$$\overline{\xi, R^*u} = \langle u, R\xi \rangle \quad \forall \xi \in \mathscr{D}(R) \quad \text{and} \quad \forall u \in \mathscr{D}(R^*). \tag{3.159}$$

Let γ be a given element of the third space and (as in (3.137)) let a be a given element of the first space.

Consider the two nonnegative operator equations

$$T^*Tx = a, \quad R^*R\xi = \gamma \tag{3.160}$$

for the determination of x and ξ respectively. The equations are independent as they stand. Each can be decomposed, as

$$T^*u = -a \quad (\alpha), \qquad Tx = -u \quad (\beta), \tag{3.161}$$

and

$$R^*v = -\gamma \quad (\alpha), \qquad R\xi = -v \quad (\beta), \tag{3.162}$$

and hence generated from the gradients of

$$L_1[x, u] = (x, T^*u) + \tfrac{1}{2}\langle u, u \rangle + (a, x), \tag{3.163}$$

$$L_2[\xi, v] = \overline{\xi, R^*v} + \tfrac{1}{2}\langle v, v \rangle + \overline{\gamma, \xi}, \tag{3.164}$$

respectively.

Theorem 3.14
Let x_0, u_0 denote any solution of (3.161), *and let ξ_0, v_0 denote any solution of* (3.162). *Then u_0 and v_0 are each unique, and simultaneous bounds*

$$L_{1\alpha} \geqslant L_{10} \geqslant L_{1\beta}, \quad L_{2\alpha} \geqslant L_{20} \geqslant L_{2\beta}$$

apply to the respective problems. Here

$$L_{10} = \tfrac{1}{2}(x_0, a), \quad L_{20} = \tfrac{1}{2}\overline{\xi_0, \gamma}$$

$$L_{1\alpha} = \tfrac{1}{2}\langle u_\alpha, u_\alpha \rangle \quad \text{for any} \quad u_\alpha \quad \text{such that} \quad T^*u_\alpha = -a,$$

$$L_{2\alpha} = \tfrac{1}{2}\langle v_\alpha, v_\alpha \rangle \quad \text{for any} \quad v_\alpha \quad \text{such that} \quad R^*v_\alpha = -\gamma,$$

$$L_{1\beta} = \frac{1}{2}\frac{(x_\beta, a)^2}{(x_\beta, T^*Tx_\beta)} = \frac{1}{2}\frac{(x_\beta, a)^2}{\langle Tx_\beta, Tx_\beta \rangle} \quad \text{for any} \quad x_\beta \in \mathscr{D}(T), \tag{3.165}$$

$$L_{2\beta} = \frac{1}{2}\frac{\overline{\xi_\beta, \gamma}^{\,2}}{\overline{\xi_\beta, R^*R\xi_\beta}} = \frac{1}{2}\frac{\overline{\xi_\beta, \gamma}^{\,2}}{\langle R\xi_\beta, R\xi_\beta \rangle} \quad \text{for any} \quad \xi_\beta \in \mathscr{D}(R). \tag{3.166}$$

Proof
$L_1[x, u]$ is strictly convex in u and weakly concave in x, so that by our standard procedures (Theorems 3.4 and 3.6) u_0 is unique, and simultaneous bounds apply. We find

$$L_{10} = (x_0, T^*u_0 + a) + \tfrac{1}{2}\langle u_0, u_0 \rangle$$
$$= -\tfrac{1}{2}\langle u_0, Tx_0 \rangle = \tfrac{1}{2}(x_0, a).$$

Since $L_1 - (x, \partial L_1/\partial x) = \tfrac{1}{2}\langle u, u \rangle$, the formula for $L_{1\alpha}$ is immediate. Since

$$L_1 - \left\langle u, \frac{\partial L_1}{\partial u} \right\rangle = -\tfrac{1}{2}\langle u, u \rangle + (a, x),$$

$L_{1\beta}$ has this value with any $u_\beta = -Tx_\beta$ for any $x_\beta \in \mathscr{D}(T)$. Maximizing the resulting inhomogeneous quadratic with respect to a scale factor in x_β gives the greatest such lower bound $L_{1\beta}$ to be that stated in the Theorem.

Exactly similar proofs apply to the independent $L_2[\xi, v]$. □

In general it is easier to find $L_{1\beta}$ than $L_{1\alpha}$, and to find $L_{2\beta}$ than $L_{2\alpha}$.

So far the problems (3.161) and (3.162) are independent, but we now describe circumstances in which they become coupled in such a way that the harder bound in one problem can be replaced by the easier bound in the other problem.

Theorem 3.15
Suppose that $a \neq 0$, $\gamma \neq 0$ and that

$$R^*Tx = -\gamma(x, a) \quad \forall x \in \mathscr{D}(T), \qquad T^*R\xi = -a\overline{\xi, \gamma} \quad \forall \xi \in \mathscr{D}(R). \tag{3.167}$$

Then the unique $u_0 = -Tx_0$ *and* $v_0 = -R\xi_0$ *associated with solutions* x_0 *and* ξ_0 *of* (3.160), *respectively, satisfy*

$$u_0 = \lambda v_0 \quad with \quad -\lambda = 2L_{10} = \frac{1}{2L_{20}} \tag{3.168}$$

and therefore $L_{10}L_{20} = \frac{1}{4}$. *The easily determined bounds*

$$\frac{1}{4L_{2\beta}} \geqslant L_{10} \geqslant L_{1\beta}, \quad \frac{1}{4L_{1\beta}} \geqslant L_{20} \geqslant L_{2\beta} \tag{3.169}$$

then replace the harder (upper) ones in Theorem 3.14.

Proof

Particular consequences of (3.167) are that

$$R^*Tx_0 = -\gamma(x_0, a) \quad \forall x_0 \quad \text{satisfying} \quad T^*Tx_0 = a \tag{3.170}$$

and that

$$T^*R\xi_0 = -a\overline{\xi_0, \gamma} \quad \forall \xi_0 \quad \text{satisfying} \quad R^*R\xi_0 = \gamma. \tag{3.171}$$

If a scale factor λ exists such that $u_0 = \lambda v_0$, then (3.161) and (3.162) imply

$$R^*Tx_0 = -R^*u_0 = -\lambda R^*v_0 = \lambda\gamma.$$

Then (3.170) implies $(\lambda + (x_0, a))\gamma = 0$ so that, since $\gamma \neq 0$, we have $\lambda = -(x_0, a) = -2L_{10}$.

Fig. 3.9. Nonnegative operator equations coupled by Theorem 3.15.

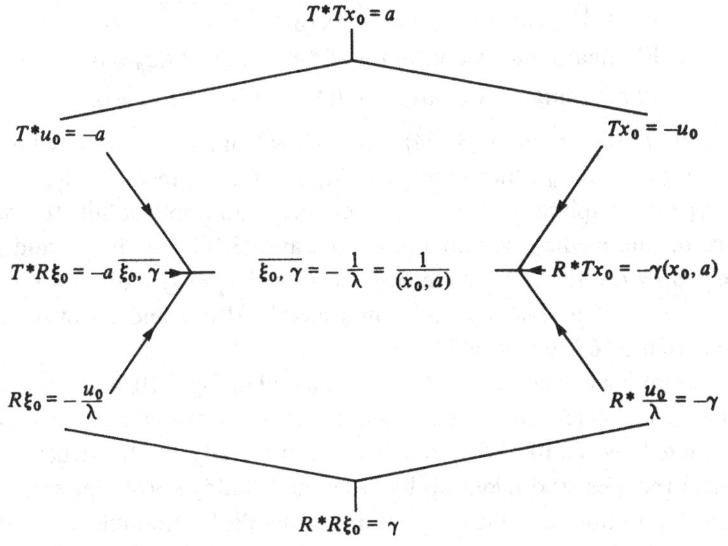

Similarly $(1/\lambda+\overline{\xi_0,\gamma})a=0$ via (3.161) so that, since $a\neq 0$, we have $1/\lambda=-\overline{\xi_0,\gamma}=-2L_{20}$. □

The coupling of (3.161) and (3.162) established by Theorem 3.15 is summarized by the interconnections in Fig. 3.9. We shall give an example of this theorem in the next subsection. Theorem 3.15 would still be true if (3.167) were satisfied by any set of x smaller than $\mathscr{D}(T)$ which still contains x_0, and by any set of ξ smaller than $\mathscr{D}(R)$ which still contains ξ_0.

Under different hypotheses from those in Theorem 3.15, we also have the following different coupling of (3.161) and (3.162).

Theorem 3.16

If $u_0=v_0$, *then* $L_{10}=L_{20}$.

$$\textit{If}\quad T^*R\xi_\beta=a \quad \textit{for some}\quad \xi_\beta\in\mathscr{D}(R), \tag{3.172}$$

then $u_\alpha=-R\xi_\beta$ *satisfies* $T^*u_\alpha=-a$ *for such* ξ_β.

$$\textit{If}\ R^*Tx_\beta=\gamma \quad \textit{for some}\quad x_\beta\in\mathscr{D}(T), \tag{3.173}$$

then $v_\alpha=-Tx_\beta$ *satisfies* $R^*v_\alpha=-\gamma$ *for such* x_β.

Then we have the alternative bounds

$$\left.\begin{array}{l} \tfrac{1}{2}\overline{\xi_\beta,R^*R\xi_\beta}=L_{1\alpha} \\[4em] \tfrac{1}{2}(x_\beta,T^*Tx_\beta)=L_{2\alpha} \end{array}\right\}\geqslant L_{10}=L_{20}\geqslant\left\{\begin{array}{l} \dfrac{(x_\beta,a)^2}{4L_{2\alpha}}, \\[4em] \dfrac{\overline{\xi_\beta,\gamma}^{\,2}}{4L_{1\alpha}}. \end{array}\right. \tag{3.174}$$

Proof

$$L_{10}=\tfrac{1}{2}(x_0,a)=\tfrac{1}{2}\langle u_0,u_0\rangle=\tfrac{1}{2}\langle v_0,v_0\rangle=\tfrac{1}{2}\overline{\xi_0,\gamma}=L_{20}.$$

Eliminating u_α, we have $a+T^*u_\alpha=a-T^*R\xi_\beta=0$.

Eliminating v_α, we have $\gamma+R^*v_\alpha=\gamma-R^*Tx_\beta=0$.

The alternative bounds (3.174) then follow from Theorem 3.14 for the particular ξ_β and x_β which satisfy (3.172) and (3.173) respectively. □

A type of coupling to which Theorem 3.16 can apply entails the *same* choice of intermediate variable in (3.161) and (3.162), i.e. $u=v$ (and not merely $u_0=v_0$, since we have also supposed $v_\beta=u_\alpha$ and $u_\beta=v_\alpha$). By contrast $u_0\neq v_0$ in Theorem 3.15, in general. Also a and γ can be zero in Theorem 3.16 but not in Theorem 3.15.

The coupling of Theorem 3.16 is summarized in Fig. 3.10, where we have omitted the suffixes. Diagrams which are variants of Fig. 3.10 have been constructed by Tonti (1975) to illustrate his study of the structure of physical theories, and taken up by Oden and Reddy (1974), for example. Many elaborations and illustrations are available, for example in terms of

Fig. 3.10. Nonnegative operator equations coupled by Theorem 3.16.

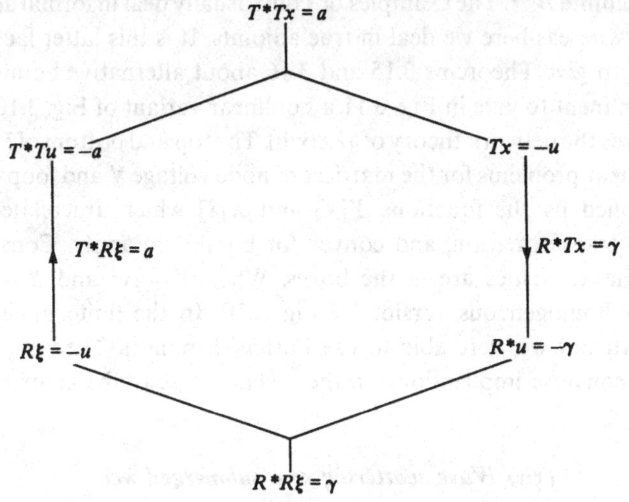

Fig. 3.11. Coupled node-branch and loop-branch network formulations.

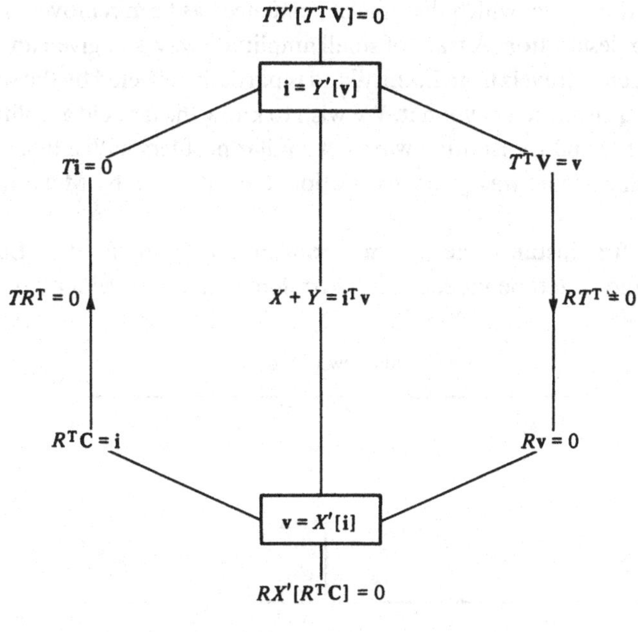

positive operators of the type $T^*B^{-1}T$ in our notation of (3.115) with positive definite B^{-1}. The examples of Tonti usually deal in formal adjoints, however, whereas here we deal in true adjoints. It is this latter fact which allows us to give Theorems 3.15 and 3.16 about alternative bounds.

It is pertinent to give in Fig. 3.11 a nonlinear variant of Fig. 3.10 which summarizes the network theory of §2.8(viii). The top and bottom of Fig. 3.11 are nonlinear problems for the matrices of node voltage $\mathbf{V}$ and loop current $\mathbf{C}$, controlled by the functions $Y[\mathbf{v}]$ and $X[\mathbf{i}]$ which are related by a Legendre transformation, and convex for bounds to apply. Constitutive branch characteristics are in the boxes. When $Y=\frac{1}{2}\mathbf{v}^2$ and $X=\frac{1}{2}\mathbf{i}^2$ we recover a homogeneous version of Fig. 3.10. In the finite dimensional network theory we were able to use Farkas' lemma in Theorem 2.20 to establish converse implications, in the reverse sense of the arrows in Fig. 3.11.

(vii) Wave scattering at a submerged weir

A horizontal layer of water of uniform depth h is bounded by a rigid bed at $t=h$, where t is distance measured downwards from the position which an equilibrium free surface would occupy at $t=0$. A submerged weir of height $h-d$ stands on the bottom, so that $0<d<h$. The idealized two dimensional side view is shown in Fig. 3.12, where the weir is a thin rigid barrier of infinite extent perpendicular to any flow. We are concerned with certain wave motions from which the time dependence has been removed from the following description. A train of small amplitude waves of given amplitude and frequency travels from infinity, and is partially reflected by the weir and partially transmitted beyond it. We wish to know the far field amplitudes of the reflected and transmitted waves. A similar problem with a thick weir of rectangular profile was given a variational formulation by Mei and Black (1969).

In the first instance the present problem may be reduced to Laplace's equation for the time independent part Φ of a complex velocity potential,

Fig. 3.12. Side view of submerged weir.

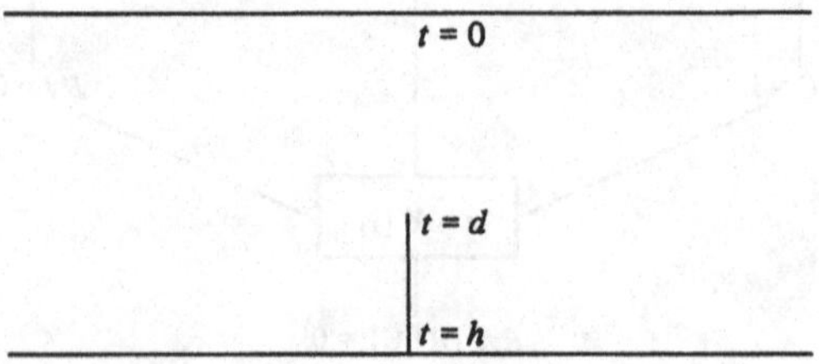

subject to $\partial\Phi/\partial n = 0$ on the bottom and on each side of the weir, and to the linearized free surface condition $\partial\Phi/\partial n = \kappa\Phi$ on $t = 0$, where n is distance along an outward normal to the fluid, and $(g\kappa)^{\frac{1}{2}}$ is the given frequency, g being the acceleration due to gravity. Appropriate asymptotic conditions at infinity are given.

Separation of variables with known horizontal dependence allows the problem to be posed in two more alternative ways, each as an integral equation, as shown by Porter (1984, unpublished). The first of these is (3.139), wherein the given functions are

$$y_i(t) = k_i^{-\frac{1}{2}}\phi_i(t) \quad \text{and} \quad a(t) = \phi_0(t) \text{ on } 0 \leqslant t \leqslant d.$$

Here the real functions on the right are

$$\phi_0(t) = \left(\frac{2\kappa}{kh + \sinh^2 k_0 h}\right)^{\frac{1}{2}} \cosh k_0(h-t),$$

$$\phi_i(t) = \left(\frac{2\kappa}{kh - \sin^2 k_i h}\right)^{\frac{1}{2}} \cos k_i(h-t), \quad i = 1, 2, \ldots, \infty,$$

where k_0 is the real root of $k \tanh kh = \kappa$ and the k_i are the positive real roots of $k \tan kh = -\kappa$. The $\phi_i(t)$ and $\phi_0(t)$ are actually defined on $0 \leqslant t \leqslant h$ and form an infinite complete orthonormal set there.

Adapting the spaces of (3.34) so that

$$(x, y) = \int_0^d x(t)y(t)\mathrm{d}t, \quad \langle u, v\rangle = \sum_{i=1}^{\infty} u_i v_i$$

with operators illustrating (3.35), the problem (3.139) for integrable $x(t)$ can be decomposed as

$$T^*u = \sum_{i=1}^{\infty} y_i u_i = -a(t), \quad (\alpha)$$

$$Tx = \{(Tx)_1, (Tx)_2, \ldots, (Tx)_\infty\} = -u, \quad (\beta) \qquad (3.175)$$

where $(Tx)_i = \int_0^d y_i(\tau)x(\tau)\mathrm{d}\tau$ for $i = 1, 2, \ldots, \infty$. This is an example of (3.161) and the generating functional is

$$L_1[x, u] = \sum_{i=1}^{\infty} u_i \int_0^d y_i(t)x(t)\mathrm{d}t + \int_0^d a(t)x(t)\mathrm{d}t + \tfrac{1}{2}\sum_{i=1}^{\infty} u_i^2.$$

This is a saddle functional strictly convex in the infinite sequences $u = \{u_i\}$ and weakly concave in the integrable functions $x(t)$. Theorem 3.14 then assures the presence of the simultaneous upper and lower bounds $L_{1\alpha} \geqslant L_{10} \geqslant L_{1\beta}$ given there.

The second alternative expression of the wave diffraction problem is in terms of integrals on the weir, instead of on the gap above it. This is the problem of finding a continuous function $\xi(t)$ on $d \leqslant t \leqslant h$ satisfying

$$\sum_{i=1}^{\infty} \eta_i(t) \int_d^h \eta_i(\tau)\xi(\tau)\mathrm{d}\tau = \gamma(t), \tag{3.176}$$

where $\gamma(t)$ and every $\eta_i(t)$ are given functions of t. Here

$$\eta_i(t) = k_i^{\frac{1}{2}}\phi_i(t) \quad \text{and} \quad \gamma(t) = \phi_0(t) \text{ on } d \leqslant t \leqslant h$$

in terms of the same $\phi_0(t)$ and $\phi_i(t)$ defined above.

Adapting the spaces of (3.34) this time so that

$$\overline{\xi, \eta} = \int_d^h \xi(t)\eta(t)\mathrm{d}t, \quad \langle u, v\rangle = \sum_{i=1}^{\infty} u_i v_i,$$

with adjoint operators R and R^* defined by a different illustration of (3.35), the problem (3.176) can be decomposed as

$$\begin{aligned} R^*v &= \sum_{i=1}^{\infty} \eta_i v_i = -\gamma(t), & (\alpha) \\ R\xi &= \{(R\xi)_1, (R\xi)_2, \ldots, (R\xi)_\infty\} = -v, & (\beta) \end{aligned} \tag{3.177}$$

where $(R\xi)_i = \int_d^h \eta_i(\tau)\xi(\tau)\mathrm{d}\tau$ for $i = 1, 2, \ldots, \infty$. This is an example of (3.162) with generating functional

$$L_2[\xi, v] = \sum_{i=1}^{\infty} v_i \int_d^h \eta_i(t)\xi(t)\mathrm{d}t + \int_d^h \gamma(t)\xi(t)\mathrm{d}t + \tfrac{1}{2}\sum_{i=1}^{\infty} v_i^2.$$

This is a saddle functional strictly convex in the infinite sequences $v = \{v_i\}$ and weakly concave in the continuous functions $\xi(t)$. Theorem 3.14 then assures the presence of the alternative simultaneous upper and lower bounds $L_{2\alpha} \geqslant L_{20} \geqslant L_{2\beta}$.

It may appear difficult, on the face of it, to find u_i which satisfy (3.175α) and hence evaluate $L_{1\alpha}$, or to find v_i which satisfy (3.177α) and hence evaluate $L_{2\alpha}$. We can avoid these difficulties by replacing $L_{1\alpha}$ by $1/4L_{2\beta}$, as in (3.169), because (3.167) holds and therefore Theorem (3.15) applies.

We only need to verify (3.167) for those integrable functions $x(\tau)$ which have the particular form

$$x(\tau) = (d-\tau)^{-\frac{1}{2}}\hat{x}(\tau) \quad \text{in} \quad 0 \leqslant \tau < d$$

where $\hat{x}(\tau)$ is a C^1 function, because it can be shown that the exact solution has this form.

$$\begin{aligned} R^*Tx &= \sum_{i=1}^{\infty} \eta_i(t) \int_0^d y_i(\tau)x(\tau)\mathrm{d}\tau \quad \text{in} \quad d < t < h \\ &= \sum_{i=1}^{\infty} \phi_i(t) \int_0^d \phi_i(\tau)x(\tau)\mathrm{d}\tau \end{aligned}$$

$$= \sum_{i=0}^{\infty} \phi_i(t) \int_0^d \phi_i(\tau)x(\tau)\mathrm{d}\tau - \phi_0(t) \int_0^d \phi_0(\tau)x(\tau)\mathrm{d}\tau$$

$$= -\gamma(a, x). \tag{3.178}$$

The last step here may be justified by the following argument. The Fourier series of a function $f(t)$ in $0<t<h$, with respect to the complete orthonormal set $\phi_i(t)$ for $i=0, 1, 2\ldots$,is

$$\sum_{i=0}^{\infty} \phi_i(t) \int_0^h f(\tau)\phi_i(\tau)\mathrm{d}\tau.$$

Therefore the Fourier series of the function defined by

$$f(t) = \begin{cases} x(t) & \text{if} \quad 0<t<d, \\ 0 & \text{if} \quad d<t<h, \end{cases}$$

is

$$\sum_{i=0}^{\infty} \phi_i(t) \int_0^d x(\tau)\phi_i(\tau)\mathrm{d}\tau \quad \text{in} \quad 0<t<h.$$

This converges to $f(t)$ for the class of $x(\tau)$ now under consideration, except at $t=d$. Hence in particular

$$0 = \sum_{i=0}^{\infty} \phi_i(t) \int_0^d x(\tau)\phi_i(\tau)\mathrm{d}\tau \quad \text{in} \quad d<t<h \tag{3.179}$$

as required in (3.178). In the absence of more rigorous arguments we are obliged to accept the convergence implicitly required in $L_1[x, u]$ and $L_2[u, v]$ above, at least for u and v near those which satisfy the constraints deployed in the proof of simultaneous bounds.

Similarly $T^*R\xi = -a\overline{\xi, \gamma}$ for any continuous $\xi(t)$ on $d \leqslant t \leqslant h$.

Therefore, using (3.165) and (3.166), the bounds (3.169) apply with

$$\left.\begin{aligned} 2L_{1\beta} &= \frac{\left[\int_0^d \phi_0(t)\psi(t)\mathrm{d}t\right]^2}{\sum_{i=1}^{\infty}\left[k_i^{-\frac{1}{2}}\int_0^d \phi_i(t)\psi(t)\mathrm{d}t\right]^2}, \\ 2L_{2\beta} &= \frac{\left[\int_d^h \phi_0(t)\psi(t)\mathrm{d}t\right]^2}{\sum_{i=1}^{\infty}\left[k_i^{\frac{1}{2}}\int_d^h \phi_i(t)\psi(t)\mathrm{d}t\right]^2}. \end{aligned}\right\} \tag{3.180}$$

Here the trial functions $x_\beta(t)$ and $\xi_\beta(t)$ have been combined into a single trial function

$$\psi(t) = \begin{cases} x_\beta(t) & \text{on} \quad 0 \leqslant t \leqslant d, \\ \xi_\beta(t) & \text{on} \quad d \leqslant t \leqslant h. \end{cases}$$

A suitable trial function to use for the case of finite depth h may be the exact solution which is known for the case of infinite depth. There is a sequence of papers in fluid mechanics on wave scattering problems of various types, which have aspects in common with those treated here. For example, the reader may gain an entry to the literature by consulting Evans and Morris (1972) or Simon (1981).

In other types of scattering problems, for example in quantum mechanics, such bounds are called Schwinger bounds. Arthurs (1980, §4.8) discusses such matters, and the reader may compare that approach with the present one.

It may be shown that the quantity bracketed by $1/4L_{2\beta} \geqslant L_{10} \geqslant L_{1\beta}$ is

$$L_{10} = \tfrac{1}{2}k_0\frac{\rho}{r},$$

where ρ and r are respectively the ratios of the far field amplitudes of the transmitted and reflected waves to that of the given incident wave, so that $\rho^2 + r^2 = 1$. Hence we bound both ρ and r, as originally required.

4

Extensions of the general approach

4.1. Introduction

The general ideas which have been developed in the first three chapters are now available to be used in a wider class of problems in applied mathematics than we have so far indicated. We can expect to go beyond the original objective, which was to find upper and lower bounds for the solution value of functionals representing, for example, an overall energy expenditure in boundary value problems.

In this chapter we start to explore such extensions.

We begin §4.2 with a discussion of bounds on pre-assigned linear functionals. This is related to the question of pointwise bounds. Some rather different viewpoints in the literature are brought together and generalized in §§4.2(i)–(iii). We carry out some preliminary detailed calculations in §§4.2(iv) and (v). Then we discuss so-called 'bivariational' bounds, which require some new hypotheses.

In §§4.3 and 4.4 we give some discussion of bounds for initial value problems. In §4.5 we return briefly to comparison methods, in order to make contact with an approach which has been influential in solid mechanics, where information about a 'hard' problem is obtained by comparison with a notional 'easy' problem.

The idea of working in a pair of inner product spaces offers some fresh viewpoints in all these contexts, which already have their own substantial literature. The purpose in this chapter is to hint at what may be achieved by the systematic development of a body of detailed and substantial examples. We give some new results, but there is ample scope for further investigation.

4.2. Bounds on linear functionals

(i) Introduction

We proved in Theorem 3.6 that, in a class of problems generated by the gradients of a quadratic functional

$$L[x,u] = -\tfrac{1}{2}(x, Ax) + \langle u, Tx\rangle + \tfrac{1}{2}\langle u, Bu\rangle + (a,x) + \langle b,u\rangle, \quad (3.59\ bis)$$

we can construct bounds on the particular linear functional

$$\tfrac{1}{2}(a, x_0) + \tfrac{1}{2}\langle b, u_0\rangle \quad (3.72\ bis)$$

of a solution x_0, u_0 of such a problem. The problem may be governed by inequalities or by equations, the generating functional needs to be a saddle functional in some sense, and there may be either simultaneous upper and lower bounds or only one of these. The bounds themselves are sometimes called energy bounds, because (3.72) may be the actual work of given forces in mechanical problems.

In other situations, or for other purposes, we may need to bound another linear functional which is defined independently of the structure of the given problem. For example, if there is a generating functional it might not be quadratic, and even when it is we may wish to estimate the value of a different linear functional from that specified by (3.72). In particular, the pointwise value of a field variable may be sought, and this might be done via a linear functional whose given 'coefficient' is, or has the same effect as, a δ function.

In this section we indicate some of the literature on such problems, and we extend it in certain directions. Notable early papers included that of Fujita (1955) for the decomposable linear equation $T^*Tx = a$ of (3.137), and the review of Diaz (1960). Barnsley and Robinson (1974, 1977) considered nonlinear equations which are not generated by saddle functionals of the unqualified or the more easily qualified kind (as in (1.58) (a)–(f) of §1.3(iii)) which we have so far considered. Under different hypotheses they constructed bounds which they called 'bivariational'. Sewell and Noble (1978) described a broad framework covering both these approaches. Cole, Mika and Pack (1984) have given further improvements on the bivariational method. Collins (W.D., 1979, 1980, 1982) and Moreau (1962, 1966b, 1971) have described a different approach which generalizes a decomposition into certain orthogonal subspaces, and which can handle inequality constraints. Much of this work is expressed in terms of Hilbert spaces, but we can establish many of the essential points in our more elementary context of inner product spaces.

A basic idea is that if we can bound the length of the position vector of the unknown solution point x_0, u_0, then we can also bound the projection of that vector onto any given direction, and thus bound a linear functional of the solution.

We begin with a simple but rather graphic example, and then proceed to generalizations.

(ii) Nonnegative operator problems

We shall exploit the identity

$$\|\tfrac{1}{2}(u_+ + u_-) - u_0\|^2 = \tfrac{1}{4}\|u_+ - u_-\|^2 + \langle u_+ - u_0, u_- - u_0\rangle, \quad (4.1)$$

which is valid for any three points u_+, u_-, u_0 in u space.

The energy bounds given for $T^*Tx = a$ in Theorem 3.14 may be written

$$\|u_\alpha\|^2 \geqslant (x_0, a) \geqslant \frac{(x_\beta, a)^2}{\|Tx_\beta\|^2},$$

where $\|\cdot\| = \langle .,. \rangle^{\frac{1}{2}}$ and the trial functions u_α and x_β satisfy

$$T^*u_\alpha = -a, \quad Tx_\beta = -u_\beta \quad (4.2)$$

respectively. That is, u_α is any solution of $T^*u_\alpha = -a$ alone, and u_β is defined by $u_\beta = -Tx_\beta$ for any $x_\beta \in \mathscr{D}(T)$. A solution x_0 of $T^*Tx = a$ satisfies both of (4.2) with unique u_0.

Suppose now that we are given any $q \in \mathscr{D}(T^*)$ and asked to bound the linear functional

$$(x_0, T^*q) = -\langle u_0, q\rangle, \quad (4.3)$$

where $u_0 = -Tx_0$. There need be no connection between a and T^*q. Fujita (1955) included the following result in a compact paper which subsumed earlier work on pointwise bounds by Diaz, Greenberg and Weinstein, Prager and Synge, and Kato. We offer a simple proof in terms of Fig. 4.1.

Theorem 4.1

The linear functional (4.3) *of a solution of $T^*Tx = a$ is bounded on both sides by each of the three inequalities*

$$\tfrac{1}{2}\|u_\alpha - u_\beta\| \, \|q\| \geqslant |\langle \tfrac{1}{2}(u_\alpha + u_\beta) - u_0, q\rangle|, \quad (4.4)$$

$$\|u_\alpha - u_\beta\| \, \|q\| \geqslant |\langle u_\alpha - u_0, q\rangle|, \quad \|u_\alpha - u_\beta\| \, \|q\| \geqslant |\langle u_\beta - u_0, q\rangle|, \quad (4.5)$$

for any u_α and u_β defined in (4.2).

Fig. 4.1. Theorem 4.1 in u-space.

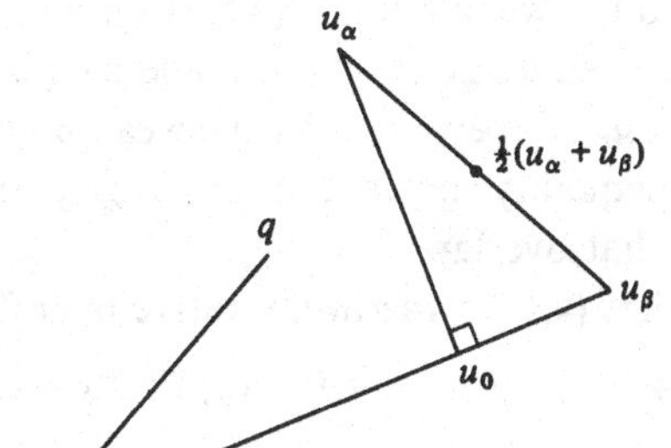

Proof

Using $u_\beta = -Tx_\beta$ with adjointness and $T^*u_0 = -a = T^*u_\alpha$ gives

$$\langle u_\beta, u_\alpha - u_0\rangle = 0.$$

This orthogonality used again, but with u_0 as a particular choice of u_β, i.e. $\langle u_0, u_\alpha - u_0\rangle = 0$, then establishes the right-angled triangle of Fig. 4.1 in the space with inner product $\langle .,.\rangle$, with hypotenuse joining u_α and u_β, and u_0 at the right angle on the join of u_β to the origin.

A hypercircle with centre $\frac{1}{2}(u_\alpha + u_\beta)$ can be drawn through these three points, and it has radius

$$\|\tfrac{1}{2}(u_\alpha + u_\beta) - u_0\| = \tfrac{1}{2}\|u_\alpha - u_\beta\|.$$

This relation also follows analytically, using (4.1), from

$$\|\tfrac{1}{2}(u_\alpha + u_\beta) - u_0\|^2 - \tfrac{1}{4}\|u_\alpha - u_\beta\|^2 = \langle u_\alpha - u_0, u_\beta - u_0\rangle = 0. \tag{4.6}$$

The projection of any given unit vector $q/\|q\|$ onto the radial vector $\frac{1}{2}(u_\alpha + u_\beta) - u_0$ must therefore be less than the length $\frac{1}{2}\|u_\alpha - u_\beta\|$ of this radius, i.e.

$$\left|\left\langle \tfrac{1}{2}(u_\alpha + u_\beta) - u_0, \frac{q}{\|q\|}\right\rangle\right| \leqslant \tfrac{1}{2}\|u_\alpha - u_\beta\|,$$

by Schwarz' inequality (3.2), hence (4.4).

Similarly the projection of $q/\|q\|$ onto either shorter side of the triangle gives, for example

$$\left|\left\langle u_\alpha - u_0, \frac{q}{\|q\|}\right\rangle\right| \leqslant \|u_\alpha - u_0\| \leqslant \|u_\alpha - u_\beta\|,$$

where the triangle inequality has also been used on the right, leading to (4.5). □

It is convenient to call (4.4) the strong bound (Fujita, *op. cit.*, equation (3.6)), in contrast to the weak bounds (4.5) (Fujita *op. cit.*, equation (3.8)), because more was given away via the triangle inequality to get the latter. The two ranges of $\langle u_0, q\rangle$ permitted by the weak bounds have an overlap, and the Schwarz inequality shows that the range permitted by the strong bound lies within that overlap.

We now generalize (4.4) to the nonnegative operator problem

$$T^*Tx + E'[x] \geqslant 0, \quad x \geqslant 0, \quad (x, T^*Tx + E'[x]) = 0. \tag{4.7}$$

A simpler version of the proof will apply to $T^*Tx + E'[x] = 0$ in (3.136), whose decomposition (3.144) is generated by (3.145) from (1.53). The decomposition of (4.7) generated by the same $L[x,u]$ from (1.55) is

$$\begin{aligned} T^*u - E'[x] \leqslant 0 \quad (\alpha), \qquad & Tx = -u \quad (\beta), \\ x \geqslant 0 \quad (\beta), \qquad & (x, T^*u - E'[x]) = 0. \end{aligned} \tag{4.8}$$

Theorem 4.2

Suppose that $E[x]$ is weakly convex and $\mathscr{D}(T) \subseteq \mathscr{D}(E)$.

(a) Simultaneous bounds $L_\alpha - (x_\alpha, \partial L/\partial x|_\alpha) \geqslant L_0 \geqslant L_\beta$ apply to (4.7). The bounds are

$$L_\alpha - (x_\alpha, \partial L/\partial x|_\alpha) = (x_\alpha, E'[x_\alpha]) - E[x_\alpha] + \tfrac{1}{2}\langle u_\alpha, u_\alpha \rangle$$

with any x_α, u_α such that $T^*u_\alpha \leqslant E'[x_\alpha]$; and

$$L_\beta = -E[x_\beta] - \tfrac{1}{2}(x_\beta, Px_\beta)$$

with any x_β such that $0 \leqslant x_\beta \in \mathscr{D}(T)$, where $P = T^*T$. Also

$$L_0 = \tfrac{1}{2}(x_0, E'[x_0]) - E[x_0]$$

where x_0 is any solution of (4.7).

(b) The unique $u_0 = -Tx_0$ derived from any solution x_0 of (4.7) satisfies

$$\| \tfrac{1}{2}(u_\alpha + u_\beta) - u_0 \|^2 \leqslant \tfrac{1}{4} \| u_\alpha - u_\beta \|^2 + C_{\alpha\beta} \tag{4.9}$$

where

$$C_{\alpha\beta} = E[x_\beta] - E[x_\alpha] + (x_\alpha, E'[x_\alpha]) - (x_\beta, T^*u_\alpha) \geqslant 0.$$

(c) Any given $q \in \mathscr{D}(T^*)$ leads to bounds

$$|\langle \tfrac{1}{2}(u_\alpha + u_\alpha) - u_0, q \rangle| \leqslant [\tfrac{1}{4} \| u_\alpha - u_\beta \|^2 + C_{\alpha\beta}]^{\frac{1}{2}} \| q \| \tag{4.10}$$

on the linear functional $-\langle u_0, q \rangle = (x_0, T^*q)$ of a solution x_0 of (4.7).

Proof

(a) Only a slight variant of the proof of Theorem 3.13 is required. If $E[x]$ is strictly convex, the uniqueness of x_0 again follows. Note that the bounded quantity L_0 is not a linear functional of the solution if $E[x]$ is not quadratic.

(b) As on the left of (4.6) we have, from (4.1) alone,

$$\| \tfrac{1}{2}(u_\alpha + u_\beta) - u_0 \|^2 - \tfrac{1}{4} \| u_\alpha - u_\beta \|^2 = \langle u_\alpha - u_0, u_\beta - u_0 \rangle.$$

The expression on the right is no longer zero, however. Instead, the properties of (4.8) give

$$\begin{aligned} \langle u_\alpha - u_0, u_\beta - u_0 \rangle &= \langle u_\alpha - u_0, T(x_0 - x_\beta) \rangle = (x_0 - x_\beta, T^*(u_\alpha - u_0)) \\ &= (x_0, T^*u_\alpha - E'[x_\alpha]) + (x_\beta, T^*u_0 - E'[x_0]) \\ &\quad + (x_0, E'[x_\alpha]) - (x_0, T^*u_0) + (x_\beta, E'[x_0]) - (x_\beta, T^*u_\alpha) \\ &\leqslant (x_0, E'[x_\alpha]) + (x_\beta - x_0, E'[x_0]) - (x_\beta, T^*u_\alpha). \end{aligned} \tag{4.11}$$

We now use the weak convexity of $E[x]$ twice in the forms

$$0 \leqslant E[x_0] - E[x_\alpha] - (x_0 - x_\alpha, E'[x_\alpha]),$$
$$0 \leqslant E[x_\beta] - E[x_0] - (x_\beta - x_0, E'[x_0]).$$

Adding these three inequalities allows us to eliminate the supposedly unknown solution x_0 from the right, giving

$$\langle u_\alpha - u_0, u_\beta - u_0 \rangle \leqslant C_{\alpha\beta}. \tag{4.12}$$

We can rearrange

$$C_{\alpha\beta} = E[x_\beta] - E[x_\alpha] - (x_\beta - x_\alpha, E'[x_\alpha]) - (x_\beta, T^* u_\alpha - E'[x_\alpha]) \geqslant 0$$

by $(x_\beta, T^* u_\alpha - E'[x_\alpha]) \leqslant 0$ and the convexity of $E[x]$. Also $C_{\alpha\beta} = L_\alpha - (x_\alpha, \partial L/\partial x|_\alpha) - L_\beta - \frac{1}{2}\| u_\alpha - u_\beta \|^2$.

(c) The Schwarz inequality gives

$$\left| \left\langle \tfrac{1}{2}(u_\alpha + u_\beta) - u_0, \frac{q}{\| q \|} \right\rangle \right| \leqslant \| \tfrac{1}{2}(u_\alpha + u_\beta) - u_0 \|,$$

whence (4.10) follows using (4.9). □

The statement of Theorem 4.2 for $T^*Tx + E'[x] = 0$ differs only in that x_β is unrestricted in sign, and $T^* u_\alpha = E'[x_\alpha]$ can be written into $C_{\alpha\beta}$. Theorem 4.2 represents the essence of a more sophisticated theory given by Collins (W.D., 1982), which is based on generalizations of vector decomposition into orthogonal subspaces, for example into so-called polar cones (Moreau, 1962, 1966*b*, 1971). Illustrations given by Collins include elastoplastic torsion, and the flow of a Bingham fluid in a pipe.

A particular example of Theorem 4.2 applies with quadratic $E[x] = \frac{1}{2}(x, Ax) - (a, x)$, when A is nonnegative. In that case (4.11) becomes

$$\langle u_\alpha - u_0, u_\beta - u_0 \rangle \leqslant -(x_\alpha - x_0, A(x_\beta - x_0)) - (x_\beta, -Ax_\alpha + T^* u_\alpha + a). \tag{4.13}$$

If A is positive we can work in a different inner product space, and bound a different linear functional without needing to give away the extra amounts involved in going from (4.11) to (4.12). Such results are indicated next.

(iii) Other saddle-generated problems

We now drop the assumption that the operator T^*T exists. The hypercircle interpretation given to the general linear problem in §1.6(iv) can be repeated verbatim for the infinite dimensional case. In particular, the orthogonality established in $(1.145)_4$ still holds, so that there is a right-angled triangle generalizing Fig. 4.1 to the space whose elements are sets $\{x, u\} = \xi$,

$\{u, v\} = \eta$ (say) with inner product

$$\overline{\xi, \eta} = \tfrac{1}{2}(x, Ay) + \tfrac{1}{2}\langle u, Bv \rangle, \tag{4.14}$$

when A and B are both positive operators. It may be that the origin no longer lies on the join of solution and β points, but that is not important. From (3.73) and (4.14) the saddle quantity $S = \| \xi_+ - \xi_- \|^2$.

Now let $p \in \mathscr{D}(T) \cap \mathscr{D}(A)$ and $q \in \mathscr{D}(T^*) \cap \mathscr{D}(B)$ be any given elements in the stated domains, normalized so that $\{p, q\} = r$ has

$$\| r \|^2 = \overline{r, r} = 1, \quad \text{i.e.} \quad \tfrac{1}{2}(p, Ap) + \tfrac{1}{2}\langle q, Bq \rangle = 1. \tag{4.15}$$

Then the following generalization of Theorem 4.1 is immediate, and associated results are summarized in Exercises 4.1.

Theorem 4.3

The linear functional $(x_0, Ap) + \langle u_0, Bq \rangle$ *of the solution of*

$$-Ax + T^*u + a = 0, \quad Tx + Bu + b = 0$$

is bounded (in the notation of (1.144)) *by the strong inequality*

$$\begin{aligned} &|(\tfrac{1}{2}(x_\alpha + x_\beta) - x_0, Ap) + \langle \tfrac{1}{2}(u_\alpha + u_\beta) - u_0, Bq \rangle| \\ &\quad \leqslant [\tfrac{1}{2}(x_\alpha - x_\beta, A(x_\alpha - x_\beta)) + \tfrac{1}{2}\langle u_\alpha - u_\beta, B(u_\alpha - u_\beta) \rangle]^{\frac{1}{2}} \\ &\quad = (L_\alpha - L_\beta)^{\frac{1}{2}} \end{aligned} \tag{4.16}$$

and by the weak inequalities

$$\begin{aligned} &|(x_\alpha - x_0, Ap) + \langle u_\alpha - u_0, Bq \rangle| \\ &\quad \leqslant 2^{\frac{1}{2}}[(x_\alpha - x_\beta, A(x_\alpha - x_\beta)) + \langle u_\alpha - u_\beta, B(u_\alpha - u_\beta) \rangle]^{\frac{1}{2}} \\ &\quad = 2(L_\alpha - L_\beta)^{\frac{1}{2}}, \end{aligned}$$

and the similar one with x_β, u_β *on the left.*

The Schwarz inequality, which is a central tool in the proofs of Theorems 4.1 and 4.3, is proved by optimizing a nonhomogeneous quadratic with respect to a scale factor. We can generalize the weak inequalities to certain nonlinear problems by returning to this more basic optimization method. Following the account of Sewell and Noble (1978), we must identify suitable nonquadratic functionals to be optimized, as follows.

Theorem 4.4

Let $L[x, u]$ *be a given saddle functional in the sense that* $S \geqslant 0$ *for every pair of points in its domain, where the saddle quantity* S *is defined by* (3.66). *Let* x_0, u_0 *denote any solution of*

$$\frac{\partial L}{\partial x} = 0 \quad (\alpha), \qquad \frac{\partial L}{\partial u} = 0 \quad (\beta). \tag{1.53 bis}$$

Then if x_β, u_β is any point satisfying (β) and if x_+, u_+ is an arbitrary point in the domain of $L[x,u]$,

$$L_+ - L_\beta - \left(x_+, \frac{\partial L}{\partial x}\bigg|_+ \right) \geqslant - \left(x_0, \frac{\partial L}{\partial x}\bigg|_+ \right). \tag{4.17}$$

Again if x_α, u_α is any point satisfying (α) and if x_-, u_- is an arbitrary point in the domain of $L[x,u]$,

$$L_\alpha - L_- + \left\langle u_-, \frac{\partial L}{\partial u}\bigg|_- \right\rangle \geqslant \left\langle u_0, \frac{\partial L}{\partial u}\bigg|_- \right\rangle. \tag{4.18}$$

Proof

Referring to (3.66), first select the minus point in $S \geqslant 0$ as a solution point, so that

$$L_+ - L_0 - \left(x_+, \frac{\partial L}{\partial x}\bigg|_+ \right) \geqslant - \left(x_0, \frac{\partial L}{\partial x}\bigg|_+ \right). \tag{4.19}$$

Adding $L_0 - L_\beta \geqslant 0$ to this gives (4.17).

Beginning again with (3.66), next select the plus point in $S \geqslant 0$ as a solution point, so that

$$L_0 - L_- + \left\langle u_-, \frac{\partial L}{\partial u}\bigg|_- \right\rangle \geqslant \left\langle u_0, \frac{\partial L}{\partial u}\bigg|_- \right\rangle. \tag{4.20}$$

Adding $L_\alpha - L_0 \geqslant 0$ to this gives (4.18). □

The quantities on the left in (4.17) and (4.18) are regarded as known, in terms of assignable x_+, u_+ and x_-, u_- and supposedly known x_α, u_α and x_β, u_β. Thus we bound the two linear functionals of x_0 and u_0 on the right of (4.17) and (4.18). Theorem 4.4 can be extended to problems like (1.50) which are governed by inequalities rather than equations (see §5(iii) of Sewell and Noble, 1978).

For $T^*Tx = a$, generated by $L[x,u] = (x, T^*u + a) + \frac{1}{2}\langle u, u\rangle$, we find that (4.19) becomes $\frac{1}{2}\langle u_+, u_+\rangle \geqslant -(x_0, T^*u_+ + \frac{1}{2}a)$, and the strong bound (4.4) is then recovered by choosing $u_+ = \frac{1}{2}(u_\alpha + u_\beta) + kq$, using the orthogonality, and optimizing with respect to the scale factor k. The same result comes from (4.20), with u_- in place of u_+.

For the nonlinear problem covered by Theorem 4.4, the extra amount given away in Theorem 4.1 via the triangle inequality to get the weak bounds is now represented by $L_0 - L_\beta \geqslant 0$ and $L_\alpha - L_0 \geqslant 0$ used to reach (4.17) and (4.18) respectively. We can generalize the weak bound $(4.5)_1$ by inserting into (4.17) the choice

$$x_+ = x_\alpha + hp, \quad u_+ = u_\alpha + kq, \tag{4.21}$$

and optimizing with respect to the two scale factors h and k, for any assigned p, q in the domain of $L[x, u]$. For the details in the general case we refer to Sewell and Noble (1978, §2(ix)). We illustrate the procedure for the example

$$L[x,u]=(x,T^*u+a)+G[u], \tag{4.22}$$

where $G[u]$ is a strictly convex functional, not necessarily quadratic, with gradient $G'[u]$.

Theorem 4.5

The linear functional $(x_0, T^*q) = -\langle G'[u_0], q\rangle$ *of a solution to the problem*

$$T^*u=-a \quad (\alpha), \qquad Tx=-G'[u] \quad (\beta), \tag{4.23}$$

is bounded by

$$[2(L_\alpha-L_\beta)\langle q, G''[u_\alpha]q\rangle]^{\frac{1}{2}} \geqslant |\langle G'[u_\alpha]-G'[u_0], q\rangle|, \tag{4.24}$$

where $G''[u_\alpha]$ *is an operator such that*

$$G[u_\alpha+kq]=G[u_\alpha]+k\langle G'[u_\alpha],q\rangle+\tfrac{1}{2}k^2\langle q,G''[u_\alpha]q\rangle+\mathrm{O}(k^3) \tag{4.25}$$

for sufficiently small k, *and* $\langle q, G''[u_\alpha]q\rangle > 0$ *for* $q \neq 0$.

Proof

The case (4.22) has convenient intermediate generality because only u_+ and not x_+ appears in (4.17), which with (4.21) becomes

$$G[u_\alpha+kq]+k(x_0,T^*q)\geqslant G[u_\beta]+(x_\beta,T^*u_\beta+a). \tag{4.26}$$

Using (4.25), when k is small enough we have a quadratic inequality to optimize by completing the square. Allowing for k to be of either sign leads to (4.24) with

$$k=\pm\left[\frac{2(L_\alpha-L_\beta)}{\langle q,G''[u_\alpha]q\rangle}\right]^{\frac{1}{2}}. \qquad \square$$

Even if such k are not small, we can insert them into (4.26) to give rigorous bounds replacing (4.24). Here

$$L_\alpha-L_\beta=G[u_\alpha]-G[u_\beta]-(x_\beta,T^*u_\beta+a).$$

We can generalize the other weak bound $(4.5)_2$ by inserting into (4.18) the choices

$$x_-=x_\beta+hp, \quad u_-=u_\beta+kq$$

instead of (4.21). Details are given by Sewell and Noble (1978).

Exercises 4.1

1. Prove Theorem 4.3.
2. The infinite dimensional version of (1.50) (cf. §3.6(i)) generated by (3.59) has

properties which can be written

$$-Ax_\alpha + T^*u_\alpha + a \leqslant 0, \qquad Tx_\beta + Bu_\beta + b \geqslant 0,$$
$$x_\beta \geqslant 0, \qquad u_\alpha \geqslant 0,$$
$$(x_0, -Ax_0 + T^*u_0 + a) = 0, \quad \langle u_0, Tx_0 + Bu_0 + b\rangle = 0.$$

The inequalities are also satisfied with the zero subscript, which signifies a solution of the whole system as before. Show that

$$\begin{aligned}(x_\alpha - x_0, A(x_\beta - x_0)) &+ \langle u_\alpha - u_0, B(u_\beta - u_0)\rangle \\ &\leqslant -(x_\beta, -Ax_\alpha + T^*u_\alpha + a) + \langle u_\alpha, Tx_\beta + Bu_\beta + b\rangle \\ &= -\left(x_\beta, \frac{\partial L}{\partial x}\bigg|_\alpha\right) + \left\langle u_\alpha, \frac{\partial L}{\partial u}\bigg|_\beta\right\rangle = K_{\alpha\beta} \quad \text{(say)}\end{aligned}$$

and that $K_{\alpha\beta} \geqslant 0$.

3. Compare the result of Exercise 2 with (4.11). Show that, in the context of (4.13), $C_{\alpha\beta} = \frac{1}{2}(x_\alpha - x_\beta, A(x_\alpha - x_\beta)) + K_{\alpha\beta}$.
4. When A and B are positive operators in Exercise 2, use the inner product (4.14) to show that
$$\|\tfrac{1}{2}(\xi_\alpha + \xi_\beta) - \xi_0\|^2 - \tfrac{1}{4}\|\xi_\alpha - \xi_\beta\|^2 = \overline{\xi_\alpha - \xi_0, \xi_\beta - \xi_0}$$
and therefore that
$$\|\tfrac{1}{2}(\xi_\alpha + \xi_\beta) - \xi_0\|^2 \leqslant \tfrac{1}{4}\|\xi_\alpha - \xi_\beta\|^2 + \tfrac{1}{2}K_{\alpha\beta}.$$
5. Hence show that if $r = \{p, q\}$ is any unit element satisfying (4.15),
$$|\overline{\tfrac{1}{2}(\xi_\alpha + \xi_\beta) - \xi_0, r}| \leqslant [\tfrac{1}{4}\|\xi_\alpha - \xi_\beta\|^2 + \tfrac{1}{2}K_{\alpha\beta}]^{\frac{1}{2}}.$$
This generalizes (4.16), giving a bound for an assigned linear functional of the solution in the presence of linear inequality constraints.
6. Show that the problem
$$T^*(Tx + c) \geqslant 0, \quad x \geqslant 0, \quad (x, T^*(Tx + c)) = 0,$$
with a given c can be expressed as a variant of (4.7) in which $C_{\alpha\beta} = -\langle u_\beta, c - u_\alpha\rangle$. Compare with equation (2.5) of Collins (1980), and also (2.5) of Collins (W.D., 1982).
7. Write $S_{\alpha\beta}$ for the value of the saddle quantity (3.66) when the + and − points are chosen as α and β points respectively, and write $B_\alpha = L_\alpha - (x_\alpha, \partial L/\partial x|_\alpha)$, $B_\beta = L_\beta - \langle u_\beta, \partial L/\partial u|_\beta\rangle$ for the values of upper and lower bounds when they apply. Show that $S_{\alpha\beta} + K_{\alpha\beta} = B_\alpha - B_\beta$ with $S_{\alpha\beta} \geqslant 0$ and $K_{\alpha\beta} \geqslant 0$, regardless of whether $L[x, u]$ is quadratic or not, and of whether the inequality constraints are present or not.
8. For the quadratic (3.59) with positive operators A and B show that $\|\xi_\alpha - \xi_\beta\|^2 = S_{\alpha\beta}$, and therefore
$$|(\tfrac{1}{2}(x_\alpha + x_\beta) - x_0, Ap) + \langle \tfrac{1}{2}(u_\alpha + u_\beta) - u_0, Bq\rangle| \leqslant [B_\alpha - B_\beta + K_{\alpha\beta}]^{\frac{1}{2}}$$
as an alternative statement of the result in Exercise 5.

(iv) Comparison problems for a cantilever beam

Here we give a specific example of another type of 'comparison problem', to be judiciously chosen to provide information about a different given problem. We also widen the ideas of strong and weak bounds mentioned in §§(ii) and (iii).

(a) Given problem

Let s denote distance from the clamped end of the thin straight cantilever shown in Fig. 4.2(*a*). The other end $s = 1$ is free, and the given piecewise continuous function $w(s)$ specifies a distributed transverse downward load per unit length. Let piecewise C^2 functions $m(s)$ and $u(s)$ denote internal bending moment and small transverse deflection respectively. These are to be found subject to the end conditions $u(0) = u'(0) = 0$, $m(1) = m'(1) = 0$, the equilibrium equation $m'' = w$, and the constitutive law $u'' = m^c$ for any given odd integer $c > 0$. For $c = 1$ we have classical elasticity, and for other c (say $c = 3$) the beam obeys a typical creep law, the moduli being normalized to unity. The transverse shear force $m'(s)$ is found from $m(s)$, which will be the

Fig. 4.2. Cantilever beam. (*a*) Given loading. (*b*) First comparison problem. (*c*) Second comparison problem.

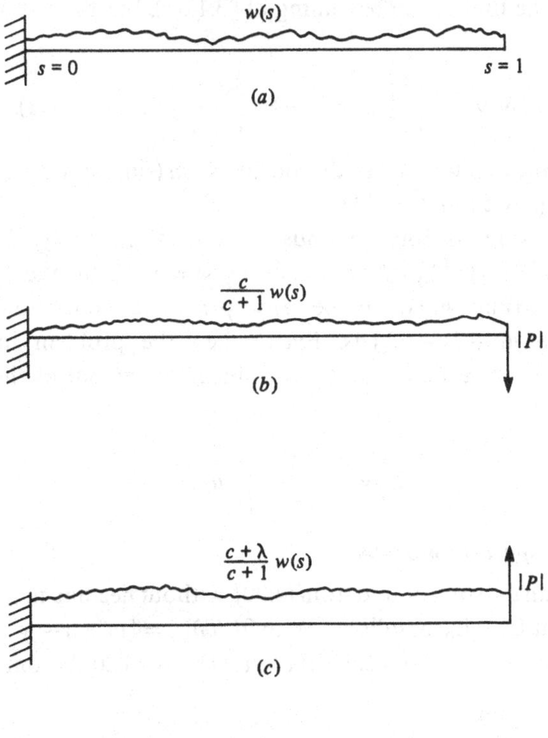

same for any c since the problem is statically determinate. In fact for constant w the exact solution is

$$m_0(s) = \tfrac{1}{2}w(s-1)^2,$$
$$u_0(s) = \frac{w^c}{2^{c+1}(2c+1)(c+1)}[(s-1)^{2c+2} - 1 + 2s(c+1)]. \qquad (4.27)$$

This will give us a check on the following calculations of bounds.

(b) Saddle functional

Our general procedure (§3.9(ii)) can rewrite the problem as

$$T^*\tilde{u} = \begin{bmatrix} u''(s) \\ u'(0) \end{bmatrix} = \begin{bmatrix} m^c \\ 0 \end{bmatrix}, \quad (\alpha)$$
$$T\tilde{m} = \begin{bmatrix} m''(s) \\ -m'(1) \end{bmatrix} = \begin{bmatrix} w \\ 0 \end{bmatrix}, \quad (\beta) \qquad (4.28)$$

where $\mathscr{D}(T)$ is the space of $\tilde{m} = [m(s) \;\; m(0)]^{\mathrm{T}}$ constructed from piecewise C^2 functions $m(s)$ satisfying $m(1) = 0$ and with $(\tilde{m}, \tilde{n}) = \int_0^1 mn\,\mathrm{d}s + m(0)n(0)$, and $\mathscr{D}(T^*)$ is the space of $\tilde{u} = [u(s) \;\; u(1)]^{\mathrm{T}}$ built from piecewise C^2 functions $u(s)$ satisfying $u(0) = 0$ and with $\langle \tilde{u}, \tilde{v} \rangle = \int_0^1 uv\,\mathrm{d}s + u(1)v(1)$. Two integrations by parts are needed to verify $(\tilde{m}, T^*\tilde{u}) = \langle \tilde{u}, T\tilde{m} \rangle$. Neither T^*T nor TT^* exist, for reasons like that at the beginning of §3.12(v), but there is a generating functional

$$L[\tilde{m}, \tilde{u}] = \int_0^1 \left[u(m'' - w) - \frac{m^{c+1}}{c+1} \right] \mathrm{d}s - u(1)m'(1). \qquad (4.29)$$

This is a saddle functional strictly concave in $\tilde{m}$ (since c is odd) and weakly convex in $\tilde{u}$ (Sewell and Noble, 1978, (6.16)).

Therefore simultaneous bounds $L_\alpha \geqslant L_0 \geqslant L_\beta$ apply, but in fact $L_0 = L_\beta = -(c+1)^{-1}\int_0^1 m_\beta^{c+1}\,\mathrm{d}s$ since $m_\beta(s) = m_0(s)$ by the static determinacy. Choosing $m_\alpha(s)$ to be any piecewise continuous function, $L_\alpha = \int_0^1 m_\alpha^c[cm_\alpha/(c+1) - m_\beta]\,\mathrm{d}s$. Even when the problem in nonlinear (i.e. when $c \neq 1$), we find that L_0 is a linear functional of the deflection solution $u_0(s)$, i.e.

$$L_0 = -\frac{1}{c+1}\int_0^1 u_0 w\,\mathrm{d}s. \qquad (4.30)$$

(c) First comparison problem

We can obtain a strong bound from (4.20), without needing to give away the extra amount $L_\alpha - L_0$, as follows. With $\partial L/\partial\tilde{u}|_- = [m_-''(s) - w \quad -m_-'(1)]^{\mathrm{T}}$ for any piecewise C^2 $m_-(s)$ such that $m_-(1) = 0$, (4.20) becomes

$$\frac{1}{c+1}\int_0^1 m_-^{c+1}\,ds \geqslant \int_0^1 u_0\left[m_-'' - \frac{cw}{c+1}\right]ds - u_0(1)m_-'(1). \qquad (4.31)$$

We can use this inequality to estimate the true downward deflection $u_0(1)$ at the free end of the given beam (Fig. 4.2(*a*)) by selecting the function $m_-(s)$, which is at our disposal, to be such that

$$-m_-'(1) = P, \quad \text{and} \quad m_-''(s) = \frac{cw(s)}{c+1} \quad \text{in} \quad 0 \leqslant s \leqslant 1. \qquad (4.32)$$

Here P is a given number, and

$$m_-(s) = \frac{c}{c+1}\int_s^1 \int_t^1 w(\tau)\,d\tau\,dt + P(1-s)$$

is interpretable as the internal bending moment which would be in equilibrium for a 'comparison problem' whose load system is a point load P at the free end together with a distributed load $cw(s)/(c+1)$ per unit length. Such $m_-(s)$ is not meant to be an approximation to the bending moment in the given problem, and so it is not being thought of as a trial function.

If P is assumed positive, this comparison loading is as shown in Fig. 4.2(*b*), with the point load acting downward, like the distributed load. We then obtain a result due to Martin (1966), that the end deflection in the given beam is bounded above by

$$u_0(1) \leqslant \frac{1}{P(c+1)}\int_0^1 m_-^{c+1}\,ds. \qquad (4.33)$$

Martin's analysis looks quite different to the present one based on (4.20), but it fits within our general scheme because it is based on potential and complementary energies (cf. also Noble 1974).

For constant w we find

$$m_-(s) = \frac{cw}{2(c+1)}(1-s)^2 + P(1-s). \qquad (4.34)$$

If $c = 1$, from (4.30) the end deflection is bounded by

$$\frac{u_0(1)}{w} \leqslant 0.127 + \frac{1}{2}\left[\left(\frac{P}{3w}\right)^{\frac{1}{2}} - \left(\frac{w}{80P}\right)^{\frac{1}{2}}\right]^2.$$

The least such bound, provided by choosing $P/w = 0.194$, is evidently quite close to the exact value $u_0(1)/w = 0.125$.

(d) Second comparison problem

There is no reason in principle why we should not explore negative P in (4.32), so that the point load in Fig. 4.2(*b*) would be reversed to act upwards, in the opposite direction to the distributed load. Then (4.31) gives

$$u_0(1) \geqslant \frac{1}{P(c+1)} \int_0^1 m_-^{c+1}\,ds, \tag{4.35}$$

instead of (4.33). Together with (4.33) and (4.34) we then find $|u_0(1)/w - \frac{1}{16}| \leqslant 240^{-\frac{1}{2}}$ for $c = 1$, but the lower bound -0.002 is not useful.

A better lower bound to the downward end deflection can be obtained using an upward point load in a second comparison problem, as Palmer (1967) showed in Martin's context. In general terms first, we begin with (4.20) and add $\lambda(L_\alpha - L_0) \geqslant 0$ to it for some scalar $\lambda > 0$ to be chosen. In deriving (4.18) we used $\lambda = 1$ to eliminate L_0, but now we can exploit

$$\lambda L_\alpha - L_- + \left\langle u_-, \frac{\partial L}{\partial u}\bigg|_-\right\rangle \geqslant \left\langle u_0, \frac{\partial L}{\partial u}\bigg|_-\right\rangle + (\lambda - 1)L_0 \tag{4.36}$$

because in the current problem L_0 is a linear functional of u_0. We may call this inequality weak compared to (4.20), because an extra amount has been given away. In place of (4.31) we reach

$$\begin{aligned} &\frac{1}{c+1}\int_0^1 m_-^{c+1}\,ds + \frac{\lambda c}{c+1}\int_0^1 m_\alpha^{c+1}\,ds - \lambda\int_0^1 u_\alpha w\,ds \\ &\geqslant \int_0^1 u_0\left[m_-'' - \frac{(c+\lambda)}{c+1}w\right]ds - m'(1)u_0(1). \end{aligned} \tag{4.37}$$

This induces the second comparison problem shown in Fig. 4.2(*c*), which leads to a family of lower bounds for $u_0(1)$ which depend on λ and on the trial functions which satisfy (4.28α) with $u_\alpha(0) = m_\alpha(1) = 0$. For example, with constant w and $c = 1$, $u_\alpha(s) = \mu s^2[s^2/12 - s/3 + \frac{1}{2}]$ in (4.37) leads to optimum $\mu = \frac{1}{2}w$, $\lambda = 1 - 5P/w$, $P = -6.25w$ and finally $u_0(1)/w \geqslant 0.063$. This is better than -0.002. Palmer performed a similar calculation for $c = 3$, and showed that $u_\alpha(s)$ could be chosen to improve the lower bound.

Summarizing, we have shown that the two different comparison problems in Figs. 4.2(*b*) and (*c*) lead to the pointwise bounds $0.063 \leqslant u_0(1)/w \leqslant 0.127$ for the end deflection $u_0(1)/w = 0.125$ in Fig. 4.2(*a*). Associated dynamical problems are discussed by Martin (1964) and Reddy (1981).

(*v*) *Other examples of pointwise bounds*

A systematic exploitation of all the general procedures in §§(ii) and (iii) is not yet available, but nevertheless they give a helpful background against which to view existing theories. Additional devices will be required in particular examples, as the cantilever beam problem illustrated. The notion of a comparison problem emerged naturally there, and an augmentation (4.36) of the general theory was also required.

Further refinements are needed when, for example, we seek to use Theorem 4.1 to get pointwise bounds via (4.3) on solutions x_0 of $T^*Tx = a$. If we wish to choose q such that T^*q has the effect of a Dirac delta function δ appropriate to the context, T^* must have a property of the type $\delta \in \mathscr{R}(T^*)$. This may call for redefinition of wider $\mathscr{D}(T^*)$ and $\mathscr{D}(T)$ than was required to write the original problem as $T^*Tx = a$, for example to allow for less smoothness in problems defined via differential equations. Such redefinition may then also induce the definition of an auxiliary problem $T^*T\xi = \delta$, where ξ represents a Green's function. In fact we can get pointwise bounds in some problems by using devices which are conceptually simpler than the Dirac δ, as suggested in (a) and (c) following.

(a) A unit spike function

A simple illustration is provided by the problem

$$-x''(s) = a(s) \quad \text{on} \quad -1 \leqslant s \leqslant +1, \qquad \text{with} \quad x(\pm 1) = 0, \tag{4.38}$$

where $a(s)$ is a given continuous function. To monitor the illustration we shall refer to the known solution

$$x_0(s) = \int_{-1}^{-1} k[s, \tau]a(\tau)\,\mathrm{d}\tau \tag{4.39}$$

where

$$k[s, \tau] = \begin{cases} \frac{1}{2}(1 - s)(1 + \tau) & \text{if} \quad \tau \leqslant s, \\ \frac{1}{2}(1 + s)(1 - \tau) & \text{if} \quad \tau \geqslant s. \end{cases}$$

This $x_0(s)$ is found from (4.38) by two integrations, and the use of

$$\frac{\mathrm{d}}{\mathrm{d}\tau}\left[\tau \int_{-1}^{\tau} a(s)\,\mathrm{d}s\right] = \int_{-1}^{\tau} a(s)\,\mathrm{d}s + \tau a(\tau)$$

(cf. the derivation of the integral equation (3.155)). By specializing (3.142) with $\Sigma_2 = 0$, we can see how to express the homogeneous Dirichlet problem for Poisson's equation as an example of $T^*Tx = a$ with $T^* = -\mathrm{div}$ and $T = \mathrm{grad}$, in any number of dimensions. For (4.38) the decomposition (4.2) becomes

$$-u_\alpha'(s) = -a(s), \qquad x_\beta'(s) = -u_\beta(s) \quad \text{with} \quad x_\beta(\pm 1) = 0. \tag{4.40}$$

Here $Tx = x'(s)$ with $\mathscr{D}(T)$ as the space of C^2 functions satisfying $x(\pm 1) = 0$ with $(x, y) = \int_{-1}^{+1} xy\,\mathrm{d}s$; and $T^*u = -u'(s)$ with $\mathscr{D}(T^*)$ as the space of C^1 functions with $\langle u, v\rangle = \int_{-1}^{+1} uv\,\mathrm{d}s$ and no end restrictions. But the δ function is not continuous and so cannot belong to the $\mathscr{R}(T^*)$ just implied.

We invoke an adaptation of Exercise (3.1.4). If we wish to bound $x_0(0)$ via Theorem 4.1, we require adjointness to be

$$(\tilde{x}, T^*u) = -\int_{-1}^{+1} x(s)u'(s)\,\mathrm{d}s - x(0)[\![u]\!]_0$$

$$= \int_{-1}^{+1} u(s)x'(s)\,\mathrm{d}s = \langle u, T\tilde{x}\rangle \qquad (4.41)$$

for all $\tilde{x}\in\mathscr{D}(T)$ and all $u(s)\in\mathscr{D}(T^*)$. This is achieved if $T\tilde{x} = x'(s)$ with $\mathscr{D}(T)$ as the space of $\tilde{x} = [x(s)\;\; x(0)]^{\mathrm{T}}$ constructed from functions $x(s)$ satisfying $x(\pm 1) = 0$, that are C^2 except that a finite gradient jump is permitted at $x = 0$, and with inner product $(\tilde{x}, \tilde{y}) = \int_{-1}^{+1} x(s)y(s)\,\mathrm{d}s + x(0)y(0)$; and if $T^*u = [-u'(s)\;\; -[\![u]\!]_0]^{\mathrm{T}}$ with $\mathscr{D}(T^*)$ as the space of functions $u(s)$ defined above except that a finite jump $[\![u]\!]_0$ at the origin is now also admitted. Then (4.41) holds and $T^*T\tilde{x}$ exists so that the given problem has been embedded in a class of wider ones associated with these redefinitions of the original T and T^*. Again we have specified true adjoints, not merely formal adjoints.

Fig. 4.3. Auxiliary problem $T^*T\tilde{\xi} = \tilde{1}$. (*a*) Green's function $\xi(s) = k[s, 0]$. (*b*) Unit step function $-\xi'(s)$. (*c*) $\tilde{1} = T^*(-\xi')$.

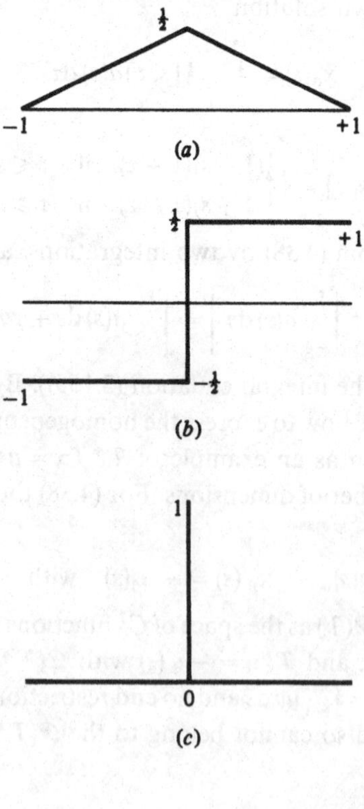

Any $u(s)$ which is constant except for a jump up of one unit at the origin, such as the Heaviside function or that shown in Fig. 4.3(b), will have $T^*u=[0\ \ -1]^{\mathrm{T}}$. We define

$$\tilde{1}=[0\ 1]^{\mathrm{T}}. \tag{4.42}$$

Geometrically $\tilde{1}$ has a spike of unit height at the origin (Fig. 4.3(c)) and is zero elsewhere (and is thus different from the Dirac δ, which has an infinite spike at the origin). The auxiliary problem $T^*T\tilde{\xi}=\tilde{1}$ has solution $\tilde{\xi}=[\xi(s)\ \xi(0)]^{\mathrm{T}}$ specified by the Green's function $\xi(s)=k[s,0]$ (obtainable from (4.39) if we do choose $a(\tau)$ as a Dirac delta in the generalized function sense – see, for example, Roach 1970).

To apply Theorem 4.1 to the original problem, we now regard it as embedded in the wider class accommodated via (4.41). Then we choose any unit step function $q(s)$ such that $T^*q=-\tilde{1}$, i.e.

$$q(s)=\begin{cases}1+b & \text{if}\quad s\geqslant 0,\\ b & \text{if}\quad s\leqslant 0,\end{cases} \tag{4.43}$$

for any constant b. We find $(x_0, T^*q)=-x_0(0)$ and $\|q\|^2=1+2b+2b^2$ (least when $b=-\frac{1}{2}$).

Now suppose a is constant. Trial functions satisfying (4.40) can be chosen as $u_\alpha=as$, $x_\beta=\frac{1}{2}\lambda(1-s^2)$ for any constant λ, giving $\|u_\alpha-u_\beta\|^2=\frac{2}{3}(a-\lambda)^2$, $\langle u_\alpha, q\rangle=\frac{1}{2}a$, $\langle u_\beta, q\rangle=\frac{1}{2}\lambda$. Then (4.4) gives

$$|\tfrac{1}{4}(a+\lambda)-x_0(0)|\leqslant\frac{1}{\sqrt{6}}|a-\lambda|[1+2b+2b^2]^{\frac{1}{2}},$$

with (4.5) simpler but less tight. The best $b=-\frac{1}{2}$ and, as we expect from the known solution $x_0(s)=\frac{1}{2}a(1-s^2)$, the best $\lambda=a$.

The purpose in giving such a simple illustration is to encourage confidence in the unfamiliar spaces which are required. They are expressed in terms of ordinary functions, and we have not actually used the Dirac δ or any other generalized function.

(b) Example of $T^*Tx+E'[x]=0$

Pointwise bounds for solutions of this equation (3.136) can be sought via a simplification of Theorem 4.2. Inequality (4.10) bounds $-\langle u_0, q\rangle=(x_0, T^*q)$, where x_0 is a solution of the problem and $u_0=-Tx_0$ is unique. We could again try to choose q such that T^*q has the effect of a δ function, but now there is an alternative approach when, as we envisaged in (3.148), the equation

$$e=E'[x]\quad\text{has inverse}\quad x=f[e] \tag{4.44}$$

under the qualifications described there. We choose $q = -Tp$ with some $p \in \mathscr{D}(T)$ so that

$$\begin{aligned}(x_0, T^*q) &= -(x_0, T^*Tp)\\ &= -(p, T^*Tx_0) = (p, E'[x_0]) = (p, e_0).\end{aligned} \tag{4.45}$$

Thus instead of choosing p to be a Green's function as we did in the last example, where T^*Tp had the effect of a δ function, we now choose p itself to have the effect of a δ function. Then (p, e_0) reduces to a pointwise solution value of e_0, and the inverse $x_0 = f[e_0]$ gives an associated approximation to x_0.

Before applying this procedure we must check whether $\delta \in \mathscr{D}(T)$. If this is not satisfied, we should need to widen the definition of $\mathscr{D}(T)$ beyond that originally required to express the given problem in the form $T^*Tx + E'[x] = 0$. For integral equations, in fact, the required spaces of integrable functions may already be wide enough to admit Dirac delta functions in either $\mathscr{D}(T)$ or $\mathscr{D}(T^*)$, as illustrated below. This contrasts with $\tilde{1} \in \mathscr{R}(T^*)$ illustrated for differential operators in example (a) above.

The bounds (4.10) simplify to

$$|\langle u_\beta - u_0, q\rangle| \leqslant (L_\alpha - L_\beta)^{\frac{1}{2}} \| q \| \tag{4.46}$$

whenever we can choose $u_\alpha = u_\beta = -Tx_\beta$ as explained in the context of (3.148). For the linear equation $T^*Tx = a - Ax$ of (3.137) we have, when $q = -Tp$ is chosen,

$$|(p, Ax_0 - a + T^*Tx_\beta)| \leqslant [\tfrac{1}{2}(x_\alpha - x_\beta, A(x_\alpha - x_\beta))]^{\frac{1}{2}} \| Tp \| \tag{4.47}$$

by (3.151).

As an example we take the integral equation

$$\int_0^t \int_s^1 x(\tau)\,d\tau\,ds = \frac{1}{m^2}[1 + (b+1)t - \tfrac{1}{2}t^2 - x(t)] \tag{3.155 bis}$$

on $0 \leqslant t \leqslant 1$. Then (4.44) applies in the simple form

$$e(t) = \frac{x(t)}{m^2} - a(t), \quad x(t) = m^2[e(t) + a(t)].$$

The spaces of integrable functions originally specified in the context of (3.155) are already wide enough to admit Dirac delta functions, if treated with care. This is because adjointness, which is now the special case

$$\begin{aligned}(x, T^*u) &= \int_0^1 x(t) \int_0^t u(s)\,ds\,dt\\ &= \int_0^1 u(s) \int_s^1 x(t)\,dt\,ds = \langle u, Tx\rangle\end{aligned} \tag{4.48}$$

of (3.31), is still satisfied when either one of $x(t)$ and $u(s)$ are delta functions.

To see this, let σ be the given place within $0<\sigma<1$ at which we wish to estimate a solution value $x_0(\sigma)$ of (3.155). Let $\delta(t-\sigma)$ denote a delta function having the defining property

$$\int_a^b \delta(t-\sigma)g(t)\,dt = \begin{cases} g(\sigma) & \text{if} \quad a<\sigma<b, \\ 0 & \text{if} \quad \sigma<a \quad \text{or} \quad \sigma>b, \end{cases} \tag{4.49}$$

for any continuous function $g(t)$, with any given a and b. Then T and T^* in (4.48) have the properties shown in Fig. 4.4. It follows, for example, that

$$(\delta(t-\sigma), T^*u) = \int_0^1 \delta(t-\sigma)\int_0^t u(s)\,ds$$
$$= \int_0^\sigma u(s)\,ds = \langle u, T\delta(t-\sigma)\rangle, \tag{4.50}$$

provided T^*u is continuous. Similarly $(x, T^*\delta(s-\sigma)) = \langle \delta(s-\sigma), Tx\rangle$ provided Tx is continuous. These arguments show that adjointness (4.48) is satisfied when either (but not both) of $x(t)$ and $u(t)$ are δ-functions. We shall not need the further details required to study (4.48) when both $x(t)$ and $u(t)$ are δ-functions. We now choose $\mathscr{D}(T)$ wide enough to admit $\delta(t-\sigma)$, but $\mathscr{D}(T^*)$ to exclude δ-functions, and to emphasize this we have used different inner product symbols in (4.48) and (4.50), instead of the single $\langle .,. \rangle$ in (3.31).

We choose p in (4.47) to be $\delta(t-\sigma)$. The trial functions used to get the energy bounds (3.157) and (3.158) were $u_\alpha(s)=u_\beta(s)=s-1$ with $x_\beta(\tau)\equiv 1$. Assuming that the solution $x_0(t)$ is continuous, the left side of (4.47) contains

$$\frac{1}{m^2}\int_0^1 \delta(t-\sigma)[1+(b+1-m^2)t-\tfrac{1}{2}t^2(1-m^2)-x_0(t)]\,dt$$
$$=\frac{1}{m^2}[1+(b+1-m^2)\sigma-\tfrac{1}{2}\sigma^2(1-m^2)-x_0(\sigma)].$$

On the right of (4.47) we have, since we are dealing with an example of the

Fig. 4.4. $Tx=\int_s^1 x(t)dt$ and $T^*u=\int_0^t u(s)dt$ applied to δ-functions.

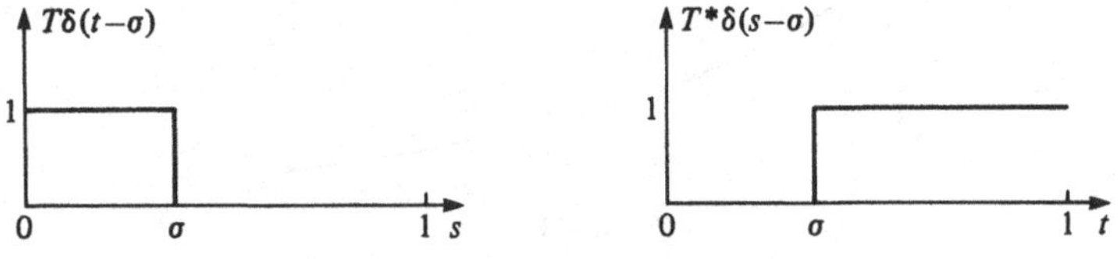

type $E[x] = \frac{1}{2}(x, Ax) - (a, x)$,

$$\begin{aligned} C_{\alpha\beta} &= E[x_\beta] - E[x_\alpha] - (x_\beta - x_\alpha, E'[x_\alpha]) - (x_\beta, T^*u_\alpha - E'[x_\alpha]) \\ &= \tfrac{1}{2}(x_\beta - x_\alpha, A(x_\beta - x_\alpha)) \\ &= \frac{1}{2m^2}\int_0^1 [\tfrac{1}{2}(m^2-1)t^2 + (b+1-m^2)t]^2\,dt \\ &= \frac{1}{2m^2}[\tfrac{1}{20}(m^2-1)^2 + \tfrac{1}{4}(m^2-1)(b+1-m^2) + \tfrac{1}{3}(b+1-m)^2]. \end{aligned}$$

Also, $\|q\| = \|T\delta(t-\sigma)\| = \sigma^{\frac{1}{2}}$ from Fig. 4.4.

Collecting these results, (4.47) gives

$$|x_0(\sigma) - [1 + (b+1-m^2)\sigma - \tfrac{1}{2}(1-m^2)\sigma^2]| \leqslant m^2(\sigma C_{\alpha\beta})^{\frac{1}{2}}. \quad (4.51)$$

As a first check, when $m = 1$ and $b = 0$ we have $C_{\alpha\beta} = 0$ and $|x_0(\sigma) - 1| \leqslant 0$, which tallies with the exact solution $x_0(\sigma) \equiv 1$ in that case. In Fig. 4.5 we plot the bounds (4.51) for two cases (a) $m = 0.25$, $b = 1$ and (b) $m = 3$, $b = 1$, and compare them with the graph of the known exact solution given in §3.7(xiii).

As elsewhere, our main purpose has been to give a quick illustration of

Fig. 4.5. Upper and lower pointwise bounds (4.51), and exact solution $x_0(\sigma)$.

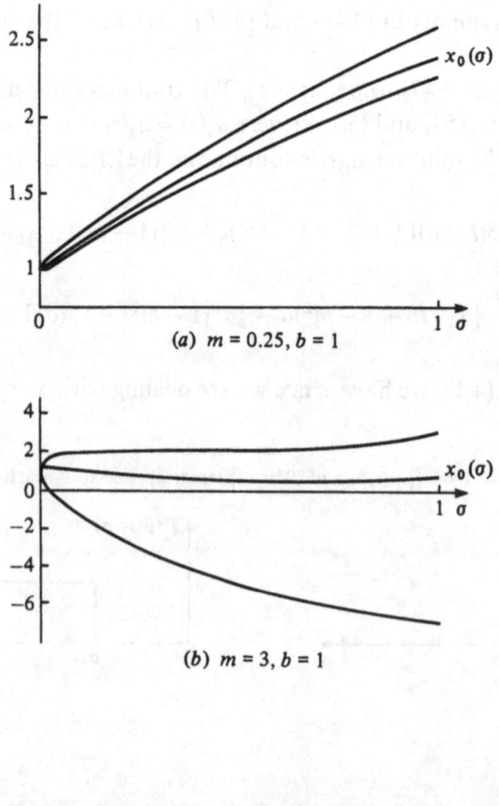

the main points of the procedure, in the present case for a theorem not previously tested in this respect. As an exercise, the reader may care to seek closer bounds by more sophisticated choice of trial functions. Also the theorem is ready for use in problems whose exact solution is not known, including problems governing by inequalities.

(c) A two dimensional problem

We now wish to estimate the value at the origin of the solution of

$$-\nabla^2\phi = a \text{ in } V, \quad \phi = 0 \text{ on } \Sigma, \tag{4.52}$$

where V is the square $|x| \leqslant b, |y| \leqslant b$ and Σ is its boundary. Here a and b are given constants. The problem is of the type $T^*T\phi = a$ as a specialization of (3.142), but it must be embedded in a wider class of such problems before Theorem 4.1 can be properly applied to estimate the solution at the origin. The following method can be adapted to deal with any other interior point similarly.

Let C be any given circle of radius $c \leqslant b$ centred on the origin (Fig. 4.6). Define $T\tilde{\phi} = \nabla\phi$ with $\mathscr{D}(T)$ as the space of $\tilde{\phi} = [\phi(V) \ \ \phi(C)]^{\mathrm{T}}$ constructed from scalar functions $\phi(V)$ of position satisfying $\phi(\Sigma) = 0$, that are C^2 except that a finite gradient jump is permitted across C, and with inner product $(\tilde{\phi}, \tilde{\psi}) = \int \phi\psi \, \mathrm{d}V + \int \phi\psi \, \mathrm{d}C$. Define $T^*\mathbf{u} = [-\nabla\cdot\mathbf{u}(V) \ \ -\mathbf{m}\cdot[\![\mathbf{u}]\!]_{\mathrm{c}}]^{\mathrm{T}}$ with $\mathscr{D}(T^*)$ as the space of vector functions $\mathbf{u}(V)$ that are C^1 except that a finite jump $[\![\mathbf{u}]\!]_{\mathrm{c}}$ is permitted across C where $\mathbf{m}$ is the unit radial vector, with $\langle \mathbf{u}, \mathbf{v} \rangle = \int \mathbf{u}\cdot\mathbf{v} \, \mathrm{d}V$. Such $\mathbf{u}$ have no boundary restrictions. Then adjointness states that, for all $\mathbf{u} \in \mathscr{D}(T^*)$ and for all $\tilde{\phi} \in \mathscr{D}(T)$,

$$(\tilde{\phi}, T^*\mathbf{u}) = -\int \phi\nabla\cdot\mathbf{u} \, \mathrm{d}V - \int \phi\mathbf{m}\cdot[\![\mathbf{u}]\!]_{\mathrm{c}} \, \mathrm{d}C = \int \mathbf{u}\cdot\nabla\phi \, \mathrm{d}V = \langle \mathbf{u}, T\tilde{\phi} \rangle. \tag{4.53}$$

The problem (4.52) is now $T^*T\tilde{\phi} = [-\nabla^2\phi \ \ -\mathbf{m}\cdot[\![\nabla\phi]\!]_{\mathrm{c}}]^{\mathrm{T}} = [a \ \ 0]^{\mathrm{T}}$.

Fig. 4.6. Circle of discontinuity of **q**.

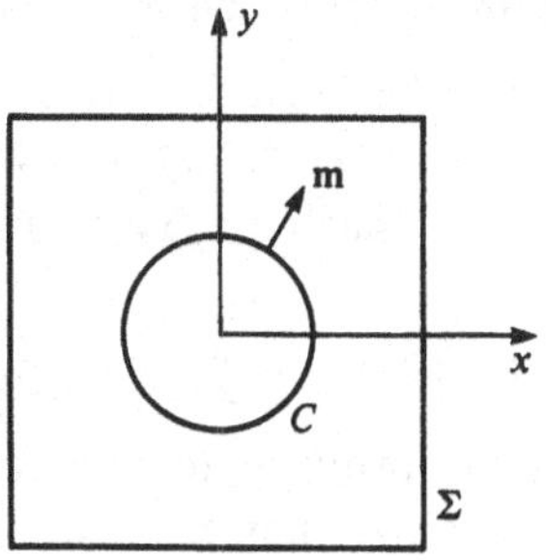

To apply Theorem 4.1 we are now in a position to choose

$$\mathbf{q}=\begin{cases}0 & \text{where } r<c,\\ \mathbf{r}/r^2 & \text{where } r\geqslant c.\end{cases}\tag{4.54}$$

Here $\mathbf{r}$ is the radius vector from the origin, $r=|\mathbf{r}|$, and we have chosen the so-called Green's vector in $r\geqslant c$. This $\mathbf{q}\in\mathscr{D}(T^*)$, and for any $\tilde{\phi}\in\mathscr{D}(T)$ (4.54) gives

$$\langle\mathbf{q},T\tilde{\phi}\rangle=-\int\frac{\phi}{r}\mathrm{d}C=-\int_0^{2\pi}\phi\,\mathrm{d}\theta\tag{4.55}$$

where θ is the polar angle. If we let $c\to 0$ we obtain the value $-2\pi\phi[0]$ at the origin, but then $\|q\|$ becomes infinite, which we must avoid in Theorem 4.1. We therefore fix c, and apply a standard mean value theorem of potential theory (Kellog, 1929, pp. 223–4) to the harmonic function $\phi_0+(a/4)r^2$, where ϕ_0 is the solution of (4.52). On $r=c$ it gives $\int_0^{2\pi}(\phi_0+(a/4)r^2)\mathrm{d}\theta=2\pi\phi_0[0]$. Hence Theorem 4.1 bounds

$$(\tilde{\phi}_0,T^*\mathbf{q})=ac^2\pi/2-2\pi\phi_0[0].\tag{4.56}$$

With fixed c we find

$$\|\mathbf{q}\|^2=8\int_0^{\pi/4}\ln\frac{b}{c\cos\theta}\mathrm{d}\theta=2\pi\ln\frac{b}{c}+0.6913.$$

The decomposition of (4.52) in the form (4.2) becomes

$$-\nabla\cdot\mathbf{u}_\alpha=-a,\quad \nabla\phi_\beta=-\mathbf{u}_\beta \text{ with } \phi_\beta(\Sigma)=0.$$

These constraints can be satisfied individually by, for example,

$$\mathbf{u}_\alpha=\tfrac{1}{2}a\mathbf{r}\quad\text{and}\quad \phi_\beta=\tfrac{1}{2}k(x^2-b^2)(y^2-b^2)$$

with any constant k. Hence

$$\|\mathbf{u}_\alpha-\mathbf{u}_\beta\|^2=\frac{a^2b^4}{9}+\frac{64b^4}{45}\left(kb^2-\frac{5a}{8}\right)^2.$$

We also find

$$\langle\mathbf{q},\mathbf{u}_\alpha\rangle=\tfrac{1}{2}a(4b^2-\pi c^2),\quad \langle\mathbf{q},\mathbf{u}_\beta\rangle=kb^2\pi(b^2-c^2),$$

since we need integrate only between the circle and square.

Particularly simple bounds are evidently achieved by choosing $c=b$ (i.e. the largest possible circle) so that $\langle\mathbf{q},\mathbf{u}_\beta\rangle=0$, and $k=5a/8b^2$. Then the strong bounds (4.4) are

$$0.2595ab^2\leqslant\phi_0[0]\leqslant 0.3062ab^2.\tag{4.57}$$

Fujita (1955) uses the same problem to illustrate his approach via δ-functions. Our bounds are better than his first approximation, but more sophisticated $\mathbf{u}_\alpha$ and ϕ_β can be readily chosen, and Fujita finally quotes $0.2939 \leqslant \phi_0[0] \leqslant 0.2953$ for the case $a = b = 1$.

If a in (4.52) had been a given function of position rather than a constant, we should have needed after (4.55) a mean value theorem for Poisson's equation itself, rather than for Laplace's equation, namely

$$\int_0^{2\pi} \phi \,\mathrm{d}\theta = 2\pi\phi[0] + \int_0^{2\pi}\int_0^c ar \ln\frac{r}{c}\,\mathrm{d}r\,\mathrm{d}\theta. \tag{4.58}$$

for any solution of $-\nabla^2\phi = a$. As before, the polar integration is round a circle of fixed radius c.

(d) Further examples

The simple illustrations which we have given can be augmented by more lengthy calculations to be found in the literature. The review of Diaz (1960) mentioned before gives a good list of the earlier references. These include a paper by Maple (1950) on pointwise bounds for the solution, and its derivatives, of a Dirichlet problem. Estimates at boundary points, as well as internal points, are considered. A series of four papers by Weber (1931, 1938, 1941, 1942) introduced some of the ideas in elasticity theory, and these have been translated into English (Noble and Sewell, 1978). More recent numerical calculations of internal estimates for Poisson's equation with mixed boundary conditions are represented by Young and Mote (1976). Another viewpoint is offered by Ponter (1975) in mechanics of solids. Robinson and Barnsley (1979) have also obtained pointwise bounds on solutions of Fredholm integral equations.

(vi) Boundedness hypotheses

Here we broach a rather different viewpoint from those exploited hitherto in this section. Recalling the infinite dimensional expression defined by

$$S = L_+ - L_- - \left(x_+ - x_-, \frac{\partial L}{\partial x}\bigg|_+\right) - \left\langle u_+ - u_-, \frac{\partial L}{\partial u}\bigg|_-\right\rangle \tag{3.66 bis}$$

for any pair of points x_+, u_+ and x_-, u_- in the domain of the functional $L[x, u]$, we now consider hypotheses that are weaker than requiring $S \geqslant 0$ for every pair of distinct points in some subdomain.

In particular we shall suppose that real numbers

$$d_x[u_+] \geqslant 0 \quad \text{and} \quad d_u[x_-] \geqslant 0 \tag{4.59}$$

exist, perhaps depending on the indicated coordinates, such that

$$|S| \leqslant \tfrac{1}{2}d_x\|x_+ - x_-\|^2 + \tfrac{1}{2}d_u\|u_+ - u_-\|^2 \tag{4.60}$$

for every pair of distinct points in some particular subdomain of $L[x, u]$. It need cause no confusion that we use the same norm symbol in different spaces.

It is possible to deduce (4.60) from other hypotheses. For example, we remarked in §3.5(ii) that second gradients of functionals could be defined to allow S to be written as a generalization of (1.28) and (1.60), at least over a rectangular domain. Generalizing Fig. 1.5 by introducing the functions

$$l(\lambda) = L[x_-, u_\lambda], \qquad u_\lambda = u_- + \lambda(u_+ - u_-),$$
$$m(\mu) = L[x_\mu, u_+], \qquad x_\mu = x_- + \mu(x_+ - x_-),$$

of single scalars λ and μ, we have

$$S = \int_0^1 (1-\lambda)l''(\lambda)\mathrm{d}\lambda - \int_0^1 \mu m''(\mu)\mathrm{d}\mu. \tag{4.61}$$

Let us now define 'mean value second derivative' linear operators by

$$\frac{1}{2}\left\langle u_+ - u_-, \overline{\frac{\partial^2 L}{\partial u^2}}(u_+ - u_-)\right\rangle = \int_0^1 (1-\lambda)l''(\lambda)\mathrm{d}\lambda,$$
$$\frac{1}{2}\left(x_+ - x_-, \overline{\overline{\frac{\partial^2 L}{\partial x^2}}}(x_+ - x_-)\right) = \int_0^1 \mu m''(\mu)\,\mathrm{d}\mu. \tag{4.62}$$

These operators depend on x_- and u_+ respectively, and act on the elements which follow them (as for $G''[u_\alpha]$ in (4.25)).

Theorem 4.6
If we suppose that

$$d_x\|x_+ - x_-\| \geqslant \left\|\overline{\overline{\frac{\partial^2 L}{\partial x^2}}}(x_+ - x_-)\right\|,$$
$$d_u\|u_+ - u_-\| \geqslant \left\|\overline{\frac{\partial^2 L}{\partial u^2}}(u_+ - u_-)\right\| \tag{4.63}$$

for any such mean values, then (4.60) *follows.*

Proof
The triangle inequality applied to (4.61), followed by the Schwarz inequality applied to (4.62), gives the result after using (4.63). □

Another way of deducing (4.60) from other hypotheses is indicated in Exercises 4.2.4 and 4.2.6.

Next let us also assume the mean value statements

$$\left.\frac{\partial L}{\partial x}\right|_+ - \left.\frac{\partial L}{\partial x}\right|_- = \overline{\overline{\frac{\partial^2 L}{\partial x^2}}}(x_+ - x_-) + \overline{\frac{\partial^2 L}{\partial x \partial u}}(u_+ - u_-),$$
$$\left.\frac{\partial L}{\partial u}\right|_+ - \left.\frac{\partial L}{\partial u}\right|_- = \overline{\overline{\frac{\partial^2 L}{\partial u \partial x}}}(x_+ - x_-) + \overline{\frac{\partial^2 L}{\partial u^2}}(u_+ - u_-) \tag{4.64}$$

where double overbars indicate evaluation at u_+ and some x_μ (not necessarily the same x_μ as in (4.62) – cf. Fig. 1.5 again), and single overbars indicate evaluation at x_- and some u_λ. Precise statements about mean value theorems for operators are given by Rall (1969, §20). Examples of second derivative operators which happen also to be constants are provided by the coefficients in (3.59), i.e.

$$\frac{\partial^2 L}{\partial x^2} = -A, \qquad \frac{\partial^2 L}{\partial x \partial u} = T^*, \qquad \frac{\partial^2 L}{\partial u \partial x} = T, \qquad \frac{\partial^2 L}{\partial u^2} = B.$$

We shall suppose, in addition to (4.59), that real numbers

$$c_x[u_+] \geqslant 0, \qquad c_u[x_-] \geqslant 0 \tag{4.65}$$

exist, perhaps depending on the indicated coordinates, such that

$$\left\| \overline{\overline{\frac{\partial^2 L}{\partial u \partial x}}}(x_+ - x_-) \right\| \geqslant c_x \| x_+ - x_- \|,$$

$$\left\| \overline{\frac{\partial^2 L}{\partial x \partial u}}(u_+ - u_-) \right\| \geqslant c_u \| u_+ - u_- \|, \tag{4.66}$$

This hypothesis is illustrated in full in Exercises 4.2. A special case of it will be invoked in the next subsection.

(vii) Embedding method

We used this approach in §§1.4(v) and 1.5(iii). There we began with the problem of maximizing a scalar function of x subject to constraints $v[x] \geqslant 0$ and $x \geqslant 0$. We introduced a second space of multipliers u, chosen to embed the problem in a larger one generated by (1.89), which with suitable assumptions has a saddle property $S \geqslant 0$.

Here we begin with a given operator $M(x)$ acting on x, taking the place of the previous $-v[x]$, and whose domain and range may belong to two different infinite dimensional spaces with inner products $(.,.)$ and $\langle .,. \rangle$ respectively, i.e. with the generality indicated for $M(x)$ in Fig. 3.1(*a*). We allow that $M(x)$ may be linear or nonlinear. It may or not be decomposable in forms such as (3.135)–(3.137). Suppose that the *ab initio* problem is to find x satisfying $M(x) = 0$. Another such problem might be to solve $M(x) \leqslant 0$.

Either case may be approached by constructing the functional

$$L[x, u] = (p, x) - \langle u, M(x) \rangle, \tag{4.67}$$

where p is an assigned member of the space to which x belongs. Suppose that $M(x)$ has a Gateaux derivative as defined by (3.48), and that this derivative has an adjoint which we shall write as $M'^*[x]$. The latter is a linear operator which may depend nonlinearly on x but acts linearly on u, such that

$$\frac{\partial L}{\partial x} = p - M'^*[x]u, \qquad \frac{\partial L}{\partial u} = -M(x).$$

An illustration of the calculation of such adjoints is given in (4.73).

Then the problem $\partial L/\partial x = \partial L/\partial u = 0$ becomes

$$p = M'^*[x]u \quad (\alpha), \qquad M(x) = 0 \quad (\beta), \tag{4.68}$$

and thus embeds the given problem in a larger one, whose solutions will be stationary points of $L[x, u]$. An inequalities version, designed to include $M(x) \leqslant 0$, would be a generalization of (1.91).

Writing x_β as any solution of $M(x) = 0$ considered alone, let us suppose that the main objective is to estimate the linear functional (p, x_β). This appears in

$$\begin{aligned} L_+ - L_- - \left\langle u_+ - u_-, \frac{\partial L}{\partial u}\bigg|_- \right\rangle &= (p, x_+ - x_-) - \langle u_+, M(x_+) - M(x_-)\rangle \\ &= (p, x_+) - \langle u_+, M(x_+)\rangle - (p, x_\beta) \end{aligned} \tag{4.69}$$

by selecting $x_- = x_\beta$, with any $x_+ \in \mathscr{D}(M)$ and any u_+.

Theorem 4.7

Suppose $\mathscr{D}(M)$ can be selected so that the operator $M(x)$ satisfies the following two hypotheses, which can be viewed as bounds on its first and second Gateaux derivatives, but expressed indirectly.

There exists a number $c_x > 0$ such that

$$\|M(x_+) - M(x_-)\| \geqslant c_x \|x_+ - x_-\| \tag{4.70}$$

for every x_+ and x_- in $\mathscr{D}(M)$.

There exists a number $d \geqslant 0$ such that, in (4.60) with (4.67)

$$d_x = \|u_+\| d. \tag{4.71}$$

Then there exists at most one solution $x_\beta \in \mathscr{D}(M)$ of $M(x) = 0$; and for any $x_+ \in \mathscr{D}(M)$ and any $u_+ \in \mathscr{D}(M^*[x_+])$,

$$\begin{aligned} &|-\langle u_+, M(x_+)\rangle + (p, x_+) - (p, x_\beta)| \\ &\quad \leqslant \frac{1}{c_x}\|M'^*[x_+]u_+ - p\| \; \|M(x_+)\| \\ &\quad + \frac{d}{2c_x{}^2}\|u_+\| \; \|M(x_+)\|^2. \end{aligned} \tag{4.72}$$

Proof.

If there were two distinct solutions of $M(x) = 0$ in $\mathscr{D}(M)$, we could choose them to be x_+ and x_-, and (4.70) would then imply that zero exceeds a positive number, which proves uniqueness (or at least that any two solutions must differ almost everywhere, if such norms as in §3.7(xvi) are envisaged).

By the triangle and Schwarz inequalities applied to (3.66),

$$\left|L_+ - L_- - \left\langle u_+ - u_-, \left.\frac{\partial L}{\partial u}\right|_- \right\rangle\right| = \left|S + \left(x_+ - x_-, \left.\frac{\partial L}{\partial x}\right|_+\right)\right|$$
$$\leqslant |S| + \|x_+ - x_-\| \left\| \left.\frac{\partial L}{\partial x}\right|_+ \right\|$$

for any pair of points in the domain of $L[x, u]$.

The linearity of (4.67) with respect to u allows us to choose $d_u = 0$ in (4.63) without loss of generality. Then (4.60) and (4.71) give

$$|S| \leqslant \tfrac{1}{2}d\|u_+\|\,\|x_+ - x_-\|^2.$$

Choosing $x_- = x_\beta$ we can invoke (4.69), and also

$$\|x_+ - x_\beta\| \leqslant \frac{\|M(x_+)\|}{c_x}$$

from (4.70). Combining all these results gives (4.72). □

The formula (4.72) can be written as a pair of inequalities bounding (p, x_β). They are given by Barnsley and Robinson (1977, inequalities (3.6)), who call them 'bivariational' bounds. The proof described here is from a different viewpoint which, in particular, shows the connection with S via (4.60). Barnsley and Robinson also indicate related results in functional analysis, including ones of Vainberg and Kantorovich and Newton's approximation. They give the following applications.

(viii) Examples

1. Bounds on a solution of an algebraic equation

Let $M(x)$ be any piecewise C^2 function of a real scalar x. Suppose that an interval containing only one solution of $M(x) = 0$ can be found by inspection. Then (4.72) can be used to bound that solution more closely, if c_x is chosen to be the least slope $M'(x)$ and d to be the greatest $M''(x)$ within that interval. The process can be repeated iteratively.

For example, $M(x) = x^4 - x - 1$ happens to be strictly convex since $M''(x) = 12x^2$, so there are at most two real roots of $x^4 - x - 1 = 0$. By inspection only one of these exists within $1 < x < 1.5$. We can estimate it by choosing $p = 1$ in (4.72) if the hypotheses of Theorem 4.7 can be validated.

We have $M'(x) = 4x^3 - 1$, and by convexity the mean slope in the interval exceeds the slope at the left hand end, so (4.70) can be satisfied by choosing $c_x = 4 - 1 = 3$. We can satisfy (4.71) by choosing $d = 27$, corresponding to the value of $12x^2$ at $x = 1.5$. Since

$$L[x, u] = x - u(x^4 - x - 1)$$

we can also arrive at this value of d by inserting $\partial^2 L/\partial x^2 = -12ux^2$ in (4.63) with (4.71), or directly from (4.60) since

$$S = u_+(x_+ - x_-)^2(x_-{}^2 + 2x_-x_+ + 3x_+{}^2).$$

Note that in fact $S \geqslant 0$ for $u_+ \geqslant 0$ (as in the proof of Theorem 1.8).

Then since $M'^*[x]u = (4x^3 - 1)u$, the required root x_β satisfies

$$|-u_+(x_+{}^4 - x_+ - 1) + x_+ - x_\beta| \leqslant \tfrac{1}{3}|(4x_+{}^3 - 1)u_+ - 1||x_+{}^4 - x_+ - 1| + \tfrac{3}{2}|u_+||x_+{}^4 - x_+ - 1|^2$$

for any x_+ in $1 < x_+ < 1.5$, and any u_+. Barnsley and Robinson choose x_+ at the middle of the interval in each iteration, and $u_+ = (4x_+{}^3 - 1)^{-1}$. Initially $x_+ = 1.25$ gives $1.217 < x_\beta < 1.234$, then $1.220\,74 < x_\beta < 1.220\,81$ at the second iteration.

2. *A nonlinear ordinary differential equation*

The temperature $1 + x(t)$ at position t in a bar occupying $0 \leqslant t \leqslant 1$, and transferring heat to its surroundings, is assumed to satisfy

$$-x'' + 4(1 + x) + (1 + x)^2 = 0 \quad \text{with} \quad x(0) = x(1) = 0,$$

as described in Bailey *et al.* (1968). We wish to bound $\int_0^1 x(t)\mathrm{d}t$ which is a linear functional of the solution and represents the heat contained in the bar. The problem is nonlinear of type (3.136), where $\mathscr{D}(T)$ is the space of piecewise C^2 functions satisfying $x(0) = x(1) = 0$ with $(x, y) = \int_0^1 xy\,\mathrm{d}t$, and $Tx = x'$. The functional

$$E[x] = \int_0^1 [2(1 + x)^2 + \tfrac{1}{3}(1 + x)^3]\mathrm{d}t$$

is convex only for $x \geqslant -3$, so that Theorem 3.13 does not apply as it stands.

Before applying Theorem 4.7 we note that at least two other approximating principles are available, although neither estimates the required linear functional directly. Firstly, any solution of the problem will make stationary the functional

$$\tfrac{1}{2}(x, T^*Tx) + E[x] = \frac{1}{2}\int_0^1 x'^2\,\mathrm{d}t + E[x]$$

for small variations with $\mathscr{D}(T)$. Secondly, we could change the variables to $y(t) = x(t) + 3$ and $\tilde{u} = [u(t)\;\; u(0)\;\; u(1)]^{\mathrm{T}}$ and consider an associated problem of type (1.55) generated by

$$L[y,\tilde{u}] = \int_0^1 [\tfrac{1}{2}u^2 - yu' - 2(y-2)^2 - \tfrac{1}{2}(y-2)^3]\,\mathrm{d}t + 3[u(1) - u(0)].$$

This is concave in y for $y \geqslant 0$, and so there is a lower bound of type $L_\beta \geqslant L_0$ (see (3.68)), but this L_0 is not a linear functional.

To illustrate (4.72), we first define the operator

$$M(x) = -x'' + 4(1+x) + (1+x)^2$$

and select $\mathscr{D}(M)$ to be the set of functions $x(t) \geqslant -1$ which belong to $\mathscr{D}(T)$ specified above. The temperature $1 + x(t) \geqslant 0$ for physical reasons. Choosing $p \equiv 1$, and $u(t)$ to be another member of $\mathscr{D}(T)$, so that $u(0) = u(1) = 0$, construct

$$L[x,u] = \int_0^1 x(t)\,\mathrm{d}t - \int_0^1 uM(x)\,\mathrm{d}t.$$

From first principles (i.e. (3.57)), after two integrations by parts,

$$\frac{\partial L}{\partial x} = 1 - 2(3+x)u + u'', \qquad \frac{\partial L}{\partial u} = -M(x), \tag{4.73}$$

thus identifying $M'^*[x]u = 2(3+x)u - u''$.

To select c_x which validates (4.70), the Schwarz inequality first gives

$$\|M(x_+) - M(x_-)\|\,\|x_+ - x_-\| \geqslant \int_0^1 [M(x_+) - M(x_-)](x_+ - x_-)\,\mathrm{d}t$$

$$= \int_0^1 (x_+ - x_-)[-(x_+ - x_-)'' + 4(x_+ - x_-)$$
$$+ (x_+ - x_-)(x_+ + x_- + 2)]\,\mathrm{d}t$$
$$\geqslant (\pi^2 + 4)\int_0^1 (x_+ - x_-)^2\,\mathrm{d}t.$$

The last step follows because $x_+ \geqslant -1, x_- \geqslant -1$ and because π^2 is the least eigenvalue of $-x'' = n^2 x$ with $x(0) = x(1) = 0$ (cf. §3.8(iv)). Hence we can choose $c_x = \pi^2 + 4$ in (4.70).

To select d which validates (4.71), we show from (3.66) and (4.73) that

$$S = \int_0^1 u_+(x_+ - x_-)^2\,\mathrm{d}t \tag{4.74}$$

for any $x_+(t)$, $x_-(t)$ and $u_+(t)$ in $\mathscr{D}(T)$. Hence

$$|S| \leqslant \|u_+\|\,\|x_+ - x_-\|^2$$

so that $d_x = 2\|u_+\|$, $d_u = 0$ in (4.60), and $d = 2$ in (4.71). In passing note that again $S \geqslant 0$ for $u_+ \geqslant 0$, as in the previous example. This means that a variant of Theorem 1.8 will apply with $v[x] = -M(x)$ and $\phi[x] = \int_0^1 x(t)dt$.

Finally evaluating (4.72), Barnsley and Robinson choose $x_+ = 1.6(t^2 - t)$, $u_+ = -1.36(t^2 - t)$ which leads to

$$-0.275 \leqslant \int_0^1 x(t)\,dt \leqslant -0.260.$$

3. A nonlinear integral equation

An equation which arises in communication theory is

$$\int_0^{\frac{\pi}{2}} \frac{\sin(t-\tau)}{\pi(t-\tau)} x(\tau)d\tau = \frac{1}{x(t)}. \tag{4.75}$$

This is a nonlinear example of (3.135) in which P has been shown to be positive, but the individual parts of a decomposition such as $P = T^*T$ have apparently not been demonstrated. Nevertheless, since $E[x] = \int_0^{\pi/2} \ln(c/x(t))dt$ with fixed c is strictly convex, Theorem 3.13 does apply, and we do not need the decomposition. The quantity bounded in the theorem is not a linear functional of the solution, but an approximation to the latter is obtained by applying the theorem (see §6.5 of Arthurs (1980)).

Barnsley and Robinson show that Theorem 4.7 also applies to this problem to give estimates for the transmission signal $\int_0^{\pi/2} x(t)\cos rt\,dt$ with any fixed $r \geqslant 0$. In particular

$$2.29 \leqslant \int_0^{\pi/2} x(t)dt \leqslant 2.31.$$

It turns out that the approximation obtained by applying Theorem 3.13 lies within this interval, although there was no advance guarantee of that.

It seems that the potentiality of Theorem 4.7, with (4.60), needs to be tested in cases where $S \geqslant 0$ is not also available in some sense, as it is in the foregoing examples (e.g. (4.74)).

Exercises 4.2

1. Show that (4.63), (4.64) and (4.66) imply

$$-d_x\|x_+ - x_-\| + c_u\|u_+ - u_-\| \leqslant \left\| \frac{\partial L}{\partial x}\bigg|_+ - \frac{\partial L}{\partial x}\bigg|_- \right\|,$$

$$c_x\|x_+ - x_-\| - d_u\|u_+ - u_-\| \leqslant \left\| \frac{\partial L}{\partial u}\bigg|_+ - \frac{\partial L}{\partial u}\bigg|_- \right\|.$$

2. Show that, if $c_x c_u - d_x d_u > 0$, the result in Exercise 1 implies

$$\|x_+ - x_-\| \leqslant (c_x c_u - d_x d_u)^{-1}\left(d_u \left\| \frac{\partial L}{\partial x}\bigg|_+ - \frac{\partial L}{\partial x}\bigg|_- \right\| + c_u \left\| \frac{\partial L}{\partial u}\bigg|_+ - \frac{\partial L}{\partial u}\bigg|_- \right\| \right),$$

$$\|u_+ - u_-\| \leqslant (c_x c_u - d_x d_u)^{-1}\left(c_x \left\| \frac{\partial L}{\partial x}\bigg|_+ - \frac{\partial L}{\partial x}\bigg|_- \right\| + d_x \left\| \frac{\partial L}{\partial u}\bigg|_+ - \frac{\partial L}{\partial u}\bigg|_- \right\| \right).$$

3. Show that the result in Exercise 2, with (4.60), implies

$$\left|L_+ - L_- - \left\langle u_+ - u_-, \frac{\partial L}{\partial u}\bigg|_-\right\rangle\right|$$
$$\leqslant \frac{\|\partial L/\partial x|_+\|}{(c_x c_u - d_x d_u)}\left(d_u\left\|\frac{\partial L}{\partial x}\bigg|_+ - \frac{\partial L}{\partial x}\bigg|_-\right\| + c_u\left\|\frac{\partial L}{\partial u}\bigg|_+ - \frac{\partial L}{\partial u}\bigg|_-\right\|\right)$$
$$+\frac{1}{2}\frac{d_x}{(c_x c_u - d_x d_u)^2}\left(d_u\left\|\frac{\partial L}{\partial x}\bigg|_+ - \frac{\partial L}{\partial x}\bigg|_-\right\| + c_u\left\|\frac{\partial L}{\partial u}\bigg|_+ - \frac{\partial L}{\partial u}\bigg|_-\right\|\right)^2$$
$$+\frac{1}{2}\frac{d_u}{(c_x c_u - d_x d_u)^2}\left(c_x\left\|\frac{\partial L}{\partial x}\bigg|_+ - \frac{\partial L}{\partial x}\bigg|_-\right\| + d_x\left\|\frac{\partial L}{\partial u}\bigg|_+ - \frac{\partial L}{\partial u}\bigg|_-\right\|\right)^2.$$

The significance of such results, given by Sewell and Noble (1978), is that the right hand sides contain first gradients, formulae for which will already be known from the statement $\partial L/\partial x = \partial L/\partial u = 0$ of the problem. The gradients are evaluated at plus and minus points which are at our disposal. They may be chosen for example, as α or β or solution points.

4. Show that if real numbers (with unrestricted sign)

$$k_x[u_+],\quad K_x[u_+],\quad k_u[x_-],\quad K_u[x_-]$$

are given, perhaps depending on the indicated coordinates, such that for each u_+ and for every pair x_+, x_-

$$K_x\|x_+ - x_-\|^2 \geqslant -\left(x_+ - x_-, \overline{\overline{\frac{\partial^2 L}{\partial x^2}}}(x_+ - x_-)\right) \geqslant k_x\|x_+ - x_-\|^2,$$

where $\|\cdot\|^2 = (.,.)$, and for each x_- and for every pair u_+, u_-

$$K_u\|u_+ - u_-\|^2 \geqslant \left\langle u_+ - u_-, \overline{\frac{\partial^2 L}{\partial u^2}}(u_+ - u_-)\right\rangle \geqslant k_u\|u_+ - u_-\|^2,$$

where $\|\cdot\|^2 = \langle .,. \rangle$, then

$$B_{+-} \geqslant S \geqslant b_{+-}$$

with the shorthand

$$B_{+-} = \tfrac{1}{2}K_x[u_+]\|x_+ - x_-\|^2 + \tfrac{1}{2}K_u[x_-]\|u_+ - u_-\|^2,$$
$$b_{+-} = \tfrac{1}{2}k_x[u_+]\|x_+ - x_-\|^2 + \tfrac{1}{2}k_u[x_-]\|u_+ - u_-\|^2.$$

The saddle property $S \geqslant 0$ is assured if $k_x[u_+] \geqslant 0$ and $k_u[x_-] \geqslant 0$ over a rectangular domain, without requiring K_x and K_u to exist, but $B_{+-} \geqslant S \geqslant b_{+-}$ applies when the bounds have either sign.

5. Use the result of Exercise 4 to show that

$$K_x[u_\alpha]\|x_\alpha - x_0\|^2 + K_u[x_0]\|u_\alpha - u_0\|^2 + K_x[u_0]\|x_\beta - x_0\|^2$$
$$+ K_u[x_\beta]\|u_\beta - u_0\|^2$$
$$\geqslant 2(L_\alpha - L_\beta)$$
$$\geqslant k_x[u_\alpha]\|x_\alpha - x_0\|^2 + k_u[x_0]\|u_\alpha - u_0\|^2 + k_x[u_0]\|x_\beta - x_0\|^2$$
$$+ k_u[x_\beta]\|u_\beta - u_0\|^2$$

for solutions of $\partial L/\partial x = \partial L/\partial u = 0$.

6. Show how (4.60) can be deduced as a special case of the result in Exercise 4.
7. Show from Exercise 4 that, in the case of a saddle functional with

$$k_x[u_\alpha] > 0 \quad \text{and/or} \quad k_u[x_\beta] > 0,$$

more can be given away to give the error estimates

$$\frac{2(L_\alpha - L_\beta)}{k_x[u_\alpha]} \geqslant \|x_\alpha - x_0\|^2 \quad \text{and/or} \quad \frac{2(L_\alpha - L_\beta)}{k_u[x_\beta]} \geqslant \|u_\beta - u_0\|^2.$$

4.3. Initial value problems

(i) Introduction

Examples in which the solution of an initial value problem is at the stationary point of an associated functional have been known for many years. In particular, such a stationary property for the heat equation was described by Morse and Feshbach (1953, p. 313) using a so-called 'mirror method'. This method requires the given initial value problem to be coupled with a suitably defined adjoint problem. A functional is constructed whose gradient is zero if both the heat equation problem and its adjoint problem are satisfied. Other examples of such stationary properties for initial value problems have been given by Gurtin (1964), Sandhu and Pister (1971), Noble (1973) and Tonti (1973).

Minimum principles for initial value problems, as distinct from stationary principles, were established by Herrera (1974). He gave a general theory from which simultaneous extremum principles for the heat equation were obtained. Herrera and Bielak (1976) extended this work to diffusion equations, and Herrera and Sewell (1978) applied it to nonnegative unsymmetric operator equations including the wave equation. Collins (1976, 1977) carried the theory further, giving results which include simultaneous extremum principles for dissipative systems.

(ii) The role of the adjoint problem

Let $E[x]$ be a given functional having a gradient $E'[x]$ in the sense of (3.50), and let P be a self-adjoint operator in the sense of (3.42).

Theorem 4.8

Any solution of the equation

$$Px + E'[x] = 0 \tag{4.76}$$

makes stationary the functional

$$\tfrac{1}{2}(x, Px) + E[x]. \tag{4.77}$$

Proof
The first variation of (4.77), in the sense of (3.51), is

$$\delta[\tfrac{1}{2}(x, Px) + E[x]] = \tfrac{1}{2}(\delta x, Px) + \tfrac{1}{2}(x, P\delta x) + \delta E$$
$$= (\delta x, Px + E'[x])$$

after using self-adjointness in the form $(x, P\delta x) = (\delta x, Px)$. □

This result is true in particular if P is also a nonnegative operator, for example $P = T^*T$ (cf. Theorem 3.1), as we have already noted in discussing Example 2 of §4.2(viii). Simultaneous bounds are then also available if $E[x]$ is convex, by Theorem 3.13. A linear operator equation which has all these properties is

$$(T^*T + A)x = a \qquad (3.137\ bis)$$

if A is nonnegative and self-adjoint.

Other linear operator equations, which are not known to involve a decomposition typified by $T^*T + A$, may require a viewpoint opposite to decomposition, namely the embedding viewpoint which we have already illustrated in §§1.4(v), 1.5(iii) and 4.2(vii). For example, we may seek to embed a given linear operator problem within a larger problem of the type

$$-Ax + T^*u + a = 0 \quad (\alpha), \qquad Tx + Bu + b = 0 \quad (\beta), \qquad (4.78)$$

for which we have already seen how to construct simultaneous bounds (Theorem 3.6).

In particular, if a given problem can be written in linear operator form as

$$Tx + b = 0. \qquad (4.79\beta)$$

where the linear operator (here T) has no known decomposition of the type explicitly displayed in (3.137), we may seek to handle it by defining an adjoint problem

$$T^*u + a = 0, \qquad (4.79\alpha)$$

and taking the two equations $(4.79\alpha + \beta)$ together as zero gradients of the bilinear functional

$$(x, T^*u) + (a, x) + \langle b, u\rangle. \qquad (4.80)$$

This is a saddle functional, and there are the following two ways of using it to obtain upper and lower bounds, as we previously remarked after (1.21).

(a) Replace (4.79) by inequality constraints, so defining an infinite dimensional version of the linear programming problem given in §1.6(i). In particular, consider

$$T^*u + a \leqslant 0 \quad (\alpha), \qquad (x, T^*u + a) = 0,$$
$$x \geqslant 0 \quad (\beta), \qquad Tx + b = 0 \quad (\beta), \qquad (4.81)$$

which is an example of (1.55). The upper and lower bounds are

$$\langle b, u_\alpha \rangle \geqslant \langle b, u_0 \rangle = (a, x_0) \geqslant (a, x_\beta)$$

in our usual suffix notation. In other words, this approach characterizes nonnegative solutions x_β of the given problem $Tx + b = 0$ by the upper bound $\langle b, u_\alpha \rangle \geqslant (a, x_\beta)$ calculated for any u_α satisfying $T^*u_\alpha + a \leqslant 0$. Without further information, in fact, we have no guarantee that negative solutions of $Tx + b = 0$ do not exist as well as (or instead of) nonnegative ones.

(b) Retain equality constraints, but change the variables from x and u, which act along the 'asymptotes' of the 'rectangular hyperbolas' $(x, T^*u) =$ *constant*, to another pair of variables which (for example) 'bisects the asymptotes'. This idea was introduced in Chapter 1 (Fig. 1.4 and Exercise 1.2.3), and we shall develop it in detail for initial value problems in what follows. The stationary point of the saddle functional is thereby characterized by upper and lower bounds.

Variable changes of the type just indicated (again see Fig. 1.4) will require that linear combinations of x and u can be defined. In higher dimensional contexts this imposes a special restriction that x and u belong to the *same* inner product space, instead of to different spaces which has been our general rule up to know.

Given that special restriction, we emphasize that no physical interpretation of the adjoint problem $T^*u + a = 0$ need be insisted upon, to mirror whatever interpretation the given problem $Tx + b = 0$ may have. We shall regard the adjoint problem, and any u which satisfies it, purely as artifacts for constructing information about solutions of the given problem. In particular, the adjoint of an initial value problem will facilitate the nomination of an arbitrary later time and a method for calculating required quantities at that time.

(iii) A simple example

Consider the problem of finding a function $x(t)$ satisfying

$$\dot{x} + m(t)x = f(t) \quad \text{in} \quad 0 \leqslant t \leqslant \tau, \qquad x(0) = 0, \tag{4.82}$$

where τ is a given real number, $m(t)$ and $f(t)$ are given functions of t, and the superposed dot denotes t-differentiation. In our composite notation (cf. (3.1)) the problem may be written $T\tilde{x} + \tilde{b} = 0$ where

$$\tilde{x} = \begin{bmatrix} x(t) \\ x(0) \\ x(\tau) \end{bmatrix}, \quad T\tilde{x} = \begin{bmatrix} \dot{x} + mx \\ x(0) \\ 0 \end{bmatrix}, \quad \tilde{b} = \begin{bmatrix} -f \\ 0 \\ 0 \end{bmatrix}. \tag{4.83}$$

There is no evident multiplicative decomposition of such T, and to construct an adjoint problem we are guided by the integration by parts formula

$$u(0)x(0)+\int_0^\tau u(\dot{x}+mx)\mathrm{d}t=x(\tau)u(\tau)+\int_0^\tau x(-\dot{u}+mu)\mathrm{d}t \qquad (4.84)$$

for any two piecewise C^1 functions $x(t)$ and $u(t)$. This leads us to introduce our second space having elements

$$\tilde{u}=\begin{bmatrix}u(t)\\u(0)\\u(\tau)\end{bmatrix} \quad \text{with} \quad T^*\tilde{u}=\begin{bmatrix}-\dot{u}+mu\\0\\u(\tau)\end{bmatrix}. \qquad (4.85)$$

Anticipating the need to form linear combinations of $\tilde{x}$ and $\tilde{u}$ we define the inner product

$$(\tilde{x},\tilde{y})=\int_0^\tau x(t)y(t)\mathrm{d}t+x(0)y(0)+x(\tau)y(\tau) \qquad (4.86)$$

and an identical inner product $(\tilde{u},\tilde{v})$. Then (4.84) expresses adjointness as $(\tilde{u},T\tilde{x})=(\tilde{x},T^*\tilde{u})$ for all $\tilde{x}\in\mathscr{D}(T)$ and for all $\tilde{u}\in\mathscr{D}(T^*)$. These domains are each the space of $\tilde{x}$ or $\tilde{u}$ having the inner product (4.86) and constructed from piecewise C^1 functions $x(t)$ or $u(t)$ respectively.

Manifestly T and T^* are not the same, even though they act within the same space, and so neither T nor T^* is self-adjoint in the sense of (3.42). Therefore even Theorem 4.8 cannot apply to the given problem (4.82) in its operator form $T\tilde{x}+\tilde{b}=0$, and to make progress we need to consider it in conjunction with an adjoint problem $T^*\tilde{u}+\tilde{a}=0$ for any assigned $\tilde{a}$. Then

$$T^*\tilde{u}+\tilde{a}=0, \quad T\tilde{x}+\tilde{b}=0 \qquad (4.87)$$

are generated by zero gradients of the bilinear functional

$$(\tilde{x},T^*\tilde{u})+(\tilde{a},\tilde{x})+(\tilde{b},\tilde{u}). \qquad (4.88)$$

This is the appropriate specialization of (4.80) for the present particular case.

Before continuing, we remark that although simpler space elements of type (3.81) and inner products of type (3.80) could have been used to write (4.84) as $\langle\tilde{u},T\tilde{x}\rangle=(\tilde{x},T^*\tilde{u})$, with the zero entries omitted from (4.83) and (4.85) as in (3.82), this would have been inadequate for two reasons. Self-adjointness (3.42) is not then defined for such T, and such $\tilde{x}$ and $\tilde{u}$ belong to different spaces so that linear combinations of them (required next) are not defined.

For the same reason, we cannot simply choose the requisite common space to be that of (say C^1) functions $x(t)$ and $u(t)$ which vanish at $t=0$, because in general $u(0) \neq 0$ even though we can choose $x(0)=0$ in (4.82) without loss of generality.

(iv) Change of variable

Leaving the simple example, we return to the generality of any inner product space with inner product $(.,.)$ and elements $x, y, \ldots, u, v, \ldots, a, b, \ldots$. Consider the problem of finding x and u such that, for given a and b,

$$T^*u + a = 0, \quad Tx + b = 0, \tag{4.89}$$

where T and T^* are given linear operators adjoint to each other in the sense that $(x, T^*u) = (u, Tx)$ for all x in $\mathscr{D}(T)$ and all u in $\mathscr{D}(T^*)$. In passing we notice that $(x, \overline{T^* - T}x) = 0$ for every x which lies in both domains, i.e. in $\mathscr{D}(T) \cap \mathscr{D}(T^*)$.

As will be clear from (4.80) (where two different inner products were admitted) or from the example (4.88), any solution of (4.89) will make stationary the bilinear functional

$$(x, T^*u) + (a, x) + (b, u). \tag{4.90}$$

The proof is an immediate consequence of (3.60). It could also be regarded as an illustration of Theorem 4.8, as we explain in §(v) below.

In what follows we work within $\mathscr{D}(T) \cap \mathscr{D}(T^*)$.

Let p, q, r, s be any fixed real numbers satisfying

$$0 \neq ps - qr = \Delta(\text{say}).$$

We can then introduce a uniquely invertible change of variables

$$\left.\begin{aligned} y &= px + qu \\ v &= rx + su \end{aligned}\right\} \Leftrightarrow \left\{\begin{aligned} x &= \frac{1}{\Delta}(sy - qv) \\ u &= \frac{1}{\Delta}(-ry + pv) \end{aligned}\right. \tag{4.91}$$

The bilinear saddle functional (4.90) can then be rewritten as

$$\begin{aligned} L[y, v] = {} & \frac{1}{\Delta^2}[-rs(y, Ty) - pq(v, Tv) + qr(y, Tv) + ps(v, Ty)] \\ & + \frac{1}{\Delta}(y, as - rb) + \frac{1}{\Delta}(v, -qa + pb). \end{aligned} \tag{4.92}$$

The domain of this functional is $\mathscr{D}(T) \cap \mathscr{D}(T^*)$.

Theorem 4.9

Any solution of (4.89) *within* $\mathscr{D}(T) \cap \mathscr{D}(T^*)$ *will make* $L[y, v]$ *stationary.*

Proof
The gradients of (4.92) are

$$\frac{\partial L}{\partial y}=\frac{s}{\Delta}\left[\frac{T^*}{\Delta}(-ry+pv)+a\right]-\frac{r}{\Delta}\left[\frac{T}{\Delta}(sy-qv)+b\right],$$

$$\frac{\partial L}{\partial v}=-\frac{q}{\Delta}\left[\frac{T^*}{\Delta}(-ry+pv)+a\right]+\frac{p}{\Delta}\left[\frac{T}{\Delta}(sy-qv)+b\right]. \tag{4.93}$$

From (4.91) we then see that

$$\frac{\partial L}{\partial y}=\frac{\partial L}{\partial v}=0 \tag{4.94}$$

if and only if (4.89) is satisfied. □

Next we introduce some stronger hypotheses than hitherto.

Theorem 4.10
(a) Suppose that T is a nonnegative operator in the sense of (3.43), *and also that*

$$rs\geqslant 0, \quad pq\leqslant 0. \tag{4.95}$$

Then $L[y,v]$ is a saddle functional concave in y and convex in v.

(b) If T is a positive operator with

$$rs>0, \quad pq<0 \tag{4.96}$$

then $L[y,v]$ is a strict saddle functional, strictly concave in y and strictly convex in v.

Proof
This is immediate, for example by inspecting the first two (unmixed quadratic) terms in (4.92). The remainder are either bilinear or linear and so will contribute an identically zero amount to the basic saddle quantity S. For any two points y_+,v_+ and y_-,v_- this is

$$\begin{aligned} S&=L_+-L_--\left(y_+-y_-,\left.\frac{\partial L}{\partial y}\right|_+\right)-\left(v_+-v_-,\left.\frac{\partial L}{\partial v}\right|_-\right)\\ &=\frac{rs}{\Delta^2}(y_+-y_-,T(y_+-y_-))-\frac{pq}{\Delta^2}(v_+-v_-,T(v_+-v_-)) \qquad (4.97)\\ &\geqslant 0. \end{aligned}$$

□

The hypotheses of Theorem 4.10 invite designation of the equivalent governing conditions (4.89) or (4.94) into α and β subsets

$$\frac{\partial L}{\partial y}=0 \quad (\alpha), \qquad \frac{\partial L}{\partial v}=0 \quad (\beta), \tag{4.98}$$

or

$$\begin{aligned} s(T^*u+a)-r(Tx+b)&=0, \quad (\alpha) \\ -q(T^*u+a)+p(Tx+b)&=0, \quad (\beta) \end{aligned} \tag{4.99}$$

These designations are different from those in (4.79). Upper and lower bounds then apply as a direct consequence of Theorem 3.6 transcribed into the y and v variables. We can express the result as follows, after transcribing back to the x and u variables.

Theorem 4.11

Let x_α, u_α denote any solution of (4.99α) *and let x_β, u_β denote any solution of* (4.99β). *Let x_0, u_0 denote any solution of* (4.99α + β) *and therefore of the pair* (4.89) *in $\mathscr{D}(T)\cap\mathscr{D}(T^*)$. Then under the hypotheses of Theorem* 4.10(*a*) *or* (*b*)

$$L_\alpha \geqslant (a, x_0) = (b, u_0) \geqslant L_\beta, \tag{4.100}$$

where

$$\begin{aligned} L_\alpha &= (u_\alpha, Tx_\alpha+b)+(a, x_\alpha), \\ L_\beta &= (u_\beta, Tx_\beta+b)+(a, x_\beta). \end{aligned}$$

Theorem 4.12

Under the hypotheses of Theorem 4.10(*b*), *if* (4.99) *has a solution x_0 with u_0, this solution pair is unique, and so therefore is a solution of* (4.89) *in $\mathscr{D}(T)\cap\mathscr{D}(T^*)$.*

Proof

The strict version of (4.96) implies

$$\left(y_+-y_-, \frac{\partial L}{\partial y}\bigg|_+ - \frac{\partial L}{\partial y}\bigg|_-\right) - \left(v_+-v_-, \frac{\partial L}{\partial v}\bigg|_+ - \frac{\partial L}{\partial v}\bigg|_-\right) > 0$$

for every pair of points in the domain of $L[y, v]$. Uniqueness follows for (4.98) by *reductio ad absurdum*, and hence for the equivalent (4.99) and (4.89). □

Notice that proof of uniqueness of solution of (4.89) cannot be based on the saddle property of (4.90) alone, because this is not a strict saddle property, and the argument by contradiction does require positive T and the change of variables with (4.96).

The information which the bounding Theorem 4.11 gives about the original problem $Tx+b=0$ will depend on how we can exploit the flexibility available in defining the adjoint problem $T^*u+a=0$. We are free to choose any a there which belongs to the range of T^*, and within this restriction (4.100) bounds any linear functional (a, x_0) of solutions of (4.89) within $\mathscr{D}(T)\cap\mathscr{D}(T^*)$. The question of pointwise bounds (cf. Fig. 4.3) suggests itself, and appears in another form (eg. see (4.114) and (4.122)).

Choices which make (4.99α) or (4.99β) homogeneous are $a = br/s$ if $s \neq 0$ or $a = bp/q$ if $q \neq 0$, respectively. With $r = s$ or $p = -q$ also, respectively, we then reach

$$(u_\alpha, Tu_\alpha) + (b, x_\alpha + u_\alpha) \geqslant (b, x_0) \quad \text{with} \quad T^*u_\alpha = Tx_\alpha \tag{4.101}$$

from (4.100), using $(u_\alpha, Tu_\alpha) = (u_\alpha, T^*u_\alpha) = (u_\alpha, Tx_\alpha)$. This bounds the 'work' done by the given b in a solution. There is also a lower bound subject to inhomogeneous constraints. Such upper and lower bounds can be regarded as approximating principles, additional to the stationary property stated after (4.90), in which the α or β constraints may be easier to satisfy than $Tx + b = 0$ itself.

We can improve (4.101) by optimizing with respect to the arbitrary scale factor in x_α and u_α, giving a result which is equivalent to one stated in (4.107).

(v) A self-adjoint representation

It is possible to discuss (4.89) and (4.90) in terms of another linear space with elements $\xi = [x\ u]^{\mathrm{T}}$, $\eta = [y\ v]^{\mathrm{T}}$ and inner product $\overline{\xi, \eta} = (x, y) + (u, v)$. Defining $\gamma = [a\ b]^{\mathrm{T}}$ we can write (4.89) as

$$\Gamma\xi + \gamma = 0 \quad \text{where} \quad \Gamma = \begin{bmatrix} 0 & T^* \\ T & 0 \end{bmatrix} \tag{4.102}$$

is a self-adjoint operator acting in this new space. For

$$\overline{\eta, \Gamma\xi} = (y, T^*u) + (v, Tx) = (x, T^*v) + (u, Ty) = \overline{\xi, \Gamma\eta}.$$

The functional in (4.90) is therefore

$$(x, T^*u) + (a, x) + (b, u) = \tfrac{1}{2}\overline{\xi, \Gamma\xi} + \overline{\gamma, \xi},$$

so that Theorem 4.8 can be applied to establish the stationary property mentioned after (4.90).

(vi) A particular change of variable

An especially simple choice of variable change (4.91) which satisfies (4.96) is just to take the sum and difference of the original variables, i.e.

$$\left.\begin{aligned} y &= x - u \\ v &= x + u \end{aligned}\right\} \Leftrightarrow \left\{\begin{aligned} x &= \tfrac{1}{2}(y + v) \\ u &= \tfrac{1}{2}(-y + v) \end{aligned}\right. \tag{4.103}$$

defined by $p = r = s = -q = 1$. In this case the designation of α and β conditions in (4.99) becomes

$$\begin{aligned} T^*u - Tx + a - b &= 0, \quad (\alpha) \\ T^*u + Tx + a + b &= 0. \quad (\beta) \end{aligned} \tag{4.104}$$

If we also choose $a = b$ in (4.104) we can give a particular consequence of Theorem 4.11 which is worth stating explicitly, as follows. This choice may be too simple to be appropriate in some problems, and further investigation of the method will be needed before a true perspective can be reached.

Theorem 4.13

Let x_α, u_α and x_β, u_β satisfy

$$T^*u = Tx \quad (\alpha), \qquad T^*u + Tx + 2b = 0 \quad (\beta), \tag{4.105}$$

respectively. Let x_0, u_0 denote any solution of (4.105$\alpha + \beta$) and therefore of the pair

$$T^*u + b = 0, \quad Tx + b = 0 \tag{4.106}$$

in $\mathscr{D}(T) \cap \mathscr{D}(T^*)$.

Then if T is a positive operator (or nonnegative but with $(u_\alpha, Tu_\alpha) > 0$ for some u_α to which we then restrict attention)

$$-\frac{(b, x_\alpha + u_\alpha)^2}{4(u_\alpha, Tu_\alpha)} \geqslant (b, x_0) \geqslant -(u_\beta, Tu_\beta) + (b, x_\beta - u_\beta). \tag{4.107}$$

Proof

We have a particular case (4.106) of (4.89) which is equivalent to (4.105), and (4.96) is satisfied with $rs = 1$ and $pq = -1$. The bounds (4.100) apply and specialize to (4.107) by deploying (4.105α) and (4.105β), and by optimizing with respect to an arbitrary scale factor in u_α. □

The choice (4.103) is made without any reference to the structure of T. It is therefore unlikely to be optimum in the sense that bisectors of the asymptotes in Fig. 1.4 is optimum. In that case the curvature of the upper and lower bounding parabolic sections L_α and L_β will be the greatest possible. Other changes of variable which reflect the structure of the nonnegative operator T, and in particular its eigenvalue structure, might therefore be more useful in some problems.

(vii) Example

Returning to the simple example of §(iii),

$$\begin{aligned}(\tilde{x}, T\tilde{x}) &= \int_0^\tau x(\dot{x} + mx)\mathrm{d}t + x(0)^2 \\ &= \int_0^\tau x(-\dot{x} + mx)\mathrm{d}t + x(\tau)^2 \\ &= \int_0^\tau mx^2\,\mathrm{d}t + \tfrac{1}{2}[x(0)^2 + x(\tau)^2].\end{aligned} \tag{4.108}$$

Hence T is positive if $m(t) > 0$, and T is nonnegative if $m(t) \geqq 0$.

To use (4.103) and (4.105) we choose

$$\tilde{y} = \tilde{x} - \tilde{u}, \quad \tilde{v} = \tilde{x} + \tilde{u}, \quad \tilde{a} = \tilde{b} \tag{4.109}$$

in (4.87), so that the (4.105) is

$$T^*\tilde{u} = T\tilde{x} \quad (\alpha), \qquad T^*\tilde{u} + T\tilde{x} + 2\tilde{b} = 0 \quad (\beta), \tag{4.110}$$

i.e.

$$\begin{bmatrix} -(\dot{u} + \dot{x}) + m(u - x) \\ -x(0) \\ u(\tau) \end{bmatrix} = 0 \quad (\alpha), \quad \begin{bmatrix} -(\dot{u} - \dot{x}) + m(u + x) - 2f \\ x(0) \\ u(\tau) \end{bmatrix} = 0 \quad (\beta). \tag{4.111}$$

Now we are in a position to evaluate the bounds (4.107).

For simplicity suppose that m is a positive constant. Then we can set $m = 1$ without loss of generality, thus normalizing the time scale. It is straightforward to show that the homogeneous (4.111α) is satisfied by

$$x_\alpha(t) = -u_\alpha(t) + \mathrm{e}^{-t}\left[2\int_0^t u_\alpha(s)\mathrm{e}^s\,\mathrm{d}s + u_\alpha(0)\right],$$

with any function $u_\alpha(t)$ satisfying $u_\alpha(\tau) = 0$. This gives flexibility to the upper bound in (4.107). The easiest choice is

$$u_\alpha = t - \tau, \quad x_\alpha = t + (\tau + 2)(\mathrm{e}^{-t} - 1),$$

leading to $(\tilde{u}_\alpha, T\tilde{u}_\alpha) = \frac{1}{2}\tau^2 + \frac{1}{3}\tau^3$. If $f(t)$ in (4.82) is a nonzero constant we choose $f = 1$ to normalize x. Then we find

$$(\tilde{b}, \tilde{x}_\alpha + \tilde{u}_\alpha) = \tau^2 + 2\tau + (2 + \tau)(\mathrm{e}^{-\tau} - 1)$$

and hence the upper bound

$$L_\alpha = -\frac{[\tau^2 + 2\tau + (2 + \tau)(\mathrm{e}^{-\tau} - 1)]^2}{2\tau^2 + \frac{4}{3}\tau^3}. \tag{4.112}$$

We can also show that, with $f(t) \equiv 1$, the nonhomogeneous (4.111β) is satisfied by

$$x_\beta(t) = u_\beta(t) + 2 - 2e^{-t}\left[1 + \tfrac{1}{2}u_\beta(0) + \int_0^t u_\beta(s)\mathrm{e}^s\,\mathrm{d}s\right],$$

with any function $u_\beta(t)$ satisfying $u_\beta(\tau) = 0$. Thus exactly the same family of arbitrary functions, vanishing at the final time, is available in choosing both lower and upper bounds. Choose

$$u_\beta = t - \tau, \quad x_\beta = -t + \tau + 4 - \mathrm{e}^{-t}(4 + \tau),$$

so that the lower bound in (4.107) becomes

$$L_\beta = -\mathrm{e}^{-\tau}(4 + \tau) + 4 - \tfrac{1}{3}\tau^3 - \tfrac{5}{2}\tau^2 - 3\tau. \tag{4.113}$$

There is scope to improve this bound.

The quantity $L_0 = (\tilde{b}, \tilde{x}_0)$ bounded by (4.107) is effectively the exact solution $x_0(t) = 1 - e^{-t}$ of (4.82) itself, since

$$L_0 = -\int_0^\tau x_0\,dt = \int_0^\tau (\dot{x}_0 - 1)dt = x_0(\tau) - \tau = 1 - e^{-\tau} - \tau. \quad (4.114)$$

The real purpose of this example is, of course, to give a quick tangible illustration of Theorem 4.13. The bounds $L_\alpha \geqslant L_0 \geqslant L_\beta$ of (4.107) just calculated are displayed in Fig. 4.7 as functions of τ.

(viii) A general first order system

Let $\mathbf{x}(t)$ be an $n \times 1$ column matrix of scalar functions satisfying

$$\dot{\mathbf{x}} + M\mathbf{x} = \mathbf{f}(t) \quad \text{in} \quad 0 \leqslant t \leqslant \tau, \quad \mathbf{x}(0) = 0, \quad (4.115)$$

where $M(t)$ and $\mathbf{f}(t)$ are given $n \times n$ and $n \times 1$ functions of t. The procedures of §(iii) can be imitated, augmented by ordinary matrix products with transposes as appropriate, for example leading to (4.108) in the revised form

$$2(\tilde{\mathbf{x}}, T\tilde{\mathbf{x}}) = \int_0^\tau \mathbf{x}^T(M + M^T)\mathbf{x}\,dt + \mathbf{x}^T(0)\mathbf{x}(0) + \mathbf{x}^T(\tau)\mathbf{x}(\tau). \quad (4.116)$$

Thus T is nonnegative if the symmetric part of M is nonnegative. This is sufficient for upper and lower bounds to be obtainable, for example as in Theorem 4.13. This result is consistent with an example (for symmetric M independent of t) given by Collins (1977) to illustrate his theory.

In the language of phase plane analysis (e.g. see Arrowsmith and Place, 1982), when $n = 2$ and $\mathbf{f} = 0$, a requirement of positive $M + M^T$ implies that the origin is a stable focus or node, never an unstable one or a saddle. Systems with positive $M + M^T$ may be called dissipative. Hamilton's

Fig. 4.7. Bounds for first order initial problem.

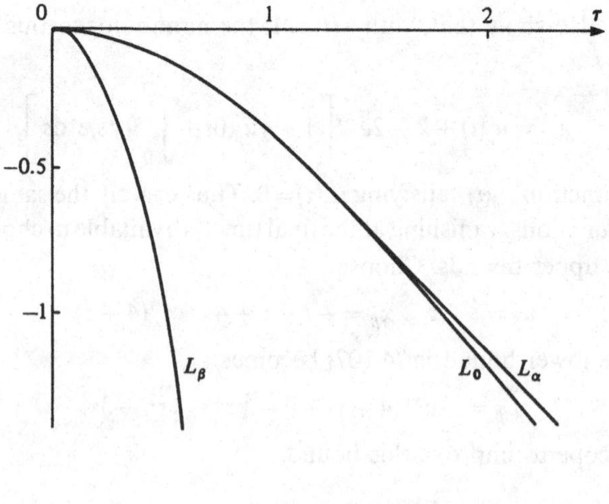

equations for $n=2$ with a quadratic Hamiltonian and non-negative $M+M^{\mathrm{T}}$ imply $M+M^{\mathrm{T}}=0$ and a centre.

(ix) A second order equation

Consider the problem of finding a scalar function $x(t)$ of $t\geqslant 0$ which satisfies

$$\begin{gathered}-\rho(t)\ddot{x}-m(t)\dot{x}+r(t)x=f(t),\\ x(0)=0,\quad \dot{x}(0)=0.\end{gathered}\tag{4.117}$$

Here $\rho(t)$, $m(t)$, $r(t)$ and $f(t)$ are given functions of t. The following treatment of this initial value problem may be contrasted with that of the boundary value problem (3.78). We introduce an intermediate variable $y=-\dot{x}$ so that (4.117) becomes

$$\begin{aligned}\dot{x}&=-y, &\quad \rho\dot{y}+my+rx&=f,\\ x(0)&=0, & y(0)&=0.\end{aligned}\tag{4.118}$$

Irrespective of whether this system is satisfied, any quartet of piecewise C^1 functions $x(t), y(t), u(t), v(t)$ defined over $0\leqslant t\leqslant\tau$ with any given τ will satisfy integration by parts in the form

$$\begin{aligned}&\int_0^\tau [u(\dot{x}+y)+v(\rho\dot{y}+my+rx)]\mathrm{d}t+[ux+v\rho y]_{t=0}\\ &\quad=\int_0^\tau [x(-\dot{u}+rv)+y(-\dot{\overline{\rho v}}+mv+u)]\mathrm{d}t+[xu+y\rho v]_{t=\tau},\end{aligned}\tag{4.119}$$

provided that $\rho(t)$ is piecewise C^1 and $m(t)$ and $r(t)$ are integrable. We can express (4.117) as a statement of adjointness in the form

$$(\tilde{\eta},T\tilde{\xi})=(\tilde{\xi},T^*\tilde{\eta})$$

if we define composite elements and operators as

$$\tilde{\xi}=\begin{bmatrix}x(t)\\ y(t)\\ x(0)\\ y(0)\\ x(\tau)\\ y(\tau)\end{bmatrix},\quad T\tilde{\xi}=\begin{bmatrix}\dot{x}+y\\ \rho\dot{y}+my+rx\\ x(0)\\ \rho(0)y(0)\\ 0\\ 0\end{bmatrix},$$

$$\tilde{\eta}=\begin{bmatrix}u(t)\\ v(t)\\ u(0)\\ v(0)\\ u(\tau)\\ v(\tau)\end{bmatrix},\quad T^*\tilde{\eta}=\begin{bmatrix}-\dot{u}+rv\\ -\dot{\overline{\rho v}}+mv+u\\ 0\\ 0\\ u(\tau)\\ \rho(\tau)v(\tau)\end{bmatrix},$$

and inner product as

$$(\tilde{\xi}, \tilde{\eta}) = \int_0^\tau (xu + yv)\mathrm{d}t + x(0)u(0) + y(0)v(0) + x(\tau)u(\tau).$$

From (4.117) we then see that

$$2(\tilde{\xi}, T\tilde{\xi}) = x(0)^2 + \rho(0)y(0)^2 + x(\tau)^2 + \rho(\tau)y(\tau)^2$$
$$+ \int_0^\tau [2(1+r)xy + (2m - \dot{\rho})y^2]\mathrm{d}t, \qquad (4.120)$$

so that the operator T is nonnegative if

$$\rho(0) \geqslant 0, \quad \rho(\tau) \geqslant 0, \quad r(t) \equiv -1, \quad 2m(t) \geqslant \dot{\rho}(t). \qquad (4.121)$$

These conditions are satisfied *a fortiori* if ρ and m are positive constants (with $r \equiv -1$). The system in the form (4.118) can then be written as

$$T\tilde{\xi} + \tilde{b} = 0 \quad \text{with} \quad \tilde{b} = [0 \;\; f \;\; 0 \;\; 0 \;\; 0 \;\; 0]^{\mathrm{T}},$$

and Theorems 4.10(a), 4.11 and (4.13) apply to it. The quantity bounded by (4.107) is the value of $(\tilde{\xi}, \tilde{b})$ taken by a solution, i.e. of

$$\int_0^\tau y_0 f\,\mathrm{d}t = -\int_0^\tau \dot{x}_0 f\,\mathrm{d}t \qquad (4.122)$$

in a solution x_0, y_0 of (4.118). If f is a constant, this becomes $-fx_0(\tau)$. Thus for constant $f \neq 0$, our procedure has the effect of bounding the solution $x_0(\tau)$ of (4.117) at any chosen time $\tau > 0$.

(x) A heat conduction problem

From several available variants we select the following version of such a problem. In a fixed two or three dimensional spatial region V with boundary Σ, find a scalar function $\phi[\mathbf{r}, t]$ of position vector $\mathbf{r}$, and of time t in a given interval $0 \leqslant t \leqslant \tau$ with fixed τ, such that

$$\frac{\partial \phi}{\partial t} - \nabla\cdot(\rho\nabla\phi) = f \quad \text{in} \quad V \quad \text{and in} \quad 0 \leqslant t \leqslant \tau,$$

$$\frac{\partial \phi}{\partial n} = g \quad \text{on} \quad \Sigma \quad \text{in} \quad 0 \leqslant t \leqslant \tau, \qquad (4.123)$$

$$\phi[\mathbf{r}, 0] = c \quad \text{in} \quad V.$$

Here ρ, f and g are given functions of $\mathbf{r}$ and t, with $\rho > 0$, c is a given function of $\mathbf{r}$, and $\partial\phi/\partial n = \mathbf{n}\cdot\nabla\phi$ is the gradient in the direction of the outward unit normal $\mathbf{n}$ on Σ.

The linear differential operator displayed in the heat equation participates in the following formula of integration by parts, for any two functions $\phi[\mathbf{r},t]$ and $\psi[\mathbf{r},t]$ (say C^2 for simplicity) regardless of whether they satisfy the system (4.123) or not.

$$\int_0^\tau\int\psi\left(\frac{\partial\phi}{\partial t}-\nabla\cdot(\rho\nabla\phi)\right)\mathrm{d}V\mathrm{d}t+\int_0^\tau\int\rho\psi\frac{\partial\phi}{\partial n}\mathrm{d}\Sigma\,\mathrm{d}t+\int\psi\phi\,\mathrm{d}V\bigg|_{t=0}$$
$$=\int_0^\tau\int\phi\left(-\frac{\partial\psi}{\partial t}-\nabla\cdot(\rho\nabla\psi)\right)\mathrm{d}V\mathrm{d}t+\int_0^\tau\int\rho\phi\frac{\partial\psi}{\partial n}\mathrm{d}\Sigma\,\mathrm{d}t+\int\phi\psi\,\mathrm{d}V\bigg|_{t=\tau}. \tag{4.124}$$

This suggests that we introduce composite elements

$$\tilde{\phi}=\begin{bmatrix}\phi[\mathbf{r},t]\\ \phi[\mathbf{r},t]\\ \phi[\mathbf{r},0]\\ \phi[\mathbf{r},\tau]\end{bmatrix},\quad \tilde{b}=\begin{bmatrix}-f\\ -g\\ -c\\ 0\end{bmatrix}\quad\text{on}\quad\begin{bmatrix}V,\text{ all }t\\ \Sigma,\text{ all }t\\ V,t=0\\ V,t=\tau\end{bmatrix}, \tag{4.125}$$

$$T\tilde{\phi}=\begin{bmatrix}\dfrac{\partial\phi}{\partial t}-\nabla\cdot(\rho\nabla\phi)\\ \dfrac{\partial\phi}{\partial n}\\ \phi[\mathbf{r},0]\\ 0\end{bmatrix},\quad T^*\tilde{\psi}=\begin{bmatrix}-\dfrac{\partial\psi}{\partial t}-\nabla\cdot(\rho\nabla\psi)\\ \dfrac{\partial\psi}{\partial n}\\ 0\\ \psi[\mathbf{r},\tau]\end{bmatrix}, \tag{4.126}$$

and an inner product

$$(\tilde{\phi},\tilde{\psi})=\int_0^\tau\int\phi\psi\,\mathrm{d}V\mathrm{d}t+\int_0^\tau\int\rho\phi\psi\,\mathrm{d}\Sigma\,\mathrm{d}t+\int\phi\psi\,\mathrm{d}V\bigg|_{t=0}+\int\phi\psi\,\mathrm{d}V\bigg|_{t=\tau}. \tag{4.127}$$

The given heat conduction problem (4.123) is then summarized as $T\tilde{\phi}+\tilde{b}=0$, and (4.124) expresses adjointness as $(\tilde{\psi},T\tilde{\phi})=(\tilde{\phi},T^*\tilde{\psi})$. There is an adjoint problem $T^*\tilde{\psi}+\tilde{a}=0$ for any given $\tilde{a}$. Since

$$(\tilde{\phi},T\tilde{\phi})=\int_0^\tau\int\rho(\nabla\phi)^2\,\mathrm{d}V\mathrm{d}t+\frac{1}{2}\int\phi^2\,\mathrm{d}V\bigg|_{t=0}+\frac{1}{2}\int\phi^2\,\mathrm{d}V\bigg|_{t=\tau}, \tag{4.128}$$

we deduce that T is a nonnegative operator. Any function uniform over V and vanishing at $t=0$ and $t=\tau$ will make $(\tilde{\phi},T\tilde{\phi})=0$. The hypotheses of Theorem 4.10(a) are thus satisfied, and upper and lower bounds can be written down.

For example, if we choose $\tilde{a}=\tilde{b}$ and the sum and difference change of variables (4.103), thus imitating (4.109), the (α) equation becomes $T^*\tilde{\psi}_\alpha=T\tilde{\phi}_\alpha$. To satisfy this requires functions $\phi_\alpha[\mathbf{r},t]$ and $\psi_\alpha[\mathbf{r},t]$ such that

$$\frac{\partial(\phi_\alpha+\psi_\alpha)}{\partial t}-\nabla\cdot(\rho\nabla(\phi_\alpha-\psi_\alpha))=0 \quad \text{in} \quad V \quad \text{for all} \quad t,$$

$$\frac{\partial(\phi_\alpha-\psi_\alpha)}{\partial n}=0 \quad \text{on} \quad \Sigma \quad \text{for all} \quad t,$$

$$\phi_\alpha[\mathbf{r},0]=0 \quad \text{and} \quad \psi_\alpha[\mathbf{r},\tau]=0 \quad \text{in} \quad V.$$

Any solution of this problem will lead to an upper bound $L_\alpha=(\tilde{\psi}_\alpha, T\tilde{\phi}_\alpha)+(\tilde{b}, \tilde{\phi}_\alpha+\tilde{\psi}_\alpha)$ by Theorem 4.11. Other methods related to the variable change (4.103) have been explored by Barrett *et al.* (1979) and Noble (1981).

(xi) Alternative approaches

Suppose the Neumann boundary condition in (4.123) is replaced by the Dirichlet boundary condition

$$\phi[\mathbf{r},t]=g \quad \text{on} \quad \Sigma \quad \text{in} \quad 0\leqslant t\leqslant\tau, \tag{4.129}$$

where $g[\mathbf{r},t]$ is a given function. This change, which might be thought innocuous, demands a different approach from that of §(x) above. There are at least two possible alternative approaches.

(a) We can seek to retain (4.124) as the guiding statement of adjointness, inducing an operator T which allows the new governing conditions to be displayed as $T\tilde{\phi}+\tilde{b}=0$, but this requires the Σ terms in (4.124) to be transferred to the opposite sides of the equation. This gives

$$\int_0^\tau\int\psi\left(\frac{\partial\phi}{\partial t}-\nabla\cdot(\rho\nabla\phi)\right)\mathrm{d}V\,\mathrm{d}t-\int_0^\tau\int\rho\frac{\partial\psi}{\partial n}\phi\,\mathrm{d}\Sigma\,\mathrm{d}t+\int\psi\phi\,\mathrm{d}V\bigg|_{t=0}$$
$$=\int_0^\tau\int\phi\left(-\frac{\partial\psi}{\partial t}-\nabla\cdot(\rho\nabla\psi)\right)\mathrm{d}V\,\mathrm{d}t-\int_0^\tau\int\rho\frac{\partial\phi}{\partial n}\psi\,\mathrm{d}\Sigma\,\mathrm{d}t+\int\phi\psi\,\mathrm{d}V\bigg|_{t=\tau}. \tag{4.130}$$

If we use the same definition (4.127) of inner product as before, we can write (4.130) as

$$(N\tilde{\psi}, T\tilde{\phi})=(N\tilde{\phi}, T^*\tilde{\psi}) \tag{4.131}$$

if we introduce *three* linear operators T, T^* and N defined by

$$T\tilde{\phi}=\begin{bmatrix}\dfrac{\partial\phi}{\partial t}-\nabla\cdot(\rho\nabla\phi)\\ \phi[\mathbf{r},t]\\ \phi[\mathbf{r},0]\\ 0\end{bmatrix},\quad T^*\tilde{\psi}=\begin{bmatrix}-\dfrac{\partial\psi}{\partial t}-\nabla\cdot(\rho\nabla\psi)\\ \psi[\mathbf{r},t]\\ 0\\ \psi[\mathbf{r},\tau]\end{bmatrix},\quad N\tilde{\phi}=\begin{bmatrix}\phi[\mathbf{r},t]\\ -\dfrac{\partial\phi}{\partial n}\\ \phi[\mathbf{r},0]\\ \phi[\mathbf{r},\tau]\end{bmatrix} \tag{4.132}$$

and acting on composite elements $\tilde{\phi}$ and $\tilde{\psi}$ as defined in (4.125).

The Dirichlet problem can then be written $T\tilde{\phi} + \tilde{b} = 0$, but at the expense of a prospective generalization in our basic definition of adjointness from $(\tilde{\psi}, T\tilde{\phi}) = (\tilde{\phi}, T^*\tilde{\psi})$ to (4.131) (for all $\tilde{\phi}$ and $\tilde{\psi}$ constructed from sufficiently smooth, say C^2, functions $\phi[\mathbf{r},t]$ and $\psi[\mathbf{r},t]$). Since also

$$(N\tilde{\phi}, T\tilde{\phi}) = \int_0^\tau \int \rho(\nabla\phi)^2 \,\mathrm{d}V\,\mathrm{d}t + \frac{1}{2}\int \phi^2\,\mathrm{d}V\bigg|_{t=0} + \frac{1}{2}\int \phi^2\,\mathrm{d}V\bigg|_{t=\tau} - 2\int_0^\tau \int \rho\phi\frac{\partial\phi}{\partial n}\,\mathrm{d}\Sigma\,\mathrm{d}t, \qquad (4.133)$$

we could adjust the definition (3.42) of positivity to allow the conclusion that T is nonnegative on the space of $\tilde{\phi}$ constructed from sufficiently smooth $\phi[\mathbf{r},t]$ which have $\partial\phi/\partial n = 0$ on Σ for all t.

Herrera's (1974) theory of upper and lower bounds for saddle functionals on spaces which have no metric, norm or inner product contains a bilinear definition of adjointness which is wide enough to include (4.131). We can, however, handle the current Dirichlet problem in another way which avoids the need to change our basic definition (3.10) of adjointness, as follows.

(b) An intermediate variable $\mathbf{u} = -\nabla\phi$ such as we used in §3.10(v) may be introduced, or may already be present in the physical formulation of the problem. The problem (4.123) but with (4.129) then becomes

$$\frac{\partial\phi}{\partial t} + \nabla\cdot(\rho\mathbf{u}) = f, \quad \mathbf{u} + \nabla\phi = 0 \quad \text{in} \quad V \quad \text{for all } t,$$
$$\phi[\mathbf{r},t] = g \quad \text{on} \quad \Sigma \quad \text{for all} \quad t, \quad \phi[\mathbf{r},0] = c \quad \text{in} \quad V. \qquad (4.134)$$

The linear operator implicit on the left sides in (4.134) suggests the following formula of integration by parts. For any two pairs of piecewise C^1 functions $\phi[\mathbf{r},t], \mathbf{u}[\mathbf{r},t]$ and $\psi[\mathbf{r},t], \mathbf{v}[\mathbf{r},t]$, each pair consisting of a scalar and a vector, and regardless of whether (4.134) is satisfied or not, we have

$$\int_0^\tau \int \left[\psi\left(\frac{\partial\phi}{\partial t} + \nabla\cdot(\rho\mathbf{u})\right) + \rho\mathbf{v}\cdot(\nabla\phi + \mathbf{u})\right]\mathrm{d}V\,\mathrm{d}t - \int_0^\tau \int \rho\mathbf{v}\cdot\mathbf{n}\phi\,\mathrm{d}\Sigma\,\mathrm{d}t + \int \psi\phi\,\mathrm{d}V\bigg|_{t=0}$$
$$= \int_0^\tau \int \left[\phi\left(-\frac{\partial\psi}{\partial t} - \nabla\cdot(\rho\mathbf{v})\right) + \rho\mathbf{u}\cdot(-\nabla\psi + \mathbf{v})\right]\mathrm{d}V\,\mathrm{d}t + \int_0^\tau \int \rho\mathbf{u}\cdot\mathbf{n}\psi\,\mathrm{d}\Sigma\,\mathrm{d}t + \int \phi\psi\,\mathrm{d}V\bigg|_{t=\tau}$$
$$(4.135)$$

We can regard this as a statement of adjointness of the type $(\tilde{y}, T\tilde{x}) = (\tilde{x}, T^*\tilde{y})$ if we introduce the composite elements and operators

$$\tilde{x} = \begin{bmatrix} \phi[\mathbf{r}, t] \\ \mathbf{u}[\mathbf{r}, t] \\ \mathbf{u}[\mathbf{r}, t] \\ \phi[\mathbf{r}, 0] \\ \phi[\mathbf{r}, \tau] \end{bmatrix}, \quad T\tilde{x} = \begin{bmatrix} \dfrac{\partial \phi}{\partial t} + \nabla\cdot(\rho\mathbf{u}) \\ \nabla\phi + \mathbf{u} \\ -\mathbf{n}\phi \\ \phi[\mathbf{r}, 0] \\ 0 \end{bmatrix} \quad \text{on} \quad \begin{bmatrix} V, \text{all } t \\ V, \text{all } t \\ \Sigma, \text{all } t \\ V, t=0 \\ V, t=\tau \end{bmatrix}, \tag{4.136}$$

$$\tilde{y} = \begin{bmatrix} \psi[\mathbf{r}, t] \\ \mathbf{v}[\mathbf{r}, t] \\ \mathbf{v}[\mathbf{r}, t] \\ \psi[\mathbf{r}, 0] \\ \psi[\mathbf{r}, \tau] \end{bmatrix}, \quad T^*\tilde{y} = \begin{bmatrix} -\dfrac{\partial \psi}{\partial t} - \nabla\cdot(\rho\mathbf{v}) \\ -\nabla\psi + \mathbf{v} \\ \mathbf{n}\psi \\ 0 \\ \psi[\mathbf{r}, \tau] \end{bmatrix} \quad \text{on} \quad \begin{bmatrix} V, \text{all } t \\ V, \text{all } t \\ \Sigma, \text{all } t \\ V, t=0 \\ V, t=\tau \end{bmatrix}, \tag{4.137}$$

together with the inner product

$$(\tilde{x}, \tilde{y}) = \int_0^\tau \int (\phi\psi + \rho\mathbf{u}\cdot\mathbf{v})\,\mathrm{d}V\,\mathrm{d}t + \int_0^\tau \int \rho\mathbf{u}\cdot\mathbf{v}\,\mathrm{d}\Sigma\,\mathrm{d}t + \int \phi\psi\,\mathrm{d}V\Big|_{t=0} + \int \phi\psi\,\mathrm{d}V\Big|_{t=\tau}. \tag{4.138}$$

The problem (4.134) is thus of the form $T\tilde{x} + \tilde{b} = 0$ where $\tilde{b} = [-f \;\; 0 \;\; \mathbf{n}g \;\; -c \;\; 0]^{\mathrm{T}}$. Since

$$(\tilde{x}, T\tilde{x}) = \int_0^\tau \int \rho\mathbf{u}^2\,\mathrm{d}V\,\mathrm{d}t + \frac{1}{2}\int \phi^2\,\mathrm{d}V\Big|_{t=0} + \frac{1}{2}\int \phi^2\,\mathrm{d}V\Big|_{t=\tau} \tag{4.139}$$

the operator T is nonnegative. Any $\tilde{x}$ constructed from functions $\phi[\mathbf{r}, t]$ and $\mathbf{u}[\mathbf{r}, t]$ such that $\phi \neq 0$ in $0 < t < \tau$ will make $(\tilde{x}, T\tilde{x}) = 0$ if $\phi[\mathbf{r}, 0] = \phi[\mathbf{r}, \tau] = 0 \equiv \mathbf{u}$.

It follows that the upper and lower bounds of Theorems 4.11 and 4.13 apply to this version of the heat conduction problem. In this way we can still use an inner product space formulation, and avoid the need to modify the definition of adjointness.

The method outlined here also applies in the presence of mixed boundary conditions. The details of an application to the transient diffusion of liquid through a permeable medium are described in §4.4.

Exercise 4.3

Let $\rho(\mathbf{r})$ be a given function C^1 of position in a fixed two or three dimensional spatial region V with boundary Σ. For a given time interval $0 \leqslant t \leqslant \tau$ with fixed τ, show that for any two sufficiently smooth functions $\phi[\mathbf{r},t]$ and $\psi[\mathbf{r},t]$,

$$\int_0^\tau\int\frac{\partial\psi}{\partial t}\left(\frac{\partial^2\phi}{\partial t^2}-\nabla\cdot(\rho\nabla\phi)\right)dV\,\mathrm{d}t+\int_0^\tau\int\rho\frac{\partial\psi}{\partial t}\frac{\partial\phi}{\partial n}\mathrm{d}\Sigma\,\mathrm{d}t$$
$$+\int\left[\frac{\partial\psi}{\partial t}\frac{\partial\phi}{\partial t}+\rho(\nabla\psi)\cdot(\nabla\phi)\right]\mathrm{d}V\bigg|_{t=0}$$
$$=\int_0^\tau\int\frac{\partial\phi}{\partial t}\left(-\frac{\partial^2\psi}{\partial t^2}+\nabla\cdot(\rho\nabla\psi)\right)\mathrm{d}V\,\mathrm{d}t-\int_0^\tau\int\rho\frac{\partial\phi}{\partial t}\frac{\partial\psi}{\partial n}\mathrm{d}\Sigma\,\mathrm{d}t$$
$$+\int\left[\frac{\partial\phi}{\partial t}\frac{\partial\psi}{\partial t}+\rho(\nabla\phi)\cdot(\nabla\psi)\right]\mathrm{d}V\bigg|_{t=\tau}.$$

Upper and lower bounds associated with certain versions of the wave equation can be obtained, starting from this formula as a statement of adjointness (cf. (4.131)) – see Herrera and Sewell (1978).

4.4. Diffusion of liquid through a porous medium

(i) Governing equations

This problem arises in the transient flow of ground water. We adopt governing equations described by Herrera and Bielak (1976), but we obtain upper and lower bounds in a different way to them.

A fixed three dimensional spatial region V with boundary $\Sigma=\Sigma_1+\Sigma_2$ is occupied by a medium possessing a symmetric positive definite permeability tensor, whose cartesian components k_{ij} are given C^1 functions of position $\mathbf{r}$ (but not time t). We use cartesian coordinates in which the components of $\mathbf{r}$ are x_i $(i=1,2,3)$. The medium is assumed to be saturated by a liquid (e.g. water or oil) which moves unsteadily with velocity components v_i under a pressure p. The functions $v_i[\mathbf{r},t]$ and $p[\mathbf{r},t]$ are to be determined from the following governing conditions, in a time interval $0\leqslant t\leqslant\tau$ with fixed τ.

Darcy's law in V for all t:	$v_i=-k_{ij}\dfrac{\partial p}{\partial x_j}.$	(4.140)
Continuity in V for all t:	$s\dfrac{\partial p}{\partial t}+\dfrac{\partial v_i}{\partial x_i}=c.$	(4.141)
Pressure assigned on Σ_1 for all t:	$p=a.$	(4.142)
Flux assigned on Σ_2 for all t:	$n_iv_i=b.$	(4.143)
Initial pressure assigned in V at $t=0$:	$p=h.$	(4.144)

Here s is a given positive function of position only, called the storage function, h is a given function of position, and a, b, and c are given functions of position and time. The components n_i of the outward unit normal to Σ are given functions of position there.

The system (4.140)–(4.144) has a structure reminiscent of (4.134) or, if we eliminate the v_i at the outset in favour of p, of (4.123). We treat it on terms similar to (4.134), since Darcy's law offers a physical choice of intermediate variable, namely $k_{ij}^{-1}v_j$ where k_{ij}^{-1} denotes the typical component of the inverse of the 3×3 matrix k_{ij} (not the inverse of the typical component). This inverse is also positive definite.

(ii) Adjointness

As recommended in the procedure (§3.9) and illustrated in the catalogue (§3.10), we begin by seeking an appropriate statement of integration by parts, being guided in this case by (4.135). Let $p[\mathbf{r},t]$, $v_i[\mathbf{r},t]$ and $q[\mathbf{r},t]$, $w_i[\mathbf{r},t]$ be any two sets of piecewise C^1 functions, each set consisting of a scalar and the components of a vector, and neither set necessarily satisfying the governing conditions. Then

$$\begin{aligned}
&\int_0^\tau \int \left[q\left(s\frac{\partial p}{\partial t} + \frac{\partial v_i}{\partial x_i} \right) + w_i\left(\frac{\partial p}{\partial x_i} + k_{ij}^{-1}v_j \right) \right] \mathrm{d}V\,\mathrm{d}t \\
&\quad - \int_0^\tau \left[\int w_i n_i p \,\mathrm{d}\Sigma_1 + \int q n_i v_i \,\mathrm{d}\Sigma_2 \right] \mathrm{d}t + \int sqp\,\mathrm{d}V \bigg|_{t=0} \\
&= \int_0^\tau \int \left[p\left(-s\frac{\partial q}{\partial t} - \frac{\partial w_i}{\partial x_i} \right) + v_i\left(-\frac{\partial q}{\partial x_i} + k_{ij}^{-1}w_j \right) \right] \mathrm{d}V\,\mathrm{d}t \\
&\quad + \int_0^\tau \left[\int v_i n_i q \,\mathrm{d}\Sigma_1 + \int p n_i w_i \,\mathrm{d}\Sigma_2 \right] \mathrm{d}t + \int spq\,\mathrm{d}V \bigg|_{t=\tau}.
\end{aligned} \tag{4.145}$$

We can regard this as a statement of adjointness of the type $(\tilde{u}, T\tilde{x}) = (\tilde{x}, T^*\tilde{u})$ if we introduce the following composite elements and operators. Here $\tilde{x}$ is unconnected with the cartesian spatial coordinates x_i. Define

$$\tilde{x} = \begin{bmatrix} p[\mathbf{r},t] \\ v_i[\mathbf{r},t] \\ v_i[\mathbf{r},t] \\ p[\mathbf{r},t] \\ p[\mathbf{r},0] \\ p[\mathbf{r},\tau] \end{bmatrix}, \quad T\tilde{x} = \begin{bmatrix} s\dfrac{\partial p}{\partial t} + \dfrac{\partial v_i}{\partial x_i} \\ \dfrac{\partial p}{\partial x_i} + k_{ij}^{-1}v_j \\ -n_i p \\ -n_i v_i \\ p \\ 0 \end{bmatrix} \quad \text{on} \quad \begin{bmatrix} V, \forall t \\ V, \forall t \\ \Sigma_1, \forall t \\ \Sigma_2, \forall t \\ V, t=0 \\ V, t=0 \end{bmatrix}, \tag{4.146}$$

$$\tilde{u}=\begin{bmatrix} q[\mathbf{r},t] \\ w_i[\mathbf{r},t] \\ w_i[\mathbf{r},t] \\ q[\mathbf{r},t] \\ q[\mathbf{r},0] \\ q[\mathbf{r},\tau] \end{bmatrix}, \quad T^*\tilde{u}=\begin{bmatrix} -s\dfrac{\partial q}{\partial t}-\dfrac{\partial w_i}{\partial x_i} \\ -\dfrac{\partial q}{\partial x_i}+k_{ij}^{-1}w_j \\ n_i q \\ n_i w_i \\ 0 \\ q \end{bmatrix} \quad \text{on} \quad \begin{bmatrix} V,\forall t \\ V,\forall t \\ \Sigma_1,\forall t \\ \Sigma_2,\forall t \\ V,t=0 \\ V,t=\tau \end{bmatrix}, \tag{4.147}$$

together with the inner product

$$(\tilde{x},\tilde{u})=\int_0^\tau\left[\int(pq+v_iw_i)\mathrm{d}V+\int v_iw_i\,\mathrm{d}\Sigma_1+\int pq\,\mathrm{d}\Sigma_2\right]\mathrm{d}t$$
$$+\int spq\,\mathrm{d}V\Big|_{t=0}+\int spq\,\mathrm{d}V\Big|_{t=\tau}. \tag{4.148}$$

Then we can see that (4.145) is $(\tilde{u},T\tilde{x})=(\tilde{x},T^*\tilde{u})$ as required.

(iii) Governing equations as $T\tilde{x}+\tilde{b}=0$

Equations (4.140)–(4.144) are equivalent to $T\tilde{x}+\tilde{b}=0$ with the T in (4.146) if we choose

$$\tilde{b}=[-c \quad 0 \quad n_ia \quad b \quad -h \quad 0]^{\mathrm{T}}. \tag{4.149}$$

The Σ_1 equation is $n_i(a-p)=0$, which is equivalent to $p=a$. Darcy's law is equivalent to $\partial p/\partial x_i+k_{ij}^{-1}v_j=0$ since the matrix k_{ij} is non-singular, being positive definite.

(iv) Nonnegative T

Since, from (4.145),

$$(\tilde{x},T\tilde{x})=\int_0^\tau\int k_{ij}^{-1}v_iv_j\,\mathrm{d}V\,\mathrm{d}t+\frac{1}{2}\int sp^2\,\mathrm{d}V\Big|_{t=0}+\frac{1}{2}\int sp^2\,\mathrm{d}V\Big|_{t=\tau} \tag{4.150}$$

the operator T is nonnegative. Any $\tilde{x}$ constructed from functions $p[\mathbf{r},t]$ and $v_i[\mathbf{r},t]$ such that $p\neq 0$ in $0<t<\tau$ will make $(\tilde{x},T\tilde{x})=0$ if $p[\mathbf{r},0]=p[\mathbf{r},\tau]=0$ and $v_i[\mathbf{r},t]\equiv 0$.

All the conditions are therefore available to establish not only a stationary property as in Theorem 4.9, but also the upper and lower bounds of Theorem 4.11. In particular the sum and difference change of variable (4.103) can be adopted, and Theorem 4.13 can be used. To put these remarks into focus we quote a particular result.

(v) Generating saddle functional

We have shown that the given problem is $T\tilde{x}+\tilde{b}=0$ and we choose $\tilde{a}=\tilde{b}$ so that the adjoint problem is $T^*\tilde{u}+\tilde{b}=0$. The generating saddle functional is then

$$\begin{aligned}L[\tilde{x},\tilde{u}] &= (\tilde{u},T\tilde{x})+(\tilde{b},\tilde{x}+\tilde{u})\\ &= \int_0^\tau\int\left[q\left(s\frac{\partial p}{\partial t}+\frac{\partial v_i}{\partial x_i}-c\right)+w_i\left(\frac{\partial p}{\partial x_i}+k_{ij}^{-1}v_j\right)-pc\right]\mathrm{d}V\,\mathrm{d}t\\ &\quad+\int_0^\tau\int n_i[av_i+(a-p)w_i]\,\mathrm{d}\Sigma_1\,\mathrm{d}t+\int_0^\tau\int[bp+q(b-n_iv_i)]\,\mathrm{d}\Sigma_2\,\mathrm{d}t\\ &\quad+\int s[ph+q(p-h)]\,\mathrm{d}V\bigg|_{t=0}\end{aligned}\tag{4.151}$$

from (4.90) and (4.145)–(4.149).

Any solution of $T\tilde{x}=-\tilde{b}=T^*\tilde{u}$ will make this functional $L[\tilde{x},\tilde{u}]$ stationary, for example by Theorem 4.9. Applications of stationary principles related to this one have been given (cf. Neumann and Witherspoon, 1971).

(vi) Upper and lower bounds

Choosing the sum and difference change of variables (4.103) rewrites the governing conditions and their adjoint problem in the form (4.105), which is as follows in the present case.

$$\begin{bmatrix} s\dfrac{\partial(p+q)}{\partial t}+\dfrac{\partial(v_i+w_i)}{\partial x_i}\\ \dfrac{\partial(p+q)}{\partial x_i}+k_{ij}^{-1}(v_j-w_j)\\ -n_i(p+q)\\ -n_i(v_i+w_i)\\ p\\ -q\end{bmatrix}=0\quad\text{on}\quad\begin{bmatrix}V,\forall t\\ V,\forall t\\ \Sigma_1,\forall t\\ \Sigma_2,\forall t\\ V,t=0\\ V,t=\tau\end{bmatrix},\tag{4.152α}$$

$$\begin{bmatrix} s\dfrac{\partial(p-q)}{\partial t}+\dfrac{\partial(v_i-w_i)}{\partial x_i}-2c\\ \dfrac{\partial(p-q)}{\partial x_i}+k_{ij}^{-1}(v_j+w_j)\\ n_i(2a+q-p)\\ 2b-n_i(v_i-w_i)\\ p-2h\\ q\end{bmatrix}=0\quad\text{on}\quad\begin{bmatrix}V,\forall t\\ V,\forall t\\ \Sigma_1,\forall t\\ \Sigma_2,\forall t\\ V,t=0\\ V,t=\tau\end{bmatrix}.\tag{4.152β}$$

A specification of α and β labelling is thereby given which allows

Theorem 4.11 to be applied, and also Theorem 4.13 if we avoid a choice of $\tilde{u}_\alpha$ which makes $(\tilde{u}_\alpha, T\tilde{u}_\alpha) = 0$. The bounds (4.107) then apply (with tildes for the present notation). We gave a simple illustration of such calculations in §4.3(vii).

4.5. Comparison methods

(i) The basic idea

If a given problem is hard to solve we may be able to introduce another *comparison problem*, for which certain quantities are more easily calculated and can be used to bound analogous quantities in the given problem.

We have already met the idea in §2.6(iii), where (2.98) proves that bifurcation in certain problems for a real elastic/plastic solid (2.84) with (2.85) cannot precede a corresponding bifurcation in a physically fictitious comparison solid (2.84) with, instead of (2.85), all $\mu_\gamma = 0$ and all λ_γ unrestricted in sign. We shall return to similar comparison solids for plasticity in §§5.8(i) and 5.10(vii).

Another type of comparison problem was employed to estimate the elastic cantilever beam deflection in §4.2(iv).

A third type of comparison problem has been stimulated by the work of Hashin and Shtrikman (1962) in elasticity, where a heterogeneous body is analysed by comparing it with a homogeneous and isotropic body. This has attracted several reformulations, in ways which look superficially different from each other, including those of Hill (1963), Walpole (1974) and Talbot and Willis (1985). Applications include estimation of the strength of polycrystals and of composite materials. Smith (1978) has developed Walpole's results for linear problems, and extended it (1981, 1985*a*, *b*) by using our saddle functional framework to nonlinear problems, including applications to Duffing's equation.

In this section we have a limited objective, namely to explain the original upper and lower bounds of Hashin and Shtrikman (1962) in a simple but generalized way.

(ii) Example of the general setting

The stationary value problem (1.53), for a general linear problem, puts to zero the gradients (1.66) and (3.60) in the finite and infinite dimensional cases respectively, for example as in (1.44). Here we consider the special cases in which $a = B = 0$. However, we enlarge the problem, and also allow for compound space elements in the sense of (3.81), for example, by considering

$$T^*\tilde{u} = \tilde{y}, \qquad T\tilde{x} = -\tilde{b}, \qquad y = Ax. \tag{4.153}$$

An illustration in a spatial region V with surface $\Sigma_1 + \Sigma_2$ is provided by vector fields $\mathbf{y}$ and $\mathbf{x}$ and a scalar field u which satisfy

$$\nabla u = \mathbf{y}, \qquad -\nabla\cdot\mathbf{x} = b, \qquad \mathbf{y} = A\mathbf{x} \quad \text{in} \quad V,$$
$$u = c \quad \text{on} \quad \Sigma_2, \qquad \mathbf{n}\cdot\mathbf{x} = a \quad \text{on} \quad \Sigma_1. \tag{4.154}$$

Here b, a and c are given scalars in V and on Σ_1 and Σ_2 respectively, and A is a given real symmetric 3×3 matrix of 'compliance' tensor components, having an inverse A^{-1} which is the matrix of 'moduli'. The problem (4.154) can represent, in particular, the electrostatic problem used by Willis (1984) as a vehicle for his applications of comparison theorems; in that case $\mathbf{x}, \mathbf{y}$ and $-u$ are the electric displacement, field and potential respectively, and A^{-1} is the dielectric matrix. We define three inner product spaces, with elements, operators and inner products as follows:

$$\tilde{\mathbf{x}} = \begin{bmatrix} \mathbf{x}(V) \\ \mathbf{x}(\Sigma_2) \end{bmatrix}, \qquad T^*\tilde{u} = \begin{bmatrix} \nabla u \\ -\mathbf{n}u \end{bmatrix},$$
$$\tilde{u} = \begin{bmatrix} u(V) \\ u(\Sigma_1) \end{bmatrix}, \qquad T\tilde{x} = \begin{bmatrix} -\nabla\cdot\mathbf{x} \\ \mathbf{n}\cdot\mathbf{x} \end{bmatrix}, \tag{4.155}$$
$$\mathbf{x} = \mathbf{x}(V) \qquad \overline{\mathbf{x}, \mathbf{y}} = \int \mathbf{x}\cdot\mathbf{y}\,\mathrm{d}V,$$
$$(\tilde{\mathbf{x}}, \tilde{\mathbf{y}}) = \int \mathbf{x}\cdot\mathbf{y}\,\mathrm{d}V + \int \mathbf{x}\cdot\mathbf{y}\,\mathrm{d}\Sigma_2,$$
$$\langle \tilde{u}, \tilde{v} \rangle = \int uv\,\mathrm{d}V + \int uv\,\mathrm{d}\Sigma_1.$$

Here $\mathbf{n}$ is the unit outward normal to Σ, as usual, and with these definitions the problem (4.154) can be written as

$$T^*\tilde{u} = \tilde{\mathbf{y}}, \qquad T\tilde{\mathbf{x}} = -\tilde{b}, \qquad \mathbf{x} = A^{-1}\mathbf{y} \tag{4.156}$$

if we choose

$$\tilde{\mathbf{y}} = \begin{bmatrix} \mathbf{y}(V) \\ -\mathbf{n}c(\Sigma_2) \end{bmatrix}, \qquad \tilde{b} = \begin{bmatrix} -b(V) \\ -a(\Sigma_1) \end{bmatrix}.$$

Equations (4.156) are obtainable by setting to zero the gradients of the functional

$$L[\tilde{\mathbf{x}}, \tilde{u}, \mathbf{y}] = (\tilde{\mathbf{x}}, T^*\tilde{u} - \tilde{\mathbf{y}}) + \langle \tilde{b}, \tilde{u} \rangle + \int U[\mathbf{y}]\mathrm{d}V$$
$$= \int [-u\nabla\cdot\mathbf{x} - \mathbf{x}\cdot\mathbf{y} + U[\mathbf{y}] - bu]\mathrm{d}V$$
$$+ \int u(\mathbf{n}\cdot\mathbf{x} - a)\,\mathrm{d}\Sigma_1 + \int \mathbf{n}\cdot\mathbf{x}c\,\mathrm{d}\Sigma_2, \tag{4.157}$$

where the given scalar function $U[\mathbf{y}] = \frac{1}{2}\mathbf{y}\cdot A^{-1}\mathbf{y}$. A similar example in elasticity is contained in Exercises 3.3.3 and 5.3.5; in that case $\mathbf{x}, \mathbf{y}$ and u correspond to stress, strain and displacement respectively. Material nonlinearities can be introduced by replacing $\frac{1}{2}\mathbf{y}\cdot A^{-1}\mathbf{y}$ by a nonquadratic strain energy density function $U[\mathbf{y}]$. The domain of $L[\tilde{\mathbf{x}}, \tilde{u}, \mathbf{y}]$ is the product of the three inner product spaces built from piecewise C^1 $\mathbf{x}$ and $\mathbf{u}$ and from integrable $\mathbf{y}$, since this is enough to justify integration by parts used in obtaining the formula

$$\delta L = (\delta\tilde{\mathbf{x}}, T^*\tilde{u} - \tilde{\mathbf{y}}) + \langle \delta\tilde{u}, T\tilde{\mathbf{x}} + \tilde{b} \rangle + \overline{\delta\mathbf{y}, U'[\mathbf{y}] - \mathbf{x}}. \qquad (4.158)$$

The gradient $U'[\mathbf{y}] = A^{-1}\mathbf{y}$ in the linear problem, and in the last term we have used the fact that the Σ_2 entry in $\tilde{\mathbf{y}}$ is a constant, so that $(\delta\tilde{\mathbf{y}}, \tilde{\mathbf{x}}) = \overline{\delta\mathbf{y}, \mathbf{x}}$.

(iii) Easy and hard problems

With the explicit example of §(ii) before us, we now return to the general linear problem (4.153), in which T and T^* are any true adjoint operators, A is a given operator which we shall regard as a matrix, and x and y are obtained from compound $\tilde{x}$ and $\tilde{y}$ by deleting boundary entries from the latter.

We imagine that an 'easy' problem can be defined in which A is denoted by an invertible symmetric A_e, having a rather simple spatial dependence (e.g. for an isotropic and homogeneous medium), so that a solution x_e, y_e, u_e can be found. The subscript e stands for 'easy', and the *easy problem* therefore has the properties

$$T^*\tilde{u}_e = \tilde{y}_e, \qquad T\tilde{x}_e = -\tilde{b}, \qquad y_e = A_e x_e. \qquad (4.159)$$

Next we suppose that we have to solve a 'hard' problem, in the same spatial region and with the same data, except the A_e is replaced by an invertible symmetric A_h, having a more complicated spatial dependence (e.g. for an anisotropic or inhomogeneous medium). The subscript h stands for 'hard', and we suppose that the *hard problem* has a solution x_h, y_h, u_h with the properties

$$T^*\tilde{u}_h = \tilde{y}_h, \qquad T\tilde{x}_h = -\tilde{b}, \qquad y_h = A_h x_h. \qquad (4.160)$$

The operators T and T^* and the data $\tilde{b}$ are the same in (4.159) and (4.160).

(iv) Difference variables

Next, extending Hashin and Shtrikman's (1962) approach for elasticity to our abstract context, we rewrite the hard problem by defining new variables

$$\tilde{u}' = \tilde{u}_h - \tilde{u}_e, \quad \tilde{y}' = \tilde{y}_h - \tilde{y}_e, \quad p = x_h - A_e^{-1} y_h, \qquad (4.161)$$

where p is called the 'polarization', and the primes denote the difference of the variables, not a derivative. We denote the difference of the hard and easy 'moduli' by

$$K = A_h^{-1} - A_e^{-1}. \tag{4.162}$$

If we also define

$$x' = p + A_e^{-1} y' = x_h - x_e \tag{4.163}$$

the hard problem (4.160) can be expressed in terms of the difference variables as

$$\begin{aligned} T^*\tilde{u}' &= \tilde{y}', \quad (\alpha) \\ T\tilde{x}' &= 0, \quad (\beta) \\ p = K(y' + y_e) \quad \text{or} \quad x' &= A_h^{-1} y' + K y_e. \quad (\beta) \end{aligned} \tag{4.164}$$

(v) Stationary and saddle properties

Next we suppose that every $\tilde{y}'$ has the property

$$(\tilde{x}', \tilde{y}') = \overline{x', y'}. \tag{4.165}$$

In practice this will happen because $\tilde{y}'$ has a zero boundary entry, as would be the case for the difference of hard and easy $\tilde{y}$ in (4.156). Our standard procedure then leads, via a generalized Hamiltonian, to the fact that the three governing conditions in (4.164) are the gradients of the functional

$$L[\tilde{x}', \tilde{u}', y'] = (\tilde{x}', T^*\tilde{u}' - \tilde{y}') + \tfrac{1}{2}\overline{y', A_h^{-1} y'} + \overline{y', K y_e}. \tag{4.166}$$

Therefore any solutions of (4.164) makes this functional stationary for unconstrained variations. In fact any pair of plus and minus points, whether or not they are close together, allow us to define the saddle quantity

$$\begin{aligned} S &= L_+ - L_- - \left(\tilde{x}_+' - \tilde{x}_-', \left.\frac{\partial L}{\partial \tilde{x}'}\right|_+ \right) \\ &\quad - \left\langle \tilde{u}_+' - \tilde{u}_-', \left.\frac{\partial L}{\partial \tilde{u}'}\right|_- \right\rangle - \overline{y_+' - y_-', \left.\frac{\partial L}{\partial y'}\right|_-} \\ &= \tfrac{1}{2}\overline{(y_+' - y_-'), A_h^{-1}(y_+' - y_-')}. \end{aligned} \tag{4.167}$$

If we now assume that

$$A_h^{-1} \text{ is positive definite} \tag{4.168}$$

it follows that $L[\tilde{x}', \tilde{u}', y']$ is a saddle functional weakly concave in $\tilde{x}'$, convex in $\tilde{u}'$ and strictly convex in y'. This justifies the α and β designations written into (4.164), and upper and lower bounds follow by our standard procedure.

Theorem 4.14
When (4.168) *applies, the value* L_0 *given to* (4.166) *by any solution* $\tilde{x}_0', \tilde{u}_0', y_0'$ *(say) of* (4.164) *is bounded by* $L_\alpha \geqslant L_0 \geqslant L_\beta$ *where*

$$L_\alpha = \tfrac{1}{2}\overline{D^*u_\alpha', A_h^{-1}D^*u_\alpha'} + \overline{D^*u_\alpha', Ky_e}, \tag{4.169}$$

$$L_\beta = -\tfrac{1}{2}\overline{(x_\beta' - Ky_e), A_h(x_\beta' - Ky_e)}. \tag{4.170}$$

Here u_α' *is any member of the domain of the formally adjoint part* D^* *of* T^*, *and* x_β' *belongs to any* $\tilde{x}_\beta'$ *satisfying* $T\tilde{x}_\beta' = 0$.

Proof
Our standard method (cf. (1.38)) applied to $S \geqslant 0$ in (4.167) shows that the upper bound will have the value of

$$L - \left(\tilde{x}', \frac{\partial L}{\partial \tilde{x}'}\right) = \tfrac{1}{2}\overline{y', A_h^{-1}y'} + \overline{y', Ky_e}$$

for any solution $\tilde{y}_\alpha' = \begin{bmatrix} D^*u_\alpha' \\ 0 \end{bmatrix}$ of (4.164α), where the zero represents the boundary entry consistent with (4.165), and D^* is the formally adjoint part of T^*, so that $y_\alpha' = D^*u_\alpha'$. Likewise (cf. (1.39)) the lower bound will have the value of

$$L - \left\langle \tilde{u}', \frac{\partial L}{\partial \tilde{u}'} \right\rangle - \overline{y', \frac{\partial L}{\partial y'}} = \tfrac{1}{2}\overline{y', A_h^{-1}y'}$$

for any solution of (4.164β), i.e. for any $y_\beta' = A_h(x_\beta' - Ky_e)$ where $T\tilde{x}_\beta' = 0$.

□

(vi) Hashin–Shtrikman functional

Hashin and Shtrikman (1962) addressed the version of equations (4.164) which applies to classical elasticity. That version is obtainable by translating the example of §(ii) above into Exercise 5.3.5. They announce, for consideration, a functional whose generalization to the context of (4.164) is

$$W[p, y'] = -\tfrac{1}{2}\overline{p, K^{-1}p} + \tfrac{1}{2}\overline{p, y'} + \overline{p, y_e}. \tag{4.171}$$

It is assumed that the inverse K^{-1} of K exists and is symmetric.

Theorem 4.15 (Stationary property)
Among the set of points which satisfy the constraints $(4.164)_1$ *and* $(4.164)_2$, $W[p, y']$ *is stationary when* $(4.164)_3$ *is also satisfied.*

Proof
Let p_+, y_+' and p_-, y_-' be any two points, not necessarily close together, in the domain of $W[p, y']$. Then

$$W[p_+,y_+'] - W[p_-,y_-'] - \overline{p_+ - p_-, y_-' + y_e' - K^{-1}p_-}$$
$$= -\tfrac{1}{2}\overline{(p_+ - p_-), K^{-1}(p_+ - p_-)} + \tfrac{1}{2}\overline{p_+ - p_-, y_+' - y_-'} + \tfrac{1}{2}\overline{p_-, y_+'}$$
$$- \tfrac{1}{2}\overline{p_+, y_-'}.$$

Using (4.165) for y_+ and y_-, and the two constraints $(4.164)_{1,2}$,

$$\overline{p_+ - p_-, y_+' - y_-'} = (\tilde{p}_+ - \tilde{p}_-, \tilde{y}_+ - \tilde{y}_-) = \langle T(\tilde{p}_+ - \tilde{p}_-), \tilde{u}_+ - \tilde{u}_- \rangle$$
$$= -\overline{(y_+' - y_-'), A_e^{-1}(y_+' - y_-')}$$

and

$$\overline{p_-, y_+'} + \overline{p_+, y_-'} = (\tilde{p}_-, T^*\tilde{u}_+') - (\tilde{p}_+, T^*\tilde{u}_-')$$
$$= \overline{y_-', A_e^{-1}y_+'} - \overline{y_+', A_e^{-1}y_-'} = 0,$$

where symmetry of A_e^{-1} is used in the last result.
Therefore

$$W_+ - W_- - \overline{p_+ - p_-, y_-' + y_e - K^{-1}p_-}$$
$$= -\tfrac{1}{2}\overline{(p_+ - p_-), K^{-1}(p_+ - p_-)} + \tfrac{1}{2}\overline{(y_+' - y_-'), A_e^{-1}(y_+' - y_-')}. \tag{4.172}$$

From this we see that, for points which are close together, the difference $W_+ - W_-$ is of second order when the minus point satisfies $(4.164)_3$ in the form $y' + y_e = K^{-1}p$. □

Theorem 4.16 (Lower bound)
Suppose that

$$\textit{both } K \textit{ and } A_e^{-1} \textit{ are positive definite.} \tag{4.173}$$

Then any value of $W[p, y']$ *for which the two constraints* $(4.164)_{1,2}$ *are satisfied is a lower bound to a value of* $W[p, y']$ *for which all of* (4.164) *are satisfied.*

Proof
This is an immediate consequence of (4.172) by choosing the minus point to satisfy all of (4.164), and the plus point to satisfy the two constraints $(4.164)_{1,2}$. □

Theorem 4.17 (Upper bound)
Suppose, instead of (4.173), *that*

$$\textit{both} -K \textit{ and } A_e^{-1} \textit{ are positive definite,} \tag{4.174}$$

together with (4.168). *Then any value of* $W[p, y']$ *for which the two constraints* $(4.164)_{1,2}$ *are satisfied is an upper bound to a value of* $W[p, y']$ *for which all of* (4.164) *are satisfied.*

Proof
First we note that, for any pair of distinct plus and minus points satisfying the constraints $(4.164)_{1,2}$, and using (4.163) and (4.165),

$$\begin{aligned}\overline{(p_+ - p_-), A_e(p_+ - p_-)} &= \overline{(x_+' - x_-'), A_e(x_+' - x_-')} \\ &\quad + \overline{(y_+' - y_-'), A_e^{-1}(y_+' - y_-')} \\ &\quad - 2(\tilde{x}_+' - \tilde{x}_-', T^*(\tilde{u}_+ - \tilde{u}_-)) \\ &> \overline{(y_+' - y_-'), A_e^{-1}(y_+' - y_-')}\end{aligned}$$

since A_e will also be positive definite.

From (4.172) we then have

$$\begin{aligned}W_+ - W_- - \overline{p_+ - p_-, y_-' + y_e - K^{-1}p_-} \\ > -\tfrac{1}{2}\overline{(p_+ - p_-), (K^{-1} + A_e)(p_+ - p_-)}.\end{aligned}$$

The theorem follows, again by choosing the minus point to satisfy all of (4.164) and the plus point to satisfy the two constraints $(4.164)_{1,2}$, because (4.174) with (4.168) is equivalent to $K^{-1} + A_e$ negative definite. This last result in matrix algebra is established in Theorem 4.18 next. □

Theorem 4.18

Let $K = P^{-1} - Q^{-1}$, where P and Q are each real symmetric positive definite $n \times n$ matrices. Then when K^{-1} exists, K is negative definite if and only if $K^{-1} + Q$ is negative definite (and similarly for positive definiteness).

Proof
Since P is positive definite we can construct, for example by spectral representation, a real and symmetric positive definite square root matrix $P^{\frac{1}{2}}$ such that $P = P^{\frac{1}{2}}P^{\frac{1}{2}}$, which itself will have an inverse $P^{-\frac{1}{2}}$. Then (using >0 to denote positive definite)

$$\begin{aligned}-K > 0 &\Leftrightarrow Q^{-1} - P^{-1} > 0 \\ &\Leftrightarrow P^{\frac{1}{2}}(Q^{-1} - P^{-1})P^{\frac{1}{2}} > 0 \\ &\Leftrightarrow P^{\frac{1}{2}}Q^{-1}P^{\frac{1}{2}} - I > 0 \\ &\Leftrightarrow I - P^{-\frac{1}{2}}QP^{-\frac{1}{2}} > 0 \\ &\Leftrightarrow P - Q > 0 \\ &\Leftrightarrow (P - Q)^{-1} > 0 \\ &\Leftrightarrow Q(P - Q)^{-1}Q > 0,\end{aligned}$$

where the hypothesis $Q > 0$ is used at the fourth and seventh steps. But

$$\begin{aligned}Q(P - Q)^{-1}Q &= Q(P - Q)^{-1}PP^{-1}Q = (P^{-1}(P - Q)Q^{-1})^{-1}P^{-1}Q \\ &= (Q^{-1} - P^{-1})^{-1}P^{-1}Q = -K^{-1}P^{-1}Q\end{aligned}$$

$$= -K^{-1}(I + (P^{-1} - Q^{-1})Q)$$
$$= -(K^{-1} + Q). \qquad \square$$

Theorems 4.15–4.17 generalize the essence of the results of Hashin and Shtrikman (1962). They have shown how their results can be used to estimate the moduli of heterogeneous materials, including electric and magnetic properties of multiphase materials, and the viscosity coefficients of flud mixtures. A review of some of these results is given by Willis (1983).

5

Mechanics of solids and fluids

5.1. Introduction

In this chapter we select topics from the mechanics of fluids and elastic and plastic solids, and use them to illustrate ideas introduced previously. In §5.2 we review some Legendre transformations of classical thermodynamics in the light of Chapter 2, exhibiting quartets and isolated singularities associated with phase changes. For compressible flow another Legendre transformation, involving a swallowtail flow stress function which plays the role of a complementary energy in the sense of Fig. 2.27, is displayed in Fig. 5.3.

Subsequently we begin each of a number of contexts with a statement of the general boundary value problem, in terms of the *ab initio* kinematic and force variables. Then we obtain, for example, stationary principles and upper and lower bound theorems at that level of generality. This is in contrast to our previous illustrations, which usually began with a differential or integral equation for a single unknown variable, presumed to have resulted from some combination and elimination of *ab initio* variables.

This chapter shows that theories of considerable algebraic complexity can be well organized within the general framework developed in Chapters 1–3. Many other examples in continuum mechanics could be given.

5.2. Thermodynamics

(i) Thermodynamic potential functions

This is the obvious classical illustration of Fig. 2.18. We use certain terminology and notation which have become familiar for the context, in order to aid the reader's comparison of our treatment with what he may already know. We can recommend the freshness of Truesdell and Toupin's review (1960, §§245–255) to any reader. Our essential concern is with the structure of certain equations; their physical interpretation will depend on what 'system' or 'substance' is subsequently associated with the equations.

To be specific, we will suppose that in the case of a moving, continuous medium, a material 'particle' can be defined and all the following variables and equations associated with it. Thus we use the language of a field theory viewpoint.

First we associate with the particle four *thermodynamic potentials*. These are scalar quantities per unit mass of the medium, namely internal energy U, enthalpy H, Gibb's free enthalpy G and Helmholtz's free energy F. Next we assume that, throughout any motion of the particle, each of these scalars is a given function (C^2 for simplicity) of two variables selected from a list of four more quantities associated with the particle, in such a way that the relationships (2.52)–(2.62) implied by regarding Fig. 5.1 as an example of Fig. 2.18 are satisfied. This second list of four quantities (called *state variables*) contains two scalars, the absolute temperature $T > 0$ and the entropy S; and two sets p and v, each containing m scalars, whose inner product will be written $\langle p, v \rangle$ ($= pv$ when $m = 1$). The sets p and v often have mechanical connotations of force-like and kinematic variables respectively. For example, in a 'simple fluid' $m = 1$, we take p as pressure, and v is the volume of unit mass (i.e. $v = 1/\rho$ where ρ is density). With $m > 1$, e.g. in thermoelasticity, the m members of v can specify the local finite deformation (e.g. via six components of a symmetric finite strain tensor), and the m members of $-p$ will be the associated stress components such that their mechanical work increment per unit mass has the value of $-\langle p, \mathrm{d}v \rangle$ in an increment $\mathrm{d}v$ of v. There may be several permissible alternative choices of v and the consequent p, and the effect of switching between such choices has been explored by Hill (1981).

Other contexts may require v and p to have electromagnetic interpretations; or to be the densities or concentrations of the constituents of a mixture; or to be themselves each subdivided into two sets (e.g. in the case of heterogeneous media). For our purpose it will be enough to leave v and p as unspecified single sets, each of $m \geqslant 1$ scalars, and to give some detailed illustrations for different simple fluids.

The equation $U = U[S, v]$ is called the *caloric* equation of state. For each given v it enables the value of the internal energy of the particle to be found when the value of the entropy S is also known (S thus acts as a specifying parameter), whatever the motion of the particle may be. By transcribing Fig. 5.1 into Fig. 2.18, with (2.52)–(2.62), we obtain the standard relations such as

$$T = \frac{\partial U}{\partial S}, \quad -p = \frac{\partial U}{\partial v} \tag{5.1}$$

between the slopes of the thermodynamic potentials and the state variables. A common alternative version of (5.1) using first differentials is

$$\mathrm{d}U = T\,\mathrm{d}S - \langle p, \mathrm{d}v \rangle \tag{5.2}$$

in the present inner product notation. Differentiating $F[T, v]$, $H[S, p]$ and $G[T, p]$ by transcribing (2.54)–(2.62) completes a quartet of pairs of relations like (5.1) which is globally valid. The theory developed in §2.5 makes it clear that each thermodynamic potential may consist of several branches joined together continuously and smoothly, but either regressively (implying global multivaluedness) or progressively (single valuedness). One may think it natural to choose, as starting point in the construction of Fig. 5.1 for a specific case, whichever potential (if any) is globally single valued.

(ii) Second derivatives of thermodynamic potentials

We next consider certain plausible assumptions about second derivatives, designed in part to distinguish strong quartets (Fig. 2.19) from other quartets which together make up the global Fig. 5.1.

A regular branch of the transformation between $U[S, v]$ and $-F[T, v]$ has the properties

$$F_{TT} = -\frac{1}{U_{SS}}, \quad F_{Tv} = \frac{U_{Sv}}{U_{SS}}, \quad F_{vv} = U_{vv} - \frac{U_{vS}U_{Sv}}{U_{SS}} \tag{5.3}$$

by (2.10) and (2.67)–(2.69). Here U_{SS}, U_{Sv}, U_{vS} and U_{vv}, respectively, denote 1×1, $1 \times m$, $m \times 1$ and $m \times m$ matrices of second derivatives indi-

Fig. 5.1. Thermodynamic potential functions related by Legendre transformations.

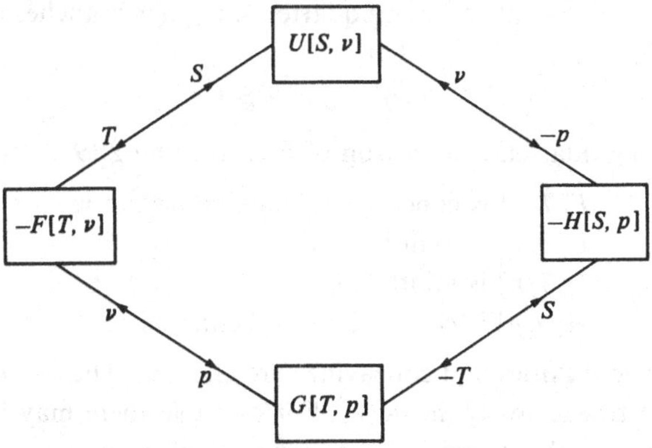

cated by the subscripts. We use this subscript notation also for the matrices of second derivatives of the other three thermodynamic potentials, e.g. of $F[T, v]$ in (5.3).

The assumption $U_{SS} > 0$ is sufficient (but not necessary) to ensure such a regular branch. It allows us to begin from $U[S, v]$ and construct $F[T, v]$ without any intervening singularity; and hence to obtain not only the *thermal* equation of state $p = -\partial F/\partial v$ (relating variables p, v and T which in principle are more easily measurable than S and U), but also the equation $S[T, v] = -\partial F/\partial T$ whose gradient $\partial S/\partial T = -F_{TT}$ appears in the definition of the *specific heat at constant v*

$$c_v = -TF_{TT} > 0. \tag{5.4}$$

By contrast, a branch of $G[T, p]$ is required to construct the equation $S[T, p] = -\partial G/\partial T$ whose gradient $\partial S/\partial T = -G_{TT}$ appears in the definition of the *specific heat at constant p*

$$c_p = -TG_{TT}. \tag{5.5}$$

On a regular branch of the transformation between $-F[T, v]$ and $G[T, p]$ we have $-F_{vv}G_{pp} = I$ by (2.10), and we can apply (2.69) in the form

$$-F_{TT} + G_{TT} = -G_{Tp}F_{vv}G_{pT} = F_{Tv}G_{pp}F_{vT}. \tag{5.6}$$

Defining $m \times m$ inverse matrices $G_{pp}^{-1} = -F_{vv}$ and $F_{vv}^{-1} = -G_{pp}$, we see that the ratio $\gamma = c_p/c_v$ of the specific heats satisfies

$$\gamma - 1 = \frac{1}{F_{TT}}G_{Tp}G_{pp}^{-1}G_{pT} = -\frac{1}{F_{TT}}F_{Tv}F_{vv}^{-1}F_{vT}. \tag{5.7}$$

By identifying the present $-F[T, v]$ with (for $n = 1, m \geqslant 1$) the $X[x_i, u_\gamma]$ of Theorem 2.15(b), we see that the conditions

$$F_{TT} < 0 \text{ with } F_{vv} \text{ positive definite} \tag{5.8}$$

are sufficient to generate a strong quartet of regular branches of the type shown in Fig. 2.19, and such that

$$c_v > 0 \quad \text{and} \quad \gamma \geqslant 1. \tag{5.9}$$

On this strong quartet, comparison of Figs. 5.1 and 2.19 shows that

$F[T, v]$ is concave in T and convex in v,
$U[S, v]$ is jointly convex,
$G[T, p]$ is jointly concave,
$H[S, p]$ is convex in S and concave in p,

where these convexities and concavities are all strict. The conditions (5.8) need not be true globally, however. For example there may be another quartet joined to the strong quartet along singularities.

(iii) Simple fluids

Here and in §5.2 we illustrate some of the foregoing ideas in the simplest case $m = 1$, when p and v become single scalars (instead of sets of scalars) representing pressure and specific volume. We assume $p \geqslant 0$ and $v > 0$.

Consider the van der Waals fluid. This is usually introduced by the familiar thermal equation of state, which in 'reduced' form is

$$\left(p + \frac{3}{v^2}\right)(v - \tfrac{1}{3}) = \tfrac{8}{3}T. \tag{5.10}$$

Here the variables have been nondimensionalized so that the 'critical point' (in physicists' phase-change language) is at $p = v = T = 1$; that is, there is a horizontal inflexion at $p = v = 1$ on the $p[v]$ isotherm defined by fixing $T = 1$ in (5.10). Assuming $v > \frac{1}{3}$ (because $v < \frac{1}{3}$ implies $p < 0$) we rewrite (5.10) as

$$-\frac{\partial F}{\partial v} = p = -\frac{3}{v^2} + \frac{8T}{3v - 1}. \tag{5.11}$$

By integration the free energy function is

$$F[T, v] = f[T] - \frac{3}{v} - \frac{8T}{3}\ln(3v - 1), \tag{5.12}$$

where $f[T]$ is an arbitrary function of T at this stage. At any given T the condition $-\partial^2 F/\partial v^2 = \partial p/\partial v = 0$ for inflexions in $F[v]$ and for stationary points of $p[v]$ is $4Tv^3 = (3v - 1)^2$. This cubic has one root in $v < \frac{1}{3}$, which we ignore because it is of no physical interest (negative $p[v]$ maximum), but in $v > \frac{1}{3}$ either two roots (if $T < 1$) or none (if $T > 1$). It is easy to show that if $\frac{27}{32} < T < 1$ the minimum of $p[v]$ is positive, and it occurs at $v > \frac{2}{3}$. The free enthalpy function $G[T, p]$ has $\partial G/\partial p = v$ and value $G = pv + F$. For fixed $T < 1$ the section $G[p]$ has two cusps which are dual to the inflexions in $F[v]$.

In Fig. 5.2, we show the qualitative shape of $p[v]$, $F[v]$ and inverted $G[p]$ for any fixed T in $\frac{27}{32} < T < 1$, via the example $T = \frac{28}{32}$ with $f(T) = -\frac{1}{2}T^2$. The liquid phase at A extends to B (or even 'metastably' to C if superheating is admitted), while the gaseous phase at F extends to E (or metastably to D if supercooling is admitted). If superheating and supercooling are not admitted, the jump in v (and thus in density $1/v$) between gas and liquid at fixed temperature takes place at the pressure value BE when the Maxwell equal areas convention for phase transition is adopted. That is, the hump D and hollow C in the van der Waals isotherm $p[v]$ are excised by the horizontal line BE having equal areas between itself and C, and between itself and D. In the corresponding energies, the tail $BCDE$ of the swallowtail-shaped function $G[p]$ is removed, leaving a single valued concave function having a jump in slope v which is the jump in specific volume, while the part $BCDE$ of $F[v]$ is removed and replaced by the dotted 'supporting hyperplane' BE whose slope is the negative of the transition pressure (Exercise 5.1.4).

When T is imagined to vary the locus of the singularities at D and C is a curve

$$p-1=-\frac{(v+2)}{v^3}(v-1)^2,\quad T-1=-\frac{(4v-1)}{4v^3}(v-1)^2 \tag{5.13}$$

in p, v, T space, from which parametric descriptions of the corresponding loci in F, v, T space and in G, p, T space can be given. We leave the reader to make detailed comparisons of the functions $p[T,v]$, $F[T,v]$ and $G[T,p]$ with the functions shown in Figs. 2.7 and 2.11. If $f[T]$ in (5.12) is chosen so that $\mathrm{d}^2f/\mathrm{d}T^2<0$ then $F_{TT}<0$. By (5.8) the segments AC and DF of $F[T,v]$ in Fig. 5.2 then belong to the same strong branch which extends, in the T direction, beyond $T>1$; there AC and DF have joined because the inflexions at C and D have disappeared after coalescing at $T=1$. Correspondingly the strong branch of $G[T,p]$ is single valued for $T>1$, but double valued for $T<1$ because of the self-intersection at BE. There is a second pair of dual, but not strong, branches swept out for varying $T<1$ by

Fig. 5.2. Van der Waals functions at fixed $T=\frac{28}{32}=0.875$.

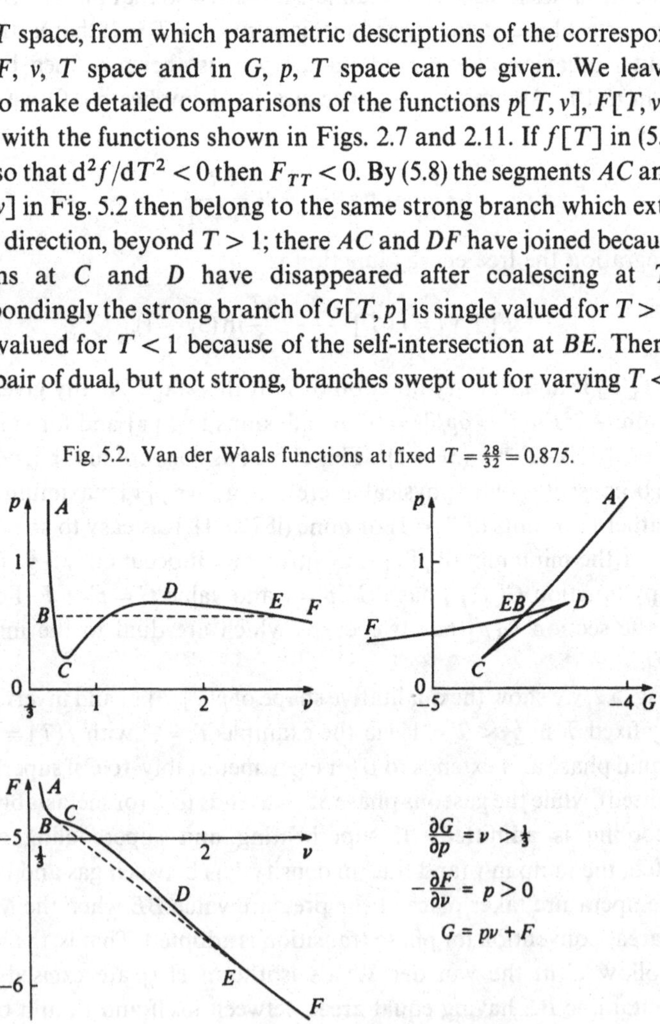

the segments DC in Fig. 5.2. The three dimensional pictures of $F[T, v]$ and $G[T, p]$ are therefore reminiscent of Fig. 2.11, but with $F_{TT} < 0$ replacing the b-linearity of Fig. 2.11, and with (5.13) specifying the shape of isolated singularities where the branches join.

The physics of phase changes has, of course, progressed far beyond van der Waals' equation (see Pippard, 1985), but this is still used as an introductory example in modern textbooks. It is not unknown for these to give the wrong curvature for the branch DC of $G[T, p]$ in Fig. 5.2 (it was correctly given, for example, by Sewell, 1977*b*, 1982).

Exercises 5.1

1. Show that (5.10) can be rewritten in terms of the density as
$$(\rho - 1)^3 + \tfrac{1}{3}[8(T-1) + p - 1](\rho - 1) + \tfrac{1}{3}[8(T-1) - 2(p-1)] = 0,$$
i.e. there is a simple change of variables which expresses van der Waals' equation as the canonical cubic equilibrium surface shown in Figs. 2.7 and 2.8 (Fowler, 1972).
2. Plot the locus of singularities (5.13) in p, v, T space, by home computer or otherwise, and show that its projection on the p, T plane has a cusp at $p = T = v = 1$.
3. Determine the qualitative graphical form of $F[T, v]$ and $G[T, p]$ for the thermal equations of state
$$p(2v-1) = T\exp(2 - 2/Tv) \quad \text{(Dieterici)},$$
$$\left(p + \frac{3}{Tv^2}\right)\left(v - \frac{1}{3}\right) = \frac{8T}{3} \quad \text{(Berthelot)}.$$
4. Prove by integration applied to $p[v]$ in Fig. 5.2 that, when equal areas are subtended by the function above and below the intercept BE, the equation for the hyperplane 'supporting' $F[v]$ emerges.
5. The *ideal gas* has thermal equation of state $pv = RT$ with constant $R > 0$. With the assumption of constant $c_v > 0$, prove by integration that $c_p - c_v = R$ (so that $R = c_v(\gamma - 1)$) and that
$$F[T, v] = -c_v[(\gamma - 1)T \ln v + T(\ln T - 1)] - sT$$
except for an arbitrary additive constant, where s is another constant. By construction of a single (strong) quartet of Legendre transformations show that
$$G[T, p] = -c_v\left[(\gamma - 1)T \ln\frac{RT}{p} + T\ln T - \gamma T\right] - sT,$$
$$U[S, v] = \frac{k}{\gamma - 1}v^{1-\gamma}, \quad k[S] = R\exp\left(\frac{S - s}{c_v}\right),$$
$$H[S, p] = \frac{\gamma}{\gamma - 1}k^{1/\gamma}p^{(\gamma-1)/\gamma}.$$

Check that the eight gradients of these thermodynamic potentials have the values of the corresponding state variables, as Fig. 5.1 requires, e.g. as in (5.1), including $p = k\rho^{\gamma}$. The constant R is the ratio of the universal gas constant ($8.31\,\mathrm{J\,mol^{-1}\,K^{-1}}$) to the molar mass (about 28.96×10^{-3} kg in the case of air).

5.3. Compressible inviscid flow

(i) Thermodynamics

Consider a particle of compressible inviscid fluid which, in subsections (i)–(iii) here, may be in any motion whatever. In particular the motion may be steady or unsteady, rotational or irrotational, with or without shocks. The pressure p, density $\rho = 1/v$, temperature T and entropy S of the particle are related throughout the motion via a given enthalpy function $H[S,p]$, such that $T = \partial H/\partial S$ and $v = \partial H/\partial p$ according to Fig. 5.1 and (2.58) specialized to any simple fluid. At each given S the function $H[p]$ has unique inverse $p[H]$ since $v > 0$; thus we can define a pressure valued function

$$p = p[H,S] \text{ with } \frac{\partial p}{\partial H} = \rho, \quad \frac{\partial p}{\partial S} = -\rho T. \tag{5.14}$$

We assume also that the second derivatives of $H[S,p]$ always satisfy

$$H_{SS} > 0, \quad H_{pp} < 0, \tag{5.15}$$

so that $H[S,p]$ is a saddle function strictly convex in S and strictly concave in p. This means that $H[S,p]$ possesses *only* a strong branch (in the language of §5.2(ii)) and that the specific heats always satisfy (5.9) (see Exercise 5.2.1). The second H-derivative of $p[H,S]$ is

$$p_{HH} = -\rho^3 H_{pp} > 0. \tag{5.16}$$

This permits the construction of a new real valued function

$$c[H,S] = \left(\frac{\partial p/\partial H}{p_{HH}}\right)^{\frac{1}{2}} > 0 \tag{5.17}$$

where we agree to take only the positive root. It can be shown by a study of wave propagation that the value c of (5.17) is the velocity of sound in the fluid. Other but equivalent formulae for c exist in terms of the partial derivatives of other functions (see Exercise 5.2.2). We also assume that $c[H,S]$ satisfies

$$\frac{\partial c}{\partial H} > 0. \tag{5.18}$$

The ideal gas (Exercise 5.1.5) provides a special illustration of (5.14)–

(5.18) in which

$$p[H,S] = k^{1/(1-\gamma)}\left[\frac{(\gamma-1)}{\gamma}H\right]^{\gamma/(\gamma-1)}, \quad c^2 = (\gamma-1)H. \tag{5.19}$$

Here $\gamma > 1$ is constant, so that (5.17) is a half-parabola $c[H]$ with S absent. The ideal gas is frequently used in numerical computation of gas flows, with $\gamma = 1.4$ for air.

All the following general results, however, are *not* restricted to the ideal gas. Instead they are based on simple general properties such as (5.14)–(5.18).

(ii) Another Legendre transformation

Let v be the speed of the fluid particle at a given instant during any motion. Define the *total energy* of the particle to be

$$h = H + \tfrac{1}{2}v^2 = U + \frac{p}{\rho} + \tfrac{1}{2}v^2 \tag{5.20}$$

in value (using $U - H = -pv$ from Fig. 5.1). Treat h as a new variable and insert the decomposition $H = h - \tfrac{1}{2}v^2$ into (5.14). This specifies a new pressure valued function

$$p[v, h, S] \tag{5.21}$$

having gradients

$$\frac{\partial p}{\partial v} = -\rho v, \quad \frac{\partial p}{\partial h} = \rho, \quad \frac{\partial p}{\partial S} = -\rho T. \tag{5.22}$$

We now regard $-p[v, h, S]$ as the starting point of a new Legendre transformation in which v is to be the active variable, and h and S the passive variables. In other words we wish to invert $(5.22)_1$ to express v as a function of h, S and the dual active variable $\rho v = Q$ (say). This scalar Q is the magnitude of the momentum density or mass flow vector $\mathbf{Q} = \rho\mathbf{v}$, where $\mathbf{v}$ is the fluid velocity vector.

The required inverse is the Q-gradient of a dual generating function

$$P[Q, h, S] \tag{5.23}$$

(say) having the values of a *flow stress*

$$P = Qv - (-p) = p + \rho v^2 \tag{5.24}$$

and gradients

$$\frac{\partial P}{\partial Q} = v, \quad \frac{\partial P}{\partial h} = \rho, \quad \frac{\partial P}{\partial S} = -\rho T, \tag{5.25}$$

by recalling the general properties of a Legendre transformation summarized, for example, in (2.52)–(2.56). Thus $(5.22)_1$ and $(5.25)_1$ are mutual inverses.

To examine possible singularities of this Legendre transformation, we use (5.14) and (5.17) to calculate second and third v-derivatives of (5.21), which are

$$p_{vv} = \rho\left(\frac{v^2}{c^2} - 1\right), \quad p_{vvv} = \frac{\rho v}{c^2}\left[3 + \frac{v^2}{c^2}\left(\frac{\partial c^2}{\partial H} - 1\right)\right]. \tag{5.26}$$

Thus we see that $p_{vv} = 0$ and $p_{vvv} > 0$ when $v = c$, invoking (5.18). We therefore say that the Legendre transformation has an isolated *sonic singularity*. When the fluid speed reaches the local sound speed, then at each h and S the function $p[v]$ has an isolated cubic-like inflexion and $P[Q]$ has a corresponding cusp (the latter was shown by Sewell 1978*b* – see Fig. 5.3).

To use this transformation in the boundary value problem (5.32)–(5.36) we need to rewrite it in vector components, by inserting $v = |\mathbf{v}|$ in (5.21). This defines a different function $p[\mathbf{v}, h, S]$ (we use the same letter for it since both functions are pressure valued), leading to another function $P[\mathbf{Q}, h, S]$ having the same values as (5.23), obtainable by inserting $Q = |\mathbf{Q}|$ into (5.23). These new functions satisfy, in particular,

$$\frac{\partial p}{\partial \mathbf{v}} = -\mathbf{Q}, \quad \frac{\partial P}{\partial \mathbf{Q}} = \mathbf{v}, \quad P = p + \mathbf{Q}\cdot\mathbf{v}, \tag{5.27}$$

and, using suffices to denote cartesian components,

$$-\frac{\partial^2 p}{\partial v_i \partial v_j} = \rho\left(\delta_{ij} - \frac{v_i v_j}{c^2}\right), \quad \rho\frac{\partial^2 P}{\partial Q_i \partial Q_j} = \delta_{ij} + \frac{v_i v_j}{c^2 - v^2}. \tag{5.28}$$

This Legendre transformation (5.21)–(5.28) was introduced by Sewell (1963*a*). We emphasize that it stems from a reinterpretation of the function $p[h - \frac{1}{2}v^2, S]$ obtained from $(5.14)_1$ with $(5.20)_1$, which depends on the velocity only via its square. Other informative Legendre transformations can be based on (5.21) or (5.23) with different choices of the active variables (e.g. see Exercise 5.2.7).

(iii) Constitutive surfaces

The functions (5.21) and (5.23) may be represented geometrically as three dimensional hypersurfaces lying in four dimensional spaces, spanned by p, v, h, S in the first case and by P, Q, h, S in the second. Equations (5.22) and (5.25) define hypersurfaces in other spaces spanned by four of Q, ρ, T, v, h, S. Each such surface conveys constitutive properties for the considered particle, i.e. as implied by the thermodynamics before the dynamical equations or jump conditions balancing mass, momentum and energy are introduced. Sewell and Porter (1980) introduced such surfaces and called them *constitutive surfaces*, and calculated the qualitative shape of some of them. Every motion of the particle induces, via the subsequent solution of the dynamical conditions in §(iv), a track on each surface. Such tracks may also jump instantaneously from one part of a constitutive surface to another (for example, when the particle traverses a shock where S jumps).

Plane sections of a constitutive surface can be drawn by fixing values of any two of its four variables. For example, if h and S are fixed in (5.21) and (5.23) the Legendre dual functions $p[v]$ and $P[Q]$ have the shapes shown in Fig. 5.3(a) and (b). There p_0 is stagnation pressure; and, for the particle having the given h and S, v_m is a supposed maximum speed $(2h)^{\frac{1}{2}}$, in which

Fig. 5.3. Constitutive functions at fixed h and S ($M = v/c$).

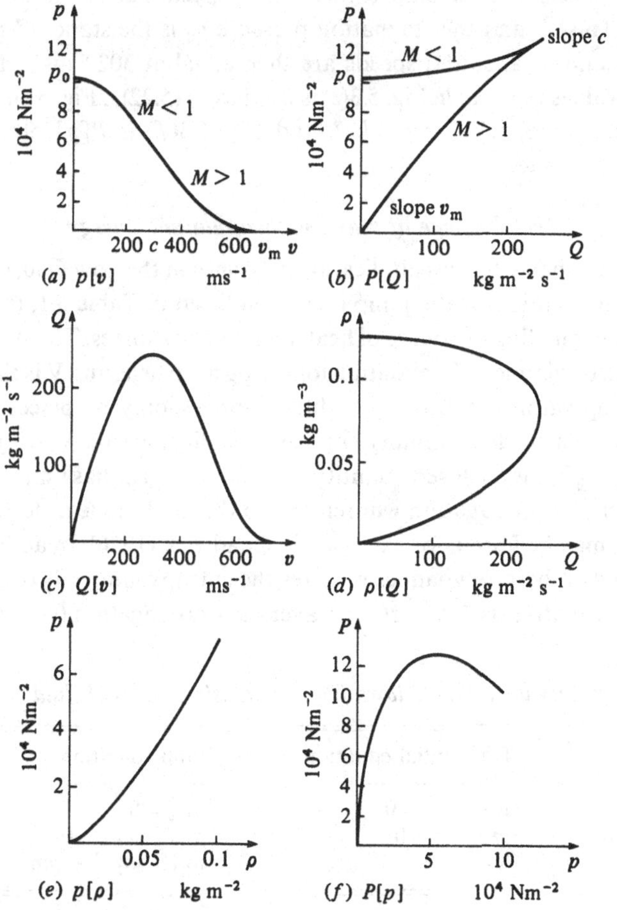

limit p and ρ vanish together. Thus we see that $-p[v]$ and $P[Q]$ each have a convex subsonic branch smoothly joined at the isolated sonic singularity to a concave supersonic branch. If either of h or S are then varied along a third axis, it can be shown under additional light assumptions that the resulting functions (say) $p[v, S]$ and $P[Q, S]$ at fixed h are shaped qualitatively like Fig. 2.11(*a*) and (*b*) in the appropriate octants.

All these qualitative properties, and in particular the shapes of Fig. 5.3(*a*) and (*b*), ensue from very light assumptions such as (5.14)–(5.18), and are not confined to the ideal gas, as we remarked at the end of §(i). Fig. 5.3 itself was calculated by Sewell (1985) for the ideal gas with $\gamma = 1.4$ and, in SI units, $k = 7.08 \times 10^4$ and $h = 2.74 \times 10^5$. This value of k corresponds to air at standard temperature 273 K and standard pressure $1.01 \times 10^5\,\mathrm{Nm}^{-2}$, and effectively fixes a value of entropy S. This value of h corresponds to standard temperature at zero speed. The maximum speed is then $740.3\,\mathrm{ms}^{-1}$, and the stagnation pressure p_0 is the standard pressure. The fluid and local sound speeds are then equal at $302.5\,\mathrm{ms}^{-1}$ for these assigned values of k and h. Fig. 5.3(*c*) is $Q[v]$ from $(5.22)_1$, Fig. 5.3(*d*) is $\rho[Q]$ from $(5.22)_2$, Fig. 5.3(*e*) is $p = k\rho^\gamma$, and Fig. 5.3(*f*) is $P[p]$ obtained by eliminating Q and v.

(iv) Balance of mass, momentum and energy

Each of these balance laws implies, at each place in the flow field, either the differential equation or the jump condition listed in Table 5.1, in the case when there is no flux or source of heat, and no body forces. The superposed dot means total time differentiation following a particle, and ∇ is the spatial gradient operator. The jump condition applies only to places through which an isolated discontinuity surface is passing, across which jumps of amount $[\![\ \ ]\!]$ in the enclosed quantity occur. The surface has unit normal **m**, and velocity of propagation $w\mathbf{m}$ relative to the fluid. A clear derivation of Table 5.1 may be found, for example, in Chadwick (1976). In addition, the second law of thermodynamics requires the entropy jump $[\![S]\!]$ of a particle to be nonnegative as the particle traverses a propagating ($w \neq 0$) surface

Table 5.1 *Pointwise balance laws for compressible inviscid fluid*

Balance	Differential equation	Jump condition
Mass	$\dot{\rho} + \rho\nabla\cdot\mathbf{v} = 0$	$[\![\rho w]\!] = 0$
Momentum	$\rho\dot{\mathbf{v}} + \nabla p = 0$	$[\![\rho w\mathbf{v} - p\mathbf{m}]\!] = 0$
Energy	$\dot{S} = 0$	$[\![\rho w(U + \frac{1}{2}v^2) - p\mathbf{m}\cdot\mathbf{v}]\!] = 0$

(see (258.4) of Truesdell and Toupin, 1960). A shock wave, for example, is defined to be a discontinuity surface across which $[\![\mathbf{m}\cdot\mathbf{v}]\!]\neq 0$. By adjoining to Table 5.1 an appropriate selection of the preceding general thermodynamical relations we reach a fully determined problem. For example, with $1/\rho=\partial H/\partial p$ from Fig. 5.1 we have four equations from which to find ρ, $\mathbf{v}$, p and S as four functions of time and spatial position, with the aid of initial and boundary conditions still to be chosen.

We can use (5.14) and (5.20)–(5.28) to construct equivalent sets of independent equations in terms of alternative lists of variables. For example, instead of p we could use H or h; instead of $\mathbf{v}$ we could use $\mathbf{Q}$, via $(5.27)_2$. Note that $(5.22)_1$ and $(5.22)_2$ are not independent of each other, since both come from $(5.14)_2$. Also, T can always be found a posteriori from $(5.14)_3$ in the absence of heat flux and sources.

There is a wide variety of stationary principles in compressible inviscid fluid mechanics, for many kinds of flow which, for example, may be steady or unsteady, and rotational or irrotational. The large literature leaves significant questions still unanswered. We have space here to consider only certain flows in which the jump conditions are satisfied identically, by seeking solutions in which ρ, $\mathbf{v}$, p and S are, for simplicity, C^1 functions of position. Then we have, by (5.14) and (5.20),

$$\nabla p=\rho\nabla(h-\tfrac{1}{2}v^2)-\rho T\nabla S \tag{5.29}$$

which we can use to rewrite the momentum equation. Any discontinuity surface now has to be treated as a boundary between two such flows.

(v) Simultaneous upper and lower bounds for steady irrotational flow

In this subsection we describe the only circumstances in which simultaneous bounds, as distinct from merely stationary principles, have been established. Throughout we consider steady flow, for which the three balance equations become

$$\nabla\cdot(\rho\mathbf{v})=0,\quad \mathbf{v}\wedge(\nabla\wedge\mathbf{v})=\nabla h-T\nabla S,\quad \mathbf{v}\cdot\nabla S=0. \tag{5.30}$$

Equation $(5.30)_3$ shows that S is constant along any streamline. A particular consequence of $(5.30)_2$ and $(5.30)_3$ is that $\mathbf{v}\cdot\nabla h=0$, i.e. that h is also constant along any streamline. This last result is called a Bernoulli integral. From these two observations it follows that we can associate with every streamline, however tortuous and in either two or three dimensional flow, a set of 15 plane diagrams which relate the six variables p, v, P, Q, ρ, T in pairs. Fig. 5.3 illustrates six of these 15 functions. All fifteen diagrams were explicitly illustrated by Sewell (1985). Each graph is a plane cross-section of a constitutive surface, and is obtainable as soon as h and S are fixed, before any specific boundary value problem is addressed. The adiabatic motion of a particle on the streamline, when subsequently found from the solution of a particular boundary value problem, will induce a time-track, perhaps to and fro, along part of each such plane graph. Also, as soon as one of the six variables has been found (perhaps numerically from a stationary principle below), the other five can be found from the graphs such as Fig. 5.3.

As a further restriction we confine attention to irrotational ($\nabla \wedge \mathbf{v} = 0$) steady flow which is homentropic ($S =$ constant everywhere), and therefore also homenergic ($h =$ constant everywhere) by $(5.30)_2$. We now treat h and S as fixed constants from the outset, and satisfy irrotationality by introducing a C^2 velocity potential ϕ such that $\mathbf{v} = \nabla\phi$. This with $(5.30)_1$ and $(5.22)_1$ makes a set of three equations to be satisfied by $\mathbf{v}$, ϕ and ρ. It is easy to verify that $(5.22)_1$, $(5.25)_1$, $(5.27)_1$ and $(5.27)_2$ are all equivalent to each other. This fact, together with the structure of $(5.30)_1$, suggests and permits the elimination of $\mathbf{v}$ and ρ in favour of $\mathbf{Q}$. The differential equations of the problem are thus reduced to $\nabla\phi = \partial P/\partial \mathbf{Q}$ with $\nabla \cdot \mathbf{Q} = 0$, for the two variables $\mathbf{Q}$ and ϕ as C^1 and C^2 functions of position respectively. Here $P[\mathbf{Q}]$ is a given function obtained by putting $Q = |\mathbf{Q}|$ in $P[Q]$, which itself is the Legendre transform of $-p[v]$. In the ideal gas illustrated in Fig. 5.3

$$p[v] = k^{1/(1-\gamma)}\left[\frac{(\gamma-1)}{\gamma}(h - \tfrac{1}{2}v^2)\right]^{\gamma/(\gamma-1)}, \tag{5.31}$$

where now k, h and γ are all fixed constants and $v = |\nabla\phi|$.

Consider flow within a fixed spatial region V in two or three dimensions, whose geometrical bounding surface $\Sigma = \Sigma_1 + \Sigma_2$ has unit outward normal $\mathbf{n}$. Let a and b be given functions of position on Σ_1 and Σ_2 respectively. We seek solutions in which $\phi = a$ on Σ_1 and $\mathbf{n}\cdot\mathbf{Q} = b$ on Σ_2. Plainly $b = 0$ on any part of Σ_2 which is a streamline, such as a rigid wall. The case $\Sigma_1 = 0$ is of most physical interest, so that $\Sigma = \Sigma_2$, and $\int b\,\mathrm{d}\Sigma = 0$ must then hold for consistency with overall mass conservation. It is little extra effort, however, to retain nonzero b and Σ_1, and this will display the structure of the theory better.

The boundary value problem so defined is frequently written in terms of ϕ alone, as

$$\nabla\cdot\left(\frac{\partial p}{\partial \nabla\phi}\right) = 0 \text{ in } V, \quad \phi = a \text{ on } \Sigma_1, \quad \mathbf{n}\cdot\frac{\partial p}{\partial \nabla\phi} = -b \text{ on } \Sigma_2, \tag{5.32}$$

where $p[\nabla\phi]$ is the function obtained by putting $v = |\nabla\phi|$ in $p[v]$, for example in (5.31) in the case of the ideal gas. The physics already gives an appropriate intermediate variable **Q** for following out the procedure for constructing bounds outlined in §§3.9 and 3.10. Using composite space elements

$$\tilde{\mathbf{Q}} = [\mathbf{Q}(V) \quad \mathbf{Q}(\Sigma_1)]^{\mathrm{T}}, \quad \tilde{\phi} = [\phi(V) \quad \phi(\Sigma_2)]^{\mathrm{T}},$$

like those of §3.10(v), the governing equations have Hamiltonian structure

$$\begin{matrix} V \\ \Sigma_1 \end{matrix} \qquad T^*\tilde{\phi} = \begin{bmatrix} \nabla\phi \\ -\mathbf{n}\phi \end{bmatrix} = \begin{bmatrix} \dfrac{\partial P}{\partial \mathbf{Q}} \\ -\mathbf{n}a \end{bmatrix} = \frac{\partial H}{\partial \tilde{\mathbf{Q}}}, \tag{5.33α}$$

$$\begin{matrix} V \\ \Sigma_2 \end{matrix} \qquad T\tilde{\mathbf{Q}} = \begin{bmatrix} -\nabla\cdot\mathbf{Q} \\ \mathbf{n}\cdot\mathbf{Q} \end{bmatrix} = \begin{bmatrix} 0 \\ b \end{bmatrix} = \frac{\partial H}{\partial \tilde{\phi}}. \tag{5.33β}$$

This H is the value of the functional which we obtain by partial integration in (5.34), and it is unconnected with the H used for enthalpy in Fig. 5.1. The integration by parts formula

$$\int \mathbf{Q}\cdot\nabla\phi\,\mathrm{d}V - \int \mathbf{Q}\cdot\mathbf{n}\phi\,\mathrm{d}\Sigma_1 = -\int \phi\nabla\cdot\mathbf{Q}\,\mathrm{d}V + \int \phi\mathbf{n}\cdot\mathbf{Q}\,\mathrm{d}\Sigma_2$$

is valid for any piecewise C^1 functions of position **Q** and ϕ, irrespective of whether they satisfy (5.33). It is the appropriate statement of adjointness $(\tilde{\mathbf{Q}}, T^*\tilde{\phi}) = \langle \tilde{\phi}, T\tilde{\mathbf{Q}} \rangle$ for the operators T and T^* defined in (5.33), with the inner products

$$(\tilde{\mathbf{Q}}_1, \tilde{\mathbf{Q}}_2) = \int \mathbf{Q}_1\cdot\mathbf{Q}_2\,\mathrm{d}V + \int \mathbf{Q}_1\cdot\mathbf{Q}_2\,\mathrm{d}\Sigma_1,$$

$$\langle \tilde{\phi}_1, \tilde{\phi}_2 \rangle = \int \phi_1\phi_2\,\mathrm{d}V + \int \phi_1\phi_2\,\mathrm{d}\Sigma_2.$$

The right sides in (5.33) are gradients of the functional

$$H[\tilde{\mathbf{Q}}, \tilde{\phi}] = \int P[\mathbf{Q}]\mathrm{d}V - \int a\mathbf{n}\cdot\mathbf{Q}\,\mathrm{d}\Sigma_1 + \int b\phi\,\mathrm{d}\Sigma_2. \tag{5.34}$$

We can now define a generating functional

$$\begin{aligned} L[\tilde{\mathbf{Q}}, \tilde{\phi}] &= (\tilde{\mathbf{Q}}, T^*\tilde{\phi}) - H[\tilde{\mathbf{Q}}, \tilde{\phi}] \\ &= \int (\mathbf{Q}\cdot\nabla\phi - P[\mathbf{Q}])\mathrm{d}V - \int (\phi - a)\mathbf{n}\cdot\mathbf{Q}\,\mathrm{d}\Sigma_1 - \int b\phi\,\mathrm{d}\Sigma_2, \end{aligned} \tag{5.35}$$

so that (5.33) can be written

$$\frac{\partial L}{\partial \tilde{\mathbf{Q}}} = \begin{bmatrix} \nabla\phi - \dfrac{\partial P}{\partial \mathbf{Q}} \\ -\mathbf{n}(\phi - a) \end{bmatrix} = 0 \quad (\alpha), \qquad \frac{\partial L}{\partial \tilde{\phi}} = \begin{bmatrix} -\nabla\cdot\mathbf{Q} \\ \mathbf{n}\cdot\mathbf{Q} - b \end{bmatrix} = 0 \quad (\beta). \tag{5.36}$$

We shall say that an *actual flow* is a pair $\mathbf{Q}$, ϕ of functions of position, respectively C^1 and C^2, which satisfy both of (5.36). From such $\mathbf{Q}$, ϕ we can calculate $\mathbf{v}$ as $\nabla\phi$ or $\partial P/\partial\mathbf{Q}$ (and such that $\nabla\wedge\mathbf{v}=0$), and ρ as Q/v. (More generally we could admit piecewise C^1 $\mathbf{Q}$ provided $\nabla\cdot\mathbf{Q}$ is continuous, and piecewise C^2 ϕ provided $\nabla\wedge\mathbf{v}$ is continuous.) However, the domain of $L[\tilde{\mathbf{Q}},\tilde{\phi}]$ is built from piecewise C^1 functions $\mathbf{Q}$ and ϕ.

Theorem 5.1 (Free stationary principle)
$L[\tilde{\mathbf{Q}},\tilde{\phi}]$ is stationary with respect to arbitrary independent small piecewise C^1 variations $\delta\mathbf{Q}$ and $\delta\phi$ of its arguments if and only if they are variations from an actual flow.

Proof
Integration by parts shows that the first variation of $L[\tilde{\mathbf{Q}},\tilde{\phi}]$ is

$$\delta L=\int\delta\mathbf{Q}\cdot\left(\nabla\phi-\frac{\partial P}{\partial\mathbf{Q}}\right)\mathrm{d}V$$
$$+\int\delta\mathbf{Q}\cdot\mathbf{n}(a-\phi)\mathrm{d}\Sigma_1+\int\delta\phi(-\nabla\cdot\mathbf{Q})\mathrm{d}V+\int\delta\phi(\mathbf{n}\cdot\mathbf{Q}-b)\mathrm{d}\Sigma_2$$
$$=\left(\delta\tilde{\mathbf{Q}},\frac{\partial L}{\partial\tilde{\mathbf{Q}}}\right)+\left\langle\delta\tilde{\phi},\frac{\partial L}{\partial\tilde{\phi}}\right\rangle.$$

Hence (5.36) implies $\delta L=0$, which proves sufficiency.

Necessity requires the fundamental lemma of the calculus of variations. That is, the hypothesis that $\delta L=0$ with one of the equations in (5.36) not satisfied locally is contradicted by the construction of a $\delta\mathbf{Q}$ or a $\delta\phi$ which is one signed in that locality and zero elsewhere (cf. §§3.7(xiv) and 3.9(iii)).

□

Theorem 5.1 can be embedded in other free stationary principles generated by functionals of more than two independent variables. We give three illustrations (with h and S still fixed).

(a) Replace the volume integrand in $L[\tilde{\mathbf{Q}},\tilde{\phi}]$ by $\mathbf{Q}\cdot\nabla\phi-\mathbf{Q}\cdot\mathbf{v}-p[v]$, with $\mathbf{Q}$, $\mathbf{v}$ and ϕ now regarded as independent of each other at the outset and $p[v]$ given by (5.21) with $(5.22)_1$ as in Fig. 5.3(a). This principle delivers $\nabla\phi=\mathbf{v}$ and $\mathbf{Q}=\rho\mathbf{v}$ as natural conditions from the coefficients of $\delta\mathbf{Q}$ and $\delta\mathbf{v}$, respectively, in place of $\nabla\phi=\partial P/\partial\mathbf{Q}$ in (5.36α), and is included in a result of Sewell (1963a). It parallels that version of Hamilton's principle in mass-point mechanics wherein the velocities and momenta are treated as *independent* variables in the action integral (e.g. see Lanczos 1949, Ch. 6, §4). Other analogues are in Exercises 3.3.3 and 5.3.5 and in §4.5(ii).

(b) Replace $p[v]$ in (a) by $\rho(h-U[1/\rho]-\frac{1}{2}v^2)$, where h is still fixed and the internal energy function $U[1/\rho]$ is regarded as given and satisfying $\partial U/\partial p = p/\rho^2$ by $(5.1)_2$. Treat **Q**, **v**, ϕ and ρ is independent variables. In this principle we start further back in the theory, and setting to zero the coefficient of $\delta\rho$ gives $p = \rho(h - U - \frac{1}{2}v^2)$, i.e. (5.20).

(c) Replace **Q** by $\rho\mathbf{v}$ in (b). This leaves a principle in which **v**, ϕ and ρ are treated as independent variables. A principle which generalizes this type to rotational unsteady flows is given in Exercise 5.2.8.

The choice of which of these free principles to use as starting point will depend on one's purpose, and there will sometimes be advantages in the presence of a single valued and algebraically explicit $p[v]$ or $U[1/\rho]$. For our purpose, now that we know explicitly the qualitative form of the double valuedness of $P[Q]$ in Fig. 5.3(*b*), it is simplest to start with Theorem 5.1 as stated. A similar issue may arise in nonlinear elasticity, where $P[Q]$ is replaced by complementary energy density, whose multivaluedness is not yet fully explored (cf. Exercise 3.3.3 and (5.150)).

We now return to the generating functional (5.35) and the free stationary principle in Theorem 5.1. Before the application of any constraints on **Q** and ϕ we note the values

$$L - \left(\tilde{\mathbf{Q}}, \frac{\partial L}{\partial \tilde{\mathbf{Q}}}\right) = \int \left[\mathbf{Q}.\frac{\partial P}{\partial \mathbf{Q}} - P[\mathbf{Q}]\right] \mathrm{d}V - \int b\phi \,\mathrm{d}\Sigma_2 \tag{5.37}$$

and

$$L - \left\langle \tilde{\phi}, \frac{\partial L}{\partial \tilde{\phi}} \right\rangle = -\int P[\mathbf{Q}]\mathrm{d}V + \int a\mathbf{n}\cdot\mathbf{Q}\,\mathrm{d}\Sigma_1 \tag{5.38}$$

of associated functionals which we shall presently show to have stationary (and later bounding) properties for an actual solution.

If either of (5.36α) or (5.36β) is imposed as a constraint on the variations in Theorem 5.1, it is self-evident that the remaining condition will be sufficient to make the constrained $L[\tilde{\mathbf{Q}}, \tilde{\phi}]$ stationary. This fact, with its converse, is a general device in the calculus of variations and was amplified in §3.7(xv). We shall assume that its converse applies here.

Theorem 5.2

The functional

$$J[\tilde{\phi}_\alpha] = -\int p[\nabla\phi_\alpha]\mathrm{d}V - \int b\phi_\alpha \mathrm{d}\Sigma_2 \tag{5.39}$$

is stationary, with respect to small variations in the class of piecewise C^2 functions ϕ_α of position which satisfy $\phi_\alpha = a$ on Σ_1, if and only if they represent variations from an actual flow. Here $p[\nabla\phi]$ is the function introduced in (5.32), *now evaluated for any ϕ_α.*

Proof

Impose (5.36α) from the outset, and use the Legendre transformation (5.27) to invert $\nabla\phi_\alpha = \partial P/\partial \mathbf{Q}_\alpha$ as $\mathbf{Q}_\alpha = -\partial p/\partial \nabla\phi_\alpha$, so that $\mathbf{Q}_\alpha$ is expressed in terms of ϕ_α. From (5.37) we see that $L[\tilde{\mathbf{Q}}_\alpha, \tilde{\phi}_\alpha]$ then becomes $J[\tilde{\phi}_\alpha]$. We have thereby solved the constraint in V explicitly and, with $\delta\phi_\alpha = 0$ on Σ_1, sufficiency of (5.36β) is evident from

$$\delta J = \int \delta\phi_\alpha \nabla\cdot\frac{\partial p}{\partial \nabla\phi_\alpha}\mathrm{d}V - \int \delta\phi_\alpha\left(\mathbf{n}\cdot\frac{\partial p}{\partial \nabla\phi_\alpha} + b\right)\mathrm{d}\Sigma_2.$$

We have assumed the function $p[\nabla\phi_\alpha]$ to be such that piecewise C^1 $\mathbf{Q}_\alpha$ requires piecewise C^2 (instead of piecewise C^1) ϕ_α (cf. §3.9(iii)). Necessity requires the fundamental lemma, showing that the natural conditions of $\delta J = 0$ are (5.36β). □

Adapting the notation of Fig. 2.26 to this infinite dimensional context, if $j[\partial L/\partial\tilde{\mathbf{Q}}, \tilde{\phi}]$ is the functional which is the Legendre dual of $-L[\tilde{\mathbf{Q}}, \tilde{\phi}]$ with active $\tilde{\mathbf{Q}}$ and passive $\tilde{\phi}$, we can interpret the $J[\tilde{\phi}_\alpha]$ defined in (5.40) as $j[0, \tilde{\phi}_\alpha]$. Recall (Fig. 2.3) that the Legendre dual of a linear segment through the origin is a single point with zero height; this is why, from this viewpoint, the terms in (5.35) which are linear in $\mathbf{Q}$ do not appear in (5.39) (cf. Exercise 5.2.4).

Theorem 5.3

The functional

$$K[\tilde{\mathbf{Q}}_\beta] = -\int P[\mathbf{Q}_\beta]\mathrm{d}V + \int a\mathbf{n}\cdot\mathbf{Q}_\beta\mathrm{d}\Sigma_1 \tag{5.40}$$

is stationary, with respect to small variations in the class of C^1 functions $\mathbf{Q}_\beta$ of position which satisfy $\nabla\cdot\mathbf{Q}_\beta = 0$ in V and $\mathbf{n}\cdot\mathbf{Q}_\beta = b$ on Σ_2, if and only if they represent variations from an actual flow.

Proof

Impose (5.36β) from the outset. From (5.38) we see that $L[\tilde{\mathbf{Q}}_\beta, \tilde{\phi}_\beta]$ then becomes $K[\tilde{\mathbf{Q}}_\beta]$. Sufficiency of (5.36α) then follows from Theorem 5.1. By contrast with Theorem 5.2, this proof does not require that we express ϕ_β in terms of $\mathbf{Q}_\beta$ (which we could not do anyway from $\nabla\cdot\mathbf{Q}_\beta = 0$), because the linearity of $L[\tilde{\mathbf{Q}}, \tilde{\phi}]$ in $\tilde{\phi}$ removes ϕ from the right side of (5.38).

This linearity also means that the use of $\partial L/\partial\tilde{\phi} = 0$ as a constraint deletes $\tilde{\phi}$ altogether from $L[\tilde{\mathbf{Q}}, \tilde{\phi}]$, so that (5.36α) (which contains ϕ) will not be necessary as it stands for a stationary value under the constraint. The difficulty is of an obvious type which can be resolved by supposing that the stationary solutions of $L[\tilde{\mathbf{Q}}, \tilde{\phi}]$ under the constraint $\partial L/\partial\tilde{\phi} = 0$ are the same as those associated with a finite Lagrange multiplier $\tilde{\mu}$ when the augmented functional

$$\bar{L}[\tilde{\mathbf{Q}}, \tilde{\phi}, \tilde{\mu}] = L[\tilde{\mathbf{Q}}, \tilde{\phi}] - \left(\tilde{\mu}, \frac{\partial L}{\partial \tilde{\phi}}\right)$$

$$= L[\tilde{\mathbf{Q}}, \tilde{\phi} - \tilde{\mu}] \tag{5.41}$$

is required to be stationary under free variations of $\tilde{\mathbf{Q}}$, $\tilde{\phi}$ and $\tilde{\mu}$. The natural conditions for this are $\partial L/\partial\tilde{\mathbf{Q}} = 0$ and $\partial L/\partial(\tilde{\phi} - \tilde{\mu}) = 0$, i.e. (5.36) but with $\phi - \mu$ replacing ϕ as the velocity potential, with arbitrary μ. Ultimately we deem the solution value of μ to be zero, as explained in §3.7(xv). □

The constraint $\nabla \cdot \mathbf{Q} = 0$ can be satisfied by introducing a vector stream function $\boldsymbol{\psi}$, such that $\mathbf{Q} = \nabla \wedge \boldsymbol{\psi}$, and single valued in a simply connected region. In plane flow such $\boldsymbol{\psi}$ reduces to a scalar ψ, which takes assigned values on Σ_2 to satisfy $\mathbf{n} \cdot \nabla \wedge \boldsymbol{\psi} = b$. Theorem 5.3 could be rephrased directly in terms of a C^2 function $\boldsymbol{\psi}$.

Theorems 5.1–5.3 are valid regardless of whether the flow is subsonic, transonic or supersonic. Our restriction to continuous flow does mean that the supersonic region must not be traversed by a shock, unless this be accommodated via extra terms at an internal boundary. But apart from this, Fig. 5.3 provides explicit knowledge (Sewell 1982) of the shape of the volume integrands in $J[\tilde{\phi}]$ and $K[\tilde{\mathbf{Q}}]$, so that application of the Theorems is no longer inhibited by uncertainty about that aspect of the sonic singularity. Fig. 5.3(*b*) shows precisely how, in applications of Theorems 5.1 and 5.3, trial values of Q can encounter double valuedness in $P[Q]$. Care will be needed to decide whether it is the subsonic or supersonic branch of $P[Q]$ (if either) which is intercepted by a chosen ordinate.

Theorem 5.4 (Simultaneous upper and lower bounds)

(a) In subsonic flow both of the functions $-p[\mathbf{v}]$ *and* $P[\mathbf{Q}]$ *are strictly convex.*

(b) In subsonic flow $L[\tilde{\mathbf{Q}}, \tilde{\phi}]$ *is a saddle functional strictly concave in* $\tilde{\mathbf{Q}}$ *and weakly convex in* $\tilde{\phi}$. *Then* (5.39) *and* (5.40) *satisfy*

$$J[\tilde{\phi}_\alpha] \geqslant L_0 \geqslant K[\tilde{\mathbf{Q}}_\beta], \tag{5.42}$$

thus providing upper and lower bounds to the value

$$L_0 = -\int p\,\mathrm{d}V - \int b\phi\,\mathrm{d}\Sigma_1 = -\int P\,\mathrm{d}V + \int a\mathbf{n}\cdot\mathbf{Q}\,\mathrm{d}\Sigma_2 \tag{5.43}$$

of L in an actual flow.

Proof

(a) The Hessians in (5.28) are both positive definite when $v < c$, which is sufficient to establish the strict convexity of $-p[\mathbf{v}]$ and $P[\mathbf{Q}]$. Note that the Hessians are indefinite when $v > c$, so that $-p[\mathbf{v}]$ and $P[\mathbf{Q}]$ are not concave in supersonic flow, even though $-p[v]$ and $P[Q]$ are then concave as Fig. 5.3 shows.

(b) The stated saddle property of $L[\tilde{\mathbf{Q}}, \tilde{\phi}]$ is evident from (5.35) when $P[\mathbf{Q}]$ is strictly convex, since the remaining terms are bilinear or linear. Alternatively, in terms of the current inner products we could introduce a saddle quantity

$$S = L_+ - L_- - \left(\tilde{\mathbf{Q}}_+ - \tilde{\mathbf{Q}}_-, \frac{\partial L}{\partial \tilde{\mathbf{Q}}}\bigg|_+\right) - \left\langle \tilde{\phi}_+ - \tilde{\phi}_-, \frac{\partial L}{\partial \tilde{\phi}}\bigg|_-\right\rangle$$

$$= \int \left[P_- - P_+ - (\mathbf{Q}_- - \mathbf{Q}_+)\cdot \frac{\partial P}{\partial \mathbf{Q}}\bigg|_+ \right] \mathrm{d}V > 0 \tag{5.44}$$

for any distinct $\mathbf{Q}_+$ and $\mathbf{Q}_-$, by the strict convexity of $P[Q]$. Proof of the upper and lower bounds follows because (5.44) justifies the application of Theorem 3.4. □

We can expect that $J[\tilde{\phi}_\alpha]$ and $K[\tilde{Q}_\beta]$ will have a stationary minimum and maximum, respectively, in an actual flow (cf. §3.9(iii)).

The integral in (5.44) could still be positive even if the integrand were negative over a limited part of the flow, and in this sense the flow could be subsonic overall without being subsonic at every point. The bounds (5.42) would still apply. There could be a limited supersonic region without violating (5.44), if it could also be shock-free to satisfy our underlying assumption at the end of §(iv). In other words, strict convexity of $P[\mathbf{Q}]$ at every point of the flow is sufficient but not necessary to ensure that $S > 0$.

The simultaneous bounds (5.42) may be alternatively called dual bounds because their integrands contain the Legendre dual functions $-p[v]$ and $P[Q]$.

Theorem 5.5 (Uniqueness)
Any actual subsonic flow has unique $\mathbf{Q}$, *and consequently unique* $\mathbf{v}, \rho, p$ *and* P.

Proof
Strict convexity of $P[\mathbf{Q}]$ implies

$$\int (\mathbf{Q}_+ - \mathbf{Q}_-)\cdot\left(\frac{\partial P}{\partial \mathbf{Q}}\bigg|_+ - \frac{\partial P}{\partial \mathbf{Q}}\bigg|_-\right)\mathrm{d}V > 0 \tag{5.45}$$

by interchanging $\mathbf{Q}_+$ and $\mathbf{Q}_-$ in (5.44) and adding. If such distinct $\mathbf{Q}_+$ and

$\mathbf{Q}_-$ could both be solutions of (5.36) in association with ϕ_+ and ϕ_- respectively, then the left side of (5.45) is

$$\begin{aligned}&\int(\mathbf{Q}_+-\mathbf{Q}_-)\cdot\nabla(\phi_+-\phi_-)\mathrm{d}V\\ &=\int(\phi_+-\phi_-)\mathbf{n}\cdot(\mathbf{Q}_+-\mathbf{Q}_-)\mathrm{d}\Sigma-\int(\phi_+-\phi_-)\nabla\cdot(\mathbf{Q}_+-\mathbf{Q}_-)\mathrm{d}V\\ &=0.\end{aligned}$$

Hence $\mathbf{Q}$ is unique by *reductio ad absurdum*; so are P, p, ρ and $\mathbf{v}$ by the single valuedness of the subsonic branches of Fig. 5.3. □

Bateman (1930) was the first to recognize a pair of simultaneous extremum principles for compressible inviscid fluid. Working in two dimensions, he gave results which can be seen as the precursors of (5.42), having already adopted (1929, quoting Hargreaves 1908) a 'guiding principle' that the integrand in one stationary principle would have the values of pressure. The incompressible versions of (5.42) characterize the mixed Dirichlet/Neumann problem (5.36) for Laplace's equation, and are obtainable starting from $p[v]=p_0-\frac{1}{2}\rho v^2$ in (5.21) or $P[Q]=p_0+\frac{1}{2}Q^2/\rho$ in (5.23) with fixed ρ and p_0. When $\Sigma_1=0$ and Σ_2 is a rigid wall ($b=0$), the actual flow minimizes the total kinetic energy, among either irrotational (α) or incompressible (β) velocities. Kelvin proved the β-principle in 1849.

Serrin (1959, §47) gave a clear review of work on compressible extremum principles up to that time, for flow in three dimensions with $\Sigma_1=0$. He mentioned the first attempts to use the principles for the numerical computation of gas flows, such as that of Lush and Cherry (1956), and for a discussion of the existence of subsonic flow. In the practical problem of flow from infinity past a finite body, such as an aerofoil or a circular cylinder, the integrals in (5.39) and (5.40) are divergent; but this difficulty can be avoided if certain additional terms representing the far field properties are added to the integrands. Fiszdon (1962*a*, *b*) gave specific illustrations of different choices of trial velocity potentials and stream functions in the application of versions of Theorems 5.2 and 5.3 for unsteady irrotational and rotational flow, with some allowance for shocks treated as internal boundaries.

Sewell (1963*a*) used the idea of convexity, explicitly for the first time, in the proofs of (5.42); and showed how the structure of these proofs, and those of versions of Theorems 5.1–5.3 for certain rotational flows, is made much clearer by conscious use of the Legendre transformation (5.21)–(5.28). These points have not been widely taken, and the large literature on stationary principles in fluid mechanics contains *ad hoc* manipulations which are awkward implicit representations of convexity and/or the Legendre transformation. This applies also to some papers which discuss the fluids version of Hamilton's principle.

A further review of numerical applications of stationary principles in compressible flow was given by Rasmussen (1974). Such a problem with free boundaries was treated by Morice (1978). One needs to be aware that stationary principles are not the only methods in use for the solution of practical flow problems. A look into the sky at an appropriate moment can indicate the enormous effort which has been invested in the problem of finding the air flow past a moving body of complicated shape. An indication of the state of the art in computational fluid mechanics can be gleaned, for example, from Peyret and Taylor (1983), and from the symposia edited by Hunt (1980), Nixon (1982) and Reynolds and MacCormack (1981).

(vi) General flows

Consider now flow which may be unsteady and rotational, through a region of space where the jump conditions are satisfied identically. The three differential balance equations in Table 5.1 relate four variables ρ, $\mathbf{v}$, S and p. A system of governing equations can be completed by adding, from (5.1), the single thermodynamical equation

$$p = \rho^2 \frac{\partial U}{\partial \rho}, \quad \text{with } U[S, 1/\rho] \tag{5.46}$$

as the internal energy function. Equivalently we could add the three equations (5.14) to provide a system of six governing equations for the six variables ρ, $\mathbf{v}$, S, p, H and T. In the case of the ideal gas (5.14) may be rearranged as

$$p = k\rho^\gamma, \quad \rho = \left[\frac{(\gamma - 1)H}{k\gamma}\right]^{1/(\gamma-1)}, \quad T = \frac{p}{\rho R}. \tag{5.47}$$

The following theorem is general, for any smooth motion as just indicated, and is not restricted to the ideal gas. Equivalence of two systems means that every set of functions satisfying one system can be converted into a set of functions satisfying the other system, and conversely.

Theorem 5.6 (Velocity representation)
Either system of governing equations just stated is equivalent to another one in which the momentum equation is replaced by the five equations

$$\mathbf{v} = \nabla\Phi + \lambda\nabla S + C_j \nabla X_j, \tag{5.48}$$

$$\dot{\Phi} = \tfrac{1}{2}v^2 - H, \quad \dot{\lambda} = T, \tag{5.49}$$

$$\dot{\mathbf{C}} = 0, \quad \dot{\mathbf{X}} = 0. \tag{5.50}$$

where the scalars Φ and λ and the vectors **C** *and* **X** *are four additional variables.*

Proof

Let **x** be the position vector at the current time t of a typical particle which, at an arbitrarily chosen initial time $t = 0$, had position vector **X** in the same reference frame. The motion is sought as a C^2 function $\mathbf{x}[\mathbf{X}, t]$, or its inverse $\mathbf{X}[\mathbf{x}, t]$ such that $\mathbf{X}[\mathbf{x}, 0] = \mathbf{x}$. The velocity $\mathbf{v} = \dot{\mathbf{x}}$ is then a C^1 function in t, and the acceleration $\dot{\mathbf{v}}$ has cartesian components satisfying

$$\dot{v}_i \frac{\partial x_i}{\partial X_j} = \overline{v_i \frac{\partial x_i}{\partial X_j}}^{\,\cdot} - \frac{\partial}{\partial X_j}(\tfrac{1}{2}v^2).$$

The C^2 property of $\mathbf{x}[\mathbf{X}, t]$ is sufficient to allow the mixed second partial derivatives with respect to **X** and t to commute, as just required. From (5.14), if p, H and S are C^1 functions of **X**,

$$\frac{\partial p}{\partial X_j} = \rho\left(\frac{\partial H}{\partial X_j} - T\frac{\partial S}{\partial X_j}\right).$$

The components of ∇p are $\partial p/\partial x_i = (\partial X_j/\partial x_i)(\partial p/\partial X_j)$ so that, since $|\partial x_i/\partial X_j| \neq 0$, the momentum equation $\rho\dot{\mathbf{v}} + \nabla p = 0$ is equivalent to

$$\overline{v_i \frac{\partial x_i}{\partial X_j}}^{\,\cdot} - \frac{\partial}{\partial X_j}(\tfrac{1}{2}v^2 - H) - T\frac{\partial S}{\partial X_j} = 0. \tag{5.51}$$

We now introduce a function $\lambda[\mathbf{X}, t]$ which is C^1 in t, and a C^2 function $\Phi[\mathbf{X}, t]$, satisfying (5.49). If we also assume that $S[\mathbf{X}, t]$ is a C^2 function, mixed second derivatives of $\Phi[\mathbf{X}, t]$ and of $S[\mathbf{X}, t]$ commute. We can then integrate, following the particle, the last version of the momentum equation to give

$$v_i \frac{\partial x_i}{\partial X_j} - \frac{\partial \Phi}{\partial X_j} - \lambda\frac{\partial S}{\partial X_j} = C_j, \tag{5.52}$$

after using $\dot{S} = 0$. Here **C** is a vector 'constant' of integration, actually an arbitrary C^1 function $\mathbf{C}[\mathbf{X}]$ of **X** only, so that $\dot{\mathbf{C}} = 0$. Writing $\mathbf{X} = \mathbf{X}[\mathbf{x}, t]$ into (5.52) to transform it into spatial (so-called Eulerian) form gives (5.48).

□

The whole of either system of governing equations can be written explicitly, if more lengthily, in spatial form via formulae such as $\dot{\Phi} = \partial\Phi/\partial t + \mathbf{v}\cdot\nabla\Phi$, where $\partial/\partial t$ is the partial t-derivative of a function of **x** and t. An alternative to $(5.49)_1$, having analogous structure to (5.48), is stated in Exercise 5.2.5.

The point of Theorem 5.6 is to make it clear that the representation (5.48) of the velocity field is general, being obtained directly by an integration following the particle in unsteady flow, subject only to the smoothness hypotheses explicitly stated in the proof. The representation is in terms of nine scalar functions, subjected to stated side conditions. Serrin (1959, §29A) gave a version of it in which **C** was chosen to be the initial velocity field. The hydrodynamical literature, from Clebsch (1859) onwards, has been much preoccupied with the fact (e.g. see Ericksen, 1960, §35) that a general vector field can be written locally as $\nabla\alpha + \beta\nabla\gamma$, i.e. in terms of only three scalar functions. But when applied to the velocity field, this shorter and apparently simpler representation has led to other difficulties which it is beyond our scope to describe here. Some discussion of related matters has been given by Benjamin (1984) and Detyna (1982).

Theorem 5.6 also offers an entry to the subject of stationary principles for general flows. An illustration is given in Exercise 5.2.8, extending the work of Bateman (1932, §2.52), Serrin (*op. cit.*, §15) and Sewell (1963*a*, §3). We do not pursue this topic further because it goes too far from our subject of maximum and minimum principles *per se*. Benjamin (*op. cit.*) reviews a number of other stationary principles in fluid mechanics.

Exercises 5.2

1. Show that (5.15) imply

$$c_v = T\left[H_{SS} - \frac{H_{Sp}{}^2}{H_{pp}}\right]^{-1} > 0, \quad \gamma - 1 = \frac{-H_{Sp}{}^2}{H_{pp}H_{SS}} \geqslant 0$$

(compare (5.4), (2.67) and Exercise 2.4.9).

2. Show that the velocity of sound c in (5.17) also satisfies

$$c^2 = \frac{\partial p}{\partial \rho} \quad \text{for} \quad p[S,\rho] \quad \text{and} \quad c^2 = v^2 U_{vv} = -\frac{v^2}{H_{pp}}.$$

3. Develop explicitly the three free stationary principles sketched under (a), (b) and (c) after Theorem 5.1.
4. Use the function $P[Q]$ of Fig. 5.3(*b*) to sketch the function

$$L[Q,\phi] = \begin{cases} Q\phi - P[Q] & Q \geqslant 0, \\ Q\phi & n < Q \leqslant 0, \end{cases}$$

for any fixed $\phi > 0$ and any finite n. Find and sketch its Legendre dual function $-j[\partial L/\partial Q, \phi]$.

5. (i) Prove that any velocity field which satisfies (5.48), $(5.49)_1$, $\dot{S} = 0$ and $\dot{\mathbf{X}} = 0$ has the property

$$h + \frac{\partial \Phi}{\partial t} + \lambda\frac{\partial S}{\partial t} + \mathbf{C}\cdot\frac{\partial \mathbf{X}}{\partial t} = 0$$

where h is the total energy defined in (5.20). This formula emphasizes that some of the potentials Φ, $\mathbf{X}$, etc., may depend on a parameter which has the same values as t, as well as on $\mathbf{x}$, even in steady motion when directly measurable variables $\mathbf{v}$, ρ, etc., depend on $\mathbf{x}$ but not t.

(ii) Hence show that any such field, which does not necessarily satisfy the equation of continuity, gives to the integrand in Exercise 8 below the values of pressure (compare Theorem 5.2).

6. For fixed S, show that the Legendre dual of $p[H]$ in (5.14) is $\rho U[1/\rho]$.
7. Use the function $p[v, h, S]$ of (5.21) to construct the example shown in Fig. 5.4 of the generalization of Fig. 2.18. That is, when a starting function depends on three sets of variables rather than two, we can choose each in turn as an active variable so that each side of the cuboid represents a Legendre transformation, between functions at the vertices whose values and gradients are given in Table 5.2 (cf. Sewell and Porter 1980). Show also that $\partial^2 p/\partial h\partial v = -\rho v/c^2$, $\partial^2 p/\partial H^2 = \rho/c^2$. Hence show that $\partial^2 p/\partial h\partial v = -\rho v/c^2$, $\partial^2 p/\partial h^2 = \rho/c^2$. Hence show that the Legendre transformation directly from $p[v, h, S]$ to $r[Q, \rho, S]$, with the first two variables simultaneously active and only S passive, is nonsingular. Show that, at any fixed S, both such generating functions are saddle-shaped at every point in the sense of (1.9), regardless of whether the flow is subsonic or supersonic.

Table 5.2. *Functions generating Fig.* 5.4

Function	Respective partial derivatives			Value
$p[v, h, S]$	$-\rho v = -Q$	ρ	$-\rho T = -\theta$ (say)	p
$P[Q, h, S]$	v	ρ	$-\theta$	$p + \rho v^2$
$r[Q, \rho, S]$	v	$-h$	$-\theta$	$\rho(\frac{1}{2}v^2 - U)$
$R[v, \rho, S]$	Q	h	θ	$\rho(\frac{1}{2}v^2 + U)$
$a[v, h, \theta]$	Q	$-\rho$	$-S$	$-p - \theta S$
$A[Q, h, \theta]$	$-v$	$-\rho$	$-S$	$-P - \theta S$
$b[Q, \rho, \theta]$	$-v$	h	$-S$	$\rho(F - \frac{1}{2}v^2)$
$B[v, \rho, \theta]$	$-Q$	$-h$	S	$-\rho(F + \frac{1}{2}v^2)$

Fig. 5.4. Cuboid representtion of Legendre transformations.

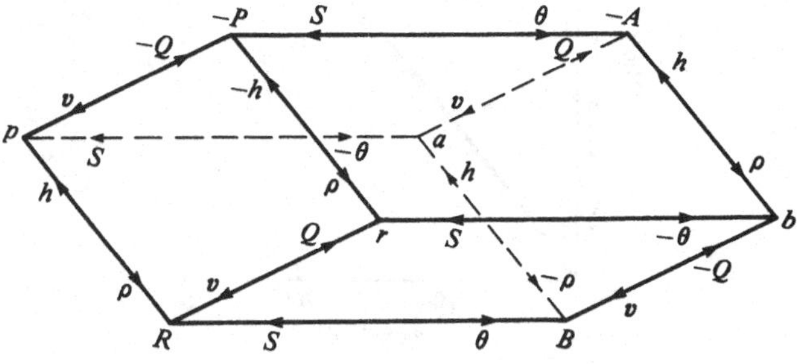

8. Consider the functional

$$I[\mathbf{v}, \rho, \Phi, S, \mathbf{X}, \lambda, \mathbf{C}]$$
$$= \iint \Big\{ \tfrac{1}{2}\rho v^2 - \rho U[S, 1/\rho] - \rho\mathbf{v}\cdot(\nabla\Phi + \lambda\nabla S + C_j\nabla X_j)$$
$$- \rho\left(\frac{\partial\Phi}{\partial t} + \lambda\frac{\partial S}{\partial t} + C_j\frac{\partial X_j}{\partial t}\right)\Big\} \mathrm{d}V\,\mathrm{d}t$$

defined over a fixed spatial region V and over an arbitrary fixed time interval $0 \leqslant t \leqslant \tau$. Show that

$$\delta I + \iint \rho\mathbf{v}\cdot\mathbf{n}(\delta\Phi + \lambda\delta S + \mathbf{C}\cdot\delta\mathbf{X})\mathrm{d}\Sigma\,\mathrm{d}t + \left[\int \rho(\delta\Phi + \lambda\delta S + \mathbf{C}\cdot\delta\mathbf{X})\mathrm{d}V\right]_0^\tau = 0$$

for unconstrained independent small variations $\delta\mathbf{v}$, δp, $\delta\Phi$, δS, $\delta\mathbf{X}$, $\delta\lambda$, $\delta\mathbf{C}$ applied at every place $\mathbf{x}$ and time t is equivalent to the system of governing equations described in Theorem 5.6, and therefore to the original system for unsteady, rotational flow without jumps such as shocks.

9. Use ideas such as those in Exercise 7 to construct alternative versions of the result in Exercise 8 in the sense of the alternatives (a), (b) and (c) which were described for Theorem 5.1. For example, replace the integrand in Exercise 8 by

$$r[\rho\mathbf{v}, -p, S] - \rho\mathbf{v}\cdot(\nabla\Phi + \lambda\nabla S + C_j\nabla X_j) - \rho(\partial\Phi/\partial t + \lambda\partial S/\partial t + C_j\partial X_j/\partial t).$$

10. Show from Table 5.1 that, when a particle crosses at right angles a shock wave which is not moving through space (standing normal shock), the quantities Q, P and h are all continuous. The representative point on Fig. 5.3(*b*) then transfers to a similar curve associated with a different value of S, and therefore of k. Show that the supersonic branch for the lower k intersects the subsonic branch for the higher k. Fig. 5.5 shows a specific example of such a transfer, from supersonic to subsonic following the arrows, and from $k = 6.70 \times 10^4$ to $k = 7.08 \times 10^4$ in SI

Fig. 5.5. Transition between flow stress functions at a standing normal shock.

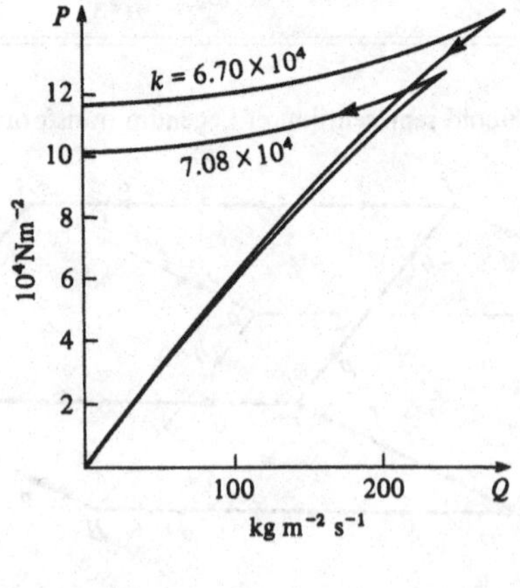

units. Fig. 5.5 was calculated by Sewell (1985) for the ideal gas with $\gamma = 1.4$ and $h = 2.74 \times 10^5\,\mathrm{J\,kg^{-1}}$, to which Fig. 5.3 applies for the higher k. Full details of such transfers may be found in Sewell and Porter (1980).

11. The standard approximation for unsteady shallow water theory ascribes to any horizontal station for all time the following variables and relations. Depth d and vertically averaged hydrostatic pressure p satisfy $p=\frac{1}{2}gd^2$, where g is acceleration due to gravity. Total energy h and horizontal fluid speed v satisfy $p=\frac{1}{2}g[\frac{1}{4}v^4 - hv^2 + h^2]$. The mass flow $Q = vd$. Develop the Legendre dual transformation which parallels (5.21)–(5.26) with S absent, i.e. starting with the stated function $p[v, h]$. Hence show that the flow stress function $P[Q, h]$ can be described either as the two-parameter family of three single valued functions

$$P = \tfrac{1}{2}gd^2 + dv^2, \quad Q = vd, \quad h = gd + \tfrac{1}{2}v^2,$$

parametrized by d and v; or equivalently as the double valued part of the swallowtail surface shown in Fig. 5.6, in which a 'subcritical' part lies above a 'supercritical' part. Here 'critical' means that the Froude number $v/(gd)^{\frac{1}{2}}$, which is the analogue of the Mach number, takes the value unity, along the cusped join of the two parts in Fig. 5.6. Sewell and Porter (1980) explained how Fig. 5.6 can be used as a basis for a geometrical description of relations between associated subcritical and supercritical 'conjugate flows', for example in channel flows, which have been discussed from other analytical viewpoints by Benjamin (1971) and co-workers.

Fig. 5.6. Swallowtail flow stress in shallow water theory.

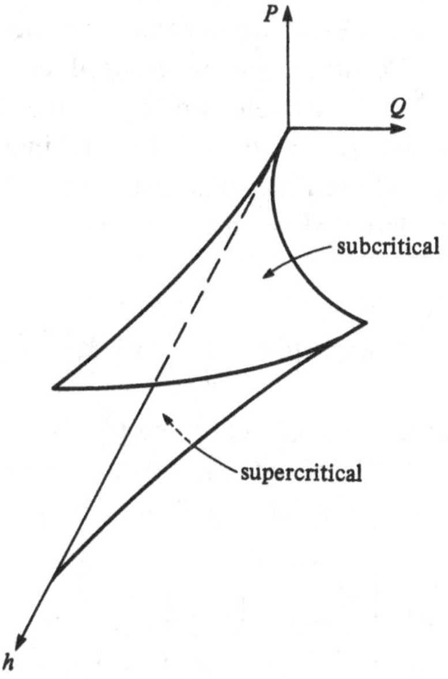

5.4. Magnetohydrodynamic pipe flow

(i) Governing equations

We study a particular linear problem which has some instructive features. Except for our remarks about networks in §2.8, we shall not describe the construction of extremum and stationary principles in electricity and magnetism generally. Hammond (1981) has made progress in this direction, and Jones (1979) also discusses the topic. Nonlinear problems of magneto-gasdynamics may require extensions of ideas in §5.3.

The nondimensionalized *ab initio* governing equations are

$$\begin{aligned} \nabla^2 w + M\frac{\partial h}{\partial y} = -1, \quad \nabla^2 b + M\frac{\partial w}{\partial y} = 0 \quad & \text{in } V, \\ w = b = 0 \qquad\qquad\qquad & \text{on } \Sigma. \end{aligned} \tag{5.53}$$

Here V is the given simply connected cross-section in the x, y plane of a straight pipe with fixed insulated boundary Σ. The pipe is filled with viscous conducting fluid whose speed normal to V is $w[x, y]$, and $b[x, y]$ is the induced magnetic field in the same direction, i.e. parallel to the pipe. The constant M is called the Hartmann number. Hunt and Shercliff (1971) give further physical background.

(ii) Decomposition method

We decompose the two second order equations into four first order equations by using ∇w and ∇b as intermediate variables. First let $\mathbf{u}$ and $\mathbf{v}$ be any vectors over V. Construct the inner product space of composite elements $\tilde{\phi} = [w \;\; \mathbf{v}]^{\mathrm{T}}$, $\tilde{\psi} = [b \;\; \mathbf{u}]^{\mathrm{T}}$, etc., which have one scalar entry and one vector entry which are integrable over V, with inner product $(\tilde{\phi}, \tilde{\psi}) = \int [wb + \mathbf{v}\cdot\mathbf{u}]\,\mathrm{d}V$. We shall need the subspace of such elements constructed from piecewise C^1 scalars and vectors such that the scalars vanish on Σ. Then

$$\int \left[b\left(M\frac{\partial w}{\partial y} + \nabla\cdot\mathbf{v} \right) + \mathbf{u}\cdot\nabla w \right] \mathrm{d}V = \int \left[-w\left(M\frac{\partial b}{\partial y} + \nabla\cdot\mathbf{u} \right) - \mathbf{v}\cdot\nabla b \right] \mathrm{d}V$$

can be regarded as a statement of adjointness $(\tilde{\psi}, T\tilde{\phi}) = (\tilde{\phi}, T^*\tilde{\psi})$ of a pair of linear operators T and T^* defined in (5.54). If the subspace is restricted to C^1 vectors and C^2 scalars, such T and T^* may be used to rewrite (5.53) as

$$T\tilde{\phi} = \begin{bmatrix} M\dfrac{\partial}{\partial y} & \nabla. \\ \nabla & 0 \end{bmatrix} \begin{bmatrix} w \\ \mathbf{v} \end{bmatrix} = \begin{bmatrix} 0 \\ \mathbf{u} \end{bmatrix} = \frac{\partial H}{\partial \tilde{\psi}}, \tag{5.54α}$$

$$T^*\tilde{\psi} = \begin{bmatrix} -M\dfrac{\partial}{\partial y} & -\nabla \\ -\nabla & 0 \end{bmatrix}\begin{bmatrix} b \\ \mathbf{u} \end{bmatrix} = \begin{bmatrix} 1 \\ -\mathbf{v} \end{bmatrix} = \frac{\partial H}{\partial \tilde{\phi}}. \qquad (5.54\beta)$$

Included here are the substitutions $\mathbf{u} = \nabla w$ and $\mathbf{v} = \nabla b$ which really begin the deduction of (5.54). On the right are the gradients of a Hamiltonian functional $H[\tilde{\psi}, \tilde{\phi}]$, such that $L[\tilde{\psi}, \tilde{\phi}] = (\tilde{\phi}, T^*\tilde{\psi}) - H$ allows (5.54) and therefore (5.53) to be again rewritten as (cf. 1.53))

$$\frac{\partial L}{\partial \tilde{\psi}} = 0 \quad (\alpha), \qquad \frac{\partial L}{\partial \tilde{\phi}} = 0 \quad (\beta). \qquad (5.55)$$

The problem is thus finally generated from a saddle functional

$$L[\tilde{\psi}, \tilde{\phi}] = \int \left[bM\frac{\partial w}{\partial y} + b\nabla\cdot\mathbf{v} + \mathbf{u}\cdot\nabla w - \tfrac{1}{2}\mathbf{u}^2 + \tfrac{1}{2}\mathbf{v}^2 - w \right] \mathrm{d}V \qquad (5.56)$$

which is concave in $\tilde{\psi}$ and convex in $\tilde{\phi}$. Uniqueness of solution, a free stationary principle (cf. Theorem 5.1) characterizing the saddle point, and simultaneous extremum principles, all follow. It must be remembered that the boundary conditions $w = b = 0$ on Σ are built into the subspace as essential constraints. Instead of the pair of spaces with different inner products $(.,.)$ and $\langle .,. \rangle$ allowed for in our general theory, we have been able to use the same space twice in this problem. We leave the reader to verify the following result.

Theorem 5.7

There is at most one solution w_0, b_0 of (5.53). It is characterized by a stationary principle based upon $0 = \delta L = (\delta\tilde{\psi}, (\partial L/\partial\tilde{\psi})) + (\delta\tilde{\phi}, (\partial L/\partial\tilde{\phi}))$. The solution value of L is $L_0 = -\frac{1}{2}\int w_0\,\mathrm{d}V$. The total fluid flux is bracketed by upper and lower bounds $L_\alpha \geqslant L_0 \geqslant L_\beta$. The solution minimizes

$$L_\alpha = \int [\tfrac{1}{2}(\mathbf{v}_\alpha{}^2 + (\nabla w_\alpha)^2) - w_\alpha]\,\mathrm{d}V \qquad (5.57)$$

over $\tilde{\phi}_\alpha = [w_\alpha \;\; \mathbf{v}_\alpha]^{\mathrm{T}}$ satisfying

$$M\,\partial w_\alpha/\partial y + \nabla\cdot\mathbf{v}_\alpha = 0 \text{ in } V, \text{ with } w_\alpha = 0 \text{ on } \Sigma. \qquad (5.58)$$

The solution maximizes

$$L_\beta = -\tfrac{1}{2}\int [\mathbf{u}_\beta{}^2 + (\nabla b_\beta)^2]\,\mathrm{d}V \qquad (5.59)$$

over $\tilde{\psi}_\beta = [b_\beta \;\; \mathbf{u}_\beta]^{\mathrm{T}}$ satisfying

$$M\,\partial b_\beta/\partial y + \nabla\cdot\mathbf{u}_\beta = -1 \text{ in } V, \text{ with } b_\beta = 0 \text{ on } \Sigma. \qquad (5.60)$$

The present decomposition (5.54) was given by Sewell (1977*b*). It is different from another decomposition of Smith (1976) who identified the same saddle functional (5.56), and obtained extremum principles for these and other boundary conditions (see also Example 3.3 of Smith (1985*a*)). There is no reason why more than one Hamiltonian representation cannot be used to characterize the same problem, or even why one saddle point cannot be viewed as belonging to more than one saddle functional.

(iii) Embedding method

As we have seen, embedding views a given problem as part of a larger problem with more variables and more equations. This device is used in the so-called mirror method (§4.3) where an adjoint problem is added to the given problem. If the latter is not characterizable by a free stationary principle, the combination of the given and adjoint problems will be so, although the combined problem will be twice the size of the given problem.

The method works *a fortiori* when the given problem does have a stationary principle, as in Theorem 5.7, and Barrett (1976) has treated (5.53) in this way. His approach can be reinterpreted in the following way. In attempting to satisfy the constraints (5.58) and (5.60) we are at liberty to choose the C^1 vector fields $\mathbf{v}_\alpha$ and $\mathbf{u}_\beta$ as, in particular, gradient fields. Thus choose C^2 scalar fields b_α and w_β (say) such that $\mathbf{v}_\alpha = \nabla b_\alpha$ in (5.58), and $\mathbf{u}_\beta = \nabla w_\beta$ in (5.60). The actual solution of (5.53) will belong to the set of w_α, w_β, b_α, b_β and will have the zero difference properties $w_\alpha - w_\beta = 0 = b_\alpha - b_\beta$. We can write (5.56) as $L[b, \mathbf{u}, w, \mathbf{v}]$, and Barrett's starting functional of b_α, w_β, w_α, b_β which generates his larger problem can be seen to be

$$
\begin{aligned}
L[b_\beta, \mathbf{u}_\beta, w_\alpha, \mathbf{v}_\alpha] &- L[b_\alpha, \mathbf{u}_\alpha, w_\beta, \mathbf{v}_\beta] \\
&= \int \left[\tfrac{1}{2}(\nabla w_\alpha)^2 + \tfrac{1}{2}(\nabla b_\alpha)^2 + b_\beta M \frac{\partial w_\alpha}{\partial y} - w_\alpha \right] \mathrm{d}V \\
&\quad - \int \left[\tfrac{1}{2}(\nabla w_\beta)^2 + \tfrac{1}{2}(\nabla b_\beta)^2 + b_\alpha M \frac{\partial w_\beta}{\partial y} - w_\beta \right] \mathrm{d}V. \qquad (5.61)
\end{aligned}
$$

In other words, a saddle functional is constructed as the difference of two other saddle functionals of C^2 scalars satisfying the boundary conditions. By requiring (5.61) to be stationary we imply the above zero difference properties, and thence (5.53).

5.5. Inner product spaces and adjoint operators in continuum mechanics

(i) Inner product spaces

In the remainder of this Chapter we shall consider several different media which are capable of transmitting shear stress, including elastic and plastic solids undergoing quasistatic incremental or large deformations, and also viscous fluids. Intrinsic in the analysis of such media are second order tensors, whose cartesian components we shall write as σ_{ij}, τ_{ij}, $e_{ij}, \ldots$ with respect to a fixed background frame, where i, $j = 1, 2, 3$ for a three dimensional case.

Up to now we have used a tilde consistently in symbols for composite elements, for example $\tilde{\phi}$, $\tilde{\mathbf{u}}$, $\tilde{\sigma}$ in §3.2, to avoid confusion with individual entries in the matrices so defined. Henceforward we omit such tildes because the meaning of the kernel letter alone will be sufficiently clear without tildes, as for σ and u in (5.62)–(5.65). The tilde has served its purpose by now, and the advent of tensor suffixes makes it desirable to simplify the notation where possible. For this reason we also omit many of the α and β suffixes such as we have regularly attached to solutions of the α and β sets of constraints. The context will make clear what is intended.

Consider a continuous body, of any material, whose reference configuration occupies a region V with external boundary Σ. Imagine $\Sigma = \Sigma_1 + \Sigma_2$ to be made up of two parts Σ_1 and Σ_2, as before, upon which two different types of boundary condition will be applied later. Let E be the linear space of composite elements σ, τ, $e, \ldots$ each defined as a set

$$\sigma = \begin{bmatrix} \sigma_{ij}(V) \\ \sigma_{ij}(\Sigma_1) \\ \sigma_{ij}(\Sigma_2) \end{bmatrix}, \quad \tau = \begin{bmatrix} \tau_{ij}(V) \\ \tau_{ij}(\Sigma_1) \\ \tau_{ij}(\Sigma_2) \end{bmatrix}, \ldots \tag{5.62}$$

of $3 \times (3 \times 3)$ component functions of position. These elements consist of three matrices each of 3×3 integrable functions over the regions, and in the order, indicated. We assign an inner product to E by the definition

$$(\sigma, \tau) = \int \sigma_{ij}\tau_{ji}\,\mathrm{d}V + \int \sigma_{ij}\tau_{ji}\,\mathrm{d}\Sigma_1 + \int \sigma_{ij}\tau_{ji}\,\mathrm{d}\Sigma_2, \tag{5.63}$$

in which the summation convention is used on the repeated subscripts, so that each integrand is the trace of a matrix product. These integrals can exist even if the integrands are products of functions which are not continuous or differentiable. The tensors can be unsymmetric, or symmetric if desired, and will represent various measures of stress, strain, etc., in due course. Alternative inner products are indicated in Exercise 5.3.5.

We can write $\sigma_{ij}\tau_{ji} = \sigma_{ij}\tau_{ij}$ in the case of symmetric tensors, which we shall be exclusively concerned with in §§5.6–5.8. It is sometimes convenient to rearrange their components into 6×1 matrices such as the $\boldsymbol{\tau}$ displayed in §2.6(ii), whence (5.62) becomes $\sigma = [\boldsymbol{\sigma}(V)\ \boldsymbol{\sigma}(\Sigma_1)\ \boldsymbol{\sigma}(\Sigma_2)]^{\mathrm{T}}$, etc., and

$$(\sigma, \tau) = \int \boldsymbol{\sigma}^{\mathrm{T}}\boldsymbol{\tau}\,\mathrm{d}V + \int \boldsymbol{\sigma}^{\mathrm{T}}\boldsymbol{\tau}\,\mathrm{d}\Sigma_1 + \int \boldsymbol{\sigma}^{\mathrm{T}}\boldsymbol{\tau}\,\mathrm{d}\Sigma_2.$$

Also intrinsic in the analysis will be various vectors (first order tensors), whose cartesian components we shall write as u_j, v_j,... Let F be the linear space of composite elements u, v,... each defined as a set

$$u = \begin{bmatrix} u_j(V) \\ u_j(\Sigma_1) \\ u_j(\Sigma_2) \end{bmatrix}, \quad v = \begin{bmatrix} v_j(V) \\ v_j(\Sigma_1) \\ v_j(\Sigma_2) \end{bmatrix}, \ldots \tag{5.64}$$

of 3×3 component integrable functions of position, over the regions and in the order indicated. We assign an inner product to F by the definition

$$\langle u, v \rangle = \int u_j v_j \mathrm{d}V + \int u_j v_j \mathrm{d}\Sigma_1 + \int u_j v_j \mathrm{d}\Sigma_2. \tag{5.65}$$

Again, at this stage the vector functions need not be continuous or differentiable. They will represent various types of force, displacement, etc.

In such definitions of inner product spaces, nothing need yet be said about the specific boundary conditions which will later be imposed over Σ_1 and Σ_2. For example, the boundary terms in (5.65) are certainly not required to take assigned values at this stage. The labels E and F for the spaces just defined replace the notations X and U of §3.3(ii), to avoid clashing with other meanings attached to X and U later.

(ii) Adjointness

We gave one illustration of this in (3.26), where σ_{ij} was not required to be symmetric. When σ_{ij} is symmetric, that illustration can be rewritten as

$$\int \sigma_{ij} \tfrac{1}{2}\left(\frac{\partial u_j}{\partial x_i} + \frac{\partial u_i}{\partial x_j}\right) \mathrm{d}V - \int \sigma_{ij} \tfrac{1}{2}(n_i u_j + n_j u_i) \mathrm{d}\Sigma_2$$
$$= -\int u_j \frac{\partial \sigma_{ij}}{\partial x_i} \mathrm{d}V + \int u_j n_i \sigma_{ij} \mathrm{d}\Sigma_1 \tag{5.66}$$

for every piecewise C^1 vector field u_j and for every piecewise C^1 symmetric tensor field σ_{ij}. Here x_i and n_j are typical cartesian components of position vector, and of unit normal to Σ, respectively. Finite jumps in the gradients of these components can be allowed across isolated internal surfaces without requiring extra terms in (5.66) to describe those jumps. We frequently require (5.66) in the sequel, and this fact prompts the following definitions.

Let E' be the subspace of E whose elements are constructed as in (5.62), but from symmetric σ_{ij}, etc., which are not merely integrable functions on V and Σ, but are also continuous and have piecewise continuous first derivatives $\partial \sigma_{ij}/\partial x_k$.

Let F' be the subspace of F whose elements are constructed as in (5.64), but from functions u_j, etc., on V and Σ which are not merely integrable, but are also continuous and have piecewise continuous first derivatives $\partial u_j/\partial x_k$.

By regarding (5.66) as an infinite dimensional example of adjointness (3.10), namely

$$(\sigma, T^*u) = \langle u, T\sigma \rangle \tag{5.67}$$

for all $\sigma \in E'$ and for all $u \in F'$, we induce in place of (3.27) the definitions

$$\begin{matrix} V \\ \\ \Sigma_1 \\ \Sigma_2 \end{matrix} \qquad T^*u = \begin{bmatrix} \frac{1}{2}\left(\frac{\partial u_j}{\partial x_i} + \frac{\partial u_i}{\partial x_j}\right) \\ 0 \\ -\frac{1}{2}(n_i u_j + n_j u_i) \end{bmatrix}, \quad T\sigma = \begin{bmatrix} -\frac{\partial \sigma_{ij}}{\partial x_i} \\ n_i\sigma_{ij} \\ 0 \end{bmatrix} \tag{5.68}$$

of a linear operator $T: \mathscr{D}(T) = F' \to E$ and its adjoint $T^*: \mathscr{D}(T^*) = E' \to F$. As indicated, the three rows correspond to V, Σ_1 and Σ_2 respectively. The inner products in (5.67) are (5.63) and (5.65). These operators were introduced by Sewell (1973*a*, *b*), where it was also shown how to accommodate the discontinuities mentioned in §(iv) below.

Any term in (5.66) could be moved to the other side of the equation, to prompt variants of the definitions (5.62)–(5.65) and (5.68). For example, if only one type of boundary condition is anticipated over the whole of Σ, all the surface integrals could be put on one side of (5.66). If this is, say, the right side of (5.66), we could write $\Sigma_2 = 0$, $\Sigma_1 = \Sigma$ to simplify (5.62)–(5.65) and (5.68) by removing the Σ_2 contributions altogether.

(iii) Virtual work

The version

$$\int \sigma_{ij}\tfrac{1}{2}\left(\frac{\partial u_j}{\partial x_i} + \frac{\partial u_i}{\partial x_j}\right) \mathrm{d}V = -\int u_j \frac{\partial \sigma_{ij}}{\partial x_i} \mathrm{d}V + \int u_j n_i \sigma_{ij} \, \mathrm{d}\Sigma \tag{5.69}$$

of (5.66) describes, regardless of boundary conditions, two alternative versions of what would be called the total internal virtual work. This name anticipates later interpretations of σ_{ij} and u_j as stress and displacement. Even if they have those meanings, the work is virtual at this stage, and not actual in any specific medium, because no relation between the σ_{ij} and u_j has yet been laid down.

Rewritten in abstract notation, (5.69) is $(\sigma, T^*u) = \langle u, T\sigma \rangle$ with

$$\begin{matrix} V \\ \\ \Sigma \end{matrix} \qquad T^*u = \begin{bmatrix} \frac{1}{2}\left(\frac{\partial u_j}{\partial x_i} + \frac{\partial u_i}{\partial x_j}\right) \\ 0 \end{bmatrix}, \quad T\sigma = \begin{bmatrix} -\frac{\partial \sigma_{ij}}{\partial x_i} \\ n_i\sigma_{ij} \end{bmatrix}.$$

An appropriate *principle of virtual work* would then be the assertion that, if g

is any member of F, not necessarily preassigned,

$$(\sigma, T^*u) = \langle g, u \rangle \tag{5.70}$$

for all $u \in F'$. If $g = [g_j(V)\ s_j(\Sigma)]^{\mathrm{T}}$ assembles body forces g_j and surface tractions s_j, (5.70) can be read as asserting that the internal virtual work is equal to the external virtual work.

Theorem 5.8
The principle of virtual work implies

$$T\sigma = g. \tag{5.71}$$

Proof
Using (5.67) with (5.70) implies $\langle u, T\sigma - g \rangle = 0$ for all $u \in F'$, i.e. for all piecewise C^1 vector fields. The result (5.71) follows in the form

$$-\frac{\partial \sigma_{ij}}{\partial x_i} = g_j \text{ in } V, \quad n_i\sigma_{ij} = s_j \text{ on } \Sigma \tag{5.72}$$

by an orthogonality argument analogous to the fundamental lemma of the calculus of variations. □

It can be seen from (5.72) that this principle of virtual work implies that the internal stresses and external forces are equilibrated. A detailed historical background to several variants of the principle of virtual work is provided by Truesdell and Toupin (1960, §232). An engineer's view of the principle, with specific applications to frame structures and to the finite element method, is described by Davies (1982). We have written a concrete case in abstract notation, in this subsection, to suggest the connection with other concrete cases such as the matrix analysis of structures, where equilibrium could again appear as (5.71), and compatibility as $T^*u = e$ (see Exercise 5.3.1).

(iv) Discontinuities of stress and displacement

We may anticipate discontinuities $[\![u_j]\!]$ of displacement and $[\![\sigma_{ij}]\!]$ of stress across isolated internal surfaces S_2 and S_1, respectively, as follows. The adjointness formula (5.66) will need to have added terms

$$-\int \sigma_{ij}\tfrac{1}{2}(n_i[\![u_j]\!] + n_j[\![u_i]\!])\,\mathrm{d}S_2 \text{ to the left side and } \int u_j n_i [\![\sigma_{ij}]\!]\,\mathrm{d}S_1$$

to the right side, simply by describing the discontinuity surfaces twice in opposite senses during the integration by parts, with n_i here the suitably oriented normal to $S_1 + S_2$. Extra terms over S_1 and S_2 can be added to the definitions (5.62)–(5.65) of spaces and inner products. We augment the adjoint operators (5.68) by adding $-\frac{1}{2}(n_i[\![u_j]\!]+n_j[\![u_i]\!])$ on S_2 and 0 on S_1 to T^*u, and by adding 0 on S_2 and $n_i[\![\sigma_{ij}]\!]$ on S_1 to $T\sigma$.

The subsequent governing conditions will impose restrictions on these extra members of T^*u and $T\sigma$. For example, equilibrium on S_1 will require $n_i[\![\sigma_{ij}]\!]=0$ there (continuous traction); incompressible material will require $n_i[\![u_i]\!]=0$ on S_2 (tangential discontinuity only). Approximation methods (e.g. finite element) will entail other conditions, as indicated, for example, by Prager (1967) and Nemat–Nasser (1972*a*). For simplicity we omit stress and displacement discontinuities in what follows, except for some remarks in §5.8.

5.6. Classical elasticity

(i) Governing equations

For classical infinitesimal elasticity we use u_j, σ_{ij} and e_{ij} to denote small displacement, true (Cauchy) stress and small strain respectively. The σ_{ij} and e_{ij} are symmetric. They are related linearly as explained in §2.6(ii), i.e. as $\boldsymbol{\sigma}=\partial U/\partial \mathbf{e}$ which we shall formally rewrite here as $\sigma_{ij}=\partial U/\partial e_{ij}$. The strain energy $U[e_{ij}]$ is a given homogeneous quadratic and strictly convex function of the six independent strains. For isotropic material this is illustrated after (2.83) and, using δ_{ij} as the Kronecker delta, leads to

$$\sigma_{ij}=2\mu e_{ij}+\lambda(e_{kk})\delta_{ij}. \tag{5.73}$$

We assume that there is no internal kinematical constraint such as incompressibility (see §5.7). Let b_j, c_j and a_j be the components of assigned vector functions of position on V, Σ_2 and Σ_1 respectively. The complete set of equations governing equilibrium is as follows.

Displacement/strain in V: $\quad \dfrac{1}{2}\left(\dfrac{\partial u_j}{\partial x_i}+\dfrac{\partial u_i}{\partial x_j}\right)=e_{ij}.\quad(\alpha)$

Assigned displacement on Σ_2: $\quad u_j=c_j.\quad(\alpha)$

Equilibrium in V: $\quad -\dfrac{\partial \sigma_{ij}}{\partial x_i}=b_j.\quad(\beta)$

Assigned traction on Σ_1: $\quad n_i\sigma_{ij}=a_j.\quad(\beta)$

Hooke's Law in V: $\quad \sigma_{ij}=\dfrac{\partial U}{\partial e_{ij}}.\quad(\beta)$

With such boundary conditions on $\Sigma_1\neq 0$ and $\Sigma_2\neq 0$, this is called the mixed boundary value problem. The subset of kinematical conditions is (5.74α), and the subset of force-like conditions is (5.74β).

(ii) Generalized Hamiltonian representation

We can use the operators in (5.68) to rewrite (5.74) as

$$T^*u = \frac{\partial H}{\partial \sigma} \quad (\alpha), \qquad T\sigma = \frac{\partial H}{\partial u} \quad (\beta), \qquad 0 = \frac{\partial H}{\partial e}, \quad (\beta) \tag{5.75}$$

where

$$\begin{matrix} V \\ \Sigma_1 \\ \Sigma_2 \end{matrix} \quad \frac{\partial H}{\partial \sigma} = \begin{bmatrix} e_{ij} \\ 0 \\ -\frac{1}{2}(n_i c_j + n_j c_i) \end{bmatrix}, \quad \frac{\partial H}{\partial u} = \begin{bmatrix} b_j \\ a_j \\ 0 \end{bmatrix}, \quad \frac{\partial H}{\partial e} = \begin{bmatrix} \sigma_{ij} - \dfrac{\partial U}{\partial e_{ij}} \\ 0 \\ 0 \end{bmatrix} \tag{5.76}$$

are the gradients of the generalized Hamiltonian functional

$$H[\sigma; u, e] = \int (\sigma_{ij} e_{ij} - U[e_{ij}] + b_j u_j)\,\mathrm{d}V - \int n_i \sigma_{ij} c_j\,\mathrm{d}\Sigma_2 + \int a_j u_j\,\mathrm{d}\Sigma_1. \tag{5.77}$$

Such gradients are defined by expressing the first variation of H as

$$\delta H = \left(\delta\sigma, \frac{\partial H}{\partial \sigma}\right) + \left\langle \delta u, \frac{\partial H}{\partial u}\right\rangle + \left(\delta e, \frac{\partial H}{\partial e}\right). \tag{5.78}$$

The Hamiltonian $H[\sigma; u, e]$ is defined over three spaces (E, F and E repeated) in the first instance, instead of the usual two. We can combine them into two, however, one of force-like variables (σ) and the other of independent kinematical variables (u and e taken together in a space with inner product $\langle .,. \rangle + (.,.)$). Then $H[\sigma; u, e]$ is a saddle functional with respect to this pair of spaces, convex in σ and concave in u with e, because of the strictly concave $-U[e_{ij}]$, the bilinear $\sigma_{ij} e_{ij}$, and the linear remaining terms.

Equations (5.75) carry some $0=0$ entries arising from the zeros in (5.68) and (5.76). These can be omitted if we use three spaces defined in a different way, as in Exercise 5.3.5, and for some purposes this is more flexible (see §4.5(ii)).

(iii) Generating saddle functional

Pursuing our approach, as already illustrated in deriving (5.36), we are led to introduce the functional

$$\begin{aligned} L[\sigma; u, e] &= (\sigma, T^*u) - H[\sigma; u, e] \\ &= \langle u, T\sigma \rangle - H[\sigma; u, e] \\ &= \int \left(-u_j \frac{\partial \sigma_{ij}}{\partial x_i} - \sigma_{ij} e_{ij} + U[e_{ij}] - b_j u_j \right) \mathrm{d}V \\ &\quad + \int u_j (n_i \sigma_{ij} - a_j)\,\mathrm{d}\Sigma_1 + \int n_i \sigma_{ij} c_j\,\mathrm{d}\Sigma_2. \end{aligned} \tag{5.79}$$

This is a saddle functional, concave in σ and convex in u with e, which is of central importance because we can expect to generate the governing equations, stationary principles and extremum principles from it by differentiation. Its gradients are

$$\frac{\partial L}{\partial \sigma} = T^*u - \frac{\partial H}{\partial \sigma}, \quad \frac{\partial L}{\partial u} = T\sigma - \frac{\partial H}{\partial u}, \quad \frac{\partial L}{\partial e} = -\frac{\partial H}{\partial e}. \tag{5.80}$$

The equilibrium problem (5.74) or (5.75) is therefore equivalent to

$$\frac{\partial L}{\partial \sigma} = 0 \quad (\alpha), \qquad \frac{\partial L}{\partial u} = \frac{\partial L}{\partial e} = 0 \quad (\beta). \tag{5.81}$$

The given data of the problem include the set of moduli (coefficients) in $U[e_{ij}]$, the functions b_j, c_j and a_j, and the shape of the body. For sufficiently smooth data, the simplest type of *actual solution* is a set of component functions σ_{ij}, u_j and e_{ij} of position, respectively C^1, C^2 and.C^1, which satisfy (5.74) or (5.81). A closer description of such details is provided by Gurtin (1972). However, the domain of $L[\sigma; u, e]$ is built from piecewise C^1 σ_{ij} and u_j and integrable e_{ij}.

Theorem 5.9 (Free stationary principle)
$L[\sigma; u, e]$ is stationary with respect to arbitrary independent small piecewise C^1 variations $\delta\sigma_{ij}$, δu_j and δe_{ij} of its arguments if and only if they are variations from an actual solution.

Proof
The first variation of $L[\sigma; u, e]$ is

$$\delta L = \left(\delta\sigma, \frac{\partial L}{\partial \sigma}\right) + \left\langle \delta u, \frac{\partial L}{\partial u}\right\rangle + \left(\delta e, \frac{\partial L}{\partial e}\right) \tag{5.82}$$

by the same calculation which produced (5.80), using adjointness (5.67). Hence (5.74), and therefore (5.81), implies $\delta L = 0$, which proves sufficiency. Necessity requires the fundamental lemma of the calculus of variations, as detailed by Gurtin (1972). See also §3.9(iii). □

Theorem 5.9 would be true even if $U[e_{ij}]$ were neither convex nor quadratic. It is a precise analogue of the variant (a) described after Theorem 5.1; σ_{ij}, u_j, e_{ij}, $U[e_{ij}]$ here correspond respectively to $\mathbf{Q}$, ϕ, $\mathbf{v}$, $-p[\mathbf{v}]$ there.

When $U[e_{ij}]$ is strictly convex, say $U = \frac{1}{2}\mathbf{e}^{\mathrm{T}}K\mathbf{e}$ with positive definite K in the notation of (2.83) as we assume henceforth, we can introduce the saddle quantity (cf. (3.66) and (5.44))

$$\begin{aligned} S &= L[\sigma_+; u_+, e_+] - L[\sigma_-; u_-, e_-] \\ &\quad - \left(\sigma_+ - \sigma_-, \left.\frac{\partial L}{\partial \sigma}\right|_+\right) - \left\langle u_+ - u_-, \left.\frac{\partial L}{\partial u}\right|_-\right\rangle - \left(e_+ - e_-, \left.\frac{\partial L}{\partial e}\right|_-\right) \\ &= \tfrac{1}{2}\int (\mathbf{e}_+ - \mathbf{e}_-)^{\mathrm{T}} K (\mathbf{e}_+ - \mathbf{e}_-)\,\mathrm{d}V > 0 \end{aligned} \tag{5.83}$$

for any pair σ_+, u_+, e_+ and σ_-, u_-, e_- in which the strains in V are distinct. From this we confirm that $L[\sigma; u, e]$ is a saddle functional, concave in σ and convex in u with e.

Theorem 5.10 (Uniqueness)
An actual solution has unique σ_{ij}, e_{ij} and, except possibly when $\Sigma_2 = 0$, unique u_j.

Proof
Interchange the plus and minus points in (5.83), and add, to give

$$-\left(\sigma_+ - \sigma_-, \frac{\partial L}{\partial \sigma}\bigg|_+ - \frac{\partial L}{\partial \sigma}\bigg|_-\right) + \left\langle u_+ - u_-, \frac{\partial L}{\partial u}\bigg|_+ - \frac{\partial L}{\partial u}\bigg|_-\right\rangle$$
$$+\left(e_+ - e_-, \frac{\partial L}{\partial e}\bigg|_+ - \frac{\partial L}{\partial e}\bigg|_-\right) = \int (\mathbf{e}_+ - \mathbf{e}_-)^{\mathrm{T}} K (\mathbf{e}_+ - \mathbf{e}_-)\mathrm{d}V > 0 \qquad (5.84)$$

for any pair of points with distinct strains. If both members of such a pair could be solutions of (5.81), the left side of (5.84) would be zero. The resulting contradiction implies uniqueness of the e_{ij} in an actual solution, and therefore uniqueness of σ_{ij} from $\sigma_{ij} = \partial U/\partial e_{ij}$. The only nonuniqueness left available to the displacement field u_j is a rigid body movement, which involves no strain, and in general this will be inconsistent with the assigned displacement over Σ_2. □

(iv) Simultaneous upper and lower bounds

Before any constraints are imposed we find from (5.79) that

$$L - \left(\sigma, \frac{\partial L}{\partial \sigma}\right) = \int (U[e_{ij}] - b_j u_j)\,\mathrm{d}V - \int a_j u_j \mathrm{d}\Sigma_1 \qquad (5.85)$$

which has the values of the total potential energy; and

$$L - \left\langle u, \frac{\partial L}{\partial u}\right\rangle - \left(e, \frac{\partial L}{\partial e}\right) = \int \left(U[e_{ij}] - e_{ij}\frac{\partial U}{\partial e_{ij}}\right)\mathrm{d}V + \int n_i \sigma_{ij} c_j \mathrm{d}\Sigma_2 \qquad (5.86)$$

whose negative can be defined, via (2.82), as having the values of the total complementary energy. Notice that, as in previous examples, the operators T and T^* do not themselves appear in (5.85) and (5.86).

Theorem 5.11 (Upper bound)
The potential energy functional

$$J[u] = \int \left(U\left[\frac{1}{2}\left(\frac{\partial u_j}{\partial x_i} + \frac{\partial u_i}{\partial x_j}\right)\right] - b_j u_j\right)\mathrm{d}V - \int a_j u_j \mathrm{d}\Sigma_1 \qquad (5.87)$$

evaluated for any piecewise C^2 displacement field which satisfies $u_j = c_j$ on Σ_2, is an upper bound to its value for an actual solution of all the conditions (5.74).

Proof
This follows from the saddle inequality (5.83) on exactly the same pattern which we first introduced in Theorem 1.4. That is (cf. (1.38)), we choose σ_+, u_+, e_+ to be any solution of the kinematical subset (5.74α) or (5.81α), and σ_-, u_-, e_- to be any solution of all of (5.74) or of (5.81). We regard (5.81) as a generalization of (1.53). The value of (5.85) under the constraint (5.81α) is an upper bound L_α, which we write as (5.87) subject to $u_j = c_j$ on Σ_2, after eliminating the e_{ij} in favour of the u_j. □

We can expect that a converse theorem is available justifying a principle of minimum potential energy (see §3.9(iii) and Gurtin, 1972, §36). The stationary character of the minimum can be inferred from the quadratic right side of (5.83). Alternatively, the device described just before Theorem 5.2 can be used here in conjunction with Theorem 5.9. Impose (5.74α) from the outset, so that e_{ij} is expressed in terms of u_j, and from (5.85), $L[\sigma; u, e]$ becomes $J[u]$. Sufficiency follows from a direct calculation of δJ exactly like that in the proof of Theorem 5.2. Necessity requires the fundamental lemma, the natural conditions now being (5.74β) in combined form with the intermediate role of σ_{ij} and e_{ij} deleted.

Theorem 5.12 (Lower bound)
The complementary energy functional

$$-K[\sigma] = \int U_c[\sigma_{ij}]\,dV - \int n_i \sigma_{ij} c_j \,d\Sigma_2 \tag{5.88}$$

evaluated for any C^1 stress field satisfying the subset (5.74β) *of force-like conditions (which include the equilibrium equations assuming continuous body force), is an upper bound to its value for an actual solution of all the conditions* (5.74); *hence* $K[\sigma]$ *is a lower bound.*

Proof
Choose σ_-, u_-, e_- in (5.83) to be any solution of (5.74β) or (5.81β). The value of (5.86) under these constraints is $K[\sigma]$ by using the Legendre transform (2.82), which in the present case defines the local complementary energy $U_c[\sigma_{ij}]$ by (2.83). Such $K[\sigma]$ is shown to be a lower bound when σ_+, u_+, e_+ in (5.83) is chosen to be any solution of all of (5.74) or (5.81). Hence $-K[\sigma]$ is an upper bound. □

We can expect a converse theorem justifying a principle of minimum complementary energy (cf. §3.9(iii)). The stationary property of this minimum can be handled by an argument like that based on the linearity property (5.41) in Theorem 5.3. Here we first deploy $\partial L/\partial e = 0$ as $\sigma_{ij} = \partial U/\partial e_{ij}$ in inverse form to eliminate $e_{ij} = \partial U_c/\partial \sigma_{ij}$ from (5.79), leaving $L[\sigma, u, e[\sigma]] = \langle u, \partial L/\partial u\rangle - K[\sigma]$. The remaining β-constraint $\partial L/\partial u = 0$ is then freed by the use of a Lagrange multiplier μ and the augmented functional

$$\begin{aligned} \bar{L}[\sigma, u, \mu, e[\sigma]] &= \langle u - \mu, \partial L/\partial u\rangle - K[\sigma] \\ &= L[\sigma, u - \mu, e[\sigma]]. \end{aligned} \tag{5.89}$$

The natural conditions for this are (5.74), but with $\partial U_c/\partial \sigma_{ij}$ in place of e_{ij} and $u_j - \mu_j$ in place of u_j, finally deeming $\mu_j = 0$.

Since $U = U_c = \frac{1}{2}\sigma_{ij}e_{ij}$ in value, the common actual solution value of $J[u]$ and $K[\sigma]$, i.e. the quantity bracketed by the upper and lower bounds, is

$$L_0 = \tfrac{1}{2}\int n_i\sigma_{ij}c_j \, d\Sigma_2 - \tfrac{1}{2}\int a_j u_j \, d\Sigma_1 - \tfrac{1}{2}\int b_j u_j \, dV.$$

This is a measure of external work which is often of considerable practical interest.

The literature is enormous and the principles associated with Theorem 5.11 and 5.12 are well known. They are often also called dual extremum principles, because of the Legendre transformation relating U and U_c. Our purpose has been to show how the bounds and the governing equations are all expressible directly in terms of a single generating saddle functional $L[\sigma; u, e]$, which was reached after identifying a generalized Hamiltonian version an as intermediate step.

A convenient review of the historical development of 'energetical principles' in the mechanics of rigid and elastic bodies, from Heraclitos of Ephesos (B.C. 550–475) to Prange (A.D. 1914), has been given by Oravas and McLean (1966). Their account reminds one that unfamiliar names, such as that of Cotterill in 1865, deserve to be associated with certain key concepts, and that the originator of an idea is not necessarily its most successful exponent or publicist. Gurtin (1972) also gives many historical references, and a discussion of technical details and of some of the many variants of stationary principles which we have not space to include here. In particular, he calls Theorem 5.9 the Hu-Washizu principle, quoting papers of 1955. However, Oravas and McLean attribute to Born, in his 1906 dissertation, the discovery of a free stationary 'minimo-maximal' principle in elasticity, using stresses and displacements as independent variables with an associated Legendre transformation; and as a way of subsuming the minimum potential energy principle into a single functional, wherein constraints and natural conditions have a reciprocal role. Gurtin indicates that work on the two extremum principles for the mixed problem treated here culminated in a paper of Trefftz in 1928. Our approach was originally stimulated by the emphasis given by Hill (1956*a*) to the role of convexity in several types of solids and fluids. Systematic study of other aspects of the structure of several stationary principles of elasticity has been carried out by Tonti (1967, 1975). Oden and Reddy (1982) review various extremum principles in theoretical mechanics.

There are many texts which illustrate the more classical applications of the extremum principles in elasticity, both in engineering structural mechanics (e.g. Gregory (1969)), and in boundary value problems (e.g. Sokolnikoff (1956) or Washizu (1968)) typically represented by estimation of torsional rigidity of a bar. In recent years extensions of the principles have appeared incorporating less smoothness than we have specified here. This is important, for example, in composite materials, where the moduli are discontinuous functions of position. In particular, bounds can be placed on the overall effective moduli for heterogeneous materials. The needs of numerical methods of solution have also stimulated consideration of discontinuous fields of stress and displacement themselves. Some of these matters are reviewed by Nemat–Nasser (1972*b*) and Willis (1982).

5.7. Slow viscous flow

(i) Governing equations

We use the problem of slow flow of a Newtonian viscous fluid to illustrate how adjoint operators can handle incompressibility, which is an example of an internal kinematical constraint. The governing equations are formally the same as those for the isotropic elastic solid (5.73) in §5.6, when subjected to this extra constraint, except that we choose $\Sigma_2 = \Sigma$ and $\Sigma_1 = 0$ because it is adequate here to deal with one type of boundary condition only.

Let v_j, σ_{ij} and $\dot{e}_{ij}$ denote cartesian components of velocity, true stress and strain-rate respectively. Let c_j be the components of an assigned vector on the fluid boundary Σ. With zero body force, and given viscosity $\mu > 0$, the governing equations are as follows.

Velocity/strain-rate in V: $$\frac{1}{2}\left(\frac{\partial v_i}{\partial x_j} + \frac{\partial v_j}{\partial x_i}\right) = \dot{e}_{ij}. \qquad (\alpha)$$

Incompressibility in V: $$-\frac{\partial v_j}{\partial x_j} = 0. \qquad (\alpha)$$

No slip on Σ: $$v_j = c_j. \qquad (\alpha) \quad (5.90)$$

Slow motion in V: $$-\frac{\partial \sigma_{ij}}{\partial x_i} = 0. \qquad (\beta)$$

Constitutive law in V: $$\sigma_{ij} - \tfrac{1}{3}(\sigma_{kk})\delta_{ij} = 2\mu \dot{e}_{ij}. \qquad (\beta)$$

The inertial forces have been neglected in comparison with the internal stresses. This will be valid if the flow is so slow and so nearly steady that the quadratic and unsteady contributions to the acceleration are each negligible. Again, the (α)-subset is that of kinematical conditions, and the (β)-subset is that of force-like conditions.

(ii) Generalized Hamiltonian representation

We need the two linear spaces E and F in §5.5 (simplified by $\Sigma_2 = \Sigma$ and $\Sigma_1 = 0$), and a third space G of elements

$$p = \begin{bmatrix} p(V) \\ p(\Sigma) \end{bmatrix}, \quad q = \begin{bmatrix} q(V) \\ q(\Sigma) \end{bmatrix}, \ldots \tag{5.91}$$

constructed from integrable scalar functions of position, so that an inner product can be assigned to G by the definition

$$\overline{p, q} = \int pq \, \mathrm{d}V + \int pq \, \mathrm{d}\Sigma. \tag{5.92}$$

In Chapter 3 we would have written $\tilde{p}$ and $\tilde{q}$ on the left in (5.91) and (5.92), but now we rely on the context to obviate any confusion which might otherwise arise from the fact that we use the same symbol for the element of G, and for the values of the function used in the definition of that element. Let G' be the subspace of G constructed from piecewise C^1 scalars. Then

$$-\int p \frac{\partial v_j}{\partial x_j} \mathrm{d}V + \int p n_j v_j \, \mathrm{d}\Sigma = \int v_j \frac{\partial p}{\partial x_j} \mathrm{d}V \tag{5.93}$$

can be interpreted as $\overline{p, R^* v} = \langle v, Rp \rangle$ to induce a pair of adjoint operators R^*: $F' \to G$ and R: $G' \to F$ with definitions

$$\begin{matrix} V \\ \\ \Sigma \end{matrix} \quad R^* v = \begin{bmatrix} -\dfrac{\partial v_j}{\partial x_j} \\ \\ n_j v_j \end{bmatrix}, \quad Rp = \begin{bmatrix} \dfrac{\partial p}{\partial x_j} \\ \\ 0 \end{bmatrix}. \tag{5.94}$$

Let r_{ij} be a symmetric second order tensor inducing an element $r = [r_{ij}(V) \; r_{ij}(\Sigma)]^{\mathrm{T}}$ of E. By using (5.68) with $\Sigma_1 = 0$, we can assemble a matrix of operators to rewrite the governing equations (5.90) as

$$\begin{bmatrix} 0 & 0 & T^* & 0 \\ 0 & 0 & R^* & 0 \\ T & R & 0 & 0 \\ 0 & 0 & 0 & 0 \end{bmatrix} \begin{bmatrix} r \\ p \\ v \\ \dot{e} \end{bmatrix} = \partial H / \partial \begin{bmatrix} r \\ p \\ v \\ \dot{e} \end{bmatrix} \quad \begin{matrix} (\alpha) \\ (\alpha) \\ (\beta) \\ (\beta) \end{matrix} \tag{5.95}$$

in terms of gradients of the generalized Hamiltonian

$$H[r,p,v,\dot{e}]=\int(r_{ij}\dot{e}_{ij}-U[\dot{e}_{ij}])\mathrm{d}V-\int n_i(r_{ij}-p\delta_{ij})c_j\,\mathrm{d}\Sigma \tag{5.96}$$

where

$$U[\dot{e}_{ij}]=\mu\dot{e}_{ij}\dot{e}_{ij},\quad \partial U/\partial\dot{e}_{ij}=2\mu\dot{e}_{ij}. \tag{5.97}$$

Explicitly, the subset (5.95β) is

$$-\frac{\partial r_{ij}}{\partial x_i}+\frac{\partial p}{\partial x_j}=0,\quad 0=r_{ij}-2\mu\dot{e}_{ij}. \tag{5.95β}$$

These equations can be identified with the force-like subset (5.90β) of governing conditions if and only if

$$p=-\tfrac{1}{3}\sigma_{kk}=\text{pressure},\quad \text{and}\quad r_{ij}=\sigma_{ij}+p\delta_{ij}=\text{stress deviator}. \tag{5.98}$$

(iii) Generating saddle functional

The composite operator matrix on the left of (5.95) is self-adjoint, in the sense that for any two 'supervectors' $[r_1\ p_1\ v_1\ \dot{e}_1]^{\mathrm{T}}$ and $[r_2\ p_2\ v_2\ \dot{e}_2]^{\mathrm{T}}$,

$$\begin{aligned}(r_2,T^*v_1)&+\overline{p_2,R^*v_1}+\langle v_2,Tr_1+Rp_1\rangle\\ &=(r_1,T^*v_2)+\overline{p_1,R^*v_2}+\langle v_1,Tr_2+Rp_2\rangle.\end{aligned} \tag{5.99}$$

From this self-adjointness we can expect to express (5.95) in terms of a single generating functional. By a natural extension of our theoretical pattern (2.112), illustrated in (5.36) and (5.79), we can introduce the functional

$$\begin{aligned}L[r,p,v,\dot{e}]&=(r,T^*v)+\overline{p,R^*v}-H\\ &=\int\left[r_{ij}\left(\frac{1}{2}\left(\frac{\partial v_i}{\partial x_j}+\frac{\partial v_j}{\partial x_i}\right)-\dot{e}_{ij}\right)+U[\dot{e}_{ij}]-p\frac{\partial v_j}{\partial x_j}\right]\mathrm{d}V\\ &\quad-\int n_i(r_{ij}-p\delta_{ij})(v_j-c_j)\,\mathrm{d}\Sigma.\end{aligned} \tag{5.100}$$

The establishment of this functional of unconstrained quantities r_{ij}, p, v_j and $\dot{e}_{ij}$ is a pivotal stage in the theory, in the sense that, as in other situations, the governing equations and stationary and extremum principles can all be obtained from it by differentiation. The strictly convex $U[\dot{e}_{ij}]$ and the bilinear or linear other terms show it to be a saddle functional concave in the stress-like variables r_{ij} and p, and convex in the kinematical variables v_j and $\dot{e}_{ij}$. The domain of (5.100) is built from piecewise C^1 σ_{ij}, p and v_j and integrable $\dot{e}_{ij}$.

Before the application of any constraints, we know from our general formalism that simultaneous extremum principles will characterise the quantities

$$L-\left(r,\frac{\partial L}{\partial r}\right)-\overline{p,\frac{\partial L}{\partial p}}=\int U[\dot{e}_{ij}]\mathrm{d}V, \tag{5.101}$$

$$L-\left\langle v,\frac{\partial L}{\partial v}\right\rangle-\left(\dot{e},\frac{\partial L}{\partial \dot{e}}\right)=\int\left(U[\dot{e}_{ij}]-\dot{e}_{ij}\frac{\partial U}{\partial \dot{e}_{ij}}\right)\mathrm{d}V$$
$$+\int n_i(r_{ij}-p\delta_{ij})c_j\,\mathrm{d}\Sigma. \tag{5.102}$$

The governing equations (5.90) or (5.95) can be rewritten

$$\frac{\partial L}{\partial r}=\frac{\partial L}{\partial p}=0 \quad (\alpha), \qquad \frac{\partial L}{\partial v}=\frac{\partial L}{\partial \dot{e}}=0 \quad (\beta). \tag{5.103}$$

The details of subsequent theoretical developments will be clear from the pattern of §5.6, and we abbreviate them in the following statement.

Theorem 5.13
An actual solution of (5.90) has unique $\dot{e}_{ij}$, r_{ij} and (unless c_j permits a rigid body motion) unique v_j (cf. Theorem 5.10). It is characterized by a free stationary principle for $L[r,p,v,\dot{e}]$ (cf. Theorem 5.9). The solution value of L is $L_0=\frac{1}{2}\int n_i\sigma_{ij}c_j\,\mathrm{d}\Sigma$, which is half the work-rate of the boundary forces (the drag in translational motion). This is also characterized by stationary simultaneous extremum principles whose upper and lower bounds in $J[v]\geqslant L_0\geqslant K[r,p]$ are as follows. The solution minimizes

$$J[v]=\int U\left[\frac{1}{2}\left(\frac{\partial v_j}{\partial x_i}+\frac{\partial v_i}{\partial x_j}\right)\right]\mathrm{d}V \tag{5.104}$$

over all incompressible C^2 velocity fields which satisfy the zero slip boundary condition. The solution maximizes

$$K[r,p]=-\int U_c[r_{ij}]\mathrm{d}V+\int n_i(r_{ij}-p\delta_{ij})c_j\,\mathrm{d}\Sigma \tag{5.105}$$

over all C^1 r_{ij} and p fields which satisfy $\partial r_{ij}/\partial x_i=\partial p/\partial x_j$. Here the Legendre dual of $U[\dot{e}_{ij}]$ in (5.97) is

$$U_c[r_{ij}]=\frac{1}{4\mu}r_{ij}r_{ij}, \quad \text{with } \frac{\partial U_c}{\partial r_{ij}}=\frac{1}{2\mu}r_{ij}=\dot{e}_{ij}. \tag{5.106}$$

(iv) Historical remarks and applications

Slow steady flows of Newtonian viscous fluid are called Stokes flows. Keller *et al.* (1967) remark that the principle minimizing (5.104) was stated by Helmholtz in 1868 and proved by Korteweg in 1883 (see also Lamb, 1952,

§344). Since $2U = r_{ij}\dot{e}_{ij}$ in value, which is $\sigma_{ij}\dot{e}_{ij}$ for incompressible flows, the quantity $J[v]$ so minimized is the total rate of energy dissipation. Although Rayleigh noted the analogy between (5.74) and (5.90), it seems that many years passed before the dual principle maximizing (5.105) was established, by Hill and Power (1956). They gave some illustrations of drag estimation on translated bodies, including the influence of containing boundaries. For instance, they deduced the bounds $\frac{40}{7} \leqslant F/(\pi\mu a w) \leqslant \frac{56}{9}$ for the drag F on a sphere of radius a moving with speed w; these bounds 5.97 ± 0.25 compare well with the known exact value 6. Bounds for the flux through tubes have been found by Gaydon and Nuttall (1959) and Jamal *et al.* (1979). Bounds for the torque needed to sustain rotation of a cylinder, and for flux in open channels applied to conveyor belt transport of viscous fluids, have been given by Nuttall (1965, 1966). Keller *et al.* (1967) gave a considerable extension of the extremum principles in the direction of more elaborate boundary conditions, including those appropriate for theories of suspensions containing particles whose motion is unknown in advance. Skalak (1969) showed how to accommodate not only rigid particles but also deformable droplets (e.g. red blood cells).

Hill (1956*a*) indicated how extremum principles for some other rheological materials could be obtained by direct integration of a convexity inequality. Apart from this, the above-mentioned authors were not concerned with the structural aspects of the theory which we have presented, namely the expression of governing conditions and extremum principles all in terms of the gradients of a single generating saddle functional. Johnson (1960, 1961), however, did give a free stationary principle as well as simultaneous extremum principles for certain non-Newtonian fluids.

Exercises 5.3

1. Consider the underdetermined problem of two equations

$$T\sigma = g, \quad T^*u = e$$

relating four quantities $\sigma \in E'$, $e \in E$, $u \in F'$, $g \in F$, none being necessarily preassigned, in the context of §5.5. Show that

$$(\sigma, e) = \langle u, g \rangle$$

is implied. Hence reformulate Theorem 5.8, and a dual version, from this symmetric starting point. Compare with a converse to the virtual work principle (Truesdell and Toupin, 1960, §234), and with Tellegen's theorem in electricity (see (2.139) and Penfield *et al.* 1970).

2. Another common type of boundary condition, different from those in (5.74), is to

assign mutually orthogonal components of the vectors of load $\mathbf{s}$ per unit area and displacement $\mathbf{u}$ at a given surface point. For example, if $\mathbf{m}$ is a given distribution of unit (not necessarily normal) vectors over the whole of Σ, $\mathbf{m}\cdot\mathbf{u}$ and $\mathbf{m}\wedge(\mathbf{s}\wedge\mathbf{m})$ could be assigned at every point of Σ. By using $s_j = n_i\sigma_{ij}$ and writing

$$\mathbf{s}\cdot\mathbf{u} = \mathbf{s}\cdot\mathbf{m}(\mathbf{u}\cdot\mathbf{m}) + \mathbf{u}\cdot\mathbf{m}\wedge(\mathbf{s}\wedge\mathbf{m})$$

construct a revised version of (5.66), and find the induced revisions of E, F, T, T^* and $L[\sigma; u, e]$. Establish the simultaneous upper and lower bounds explicitly (cf. (3.4) of Sewell, 1969).

3. In the 6×1 column matrix notation of (2.82), the Newtonian viscous fluid is $\mathbf{r} = 2\mu\dot{\mathbf{e}}$. Letting $|\dot{\mathbf{e}}| = (\dot{\mathbf{e}}^{\mathrm{T}}\dot{\mathbf{e}})^{\frac{1}{2}}$, make the changes required in §5.7 when

$$\mathbf{r} = 2\mu|\dot{\mathbf{e}}|^{(n-1)/2}\dot{\mathbf{e}} \quad \text{for any fixed } n > 0.$$

In particular determine associated functions $U[|\dot{\mathbf{e}}|]$ and $U_c[|\mathbf{r}|]$. Such constitutive equations have been used for secondary creep of metals and slow flow of ice (Hill, 1956*a*). The case $n = 0$ formally defines a Lévy–Mises rigid-plastic solid (cf. §5.8).

4. Repeat Exercise 3 for

$$\mathbf{r} = 2(\mu + k|\dot{\mathbf{e}}|^{-1})\dot{\mathbf{e}},$$

where $k > 0$ is a given constant. This is associated with the viscoplastic Bingham solid, for which the extremum principles were obtained by Prager (1954) by the Schwarz inequality.

5. Show that the same conclusions emerge from §5.5, and are generated from the same functional (5.79), by using an abbreviated version which omits the zeros from expressions like (5.68) and (5.76), as first suggested in the contrast between (3.21) and (3.12)–(3.15) (cf. Exercise 3.3.3 and §4.5(ii)). In other words, supply the details which validate the use of

$$\sigma = \begin{bmatrix} \sigma_{ij}(V) \\ \sigma_{ij}(\Sigma_2) \end{bmatrix}, \quad T^*u = \begin{bmatrix} \dfrac{1}{2}\left(\dfrac{\partial u_j}{\partial x_i} + \dfrac{\partial u_i}{\partial x_j}\right) \\ -\frac{1}{2}(n_i u_j + n_j u_i) \end{bmatrix},$$

$$u = \begin{bmatrix} u_j(V) \\ u_j(\Sigma_1) \end{bmatrix}, \quad T\sigma = \begin{bmatrix} -\dfrac{\partial \sigma_{ij}}{\partial x_i} \\ n_i\sigma_{ij} \end{bmatrix},$$

$$e = e_{ij}(V), \qquad \overline{e, f} = \int e_{ij} f_{ji}\,\mathrm{d}V,$$

$$(\sigma, \tau) = \int \sigma_{ij}\tau_{ji}\,\mathrm{d}V + \int \sigma_{ij}\tau_{ji}\,\mathrm{d}\Sigma_2,$$

$$\langle u, v\rangle = \int u_j v_j\,\mathrm{d}V + \int u_j v_j\,\mathrm{d}\Sigma_1,$$

$$\frac{\partial L}{\partial \sigma}=\begin{bmatrix}\frac{1}{2}\left(\frac{\partial u_j}{\partial x_i}+\frac{\partial u_i}{\partial x_j}\right)-e_{ij}\\ \frac{1}{2}n_i(c_j-u_j)+\frac{1}{2}n_j(c_i-u_i)\end{bmatrix},\quad \frac{\partial L}{\partial u}=\begin{bmatrix}-\frac{\partial \sigma_{ij}}{\partial x_i}-b_j\\ n_i\sigma_{ij}-a_j\end{bmatrix},$$

$$\frac{\partial L}{\partial e}=\frac{\partial U}{\partial e_{ij}}-\sigma_{ij}.$$

5.8. Rigid/plastic yield point problem

(i) Governing equations

The *rigid/plastic solid* is a useful idealization of certain behaviour in metals (and also soils) in which the permanent deformation dominates the elastic or recoverable deformation. The latter is neglected, so that no deformation at all is imagined to take place while the true stress σ_{ij} satisfies $m \geqslant 1$ strict inequalities $f_\gamma[\sigma_{ij}]<0$ $(\gamma=1,\ldots,m)$, where the $f_\gamma[\sigma_{ij}]$ are m given convex C^1 functions of stress called 'yield functions', whose form will have been decided by the previous history of deformation. For example (see Hill, 1950), the Mises yield surface $f[\sigma_{ij}]=0$ in stress space has $m=1$, where $f[\sigma_{ij}]=\frac{1}{2}r_{ij}r_{ij}-k^2$ and $r_{ij}=\sigma_{ij}-\frac{1}{3}(\sigma_{kk})\delta_{ij}$, and k is a given constant; the Tresca yield surface is made up of facets from $m=6$ linear functions of r_{ij}. Cross-sections of these two surfaces are illustrated in Fig. 2.21 (the unspecified τ_{ij} there is now to be chosen as the true stress σ_{ij} for the purpose of this section). When $f_\gamma[\sigma_{ij}]=0$ at some point of the body for one or more values of γ, quasistatic deformation becomes possible at that point according to a so-called plastic flow rule. This relates the permitted quasistatic velocity v_j (say) to the current stress. To express the flow rule we need to introduce m scalar multipliers λ_γ as auxiliary position-dependent variables (as we did in §2.6(iii)). If we use the same boundary conditions on Σ_1 and Σ_2 as in (5.74), and the same body forces in V, the boundary value problem of quasistatic plastic yielding can be written as follows.

Flow rule in V:	$\frac{1}{2}\left(\frac{\partial v_j}{\partial x_i}+\frac{\partial v_i}{\partial x_j}\right)=\lambda_\gamma\frac{\partial f_\gamma}{\partial \sigma_{ij}}.$	(α)	
Assigned velocity on Σ_2:	$v_j=c_j.$	(α)	
Equilibrium in V:	$-\frac{\partial \sigma_{ij}}{\partial x_i}=b_j.$	(β)	(5.107)
Assigned traction on Σ_1:	$n_i\sigma_{ij}=a_j$	(β)	
Admissible stresses in V:	$f_\gamma[\sigma_{ij}]\leqslant 0,$	(β)	
Admissible multipliers in V:	$\lambda_\gamma \geqslant 0.$	(α)	
Yield/flow compatibility in V:	$\lambda_\gamma f_\gamma[\sigma_{ij}]=0.$		

Recall that summation is implied by a repeated γ, from $\gamma = 1$ to $\gamma = m$. In the Mises example $\partial f/\partial\sigma_{ij} = r_{ij}$, and k is interpretable as the yield stress in pure shear. The type of flow rule here is said to be 'associated' with the yield functions, and was already foreshadowed in $(2.87)_1$, with $M = 0$ and $\dot{e}_{ij} = \frac{1}{2}(\partial v_i/\partial x_j + \partial v_j/\partial x_i)$ there. At this stage in the yield point problem, however, we do *not* lay down relations between the λ_γ and the stress-rates as we did via $(2.87)_2$ and (2.85) (such relations enter the boundary value problems of §5.10). Other types of *non*associated flow rules $\dot{e}_{ij} = \lambda_\gamma \partial g_\gamma/\partial\sigma_{ij}$ are needed for some materials, where the 'plastic potentials' $g_\gamma[\sigma_{ij}]$ are different from the yield functions $f_\gamma[\sigma_{ij}]$, but we shall not consider them here.

As before, the (α) and (β) subsets in (5.107) consist of kinematical and force-like conditions respectively. The last condition in (5.107) is assigned to neither subset – it is an orthogonality condition of the kind which we first introduced in (1.30).

To facilitate part of the discussion we define an intermediate auxiliary device called the *comparison solid*. This is a physically fictitious body of material whose governing conditions are obtained from (5.107) by replacing the last trio there by $f_\gamma[\sigma_{ij}] = 0$ for every γ, with no restriction on the λ_γ. The rigid/plastic and comparison solids will have solutions in common where all $\lambda_\gamma \geqslant 0$ in the comparison solid, provided that elsewhere a stress field in the rigid/plastic solid can be found to satisfy (5.107β).

(ii) Generalized Hamiltonian representation

Again we need a third linear space G, different from that of §5.7(ii), and consisting here of sets

$$\lambda = \{\lambda_\gamma(V)\}, \quad f = \{f_\gamma(V)\}, \ldots \tag{5.108}$$

of m scalar integrable functions of position in V, with an inner product defined by

$$\overline{\lambda, f} = \int \lambda_\gamma f_\gamma \, \mathrm{d}V. \tag{5.109}$$

Using the T and T^* of (5.68), the governing equations for the comparison solid can be assembled as

$$\begin{bmatrix} 0 & T^* & 0 \\ T & 0 & 0 \\ 0 & 0 & 0 \end{bmatrix} \begin{bmatrix} \sigma \\ v \\ \lambda \end{bmatrix} = \partial H/\partial \begin{bmatrix} \sigma \\ v \\ \lambda \end{bmatrix} \quad \begin{matrix} (\alpha) \\ (\beta) \\ (\beta) \end{matrix} \tag{5.110}$$

in terms of the gradients of the generalized Hamiltonian

$$H[\sigma; v, \lambda] = \int (\lambda_\gamma f_\gamma[\sigma_{ij}] + b_j v_j) \mathrm{d}V - \int n_i \sigma_{ij} c_j \mathrm{d}\Sigma_2 + \int a_j v_j \mathrm{d}\Sigma_1, \qquad (5.11)$$

which is defined over the three spaces E', F' and G. These gradients are defined (cf. (5.78)) by expressing the first variation of H as

$$\delta H = \left(\delta\sigma, \frac{\partial H}{\partial \sigma} \right) + \left\langle \delta v, \frac{\partial H}{\partial v} \right\rangle + \overline{\delta\lambda, \frac{\partial H}{\partial \lambda}}. \qquad (5.112)$$

The last equation $\partial H/\partial\lambda = 0$ of (5.110) must be replaced, for the rigid/plastic solid itself, by the trio

$$\frac{\partial H}{\partial \lambda} \leqslant 0 \quad (\beta), \qquad \lambda \geqslant 0 \quad (\alpha), \quad \overline{\lambda, \frac{\partial H}{\partial \lambda}} = 0 \qquad (5.113)$$

which are equivalent to the last trio in (5.107).

(iii) Generating saddle functional

By imitating (5.79) we introduce the functional

$$\begin{aligned} L[\sigma; v, \lambda] &= \langle v, T\sigma \rangle - H[\sigma; v, \lambda] \\ &= \int \left(-v_j \frac{\partial \sigma_{ij}}{\partial x_i} - \lambda_\gamma f_\gamma[\sigma_{ij}] - b_j v_j \right) \mathrm{d}V \\ &\quad + \int v_j (n_i \sigma_{ij} - a_j) \mathrm{d}\Sigma_1 + \int n_i \sigma_{ij} c_j \mathrm{d}\Sigma_2 \end{aligned} \qquad (5.114)$$

whose gradients are

$$\frac{\partial L}{\partial \sigma} = T^* v - \frac{\partial H}{\partial \sigma}, \quad \frac{\partial L}{\partial v} = T\sigma - \frac{\partial H}{\partial v}, \quad \frac{\partial L}{\partial \lambda} = -\frac{\partial H}{\partial \lambda}. \qquad (5.115)$$

A free stationary principle characterizing solutions of (5.110), i.e. of (5.115) all put to zero, can be established for the comparison solid, by analogy with Theorem 5.9. This could be used to explore *some* of the solutions of the rigid/plastic itself. However, the governing conditions (5.107) of the latter are equivalent to

$$\begin{aligned} &\frac{\partial L}{\partial \sigma} = 0, \quad \lambda \geqslant 0, \quad (\alpha) \\ &\frac{\partial L}{\partial v} = 0, \quad \frac{\partial L}{\partial \lambda} \geqslant 0, \quad (\beta) \\ &\overline{\lambda, \frac{\partial L}{\partial \lambda}} = 0. \end{aligned} \qquad (5.116)$$

The convexity of the yield functions suggests that $L[\sigma; v, \lambda]$ will be, in some sense, a saddle functional concave in σ, and convex in v with λ. To explore this question we define the saddle quantity

$$S = L[\sigma_+; v_+, \lambda_+] - L[\sigma_-; v_-, \lambda_-]$$
$$-\left(\sigma_+ - \sigma_-, \frac{\partial L}{\partial \sigma}\bigg|_+\right) - \left\langle v_+ - v_-, \frac{\partial L}{\partial v}\bigg|_-\right\rangle - \overline{\lambda_+ - \lambda_-, \frac{\partial L}{\partial \lambda}\bigg|_-}$$
$$= \int \lambda_{\gamma+}\left[f_\gamma[\sigma_{ij-}] - f_\gamma[\sigma_{ij+}] - (\sigma_{ij-} - \sigma_{ij+})\frac{\partial f_\gamma}{\partial \sigma_{ij}}\bigg|_+\right]\mathrm{d}V \qquad (5.117)$$

for any pair of plus and minus points in the domain of (5.114). The convexity inequality (1.4) applied to each $f_\gamma[\sigma_{ij}]$ shows that $S \geqslant 0$ if each $\lambda_{\gamma+} \geqslant 0$. In other words, within the pyramid $\lambda \geqslant 0$ $L[\sigma; v, \lambda]$ is a saddle functional concave in σ and convex in v with λ; but within $\lambda \leqslant 0$ it has the opposite saddle quality, convex in σ and concave in v with λ. We shall only need the former of these cases. In the language of §1.3(iii) we have a more restricted saddle functional than in (1.58), more reminiscent of that in the proof of Theorem 1.8(a). There is a close relation between the present structure and that of non-linear programming.

(iv) Uniqueness of stress

The simplest type of *actual solution* is a set of functions σ_{ij}, v_j and λ_γ of position. respectively C^1, C^2 and C^1, which satisfy all of (5.107), or equivalently (5.116). For simplicity we confine attention to these, but the extensive theory of slip-line fields, reviewed by Collins (I.F., 1982), for example, demonstrates the technological importance of solutions with isolated velocity and stress discontinuities. In §5.5(iv) we have indicated how to approach these within the present framework. Texts such as Johnson and Mellor (1973) give many illustrations of the role of discontinuities.

Under the given data in (5.107) we cannot prove uniqueness of the v_j or λ_γ in an actual solution. The *deformable* region is defined to be that part of V where an actual solution permits $\lambda_\gamma > 0$ for some γ; elsewhere every $\lambda_\gamma = 0$ in every actual solution and the material remains rigid. In the deformable region it is necessary that the stress satisfies $f_\gamma[\sigma_{ij}] = 0$ for one or more values of γ.

Theorem 5.14 (Uniqueness)
The stress is unique in the deformable region, to the extent that the convexity of $f_\gamma[\sigma_{ij}]$ is strict for the appropriate γ.

Proof
Interchanging the plus and minus labels in (5.117), and adding the results,

gives

$$-\left(\sigma_+ - \sigma_-, \frac{\partial L}{\partial \sigma}\bigg|_+ - \frac{\partial L}{\partial \sigma}\bigg|_-\right) + \left\langle v_+ - v_-, \frac{\partial L}{\partial v}\bigg|_+ - \frac{\partial L}{\partial v}\bigg|_-\right\rangle$$
$$\overline{+\lambda_+ - \lambda_-, \frac{\partial L}{\partial \lambda}\bigg|_+ - \frac{\partial L}{\partial \lambda}\bigg|_-}$$
$$= \int\left\{\lambda_{\gamma+}\left(f_\gamma[\sigma_{ij-}] - f_\gamma[\sigma_{ij+}] - (\sigma_{ij-} - \sigma_{ij+})\frac{\partial f_\gamma}{\partial \sigma_{ij}}\bigg|_+\right)\right.$$
$$\left. + \lambda_{\gamma-}\left(f_\gamma[\sigma_{ij+}] - f_\gamma[\sigma_{ij-}] - (\sigma_{ij+} - \sigma_{ij-})\frac{\partial f_\gamma}{\partial \sigma_{ij}}\bigg|_-\right)\right\} \mathrm{d}V. \tag{5.118}$$

Suppose σ_+, v_+, λ_+ and σ_-, v_-, λ_- are two possible actual solutions. Then the left side of (5.118) is $\leqslant 0$, by (5.116). The right side will be >0 for a pair of stresses on a strictly convex segment of $f_\gamma(\sigma_{ij})$ where the corresponding $\lambda_\gamma > 0$. The ensuing contradiction establishes the result. □

In the case of the Mises or the Tresca yield function of deviatoric stress, for example, the stress turns out to be unique to within an all-round pressure (this indeterminacy is removed if the deformable region has points in common with Σ_1). Theorem 5.14 was originally proved by Hill (1948; see 1950, p. 59) in this case. More data (cf. §5.10) is required before uniqueness of v_j or λ_γ can be discussed.

(v) Simultaneous upper and lower bounds

Before any constraints are imposed we find from (5.114) that

$$L - \left(\sigma, \frac{\partial L}{\partial \sigma}\right) = \int\left[\lambda_\gamma\left(\sigma_{ij}\frac{\partial f_\gamma}{\partial \sigma_{ij}} - f_\gamma[\sigma_{ij}]\right) - b_j v_j\right]\mathrm{d}V - \int a_j v_j \,\mathrm{d}\Sigma_1, \tag{5.119}$$

$$L - \left\langle v, \frac{\partial L}{\partial v}\right\rangle - \overline{\lambda, \frac{\partial L}{\partial \lambda}} = \int n_i \sigma_{ij} c_j \,\mathrm{d}\Sigma_2. \tag{5.120}$$

We can expect, from the general structure of our theory, that simultaneous extremum principles will characterize quantities having the values of (5.119) and (5.120). We shall only give the proofs of sufficiency in Theorems 5.15 and 5.16.

We notice in passing that the right side of (5.119) contains the values of the Legendre duals of the yield functions. Notice also that the right side of (5.120) is the virtual work done by *any* boundary traction $n_i\sigma_{ij}$ on Σ_2 in the assigned displacement there.

We shall say that a set of fields σ_{ij}, v_j and λ_γ is *kinematically admissible* if it satisfies all the conditions (5.107α), and *statically admissible* if it satisfies all the conditions (5.107β).

Let L_0 be the actual value assigned to (5.114), and also to (5.119) and (5.120), by an actual solution of (5.107) or equivalently (5.116). Then

$$L_0 = \int n_i \sigma_{ij} c_j \, d\Sigma_2 \tag{5.121}$$

is the actual work of the boundary traction needed to enforce the assigned displacement, with σ_{ij} *here* being the solution stress.

Theorem 5.15 (Upper bound)
L_0 is bounded above by the value L_α (say) assigned to (5.119) by any kinematically admissible set of fields, i.e.

$$L_\alpha \geqslant L_0. \tag{5.122}$$

Proof
We follow again the pattern of proof in Theorem 1.4, with similar use of the α and 0 subscripts. Choose σ_+, v_+, λ_+ in (5.117) to be any solution of (5.116α), so that in particular each $\lambda_{\gamma+} \geqslant 0$. Then, by the convexity of the yield functions, $S \geqslant 0$ for any σ_-, v_-, λ_-. Now choose this minus point to be any actual solution of the whole problem (5.116). From the left side of (5.117) we see that

$$L_\alpha - L_0 \geqslant \overline{\lambda_\alpha, \left.\frac{\partial L}{\partial \lambda}\right|_0} \geqslant 0. \tag{5.123}$$

□

The total amount given away to get the upper bound inequality is the sum of the right sides of (5.117) and (5.123). The former will be quadratic in $\sigma_{ij-} - \sigma_{ij+}$ (cf. (5.83)) if the yield functions are smooth enough; but the latter is $-\int \lambda_{\gamma+} f_\gamma[\sigma_{ij-}] \, dV$, which is not quadratic or higher order in the differences of plus and minus fields (and might not be zero unless every point of the body is deformable). Therefore the minimum in (5.122) is not in general stationary.

The upper bound can be written as

$$L_\alpha = \int \left(\sigma_{ij} \frac{\partial v_j}{\partial x_i} - \lambda_\gamma f_\gamma[\sigma_{ij}] - b_j v_j \right) dV - \int a_j v_j \, d\Sigma_1 \tag{5.124}$$

by inserting $(5.107)_1$. For any chosen σ_{ij} distribution, this particular α-constraint can be regarded as a system of differential equations which, in certain cases, can be solved to express the v_j and λ_γ as explicit families of functions of position (see Exercise 5.4.1). It seems to be conventional to omit the $\lambda_\gamma f_\gamma$ term from (5.124), which can be done by imposing either $(5.107)_7$ or $(5.107)_5$ or both as optional extra constraints on the plus point in the proof of Theorem 5.15 (cf. the supplementary constraints mentioned after Theorem 2.16).

Theorem 5.16 (Lower bound: principle of maximum plastic work)
L_0 is bounded below by the value assigned to (5.120) by any statically admissible set of fields, i.e.

$$L_0 \geqslant L_\beta - \overline{\lambda_\beta, \left.\frac{\partial L}{\partial \lambda}\right|_\beta}. \tag{5.125}$$

Proof
Choose σ_+, v_+, λ_+ in (5.117) to be any actual solution of the whole problem (5.116). Again each $\lambda_{\gamma+} \geqslant 0$, and $S \geqslant 0$ for any σ_-, v_-, λ_-. Choose the latter this time to be any solution of the subset (5.116β). Using the subscripts 0 and β as in Theorem 1.4, we find from the left side of (5.117) that

$$L_0 - L_\beta + \overline{\lambda_\beta, \left.\frac{\partial L}{\partial \lambda}\right|_\beta} \geqslant \overline{\lambda_0, \left.\frac{\partial L}{\partial \lambda}\right|_\beta} \geqslant 0. \tag{5.126}$$

□

The amount given away on the right of (5.126) is not quadratic or higher order in the difference fields, by a similar argument to that given previously, and so the maximum in (5.125) is not in general stationary. Because the lower bound in (5.125) is, by (5.120), the value of $\int n_i \sigma_{ij} c_j \mathrm{d}\Sigma_2$ for any equilibriated stress field which satisfies the traction boundary condition and the yield restrictions $f_\gamma[\sigma_{ij}] \leqslant 0$, the optimization of this bound is called the *principle of maximum plastic work*.

Certain particular cases of Theorems 5.15 and 5.16 were proved by Markov (1947) and Hill (1948) respectively, independently of earlier work by Gvozdev (1960, translation of a 1936 paper). The subject received its real stimulus with definitive accounts of both extremum principles for the Mises material, published by Hill (1950, pp. 66–7 – see also 1951) and Prager and Hodge (1951, §37). The approach via the Hamiltonian (5.111) presented here was introduced by Sewell (1969, 1973*a*). Many applications and variants of the upper and lower bounds have been developed. An early summary was given by Hill (1956*b*). Several books on engineering applications have since appeared, in both the plastic theory of structures (e.g. Horne, 1971) and in metal forming processes (e.g. Johnson and Mellor, 1973; and Johnson *et al.*, 1982). These contain numerous references to the literature and many explicit calculations. Upper bounds are much more common than lower bounds. This reflects the fact that it has proved easier to find kinematically admissible rather than statically admissible fields, and that knowledge of an upper bound to the load needed in a metal forming process will be sufficient to effect that process. Tangential velocity discontinuities in incompressible material have acquired an important role in plane strain slip-line field applications of Theorem 5.15, and the book by Johnson *et al.* (1982) is entirely devoted to this. The nonstationary character of the bounds in both Theorems has led to an exploitation of established techniques of mathematical programming and complementarity theory. For example, Charnes, Lemke and Zienkiewicz (1959) expressed the static and kinematic plastic collapse principle for frames as dual linear programming problems. This approach is conveyed in the volume edited by Cohn and Maier (1979). Another line of development, much more mathematical in tone, is represented by the rigorous work of Temam (1978).

Exercises 5.4

1. Suppose that there is only one yield function $f[\sigma_{ij}]$ at the local stress point, whose normal $\partial f/\partial\sigma_{ij}$ has principal components proportional to constant l, m, n over V. Deduce that the most general continuous velocity solution of $(5.107)_1$ is, if $lmn \neq 0$,

$$v_1 = \tfrac{1}{2}a(lx_1{}^2 - mx_2{}^2 - nx_3{}^2) + lbx_2x_1 + lcx_3x_1 + ldx_1,$$
$$v_2 = max_1x_2 + \tfrac{1}{2}b(mx_2{}^2 - nx_3{}^2 - lx_1{}^2) + mcx_3x_2 + mdx_2,$$
$$v_3 = nax_1x_3 + nbx_2x_3 + \tfrac{1}{2}c(nx_3{}^2 - lx_1{}^2 - mx_2{}^2) + ndx_3,$$

or if $m = 0$, $l|n| = -n|l| \neq 0$, is

$$v_1 = l|l|^{-\frac{1}{2}}(\phi - \psi) + lbx_1x_2,$$
$$v_2 = -\tfrac{1}{2}b(lx_1{}^2 + nx_3{}^2),$$
$$v_3 = n|n|^{-\frac{1}{2}}(\phi + \psi) + nbx_3x_2,$$

where a, b, c, d are arbitrary constants and $\phi(|n|^{\frac{1}{2}}x_3 + |l|^{\frac{1}{2}}x_1)$ and $\psi(|n|^{\frac{1}{2}}x_3 - |l|^{\frac{1}{2}}x_1)$ are arbitrary functions of the indicated arguments. These fields were deduced (Sewell, 1963*b*) by extending Prager's derivation from the incompressible ($l + m + n = 0$) to the compressible case, so that they can apply to soils as well as metals. They will often apply in regions of uniform stress, and the same approach can work for certain nonuniformly stressed bodies (Miles, 1969).
2. In the notation of §5.5(iv), suppose that isolated internal surfaces S_2 and S_1 are present where discontinuities $[\![v_j]\!]$ and $n_i[\![\sigma_{ij}]\!]$ respectively may occur, but are to be subject to assigned values $\bar{c}_j$ and $\bar{a}_j$ ($\bar{a}_j = 0$ for equilibrium). Construct a precise statement of subspaces allowing appropriate augmentation of adjointness (5.67) (see Sewell 1973*a*). Show that, before any governing conditions are imposed

$$\int v_j(n_i[\![\sigma_{ij}]\!] - \bar{a}_j)\mathrm{d}S_1 + \int n_i\sigma_{ij}\bar{c}_j\mathrm{d}S_2$$

must be added to the generating functional (5.114), and that $-\int \bar{a}_j v_j \mathrm{d}S_1$ and $\int n_i\sigma_{ij}\bar{c}_j\mathrm{d}S_2$ must be added to (5.119) and (5.120) respectively. The quantity bounded, in place of (5.121), is

$$L_0 = \int n_i\sigma_{ij}c_j \mathrm{d}\Sigma_2 + \int n_i\sigma_{ij}\bar{c}_j \mathrm{d}S_2,$$

wherein σ_{ij} is the actual stress.

3. Show that in the adjointness formula (§5.5(iv)) for the present context we can write

$$-\int \sigma_{ij}\tfrac{1}{2}(n_i[\![v_j]\!] + n_j[\![v_i]\!])\mathrm{d}S_2 = \int \tau[\![v]\!]\mathrm{d}S_2,$$

where τ is the traction component in the direction of a velocity jump of magnitude $[\![v]\!]$. It may be unrealistic to assign the value of a velocity jump. Instead of this, repeat Exercise 2 showing how to accommodate a velocity jump on S_2 by adjoining

$$\tau - k \leqslant 0 \quad (\beta), \qquad [\![v]\!] \geqslant 0 \quad (\alpha), \qquad [\![v]\!](\tau - k) = 0$$

to (5.107) (cf. Hill 1951). Verify that with these additions to the conditions defining kinematically and statically admissible fields, Theorems 5.15 and 5.16 apply as they stand when $S_1 = 0$, except that $\int k[\![v]\!]\mathrm{d}S_2$ must be added to the upper bound (5.124).

4. The foregoing condition on S_2 may be adopted as a tangential boundary condition on the ends of a cylindrical block being compressed under dies which assign the normal velocity on the ends. Use the fields in Exercise 1 with $a = b = c = 0$ to obtain an upper bound for the die pressure (see Johnson and Mellor, 1973, §16.5.1). In fact a number of different frictional boundary conditions have been considered to have physical significance, some being associated with nonself-adjointness or with a 'duality gap' between the bounds (see Collins 1969).

5. The governing conditions of an initial motion problem which generalize those of §1.7(iv) and Exercise 3.2.4 to a wide variety of incompressible media are as follows.

V	$\rho\dot{\mathbf{v}} + \nabla p = \mathbf{z},$	(α)
V	$\nabla\cdot\dot{\mathbf{v}} \geqslant \nabla\cdot[\mathbf{v}\cdot\nabla\mathbf{v}],$	(β)
V	$p \geqslant q,$	(α)
V	$(p-q)(\nabla\cdot\dot{\mathbf{v}} - \nabla\cdot[\mathbf{v}\cdot\nabla\mathbf{v}]) = 0,$	
Σ_1	$p = c,$	(α)
Σ_2	$\mathbf{n}\cdot\dot{\mathbf{v}} \leqslant k,$	(β)
Σ_2	$p \geqslant q,$	(α)
Σ_2	$(p-q)(k - \mathbf{n}\cdot\dot{\mathbf{v}}) = 0.$	

Here the medium occupies a fixed spatial region V with surface $\Sigma_1 + \Sigma_2$, and the basic unknowns at an assigned instant in unsteady motion are the spatial fields of pressure p and acceleration $\dot{\mathbf{v}}$, where $\mathbf{v}$ is a supposedly known velocity field.

Assigned scalar fields are ρ (density) and q (vaporization pressure) in V, c on Σ_1 and k on Σ_2. The vector field $\mathbf{z}$ is assigned in V. For example, in Newtonian fluid with given viscosity μ and body force $\mathbf{b}$, $\mathbf{z}=\mu\nabla^2\mathbf{v}+\mathbf{b}$.

Explain how the last trio of conditions in V can represent the onset of cavitation in the incompressible medium *per se* (as distinct from a limiting process in a compressible medium), via incipient breakdown of the continuity equation $\nabla\cdot\mathbf{v}=0$. Give an associated interpretation of the conditions on Σ_2.

Show how the system may be expressed in terms of the gradients of a saddle functional

$$L=\int\left\{(p-q)(\nabla\cdot\dot{\mathbf{v}}-\nabla\cdot[\mathbf{v}\cdot\nabla\mathbf{v}])-\tfrac{1}{2}\rho\dot{\mathbf{v}}^2+\dot{\mathbf{v}}(\mathbf{z}-\nabla q)\right\}\mathrm{d}V$$

$$-\int(c-q)\mathbf{n}\cdot\dot{\mathbf{v}}\,\mathrm{d}\Sigma_1+\int(k-\mathbf{n}\cdot\dot{\mathbf{v}})(p-q)\,\mathrm{d}\Sigma_2,$$

which is strictly concave in $\dot{\mathbf{v}}$ and weakly convex in p.

Hence show that if p_α is any pressure field satisfying only the α conditions, and if $\dot{\mathbf{v}}_\beta$ is any acceleration field satisfying the β conditions, the quantities

$$\int\left\{\tfrac{1}{2}\rho(\mathbf{z}-\nabla p_\alpha)^2-(p_\alpha-q)\nabla\cdot[\mathbf{v}\cdot\nabla\mathbf{v}]\right\}\mathrm{d}V+\int k(p_\alpha-q)\,\mathrm{d}\Sigma_2$$

and

$$\int\{-\tfrac{1}{2}\rho\dot{\mathbf{v}}_\beta{}^2+\dot{\mathbf{v}}_\beta\cdot(\mathbf{z}-\nabla q)\}\,\mathrm{d}V-\int(c-q)\mathbf{n}\cdot\dot{\mathbf{v}}_\beta\,\mathrm{d}\Sigma_1$$

bound, from above and below respectively, the common value which they take for a solution of the whole system (Sewell, 1969).

Similar results can be given for media having internal constraints which are more general linear functions of velocity than $\nabla\cdot\mathbf{v}=0$, for example

$$A_{ij}v_{j,i}+B=0$$

in cartesian components, where a comma signifies spatial differentiation. The stress in such a medium is taken as $\sigma_{ij}=r_{ij}-pA_{ij}$. The equations of motion and of time-differentiated constraint for finding $\dot{v}_j$ and p are then

$$\rho\dot{v}_j+(pA_{ij})_{,i}=z_j,\quad(\alpha)$$

$$A_{ij}\dot{v}_{j,i}=s,\quad(\beta)$$

Here $z_j=r_{ij,i}+b_j$ and $s=-(\dot{A}_{ij}v_{j,i}+\dot{B})$ are treated as known, because A_{ij}, B, r_{ij}, b_i and v_j are supposed known at the considered instant. Upper and lower bounds apply when suitable boundary conditions are appended.

Reddy (1981) has applied these ideas to the rigid/plastic solid.

5.9. Finite elasticity

(i) Strain measures

Let a typical material particle of the body have position vector $\mathbf{X}$ in a specified initial configuration at time $t=0$, and $\mathbf{x}$ in the current configuration at time t, as in the proof of Theorem 5.6. Again the motion may be sought as a function $\mathbf{x}[\mathbf{X}, t]$, or as a displacement $\mathbf{u}[\mathbf{X}, t] = \mathbf{x} - \mathbf{X}$ which may now be large, whereas it was not in §5.6. Equilibrium problems can be described by reinterpreting t as any parameter ordering a sequence of quasistatic displacements. We express vectors and tensors in terms of their components with respect to the same cartesian frame of reference throughout.

We define 3×3 unsymmetric matrices with components

$$F_{ij} = \frac{\partial x_i}{\partial X_j} \quad \text{and} \quad d_{ij} = \frac{\partial u_i}{\partial X_j} \tag{5.127}$$

called deformation gradient and displacement gradient respectively. Let

$$e_{ij} = \tfrac{1}{2}(F_{ki}F_{kj} - \delta_{ij}) = \tfrac{1}{2}(d_{ij} + d_{ji} + d_{ki}d_{kj}) \tag{5.128}$$

be the symmetric (George) Green strain components. It is the quadratic terms in the latter which were neglected in $(5.74)_1$, and it was consistent with that approximation that we were able to use the x_j instead of the X_j as independent variables in §5.6. No such small strain approximation is assumed to be available here. We seek an exact theory valid for large strains and rotations.

We may assume without loss of generality that the determinant $|F_{ij}| > 0$. The polar decomposition theorem permits any nonsingular matrix to be written uniquely as the product of an orthogonal matrix with a positive definite symmetric matrix. A proof is given by Chadwick (1976, p. 33). Here we write

$$F_{ij} = R_{ik}\Lambda_{kj} \tag{5.129}$$

where R_{ik} is proper orthogonal, and the symmetric Λ_{kj} is called the 'right stretch' matrix. The stretch, say λ, of any material fibre is its current length divided by its initial length. Understanding of the decomposition (5.129) is helped by an explicit computation (Fig. 5.7). Given F_{ij}, form the symmetric $F_{ki}F_{kj} = \Lambda_{ki}\Lambda_{kj}$ and find its eigenvalues, say $\lambda_{(1)}{}^2$, $\lambda_{(2)}{}^2$, $\lambda_{(3)}{}^2$, and orthogonal eigenvectors, say $p_{k(1)}, p_{k(2)}, p_{k(3)}$. Then construct Λ_{kj} by spectral representation to have the same eigenvectors, and positive eigenvalues $\lambda_{(1)}, \lambda_{(2)}, \lambda_{(3)}$, so that

$$\Lambda_{kj}p_{j(1)} = \lambda_{(1)}p_{k(1)}, \text{ etc.} \tag{5.130}$$

Finally obtain R_{ik} by applying the inverse of Λ_{kj} to (5.129). Any infinitesimal fibre with components $\mathrm{d}X_j$ becomes $\mathrm{d}x_i = F_{ij}\mathrm{d}X_j$. Thus (5.129) with (5.130) shows that the current local deformation determines a triad of *principal* fibres, orthogonal initially and currently, which initially had directions $p_{k(1)}$, etc., and which have been rotated to current directions $R_{ik}p_{k(1)}$, etc., and stretched by amounts $\lambda_{(1)}$, etc., called the principal stretches. In two (or three) dimensions the four (or nine) independent components of F_{ij} are expressed by (5.129) in terms of three (or six) independent components of Λ_{kj} and one (or three) finite angles.

The two symmetric matrices of pure strain considered here are related by

$$e_{ij} = \tfrac{1}{2}(\Lambda_{ik}\Lambda_{kj} - \delta_{ij}). \tag{5.131}$$

Hill (1978) reviews a 'principal axes technique' which shows how the strain measures selected here fit into a coherent family.

(ii) Stress measures

Consider a material area element which initially had magnitude $d\Sigma$ and whose unit normal had components n_i then. Let the current load vector

Fig. 5.7. Plane strain example of polar decomposition.

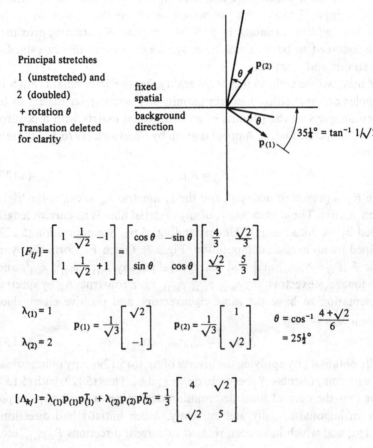

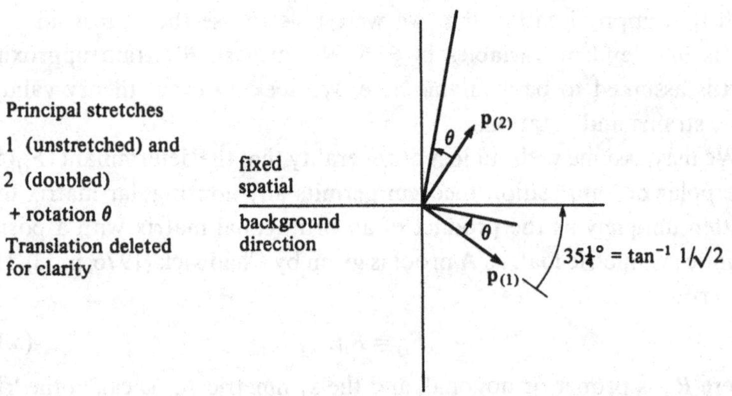

$$[F_{ij}] = \begin{bmatrix} 1 & \frac{1}{\sqrt{2}} - 1 \\ 1 & \frac{1}{\sqrt{2}} + 1 \end{bmatrix} = \begin{bmatrix} \cos\theta & -\sin\theta \\ \sin\theta & \cos\theta \end{bmatrix} \begin{bmatrix} \frac{4}{3} & \frac{\sqrt{2}}{3} \\ \frac{\sqrt{2}}{3} & \frac{5}{3} \end{bmatrix}$$

$$\lambda_{(1)} = 1 \qquad \mathbf{p}_{(1)} = \frac{1}{\sqrt{3}} \begin{bmatrix} \sqrt{2} \\ -1 \end{bmatrix} \qquad \mathbf{p}_{(2)} = \frac{1}{\sqrt{3}} \begin{bmatrix} 1 \\ \sqrt{2} \end{bmatrix} \qquad \theta = \cos^{-1}\frac{4+\sqrt{2}}{6}$$

$$\lambda_{(2)} = 2 \qquad\qquad = 25\tfrac{1}{2}°$$

$$[\Lambda_{kj}] = \lambda_{(1)}\mathbf{p}_{(1)}\mathbf{p}_{(1)}^T + \lambda_{(2)}\mathbf{p}_{(2)}\mathbf{p}_{(2)}^T = \frac{1}{3}\begin{bmatrix} 4 & \sqrt{2} \\ \sqrt{2} & 5 \end{bmatrix}$$

acting on the area be decomposed as

$$n_i s_{ij} \mathbf{i}_j \mathrm{d}\Sigma = n_i \tau_{ik} \frac{\partial \mathbf{x}}{\partial X_k} \mathrm{d}\Sigma, \tag{5.132}$$

where $\mathbf{i}_j = \partial \mathbf{X}/\partial X_j$ are the cartesian unit base vectors. This decomposition defines components of *nominal* stress s_{ij} and Kirchhoff stress τ_{ij}, related by

$$s_{ij} = \tau_{ik} F_{jk} = \tau_{ik}(\delta_{jk} + d_{jk}). \tag{5.133}$$

Thus s_{ij} is the j-component of current load, per unit initial area, acting upon an area element which initially faced in the i-direction. This direct physical interpretation of nominal stress also facilitates the simple form $(5.141)_3$ below of the equilibrium equations of finite strain, as recognized by Hill (1957). This simplicity is associated with the use of the X_j as independent variables, so allowing the boundary value problem for finite strain to be posed over a fixed region, namely the region initially occupied by the body. Other authors call (the transpose of) s_{ij} and τ_{ij} the first and second Piola–Kirchhoff tensors (see Truesdell and Toupin, 1960, §210). It may be shown that current true (Cauchy) stress σ_{ij} satisfies

$$\tau_{ij} = \frac{\rho[\mathbf{X}, 0]}{\rho[\mathbf{X}, t]} \sigma_{ij}, \tag{5.134}$$

where $\rho[\mathbf{X}, t]$ and $\rho[\mathbf{X}, 0]$ are current and initial densities at the considered particle. Local moment equilibrium ensures that σ_{ij} is symmetric, and so therefore is τ_{ij} but not s_{ij}, which must satisfy $s_{ik} \partial X_j/\partial x_k = s_{jk} \partial X_i/\partial x_k$.

Another symmetric stress measure, attributed by Hill (1978) to Biot (1965), is defined by

$$\begin{aligned} t_{ij} &= \tfrac{1}{2}(s_{ik} R_{kj} + s_{jk} R_{ki}) \\ &= \tfrac{1}{2}(\tau_{ik} \Lambda_{kj} + \tau_{jk} \Lambda_{ki}). \end{aligned} \tag{5.135}$$

We take note of the local inner product relations

$$s_{ij} F_{ji} = t_{ij} \Lambda_{ij} = \tau_{ij}(2e_{ij} + \delta_{ij}) = s_{ij}(d_{ji} + \delta_{ji}). \tag{5.136}$$

Their value is the current internal virtual work, per unit initial volume, associated with the current position vector x_j (not the displacement). The current virtual work-rate, associated with the current velocity $\dot{x}_j$, is

$$s_{ij} \dot{F}_{ji} = t_{ij} \dot{\Lambda}_{ij} = \tau_{ij} \dot{e}_{ij} = s_{ij} \dot{d}_{ji}, \tag{5.137}$$

where we have used $\dot{R}_{ki} R_{kj} = -\dot{R}_{kj} R_{ki}$. The formulae in this and the previous subsection apply to the large deformation of any medium, which may be, but does not have to be, an elastic body.

(iii) Hyperelastic solid

A hyperelastic solid can be defined by the property

$$\tau_{ij} = \frac{\partial U}{\partial e_{ij}}, \tag{5.138}$$

where $U[e_{ij}]$ is any given single valued symmetrized scalar function of the e_{ij}. From this starting point it is easy to see that different functions $U[F_{ji}]$ and symmetrized $U[\Lambda_{ij}]$ are implied by (5.128) and (5.131) respectively, such that

$$s_{ij} = \frac{\partial U}{\partial F_{ji}} \quad \text{and} \quad t_{ij} = \frac{\partial U}{\partial \Lambda_{ij}} \tag{5.139}$$

from (5.133) and (5.135). For simplicity we are considering the case in which there are no internal kinematical constraints such as incompressibility. The hyperelastic solid is a useful model of the elastic behaviour of many rubber-like materials, in which (5.137) takes the values of $\dot{U}$ so that U is interpretable as the stored 'strain energy' per unit initial volume.

A review of restrictions which have been proposed for hyperelastic strain energy functions was given by Truesdell and Noll (1965, §52). Progress on this so-called '*hauptproblem*' of Truesdell (1956) has continued since then. In particular, Ogden (1978) has concluded that it is reasonable to suppse that $U[\Lambda_{ij}]$ is strictly convex for all deformations of practical interest in rubberlike materials. A special example is the so-called semilinear material, defined by writing $\Lambda_{ij} - \delta_{ij}$ in place of the small strain in (2.83), as explained by Koiter (1976*a*); this belongs to a wider class of materials for which boundary value problems have been solved by Ogden and Isherwood (1978), for example that in plane strain ($\lambda_{(3)} \equiv 1$)

$$\begin{aligned} U &= f[\lambda_{(1)} + \lambda_{(2)} - 2] + \tfrac{1}{2}\mu(\lambda_{(1)} - \lambda_{(2)})^2 \\ &= f[\Lambda_{11} + \Lambda_{22} - 2] + \mu[\tfrac{1}{2}(\Lambda_{11} - \Lambda_{22})^2 + \Lambda_{12}{}^2 + \Lambda_{21}{}^2]. \end{aligned} \tag{5.140}$$

Here $f[.]$ is a given function of $\lambda_{(1)} + \lambda_{(2)} - 2$, and μ is a given constant, and it may be shown that $U[\Lambda_{ij}]$ is strictly convex if $\mu > 0$ and the second derivative $f''[.] > 0$ (see Isherwood and Ogden, 1977). The semilinear example is discussed explicitly in Exercise 5.5.4. A review of strain energy functions which have been used for incompressible and compressible rubberlike materials was given by Ogden (1982), oriented somewhat towards data from biaxial deformations. Even if $U[e_{ij}]$ or $U[\Lambda_{ij}]$ were arbitrary functions, $U[F_{ji}]$ could not be arbitrary because we must exclude the possibility that pure rotation could change the true stress. Neither could $U[F_{ji}]$ be convex without restriction because, as pointed out by Hill (1957), such convexity would imply unique large displacement in problems where bifurcations are known to intervene, for example by buckling in compression and by necking in tension. In fact convex $U[\Lambda_{ij}]$ does not imply convex $U[F_{ji}]$.

The hyperelastic solid also describes certain so-called 'deformation theories' of plasticity (see Neale, 1981, §4.4), which apply to the plastic behaviour of metals under restricted loading programs.

(iv) Governing equations

Suppose that the body initially occupies a region V with external boundary Σ. Let b_j, a_j and c_j be the components of currently assigned vector functions of position X_j on V, Σ_1 and Σ_2, respectively, with $\Sigma = \Sigma_1 + \Sigma_2$. We satisfy moment equilibrium automatically by assuming that U is any given single valued symmetrized function $U[e_{ij}]$ from the outset, which makes $\partial U/\partial F_{ji} = F_{jk}\partial U/\partial e_{ki}$ available for use below. Then the following set of equations govern current equilibrium of a hyperelastic solid after finite strain from the initial configuration.

$$
\begin{array}{llll}
V & \dfrac{\partial x_i}{\partial X_j} = F_{ij}, & (\alpha) & \\[2mm]
\Sigma_2 & x_j = c_j, & (\alpha) & \\[2mm]
V & -\dfrac{\partial s_{ij}}{\partial X_i} = b_j, & (\beta) & \qquad (5.141) \\[2mm]
\Sigma_1 & n_i s_{ij} = a_j, & (\beta) & \\[2mm]
V & s_{ij} = \dfrac{\partial U}{\partial F_{ji}}. & (\beta) &
\end{array}
$$

As with (5.74), we have divided (5.141) into a subset of kinematical (α) conditions and a subset of force-like (β) conditions. The simplest type of actual solution is a set of functions s_{ij}, x_j, F_{ij} of the X_k which satisfy (5.141) and are C^1, C^2, C^1 respectively.

(v) Stationary principles

Actual solutions of (5.141) can be characterized by a free stationary principle, uninhibited by the lack of convexity of $U[F_{ji}]$, and without requiring convexity of $U[\Lambda_{ij}]$ or of $U[e_{ij}]$ either. To apply our previous approach, replace (5.66) by

$$\int s_{ij}\frac{\partial x_j}{\partial X_i}\mathrm{d}V - \int s_{ij}n_i x_j \,\mathrm{d}\Sigma_2 = -\int x_j\frac{\partial s_{ij}}{\partial X_i}\mathrm{d}V + \int x_j n_i s_{ij}\,\mathrm{d}\Sigma_1, \quad (5.142)$$

rewrite this as $(s, T^*x) = \langle x, Ts \rangle$, and redefine T in (5.68) so that

$$T^*x = [\partial x_j/\partial X_i \quad 0 \quad -n_i x_j]^{\mathrm{T}}, \quad Ts = [-\partial s_{ij}/\partial X_i \quad n_i s_{ij} \quad 0]^{\mathrm{T}}. \tag{5.143}$$

This allows (5.141) to be written as in (5.75) but with $H = H[s; x, F]$. We are using s, F and x here to denote matrices of type (5.62) and (5.64). It can then be verified that (5.141) can be expressed as

$$\frac{\partial L}{\partial s} = 0 \quad (\alpha), \qquad \frac{\partial L}{\partial x} = \frac{\partial L}{\partial F} = 0 \quad (\beta), \tag{5.144}$$

where the single generating functional replacing (5.79) is

$$L[s; x, F] = \int \left(-x_j \frac{\partial s_{ij}}{\partial X_i} - s_{ij} F_{ji} + U[F_{ji}] - b_j x_j \right) \mathrm{d}V + \int x_j (n_i s_{ij} - a_j) \mathrm{d}\Sigma_1 + \int n_i s_{ij} c_j \, \mathrm{d}\Sigma_2. \tag{5.145}$$

The domain of $L[s; x, F]$ may be regarded as built from piecewise C^1 s_{ij} and x_j and integrable F_{ji}. As remarked in §3.9(iii) it may be convenient to work with more smoothness when seeking approximations to an actual solution, as in the next theorem.

Theorem 5.17 (Free stationary principle)
$L[s; x, F]$ is stationary with respect to arbitrary independent small piecewise C^1 variations δs_{ij}, δx_j and δF_{ji} of its arguments if and only if they are variations from an actual solution.

Proof
Sufficiency follows at once from the first variation

$$\begin{aligned}\delta L &= \left(\delta s, \frac{\partial L}{\partial s} \right) + \left\langle \delta x, \frac{\partial L}{\partial x} \right\rangle + \left(\delta F, \frac{\partial L}{\partial F} \right) \\ &= \int \left[\delta s_{ij} \left(\frac{\partial x_j}{\partial X_i} - F_{ji} \right) - \delta x_j \left(\frac{\partial s_{ij}}{\partial X_i} + b_j \right) - \delta F_{ji} \left(s_{ij} - \frac{\partial U}{\partial F_{ji}} \right) \right] \mathrm{d}V \\ &\quad + \int \delta x_j (n_i s_{ij} - a_j) \mathrm{d}\Sigma_1 + \int n_i \delta s_{ij} (c_j - x_j) \mathrm{d}\Sigma_2 \end{aligned} \tag{5.146}$$

and necessity depends on the fundamental lemma of the calculus of variations. □

Theorem 5.17 can be regarded as a generalization of Theorem 5.9, and is also a precise analogue of the variant (a) described after Theorem 5.1. Here s_{ij}, x_j, F_{ji}, $U[F_{ji}]$ correspond respectively to $\mathbf{Q}$, ϕ, $\mathbf{v}$, $-p[\mathbf{v}]$ there

Before any constraints are imposed we find (cf. (5.85) and (5.86)) that

$$L-\left(s,\frac{\partial L}{\partial s}\right)=\int\left(U[F_{ji}]-b_jx_j\right)\mathrm{d}V-\int a_jx_j\,\mathrm{d}\Sigma_1, \qquad (5.147)$$

$$L-\left\langle x,\frac{\partial L}{\partial x}\right\rangle-\left(F,\frac{\partial L}{\partial F}\right)=\int\left(U[F_{ji}]-F_{ji}\frac{\partial U}{\partial F_{ji}}\right)\mathrm{d}V+\int n_is_{ij}c_j\,\mathrm{d}\Sigma_2. \qquad (5.148)$$

We can see that (5.147) has the values of the total potential energy, with respect to a suitable datum level. We define the negative of (5.148) to be the total complementary energy, for reasons given before Theorem 5.19.

Theorem 5.18 (Stationary potential energy)
The potential energy functional

$$J[x]=\int\left(U\left[\frac{\partial x_j}{\partial X_i}\right]-b_jx_j\right)\mathrm{d}V-\int a_jx_j\,\mathrm{d}\Sigma_1 \qquad (5.149)$$

is stationary, with respect to small piecewise C^2 variations of the positional field x_j satisfying $x_j=c_j$ on Σ_2, if and only if they are variations from an actual solution.

Proof
We can follow steps similar to those in Theorems 5.2 and 5.11. Impose (5.141α) on the variations in Theorem 5.17, so that F_{ji} is expressed in terms of x_j. Then from (5.147), $L[s;x,F]$ becomes $J[x]$ and sufficiency follows.

Sufficiency also follows from a direct calculation of δJ in terms of δx_j (vanishing on Σ_2), the natural conditions being (5.141β) in combined form with the intermediate role of s_{ij} and F_{ji} deleted. Necessity requires the fundamental lemma. □

The literature reveals some debate about what is an appropriate definition of complementary energy for hyperelastic solids, reflecting in part the fact that different types of stress measure present themselves. Here we choose a definition of local complementary energy to have the values of

$$F_{ji}\frac{\partial U}{\partial F_{ji}}-U[F_{ji}]=\Lambda_{ij}\frac{\partial U}{\partial \Lambda_{ij}}-U[\Lambda_{ij}]=U_c \quad \text{(say)} \qquad (5.150)$$

per unit initial volume. The reason stems from $(5.136)_1$, whose kinematic parallel is used in (5.150). Both variants of (5.150) have the structure needed to anticipate a Legendre transformation. This allows the inverses of (5.139), when they exist, and even if they are not single valued, to be written as

$$F_{ji} = \frac{\partial U_c}{\partial s_{ij}} \quad \text{and} \quad \Lambda_{ij} = \frac{\partial U_c}{\partial t_{ij}},$$

respectively, in terms of different functions $U_c[s_{ij}]$ and $U_c[t_{ij}]$. Ogden (1977) shows that, even when $U_c[t_{ij}]$ is single valued as a consequence of the convexity of $U[\Lambda_{ij}]$, $U_c[s_{ij}]$ may be multivalued, and an illustration of this is provided in Exercise 5.5.4. In the compressible fluid mechanics analogue, the role of U_c is taken by the double valued $P[Q]$ in Fig. 5.3(*b*).

Theorem 5.19 (Stationary complementary energy)
The complementary energy functional

$$-K[s] = \int U_c[s_{ij}]\,\mathrm{d}V - \int n_i s_{ij} c_j \,\mathrm{d}\Sigma_2$$

is stationary, with respect to small C^1 variations of s_{ij} which satisfy $-\partial s_{ij}/\partial X_i = b_j$ in V and $n_i s_{ij} = a_j$ on Σ_1, if and only if they represent variations from an actual solution.

Proof
We follow steps in Theorems 5.3 and 5.12. Impose (5.141β) on the variations in Theorem 5.17. From (5.148) we see that $L[s; x, F]$ then becomes $-K[s]$, and sufficiency follows.

To handle necessity, we first deploy $s_{ij} = \partial U/\partial F_{ji} \Leftrightarrow \partial L/\partial F = 0$ in inverse form to eliminate $F_{ji} = \partial U_c/\partial s_{ij}$ from (5.145), leaving $L[s, x, F[s]] = \langle x, \partial L/\partial x\rangle - K[s]$. This is linear in x, and the remaining β-constraint $\partial L/\partial x = 0$, which comprises exactly the constraints stated in the current theorem, would delete x altogether from $L[s, x, F[s]]$. We suppose that the stationary solutions under this constraint are the same as those associated with a Lagrange multiplier μ when the augmented functional

$$\begin{aligned} \bar{L}[s, x, \mu, F[s]] &= \langle x - \mu, \partial L/\partial x\rangle - K[s] \\ &= L[s, x - \mu, F[s]] \end{aligned} \qquad (5.151)$$

is required to be stationary under free variations of s_{ij}, x_j and μ_j. The natural conditions for this are (5.141), but with $\partial U_c/\partial s_{ij}$ in place of F_{ji} and $x_j - \mu_j$ in place of x_j, and arbitrary μ_j finally deemed to be zero. □

We recall that $U_c[s_{ij}]$ in $K[s]$ may be multivalued, although a full description of even the qualitative nature of this multivaluedness is not currently available, unlike the situation for the analogous $P[Q]$ already described. Recall also that moment equilibrium has been satisfied from the outset here, by requiring $U_c[s_{ij}]$ to have been obtained via a strain energy such as $U[e_{ij}]$ or $U[\Lambda_{ij}]$ as starting point, rather than an arbitrary function $U[F_{ji}]$. Variants of Theorem 5.19 are a subject of current research; for example, in place of $K[s]$, the functional to be made stationary may be $\int U_c[t_{ij}]\mathrm{d}V - \int n_i s_{ij} c_j \,\mathrm{d}\Sigma_2$, where $U_c[t_{ij}]$ obtained via (5.150) is single valued but depends, via (5.135), on both s_{ij} and R_{kj} (see Ogden, 1975, 1978; and also Exercise 5.5.4 here).

Some extremum principles deriving from the convexity of $U[\Lambda_{ij}]$ are described by Ogden (1978, 1984).

Reviews of the literature on stationary principles for hyperelastic solids have been given by Truesdell and Noll (1965, §88), and by Nemat-Nasser (1972*a*, *b*) who also mentioned applications, including the role of discontinuities. Early treatments such as that of Hellinger (1914) were ambiguous about boundary conditions, which began to be clarified by Reissner (1953). Koiter (1973, 1976*a*, *b*) gave special attention to the complementary energy principle and its applications, as have Levinson (1965), Nemat-Nasser (1977), Lee and Shield (1980*a*, *b*) and Ogden (*op. cit.*). The approach to a single generating functional (5.145) via a pair of adjoint operators was described by Sewell (1969, 1973*a*, *b*), where the structure of (5.147) and (5.148), relating L to the energy and complementary energy, was first indicated.

Exercises 5.5

1. Hydrostatic pressure loading is defined to be such that the current load vector (5.132) always acts along the inward normal to the current area element, with assigned uniform current magnitude $p > 0$ per unit current area. Show that the nominal traction is thus assigned to be a function

$$n_i s_{ik} = -p\tfrac{1}{2}\varepsilon_{ijk}\varepsilon_{pqr}n_r F_{ip}F_{jq}$$
$$= -p[n_k + n_k d_{jj} - n_j d_{jk} + \tfrac{1}{2}\varepsilon_{ijk}\varepsilon_{pqr}n_r d_{ip}d_{jq}]$$

of deformation or displacement gradient, where $\varepsilon_{ijk} = +1, -1$ or 0 if the sequence ijk is cyclic, acyclic or has two equal members. This is an example of configuration-dependent loading, where the nominal traction, instead of being an assigned vector from the outset as in $(5.141)_4$, senses the deformation of the surface and 'follows' it.

2. Let pressure loading be applied to a part Σ_1 of Σ which either (a) occupies the whole of Σ ($\Sigma_2 = 0$), or (b) is terminated by a closed curve on which displacement is assigned (for example to be continuous with assigned or zero displacement on Σ_2 contiguous with Σ_1). Prove that the incremental work of the pressure loading, integrated over Σ_1, is the first variation of the 'potential'

$$-\frac{1}{6}p\varepsilon_{ijk}\varepsilon_{pqr}\int n_r x_k \frac{\partial x_i}{\partial X_p}\frac{\partial x_j}{\partial X_q}\mathrm{d}\Sigma_1$$

or of

$$-p\iint\left[n_k u_k+\frac{1}{2}\left(n_k\frac{\partial u_j}{\partial X_j}-n_j\frac{\partial u_j}{\partial X_k}\right)u_k+\frac{1}{6}\varepsilon_{ijk}\varepsilon_{pqr}u_k n_r\frac{\partial u_i}{\partial X_p}\frac{\partial u_j}{\partial X_q}\right]\mathrm{d}\Sigma_1.$$

3. Hence construct a free stationary principle characterizing pressure loading on Σ_1, in place of the assigned traction term in (5.145). Prove also that pressure loading is conservative. Detailed answers to the foregoing questions may be found in Sewell (1965, 1967). Relate the pressure loading potential to the volume swept out by the particles which began on Σ_1 (see also Pearson, 1956, equation (25); and Truesdell and Noll, 1965, equation (88.14)).
4. (i) Show that, for given $\mu > 0$ and $\lambda > -\mu$, the semilinear example

$$U[\Lambda_{ij}]=\tfrac{1}{2}(\lambda+\mu)(\Lambda_{11}+\Lambda_{22}-2)^2+\mu[\tfrac{1}{2}(\Lambda_{11}-\Lambda_{22})^2+\Lambda_{12}^2+\Lambda_{21}^2]$$

of (5.140) can be written as

$$U[e_{ij}]=(\lambda+\mu)[\{1+e_{11}+e_{22}+((1+2e_{11})(1+2e_{22})-4e_{12}e_{21})^{\frac{1}{2}}\}^{\frac{1}{2}}-2^{\frac{1}{2}}]^2$$
$$+\mu[1+e_{11}+e_{22}-((1+2e_{11})(1+2e_{22})-4e_{12}e_{21})^{\frac{1}{2}}].$$

(ii) Show that this $U[\Lambda_{ij}]$ implies the Legendre dual function

$$U_c[t_{ij}]=t_{11}+t_{22}+\frac{(t_{11}+t_{22})^2}{8(\lambda+\mu)}+\frac{(t_{11}-t_{22})^2}{8\mu}+\frac{t_{12}^2}{4\mu}+\frac{t_{21}^2}{4\mu}.$$

It can be shown, using the isotropy (see Isherwood and Ogden, 1977), that $t_{11}+t_{22}=\pm[(s_{11}+s_{22})^2+(s_{12}-s_{21})^2]^{\frac{1}{2}}$. This leads to the double valued function

$$U_c[s_{ij}]=\pm[(s_{11}+s_{22})^2+(s_{12}-s_{21})^2]^{\frac{1}{2}}+\frac{(s_{11}+s_{22})^2+(s_{12}-s_{21})^2}{8(\lambda+\mu)}$$
$$+\frac{(s_{11}-s_{22})^2+(s_{12}+s_{21})^2}{8\mu}.$$

5.10. Incremental elastic/plastic distortion

(i) Governing equations

The *elastic/plastic solid* is the theoretical model for the behaviour of a metal in which permanent deformation does *not* dominate elastic deformation, in contrast to the rigid/plastic solid of §5.8. The body may contain an elastic region where the stress has not reached the level needed to satisfy any yield function, and a plastic region where the stress has reached such a 'yield point' level. Such regions are supposed known here, and we are concerned only with the quasi-static increment of distortion from a known configuration of such a material. In the elastic region this incremental response is reversible and independent of the direction of additional stressing. In the plastic region, depending on the direction of further stressing, the consequent further distortion either may contain an irreversible component (induced by 'plastic loading'), or may be fully reversible (e.g. after 'elastic unloading').

A number of different theories of this incremental response are available, and reviews of them has been given by Sewell (1972), Hutchinson (1974), Hill (1978) and Neale (1981). We select just one of these theories, having an analytical structure which has been an insistent guide during the refinements which have been added to plasticity theory over the years. One of our aims here is to explain how this structure is intimately related to the ideas of saddle functions and Legendre dual transformations (Fig. 2.20) described earlier in this book.

It is most convenient to give the full list of governing conditions at the outset, as now following, and then explain them. Recall that Greek suffixes have the range $\gamma = 1, \ldots, m$. The value of m now describes the number of local yield surfaces attained by the stress, and hence the number of facets of a yield vertex ($m = 1$ at a smooth yield surface). This m is regarded as an assigned function of position in the plastic region. In the elastic region we may formally put $m = 0$, to delete quantities and equations having Greek suffixes in the following.

$$\frac{\partial \dot{u}_j}{\partial X_i} = \dot{d}_{ji} \quad \text{in} \quad V, \qquad \dot{u}_j = c_j \quad \text{on} \quad \Sigma_2, \qquad (\alpha)$$

$$\frac{1}{2}\left(\dot{d}_{ij} + \dot{d}_{ji} + \dot{d}_{ki}\frac{\partial u_k}{\partial X_j} + \frac{\partial u_k}{\partial X_i}\dot{d}_{kj}\right) = \frac{\partial X}{\partial \dot{\tau}_{ij}}, \qquad (\alpha)$$

$$-\frac{\partial \dot{s}_{ij}}{\partial X_i} = b_j \quad \text{in} \quad V, \qquad n_i \dot{s}_{ij} = a_j \quad \text{on} \quad \Sigma_1, \qquad (\beta)$$

$$\dot{s}_{ij} = \dot{\tau}_{ij} + \dot{\tau}_{ik}\frac{\partial u_j}{\partial X_k} + \tau_{ik}\dot{d}_{jk}, \qquad (\beta) \qquad (5.152)$$

$$\frac{\partial X}{\partial \lambda_\gamma} \leqslant 0, \qquad (\beta)$$

$$\lambda_\gamma \geqslant 0, \qquad (\alpha)$$

$$\lambda_\gamma \frac{\partial X}{\partial \lambda_\gamma} = 0.$$

(ii) Strain-rate

Strain measures such as those described in §5.9(i) are again available, but the parameter t in the function $\mathbf{x}[\mathbf{X}, t]$ is not here regarded as real time. Instead it can be any scalar ordering parameter whose monotonic notional variation defines a sequence of equilibrium configurations. Examples of such t include the magnitude of a load whose direction is fixed, or a modulus, or a dimension. We are focussing on the next increment from a known configuration in such a sequence. There is no connection between the symbols $\mathbf{X}$ for the initial position vector and X for the value of the given function $X[\dot{\tau}_{ij}, \lambda_\gamma]$.

Partial derivatives with respect to t at fixed $\mathbf{X}$, i.e. following the particle, are denoted by a superposed dot. The quasistatic velocity is therefore $\dot{\mathbf{x}} = \dot{\mathbf{u}}[\mathbf{X}, t]$, and each different deformation measure in §5.9(i) has a corresponding rate of change. In particular

$$\dot{e}_{ij} = \tfrac{1}{2}(\dot{d}_{ij} + \dot{d}_{ji} + \dot{d}_{ki}d_{kj} + d_{ki}\dot{d}_{kj}) \tag{5.153}$$

is sometimes called the Lagrangian strain-rate based on the initial configuration. When the current instant is taken to be the initial instant we can put $t = 0$ and $\mathbf{u}[\mathbf{X}, 0] = 0$ after differentiation, and the value of $\dot{e}_{ij}$ is then $\frac{1}{2}(\dot{d}_{ij} + \dot{d}_{ji})$. The latter quantity when thus based on the current configuration is called the Eulerian strain-rate, i.e. the same as $(5.90)_1$ familiar in fluid mechanics, and the direct analogue of the small strain $(5.74)_1$ of classical elasticity.

It is the Lagrangian strain-rate (5.153) whose values appear on the left of $(5.152)_2$, the current displacement function $\mathbf{u}[\mathbf{X}, t]$ being known, along with its gradient $d_{ij} = \partial u_i/\partial X_j$.

The initial configuration of the body occupies a region V, with external boundary $\Sigma_1 + \Sigma_2$. Equations $(5.152)_1$ merely relate velocity to displacement gradient-rate, allowing the latter to be eliminated whenever desired; for simplicity we specify boundary conditions of assigned velocity on Σ_2, the c_j being components of an assigned vector over Σ_2.

(iii) Stress-rate

The rate of change of nominal and Kirchhoff stresses are related by the derivative of (5.133), which is $(5.152)_4$. These $\dot{\tau}_{ij}$ components are objective measures of stress-rate in the material, in the sense that they will be zero when the body is undergoing pure rotation without distortion. This property is not then possessed by the $\dot{s}_{ij}$ when the body is under non-zero stress, as can be seen from the last term of $(5.152)_4$. The $\dot{s}_{ij}$ form a non-symmetric matrix, but the $\dot{\tau}_{ij}$ form a symmetric matrix.

(iv) Constitutive rate equations

We adopt the structure of constitutive rate equations laid out in §2.6(iii) for an elastic/plastic solid. The precise definitions of $\dot{\boldsymbol{\tau}}$ and $\dot{\mathbf{e}}$ were left unspecified there. We now choose these to be the 6 × 1 column matrices constructed from the six independent cartesian components $\dot{\tau}_{ij}$ and from the $\dot{e}_{ij}$, respectively, as defined in the preceding two subsections. Such $\dot{\tau}_{ij}$ have the values of what are called, in the specialist literature, the convected derivative of contravariant Kirchhoff stress components, with respect to embedded coordinates which are here initially cartesian.

Thus we hypothesize (2.84) with (2.85), in terms of a given single valued function $X[\dot{\tau}, \lambda_\gamma]$, homogeneous of degree two, of which a straightforward example is the homogeneous quadratic (2.86). If we eliminate $\dot{e}_{ij}$ at once from $(2.84)_1$ and (5.153), this leaves $(5.152)_2$. Of course, the elements of the matrix $\partial X/\partial\dot{\tau}$ are rewritten $\partial X/\partial\dot{\tau}_{ij}$ here; to be formally correct we should note that this is a shorthand for $\frac{1}{2}(\partial X/\partial\dot{\tau}_{ij} + \partial X/\partial\dot{\tau}_{ji})$, computed after symmetrizing $X[\dot{\tau}_{ij}, \lambda_\gamma]$ by writing $\frac{1}{2}(\dot{\tau}_{ij} + \dot{\tau}_{ji})$ in place of $\dot{\tau}_{ij}$ and of $\dot{\tau}_{ji}$. If we eliminate the μ_γ at once from $(2.84)_2$ and (2.85) this leaves the last trio of (5.152).

Examples of the gradients of $X[\dot{\tau}_{ij}, \lambda_\gamma]$ are provided by the right sides of (2.87), which in suffix notation are

$$\frac{\partial X}{\partial \dot{\tau}_{ij}} = M_{ijkl}\dot{\tau}_{kl} + v_{ij\gamma}\lambda_\gamma, \qquad \frac{\partial X}{\partial \lambda_\gamma} = \dot{\tau}_{ij}v_{ij\gamma} - h_{\gamma\delta}\lambda_\delta. \tag{5.154}$$

The constitutive system $(5.152)_{2,5,6,7}$ is then a system of linear conditions. The 36 given elastic compliance components M_{ijkl} are symmetric with respect to the interchanges $i \leftrightarrow j$, $k \leftrightarrow l$ and $ij \leftrightarrow kl$. The $v_{ij\gamma} = v_{ji\gamma}$ and $h_{\gamma\delta} = h_{\delta\gamma}$ are also regarded as given, all as part of a known current stress state. In practice these parameters will have been decided by the previous history of distortion. To begin with we desist from imposing sign-definiteness hypotheses for the M_{ijkl} and the $h_{\gamma\delta}$ of the type which were discussed in §2.6(iii), although we shall need to introduce them later.

We defer until §(xii) below the conventional *a priori* elimination of the λ_γ in favour of $\dot{\tau}_{ij}$, for example as in (2.92). We explore first the consequences of leaving the constitutive equations in the form $(5.152)_{2,5,6,7}$ generated by one single valued explicit function $X[\dot{\tau}_{ij}, \lambda_\gamma]$. This contrasts with the multibranched, possibly multivalued stress-rate potential $V_c[\dot{\tau}_{ij}]$, which is only given implicitly when $m > 1$ until we enumerate cases (cf. Exercise 2.5.5).

(v) Continuing equilibrium

Although the nominal stress-rate $\dot{s}_{ij}$ is not objective, it offers an especially simple preliminary way of writing the equations governing continuing equilibrium. This simplicity stems from the direct physical interpretation of current nominal stress in terms of the initial reference configuration described after (5.133). The equations of continuing current equilibrium are $(5.152)_3$ in V, as if just by time differentiation of $(5.141)_3$. Here the b_j are the components of a currently assigned body force rate vector function of position X_j in V. Continuing moment equilibrium is satisfied by requiring the symmetry of $\dot{\tau}_{ij}$ imposed above.

In $(5.152)_3$ we also specify boundary conditions of assigned loading rate on Σ_1, the a_j being assigned functions of position on Σ_1.

This completes the explanation of the list (5.152) of governing conditions.

(vi) Adjointness

We can reduce the number of conditions in this list by deploying $(5.152)_{1,4}$ to eliminate the $\dot{d}_{ij}$ and $\dot{s}_{ij}$, whenever desired, leaving a system from which to determine the $\dot{\tau}_{ij}, \dot{u}_j$ and λ_γ as functions of the initial position X_i.

To begin with, we desist from this elimination in order to note the especially simple nature of adjoint operators which appear in $(5.152)_{1,3}$. An appropriate integration by parts formula, replacing (5.66) and (5.142), is

$$\int \dot{s}_{ij} \frac{\partial \dot{u}_j}{\partial X_i} dV - \int \dot{s}_{ij} n_i \dot{u}_j d\Sigma_2 = -\int \dot{u}_j \frac{\partial \dot{s}_{ij}}{\partial X_i} dV + \int \dot{u}_j n_i \dot{s}_{ij} d\Sigma_1 \quad (5.155)$$

for any piecewise C^1 functions $\dot{u}_j$ and $\dot{s}_{ij}$ of the X_i. This induces definitions of inner product spaces in terms of which (5.155) can be rewritten

$$(\dot{s}, T^*\dot{u}) = \langle \dot{u}, T\dot{s} \rangle \quad (5.156)$$

where the adjoint linear operators have the simple definitions

$$
\begin{array}{c} V \\ \Sigma_1 \\ \Sigma_2 \end{array}
\quad
T^*\dot{u} = \begin{bmatrix} \dfrac{\partial \dot{u}_j}{\partial X_i} \\ 0 \\ -n_i \dot{u}_j \end{bmatrix}, \quad
T\dot{s} = \begin{bmatrix} -\dfrac{\partial \dot{s}_{ij}}{\partial X_i} \\ n_i \dot{s}_{ij} \\ 0 \end{bmatrix}, \quad (5.157)
$$

which replace (5.68). These definitions summarize only the left sides of the equations in $(5.152)_{1,3}$, and are independent of the constitution of the body, of whether equilibrium is sustained, and of the particular boundary conditions implied by the right sides of those equations.

(vii) Generating functional

The two inner product spaces implied by (5.156), say E constructed from second order tensors and F constructed from vectors, are augmented by a third G constructed from the multiples of scalars such as we have already used in (5.108) and (5.109).

Then a generalized Hamiltonian representation can be given (see

Exercise 5.6.1) along the lines of (5.75) or (5.110) with (5.113), extended to account for the additional variables present now, and containing the operators T and T^* of (5.157). We telescope this stage, in view of the previous examples. It leads, after deducing the Hamiltonian by integration and subtracting it from either of (5.156), to the generating functional (Sewell 1973*b*, 1982)

$$L[\dot{s}, \dot{\tau}, \dot{u}, \dot{d}, \lambda] = \int \left[\dot{s}_{ij} \frac{\partial \dot{u}_j}{\partial X_i} - \left(\dot{s}_{ij} - \dot{\tau}_{ij} - \dot{\tau}_{ik} \frac{\partial u_j}{\partial X_k} \right) \dot{d}_{ji} + \tfrac{1}{2} \tau_{ik} \dot{d}_{ji} \dot{d}_{jk} \right.$$
$$\left. - X[\dot{\tau}_{ij}, \lambda_\gamma] - b_j \dot{u}_j \right] \mathrm{d}V - \int a_j \dot{u}_j \mathrm{d}\Sigma_1 + \int n_i \dot{s}_{ij} (c_j - \dot{u}_j) \, \mathrm{d}\Sigma_2 . \tag{5.158}$$

This is defined on a subspace of $E \times E \times F \times E \times G$, constructed from sufficiently smooth (say piecewise C^1) tensors, vectors and scalars (recall §3.9(iii) again; in the remaining theorems we shall not state the weaker admissible smoothness, such as we have done previously, preferring instead to concentrate on other details of the theory).

Theorem 5.20

The governing conditions (5.152) *can be expressed in terms of the gradients of the functional* (5.158) *as, respectively,*

$$\begin{aligned}
\frac{\partial L}{\partial \dot{s}} &= 0, \quad (\alpha) \\
\frac{\partial L}{\partial \dot{\tau}} &= 0, \quad (\alpha) \\
\frac{\partial L}{\partial \dot{u}} &= 0, \quad (\beta) \\
\frac{\partial L}{\partial \dot{d}} &= 0, \quad (\beta) \\
\frac{\partial L}{\partial \lambda} &\geqslant 0, \quad (\beta) \\
\lambda &\geqslant 0, \quad (\alpha) \\
\overline{\lambda, \frac{\partial L}{\partial \lambda}} &= 0.
\end{aligned} \tag{5.159}$$

Proof

The gradients can be identified by a direct computation of the first variation in the form

$$\delta L = \left(\delta \dot{s}, \frac{\partial L}{\partial \dot{s}}\right) + \left(\delta \dot{\tau}, \frac{\partial L}{\partial \dot{\tau}}\right) + \left\langle \delta \dot{u}, \frac{\partial L}{\partial \dot{u}}\right\rangle + \left(\delta \dot{d}, \frac{\partial L}{\partial \dot{d}}\right) + \overline{\delta \lambda, \frac{\partial L}{\partial \lambda}}. \tag{5.160}$$

The reader may verify this by elementary methods, i.e. by inserting first variations $\delta \dot{s}_{ij}, \delta \dot{\tau}_{ij}, \delta \dot{u}_j, \delta \dot{d}_{ij}, \delta \lambda_\gamma$ into the right side of (5.158), and performing the classical manipulations of integration by parts (Exercise 5.6.3). □

As in previous sections, the subset of kinematical conditions is labelled (α), and the subset of force-like conditions is labelled (β).

As in the rigid/plastic yield point problem (§5.8(i)), it is convenient for certain purposes to introduce an auxiliary *comparison solid*. Here it is defined by replacing the trio $(5.152)_{5,6,7}$ in the plastic region by $\partial X/\partial \lambda_\gamma = 0$ for every λ_γ, with no sign restriction on the λ_γ. The real and comparison solids then have solutions in common if and only if it turns out that all $\lambda_\gamma \geqslant 0$ in the plastic region of the comparison solid. In the elastic region, the real and comparison solids have the same purely elastic response. This is governed by using $X[\dot{\tau}_{ij}, 0]$ in place of $X[\dot{\tau}_{ij}, \lambda_\gamma]$, or equivalently by choosing $m = 0$ as explained in §(i) above.

The simplest type of *actual solution* in the real solid is a set of functions $\dot{s}_{ij}, \dot{\tau}_{ij}, \dot{u}_j, \dot{d}_{ij}, \lambda_\gamma$ of initial position X_j, respectively, C^1, C^1, C^2, C^1 and C^1, which satisfy all of (5.152), or equivalently (5.159), in the plastic region, and which satisfy their stated modifications in the elastic region.

The following Theorem characterizes solutions in the comparison solid, and in particular some actual solutions in the real solid.

Theorem 5.21 (Free stationary principle)
$L[\dot{s}, \dot{\tau}, \dot{u}, \dot{d}, \lambda]$ is stationary with respect to arbitrary small variations of its arguments if and only if they are variations from an actual solution in the comparison solid.

Proof
Sufficiency is immediate, since (5.159) imply $\delta L = 0$ in (5.160). Necessity depends on the fundamental lemma of the calculus of variations. It is sufficient, if not necessary, to require the class of variations to have the same smoothness as that required for an actual solution. □

Various specializations of Theorem 5.21 can be constructed by imposing some of (5.159) as constraints at the outset. For example, the $\dot{s}_{ij}$ could be eliminated from the beginning by imposing $\partial L/\partial \dot{d} = 0$ i.e. $(5.152)_4$. Also, the $\dot{d}_{ij}$ could be eliminated at will by imposing $(5.152)_1$.

Before introducing these or any other modifications, however, the division of (5.159) into kinematical (α) and force-like (β) subsets tells us, as a consequence of the general theoretical structure developed in this book, that special significance will attach to the following two functionals generated by L and its gradients.

$$L-\left(\dot{s},\frac{\partial L}{\partial \dot{s}}\right)-\left(\dot{\tau},\frac{\partial L}{\partial \dot{\tau}}\right)$$
$$=\iint\left[\dot{\tau}_{ij}\frac{\partial X}{\partial \dot{\tau}_{ij}}-X[\dot{\tau}_{ij},\lambda_\gamma]+\tfrac{1}{2}\tau_{ik}\dot{d}_{ji}\dot{d}_{jk}-b_j\dot{u}_j\right]\mathrm{d}V-\int a_j\dot{u}_j\,\mathrm{d}\Sigma_1. \quad (5.161)$$

$$L-\left\langle\dot{u},\frac{\partial L}{\partial \dot{u}}\right\rangle-\left(\dot{d},\frac{\partial L}{\partial \dot{d}}\right)-\overline{\lambda,\frac{\partial L}{\partial \lambda}}$$
$$=\iint\left[\lambda_\gamma\frac{\partial X}{\partial \lambda_\gamma}-X[\dot{\tau}_{ij},\lambda_\gamma]-\tfrac{1}{2}\tau_{ik}\dot{d}_{ji}\dot{d}_{jk}\right]\mathrm{d}V+\int n_i\dot{s}_{ij}c_j\,\mathrm{d}\Sigma_2. \quad (5.162)$$

These expressions (Sewell, 1973*b*, 1982) may be viewed as generalizations of the total potential energy (5.85) and negative complementary energy (5.86) for a classical elastic solid.

The Legendre transformation properties in Fig. 2.20 show that the first two terms in these integrands have the values of

$$\dot{\tau}_{ij}\frac{\partial X}{\partial \dot{\tau}_{ij}}-X=Y[\dot{e}_{ij},\lambda_\gamma], \quad (5.163)$$

$$\lambda_\gamma\frac{\partial X}{\partial \lambda_\gamma}-X=Z[\dot{\tau}_{ij},\mu_\gamma], \quad (5.164)$$

whenever the functions on the right exist.

(viii) Upper and lower bounds

We can anticipate that, if $L[\dot{s},\dot{\tau},\dot{u},\dot{d},\lambda]$ can be shown to be a saddle functional, concave in the force-like variables $\dot{s},\dot{\tau}$ and convex in the kinematical variables $\dot{u},\dot{d},\lambda$, then (5.161) will have the value of an upper bound and (5.162) that of a lower bound. This leads us to define the *saddle quantity*

$$S=L[\dot{s}_+,\dot{\tau}_+;\dot{u}_+,\dot{d}_+,\lambda_+]-L[\dot{s}_-,\dot{\tau}_-;\dot{u}_-,\dot{d}_-,\lambda_-]$$
$$-\left(\dot{s}_+-\dot{s}_-,\left.\frac{\partial L}{\partial \dot{s}}\right|_+\right)-\left(\dot{\tau}_+-\dot{\tau}_-,\left.\frac{\partial L}{\partial \dot{\tau}}\right|_+\right)$$
$$-\left\langle\dot{u}_+-\dot{u}_-,\left.\frac{\partial L}{\partial \dot{u}}\right|_-\right\rangle-\left(\dot{d}_+-\dot{d}_-,\left.\frac{\partial L}{\partial \dot{d}}\right|_-\right)$$
$$-\overline{\lambda_+-\lambda_-,\left.\frac{\partial L}{\partial \lambda}\right|_-} \quad (5.165)$$

for all pairs of points in the domain of (5.158). This is the generalization of (1.57) which is appropriate to the current circumstances – compare (5.117) for the rigid/plastic yield point problem. After inserting the gradients we find that

$$S = \frac{1}{2}\int \tau_{ik}(\dot{d}_{ji+} - \dot{d}_{ji-})(\dot{d}_{jk+} - \dot{d}_{jk-})\mathrm{d}V$$

$$+ \int\Bigg[X[\dot{\tau}_{ij-}, \lambda_{\gamma-}] - X[\dot{\tau}_{ij+}, \lambda_{\gamma+}]$$

$$- (\dot{\tau}_{ij-} - \dot{\tau}_{ij+})\frac{\partial X}{\partial \dot{\tau}_{ij}}\Bigg|_{+} - (\lambda_{\gamma-} - \lambda_{\gamma+})\frac{\partial X}{\partial \lambda_{\gamma}}\Bigg|_{-} \Bigg]\mathrm{d}V. \qquad (5.166)$$

As in other examples, the bilinear and linear terms in L do not contribute to S.

We now recall that in §1.3(iii) we gave a list of free or constrained hypotheses (1.58) (a)–(f) to illustrate how saddle or convexity statements could be expressed in terms of $S > 0$ or $S \geq 0$, with consequent implications (Fig. 1.11) for extremum principles or bounds. In what follows we give selected results from a similar program in terms of (5.166).

Theorem 5.22 (Weak saddle functional)
A set of sufficient conditions for $L[\dot{s}, \dot{\tau}; \dot{u}, \dot{d}, \lambda]$ to be a weak saddle functional, concave in the stress-rates $\dot{s}, \dot{\tau}$ and convex in the kinematical variables $\dot{u}, \dot{d}, \lambda$, in the sense that

$$S \geq 0 \qquad (5.167)$$

for every pair of distinct points in the domain of L, is that
   (a) the current principal stresses are nonnegative, and
   (b) $X[\dot{\tau}_{ij}, \lambda_{\gamma}]$ is a saddle function, convex in the $\dot{\tau}_{ij}$ and concave in the λ_{γ} (at least weakly so in both).

Proof
This is an immediate consequence of (5.166), because (a) and (b) ensure that the first and second integrands there are nonnegative respectively. ☐

Theorem 5.23 (Simultaneous upper and lower bounds)
When (5.167) holds, the bounds

$$L_{\alpha} \geq L_0 \geq L_{\beta} - \overline{\lambda_{\beta}, \frac{\partial L}{\partial \lambda}}\Bigg|_{\beta} \qquad (5.168)$$

apply, where L_{α} is the value of (5.161) for any solution of the subset (5.159α), $L_{\beta} - \overline{\lambda_{\beta}, \partial L/\partial \lambda}|_{\beta}$ is the value of (5.162) for any solution of the subset (5.159β), and L_0 is the value of L for any solution of all the conditions (5.159).

Proof
The method of proof is a direct extension of Theorem 1.4 or 1.6(c) as in Theorem 3.4, but with the present $S \geqslant 0$. Thus, as (1.38) suggests, we first choose $\dot{s}_+, \dot{t}_+, \dot{u}_+, \dot{d}_+, \lambda_+$ to be any solution of (5.159α) and $\dot{s}_-, \dot{t}_-, \dot{u}_-, \dot{d}_-, \lambda_-$ to be any solution of all of (5.159). This gives the upper bound property, from (5.165). Then again, as (1.39) suggests, we choose $\dot{s}_-, \dot{t}_-, \dot{u}_-, \dot{d}_-, \lambda_-$ to be any solution of (5.159β) and $\dot{s}_+, \dot{t}_+, \dot{u}_+, \dot{d}_+, \lambda_+$ to be any solution of all of (5.159). This gives the lower bound property. □

We now seek to deploy the kinematical constraints (5.159α), eliminating variables where possible, to give an explicit formula for the upper bound L_α from (5.161). In suffix form the set (5.159α) consists of (5.152α). When $X[\dot{t}_{ij}, \lambda_\gamma]$ is strictly convex with respect to $\dot{t}_{ij}$ we can impose $(5.152)_2$ and rewrite it as $\dot{t}_{ij} = \partial Y/\partial \dot{e}_{ij}$ in terms of (5.163). Using the first of $(5.152)_1$ to eliminate the $\dot{d}_{ij}$, we find

$$L_\alpha = \int\left[Y[\dot{e}_{ij}, \lambda_\gamma] + \frac{1}{2}\tau_{ik}\frac{\partial \dot{u}_j}{\partial X_i}\frac{\partial \dot{u}_j}{\partial X_k} - b_j\dot{u}_j\right]\mathrm{d}V - \int a_j\dot{u}_j\,\mathrm{d}\Sigma_1, \quad (5.169)$$

wherein $\dot{e}_{ij}$ is to be regarded merely as a shorthand symbol for

$$\dot{e}_{ij} = \frac{1}{2}\left(\frac{\partial \dot{u}_j}{\partial X_i} + \frac{\partial \dot{u}_i}{\partial X_j} + \frac{\partial \dot{u}_k}{\partial X_i}\frac{\partial u_k}{\partial X_j} + \frac{\partial u_k}{\partial X_i}\frac{\partial \dot{u}_k}{\partial X_j}\right) \quad (5.170)$$

from (5.153). In this way we obtain a functional L_α of the $\dot{u}_j$ and λ_γ alone, which is an upper bound, to be minimized subject to the remaining α constraints, which are $\dot{u}_j = c_j$ on Σ_2 and all $\lambda_\gamma \geqslant 0$.

To give an explicit formula for the lower bound $L_\beta - \overline{\lambda_\beta, \partial L/\partial\lambda|_\beta}$ from (5.162), we deploy the set (5.159β) of force-like constraints, or equivalently (5.152β). We can use $(5.152)_4$ at the outset to eliminate the $\dot{s}_{ij}$ from $(5.152)_3$ and from the Σ_2 integral in (5.162). When $X[\dot{t}_{ij}, \lambda_\gamma]$ is strictly concave with respect to the λ_γ we can incorporate the definitions $\mu_\gamma = \partial X/\partial\lambda_\gamma$ of the slack variables via $\lambda_\gamma = \partial Z/\partial\mu_\gamma$ in terms of (5.164). This permits $(5.152)_5$ to be rewritten in the more tangible form that all $\mu_\gamma \leqslant 0$. Hence we find

$$L_\beta - \overline{\lambda_\beta, \frac{\partial L}{\partial\lambda}\bigg|_\beta} = \int [Z[\dot{t}_{ij}, \mu_\gamma] - \tfrac{1}{2}\tau_{ik}\dot{d}_{ji}\dot{d}_{jk}]\mathrm{d}V$$
$$+ \int n_i\left(\dot{t}_{ij} + \dot{t}_{ik}\frac{\partial u_j}{\partial X_k} + \tau_{ik}\dot{d}_{jk}\right)c_j\,\mathrm{d}\Sigma_2. \quad (5.171)$$

In this way we obtain a functional of the $\dot{t}_{ij}, u_\gamma$ and $\dot{d}_{ij}$ which is a lower bound, to be maximized subject to all $\mu_\gamma \leqslant 0$ and to the nontrivial constraints $(5.152)_3$ (expressed via $(5.152)_4$ in terms of $\dot{t}_{ij}$ and $\dot{d}_{ij}$).

In résumé, the sufficient conditions (a) and (b) in Theorem 5.22 lead to simultaneous upper and lower bounds (5.168); if $X[\dot{\tau}_{ij}, \lambda_\gamma]$ is strictly convex in $\dot{\tau}_{ij}$ and weakly concave in λ_γ, the upper bound can be expressed as (5.169); if $X[\dot{\tau}_{ij}, \lambda_\gamma]$ is weakly convex in $\dot{\tau}_{ij}$ and strictly convex in λ_γ, the lower bound can be expressed as (5.171). Both such expressions are available if $X[\dot{\tau}_{ij}, \lambda_\gamma]$ is strictly convex in $\dot{\tau}_{ij}$ and strictly concave in λ_γ; if, also, the principal stresses are positive then, we see from (5.166) that $S > 0$ for all pairs of points which have distinct values of $\dot{d}_{ij}$, $\dot{\tau}_{ij}$ and λ_γ. In this case it can be shown that actual solutions of (5.152) will have unique values of these three quantities.

The quadratic examples (2.86)–(2.89), rewritten in suffix notation, are

$$X[\dot{\tau}_{ij}, \lambda_\gamma] = \tfrac{1}{2}M_{ijkl}\dot{\tau}_{ij}\dot{\tau}_{kl} + \dot{\tau}_{ij}\nu_{ij\gamma}\lambda_\gamma - \tfrac{1}{2}h_{\gamma\delta}\lambda_\gamma\lambda_\delta, \tag{5.172}$$

$$Y[\dot{e}_{ij}, \lambda_\gamma] = \tfrac{1}{2}K_{ijkl}\dot{e}_{ij}\dot{e}_{kl} - \lambda_\gamma\nu_{ij\gamma}K_{ijkl}\dot{e}_{kl} + \tfrac{1}{2}g_{\gamma\delta}\lambda_\gamma\lambda_\delta, \tag{5.173}$$

$$Z[\dot{\tau}_{ij}, \mu_\gamma] = -\tfrac{1}{2}M_{ijkl}\dot{\tau}_{ij}\dot{\tau}_{kl} - \tfrac{1}{2}h^{-1}_{\gamma\delta}(\mu_\gamma - \nu_{ij\gamma}\dot{\tau}_{ij})(\mu_\delta - \nu_{kl\delta}\dot{\tau}_{kl}). \tag{5.174}$$

The saddle quantity (5.166) then becomes

$$S = \tfrac{1}{2}\int [M_{ijkl}(\dot{\tau}_{ij+} - \dot{\tau}_{ij-})(\dot{\tau}_{kl+} - \dot{\tau}_{kl-}) + h_{\gamma\delta}(\lambda_{\gamma+} - \lambda_{\gamma-})(\lambda_{\delta+} - \lambda_{\delta-})$$
$$+ \tau_{ik}(\dot{d}_{ji+} - \dot{d}_{ji-})(\dot{d}_{jk+} - \dot{d}_{jk-})]\,\mathrm{d}V. \tag{5.175}$$

This quadratic $X[\dot{\tau}_{ij}, \lambda_\gamma]$ is strictly convex in the $\dot{\tau}_{ij}$ if the M_{ijkl} comprise a positive definite matrix, which then has a positive definite inverse with components K_{ijkl}; and (5.172) is strictly concave in the λ_γ if the $h_{\gamma\delta}$ comprises a positive definite matrix, which then has a positive definite inverse with components $h^{-1}_{\gamma\delta}$. Sufficient but not necessary conditions that $g_{\gamma\delta} = h_{\gamma\delta} + K_{ijkl}\nu_{ij\gamma}\nu_{kl\delta}$ be positive definite are that both $h_{\gamma\delta}$ and K_{ijkl} are so.

(ix) Weaker hypotheses

The requirement (a) in Theorem 5.22 that the principal stresses be non-negative is a very strong condition since it excludes compressive states of stress. We can drop it, and still find that $S > 0$ or $S \geqslant 0$ over *restricted* pairs of points, in the sense first indicated in cases (1.58) (c)–(f) of §1.3(iii). There follow individual but not simultaneous bounds.

(x) Kinematical constraints

In this subsection the kinematical conditions (5.159α), or equivalently (5.152α), are imposed as constraints from the outset.

Theorem 5.24 (Upper bound)
Suppose that $S \geqslant 0$ for every pair of distinct points in the domain of L which also satisfy the kinematical conditions (5.159α). *Then the upper bound*

property

$$L_\alpha \geqslant L_0 \tag{5.176}$$

holds, in the notation of Theorem 5.23, *but not necessarily the lower bound.*

Proof

This is the same as the first part of the proof of Theorem 5.23, but now calling only upon the weaker hypothesis of the present Theorem. □

We must now assess this weaker hypothesis explicitly. That is, we must provide the counterpart of Theorem 5.22 allowing, in particular, that the principal stresses may now be negative. This assessment is expressed in terms of uniqueness criteria, towards which the emphasis now shifts from the upper bound (5.176).

Henceforward we shall suppose that $X[\dot{\tau}_{ij}, \lambda_\gamma]$ is strictly convex in the $\dot{\tau}_{ij}$. This is *not* likely to be restrictive – for example, in the quadratic case (5.172) it corresponds to requiring the matrix of elastic compliances M_{ijkl} to be positive definite. Then the eliminations described in the derivation of (5.169) all apply. We can now rewrite the saddle quantity (5.166) as

$$S = \frac{1}{2}\int \tau_{ik}\frac{\partial(\dot{u}_{j+} - \dot{u}_{j-})}{\partial X_i}\frac{\partial(\dot{u}_{j+} - \dot{u}_{j-})}{\partial X_k}\mathrm{d}V + \int\Bigg[Y[\dot{e}_{ij+}, \lambda_{\gamma+}] - Y[\dot{e}_{ij-}, \lambda_{\gamma-}]$$
$$- (\dot{e}_{ij+} - \dot{e}_{ij-})\frac{\partial Y}{\partial \dot{e}_{ij}}\bigg|_- - (\lambda_{\gamma+} - \lambda_{\gamma-})\frac{\partial Y}{\partial \lambda_\gamma}\bigg|_-\Bigg]\mathrm{d}V. \tag{5.177}$$

Here $\dot{e}_{ij}$ is merely a shorthand symbol for (5.170), as in the expression (5.169) of the upper bound. Therefore (5.177) is a functional of the values of $\dot{u}_j$ and λ_γ at the plus and minus points.

In the case when $Y[\dot{e}_{ij}, \lambda_\gamma]$ is any homogeneous quadratic, in particular (5.173), we can write

$$S = F[\dot{u}_{j+} - \dot{u}_{j-}, \lambda_{\gamma+} - \lambda_{\gamma-}] \tag{5.178}$$

in terms of the functional defined by

$$F[\dot{u}_j, \lambda_\gamma] = \int\left[Y[\dot{e}_{ij}, \lambda_\gamma] + \tfrac{1}{2}\tau_{ik}\frac{\partial \dot{u}_j}{\partial X_i}\frac{\partial \dot{u}_j}{\partial X_k}\right]\mathrm{d}V. \tag{5.179}$$

Theorem 5.25 (Uniqueness of velocity and multipliers)

With reference to (5.177), *or* (5.178) *in the quadratic case, suppose that* $S > 0$ *for every pair* $\dot{u}_{j+}, \lambda_{\gamma+}$ *and* $\dot{u}_{j-}, \lambda_{\gamma-}$ *of distinct points which satisfy the remaining kinematical conditions* $\dot{u}_j = c_j$ *on* Σ_2 *and all* $\lambda_\gamma \geqslant 0$. *Then any actual solution of* (5.159) *has unique* $\dot{u}_j$ *and* λ_γ, *and Theorem* 5.24 *applies a fortiori.*

Proof

Interchanging plus and minus labels in (5.165), and adding the two versions of S so obtained, allows the present hypothesis to be expressed as

$$-\left(\dot{s}_+ - \dot{s}_-, \left.\frac{\partial L}{\partial \dot{s}}\right|_+ - \left.\frac{\partial L}{\partial \dot{s}}\right|_-\right) - \left(\dot{\tau}_+ - \dot{\tau}_-, \left.\frac{\partial L}{\partial \dot{\tau}}\right|_+ - \left.\frac{\partial L}{\partial \dot{\tau}}\right|_-\right)$$

$$+ \left\langle \dot{u}_+ - \dot{u}_-, \left.\frac{\partial L}{\partial \dot{u}}\right|_+ - \left.\frac{\partial L}{\partial \dot{u}}\right|_-\right\rangle + \left(\dot{d}_+ - \dot{d}_-, \left.\frac{\partial L}{\partial \dot{d}}\right|_+ - \left.\frac{\partial L}{\partial \dot{d}}\right|_-\right)$$

$$\overline{+ \lambda_+ - \lambda_-, \left.\frac{\partial L}{\partial \lambda}\right|_+ - \left.\frac{\partial L}{\partial \lambda}\right|_-} > 0 \qquad\qquad (5.180)$$

for the stated pairs. But if plus and minus points could both satisfy all the conditions (5.159), the left side of (5.180) would be

$$-\overline{\lambda_+, \left.\frac{\partial L}{\partial \lambda}\right|_-} - \overline{\lambda_-, \left.\frac{\partial L}{\partial \lambda}\right|_+} \leqslant 0.$$

Uniqueness of $\dot{u}_j$ and of λ_γ in a solution of (5.159) follows by *reductio ad absurdum*.

The upper bound (5.159) applies *a fortiori* simply because $S > 0$ is stronger than $S \geqslant 0$. $\qquad\qquad \square$

Theorem 5.26 (Exclusion functional)
When $Y[\dot{e}_{ij}, \lambda_\gamma]$ is a homogeneous quadratic, a sufficient condition that any actual solution of (5.159) has unique $\dot{u}_j$ and λ_γ is that

$$F[\dot{u}_j, \lambda_\gamma] > 0 \qquad\qquad (5.181)$$

for all $\dot{u}_j$ which vanish on Σ_2, and for all real numbers λ_γ irrespective of sign.

Proof
Any $\dot{u}_j$ which vanishes on Σ_2 may be represented as the difference of a pair $\dot{u}_{j+}$ and $\dot{u}_{j-}$ which each have the same assigned values c_j on Σ_2, and vice versa. Any real λ_γ, having either sign, may be represented as the difference of a nonnegative pair $\lambda_{\gamma+}$ and $\lambda_{\gamma-}$, and vice versa. Therefore the class of admissible fields in this Theorem is equivalent to the class of difference fields in Theorem 5.25. By (5.178) it follows that (5.181) implies $S > 0$, and the uniqueness follows from Theorem 5.25. $\qquad\qquad \square$

The functional $F[\dot{u}_j, \lambda_\gamma]$ may be called an 'exclusion functional' because it is used as a device to exclude nonuniqueness and bifurcation in the original problem. With another hypothesis we can construct another exclusion functional, of $\dot{u}_j$ alone, as follows.

Theorem 5.27
When $Y[\dot{e}_{ij}, \lambda_\gamma]$ is strictly convex in the λ_γ, then (5.181) is satisfied if

$$E[\dot{u}_j] = \int \left[-W[\dot{e}_{ij}, 0] + \tfrac{1}{2}\tau_{ik}\frac{\partial \dot{u}_j}{\partial X_i}\frac{\partial \dot{u}_j}{\partial X_k}\right] dV > 0 \qquad (5.182)$$

for all $\dot{u}_j$ which vanish on Σ_2, where $W[\dot{e}_{ij}, \mu_\gamma]$ is the Legendre dual of $Y[\dot{e}_{ij}, \lambda_\gamma]$ in the sense of Fig. 2.20.

When $Y[\dot{e}_{ij}, \lambda_\gamma]$ is also a homogeneous quadratic, an actual solution of (5.159) has unique $\dot{u}_j$ when (5.182) holds.

Proof

The condition (5.181) holds for all λ_γ if $F[\dot{u}_j, \lambda_\gamma]$ has a positive minimum with respect to the λ_γ, for each admissible $\dot{u}_j$. This entails minimizing $Y[\dot{e}_{ij}, \lambda_\gamma]$. By the strict convexity and the equations

$$-\mu_\gamma = \frac{\partial Y}{\partial \lambda_\gamma}, \quad -W[\dot{e}_{ij}, \mu_\gamma] = \mu_\gamma \lambda_\gamma + Y[\dot{e}_{ij}, \lambda_\gamma],$$

implied by the Legendre transformation, the minimum is stationary and achieved when all $\mu_\gamma = 0$. Hence $E[\dot{u}_j]$ is the required minimum of $F[\dot{u}_j, \lambda_\gamma]$. Hence (5.182) implies (5.181).

The uniqueness of $\dot{u}_j$ then follows from Theorem 5.26 when Y $[\dot{e}_{ij}, \lambda_\gamma]$ is homogeneous quadratic. □

The strict convexity of $Y[\dot{e}_{ij}, \lambda_\gamma]$ in the example (5.173) requires that the $m \times m$ matrix with components

$$g_{\gamma\delta} \quad \text{is positive definite,} \tag{5.183}$$

whence, from (2.90) written in suffix notation, (5.182) contains

$$-W[\dot{e}_{ij}, 0] = \tfrac{1}{2} K_{ijkl} \dot{e}_{ij} \dot{e}_{kl} - \tfrac{1}{2} g_{\gamma\delta}^{-1} v_{pq\gamma} K_{pqij} v_{rs\delta} K_{rskl} \dot{e}_{ij} \dot{e}_{kl}. \tag{5.184}$$

The exclusion functional (5.182) for this case $m > 1$ was obtained by Sewell (1972, equation (4.12)) via a different approach ((3.60) there) from that given here. The pioneering use of the exclusion functional, in the case $m = 1$, is due to Hill (1958). The name 'exclusion functional' appeared later (Hill, 1978).

We now briefly explain how to use the exclusion functional $E[\dot{u}_j]$ to describe the range of stress over which the boundary value problem (5.152) has no more than one solution for the velocity, for the real elastic/plastic solid with quadratic $X[\dot{\tau}_{ij}, \lambda_\gamma]$ given by (5.172) with positive definite M_{ijkl} and $g_{\gamma\delta}$. Introduce an associated *homogeneous* boundary value problem by setting $c_j = a_j = b_j = 0$ in $(5.152)_{1,2,3,4}$. Seek solutions of this homogeneous problem for the comparison solid defined after Theorem 5.20, i.e. with $(5.152)_{5,6,7}$ replaced by $\partial X/\partial \lambda_\gamma = 0$ for all λ_γ. The geometrical and material data for this homogeneous problem in the comparison solid are otherwise the same as for the original inhomogeneous problem in the real elastic/plastic solid.

There will be an upper bound

$$L_\alpha = F[\dot{u}_j, \lambda_\gamma]$$

for the homogeneous problem in the comparison solid, by setting

$b_j = a_j = 0$ in (5.169) and identifying it with (5.178). The free stationary principle in Theorem 5.21 applies. Taking account of the fact that constraints have already been used to eliminate $\dot{s}_{ij}$, $\dot{t}_{ij}$ and $\dot{d}_{ij}$, and of the fact that minimization with respect to the λ_γ is provided by $\partial Y/\lambda_\gamma = 0$, i.e. by

$$\lambda_\gamma = g_{\gamma\delta}^{-1} v_{ij\delta} K_{ijkl} \dot{e}_{kl}, \tag{5.185}$$

it remains only to find the stationary minimum of

$$L_\alpha = E[\dot{u}_j],$$

with respect to $\dot{u}_j$ vanishing on Σ_2. This reduces the problem to minimization of a homogeneous quadratic functional. Let w_j be the values of $\dot{u}_j$ which provide this minimum with the value zero, for example after recasting $E[\dot{u}_j] > 0$ in the form of a Rayleigh quotient which itself is then minimized. These w_j are a type of eigenmode in the comparison solid. By adding them to the previously unique mode in the inhomogeneous problem for the real elastic/plastic solid, a class of bifurcation modes can be described for that problem which, when inserted on the right of (5.185), enforce all $\lambda_\gamma > 0$ throughout the body in the combined modes. This is achieved *a posteriori* by choosing a range of boundary data (e.g. of c_j and a_j) which are sufficiently distant from zero, so that real and comparison solids have the same behaviour, at least for that restricted range of boundary values. For example, a column which bifurcates by a combination of bending and axial compression will find that even its extreme fibre on the convex side will still be compressing if the axial compression rate on the ends of the column is large enough (Shanley 1946, 1947; Hill and Sewell, 1960*a*, *b*, 1962).

(xi) Equilibrium constraints

In this subsection the force-like conditions (5.159β), or equivalently (5.152β), are imposed as constraints from the outset. We shall not here give such explicit results as those of the previous subsection, because of the difficulty of satisfying the continuing equilibrium equations in $(5.152)_3$ *a priori*; cf. the remarks of Truesdell and Toupin (1960, §227) about general solutions of equilibrium equations.

Theorem 5.28 (Lower bound)
Suppose that $S \geqslant 0$ for every pair of distinct points in the domain of L which also satisfy the force-like conditions (5.159β). Then the lower bound property

$$L_0 \geqslant L_\beta - \overline{\lambda_\beta, \frac{\partial L}{\partial \lambda}}\bigg|_\beta \tag{5.186}$$

holds, in the notation of Theorem 5.23, but not necessarily the upper bound.

Proof.
This is the same as the second part of the proof of Theorem 5.23, but this time calling upon only the weaker hypothesis of the present Theorem. This is a different hypothesis from that of Theorem 5.24. □

To assess the hypothesis of Theorem 5.28, suppose that $X[\dot{\tau}_{ij}, \lambda_\gamma]$ is strictly concave in the λ_γ. This will be at least mildly restrictive – for example, in the quadratic case (5.172) it corresponds to requiring the matrix of hardening coefficients $h_{\gamma\delta}$ to be positive definite, which is a stronger hypothesis than (5.183). The eliminations described in the derivation of (5.171) now all apply. The function $Z[\dot{\tau}_{ij}, \mu_\gamma]$ now exists, by Fig. 2.20, having a single branch which is strictly concave in the μ_γ (for example, see (5.174)). We can now rewrite the saddle quantity (5.166) as

$$S = \frac{1}{2}\int \tau_{ik}(\dot{d}_{ji+} - \dot{d}_{ji-})(\dot{d}_{jk+} - \dot{d}_{jk-})\mathrm{d}V + \int\bigg[Z[\dot{\tau}_{ij+}, \mu_{\gamma+}] - Z[\dot{\tau}_{ij-}, \mu_{\gamma-}] - (\dot{\tau}_{ij+} - \dot{\tau}_{ij-})\frac{\partial Z}{\partial \dot{\tau}_{ij}}\bigg|_+ - (\mu_{\gamma+} - \mu_{\gamma-})\frac{\partial Z}{\partial \mu_\gamma}\bigg|_+\bigg]\mathrm{d}V. \tag{5.187}$$

In the case when $Z[\dot{\tau}_{ij}, \mu_\gamma]$ is any homogeneous quadratic, in particular (5.174), we can write

$$S = F_c[\dot{\tau}_{ij+} - \dot{\tau}_{ij-}, \mu_{\gamma+} - \mu_{\gamma-}, \dot{d}_{ji+} - \dot{d}_{ji-}] \tag{5.188}$$

in terms of the functional defined by

$$F_c[\dot{\tau}_{ij}, \mu_\gamma, \dot{d}_{ji}] = \int[-Z[\dot{\tau}_{ij}, \mu_\gamma] + \tfrac{1}{2}\tau_{ik}\dot{d}_{ji}\dot{d}_{jk}]\mathrm{d}V. \tag{5.189}$$

Theorem 5.29 (Uniqueness of $\dot{\tau}_{ij}, \mu_\gamma$ and $\dot{d}_{ji}$)
With reference to (5.187), *or* (5.188) *in the quadratic case, suppose that* $S > 0$ *for every pair* $\dot{\tau}_{ij+}, \mu_{\gamma+}, \dot{d}_{ji+}$ *and* $\dot{\tau}_{ij-}, \mu_{\gamma-}, \dot{d}_{ji-}$ *of distinct points which satisfy* (5.152β), *Then any actual solution of* (5.152) *has unique* $\dot{\tau}_{ij}, \mu_\gamma$ *and* $\dot{d}_{ji}$, *and Theorem 5.28 applies a fortiori.*

Proof
The hypothesis is that (5.180) holds for all pairs now stated, instead of those in Theorem 5.25. The argument in that proof can be repeated, this time establishing uniqueness of $\dot{\tau}_{ij}, \mu_\gamma$ and $\dot{d}_{ji}$ in a solution of (5.152).

□

Theorem 5.30 (Complementary exclusion functional)
When $Z[\dot{\tau}_{ij}, \mu_\gamma]$ *is a homogeneous quadratic, a sufficient condition ensuring that any actual solution of* (5.152) *has unique* $\dot{\tau}_{ij}, \mu_\gamma$ *and* $\dot{d}_{ji}$ *is that*

$$F_c[\dot{\tau}_{ij}, \mu_\gamma, \dot{d}_{ji}] > 0 \tag{5.190}$$

for all $\dot{\tau}_{ij}$ *and* $\dot{d}_{ji}$ *which satisfy*

$$\frac{\partial}{\partial X_i}\left[\dot{\tau}_{ij}+\dot{\tau}_{ik}\frac{\partial u_j}{\partial X_k}+\tau_{ik}\dot{d}_{jk}\right]=0 \quad in \quad V, \tag{5.191}$$

$$n_i\left[\dot{\tau}_{ij}+\dot{\tau}_{ik}\frac{\partial u_j}{\partial X_k}+\tau_{ik}\dot{d}_{jk}\right]=0 \quad on \quad \Sigma_1, \tag{5.192}$$

and for all real μ_γ *irrespective of sign.*

Proof

Any $\dot{\tau}_{ij}$ and $\dot{d}_{jk}$ which satisfy the linear homogeneous equations (5.190) and (5.191) may be represented as the difference of pairs $\dot{\tau}_{ij+}, \dot{d}_{jk+}$ and $\dot{\tau}_{ij-}, \dot{d}_{jk-}$ which each satisfy the inhomogeneous conditions $(5.152)_{3,4}$, and vice versa. Any real μ_γ, having either sign, may be represented as the difference of a nonpositive pair $\mu_{\gamma+}$ and $\mu_{\gamma-}$, and vice versa. Therefore the class of admissible fields in this Theorem is equivalent to the class of difference fields in Theorem 5.29. By (5.188) it follows that (5.190) implies $S>0$, and the uniqueness follows from Theorem 5.29. □

We can also call $F_c[\dot{\tau}_{ij}, \mu_\gamma, \dot{d}_{ji}]$ an exclusion functional. It is used in Theorem 5.30 as a device to exclude uniqueness of $\dot{\tau}_{ij}, \mu_\gamma$ and $\dot{d}_{ji}$.

Theorem 5.31

Since $X[\dot{\tau}_{ij}, \lambda_\gamma]$ *is strictly concave in the* λ_γ, (5.190) *is satisfied if*

$$E_c[\dot{\tau}_{ij}, \dot{d}_{ji}] = \int [X[\dot{\tau}_{ij}, 0] + \tfrac{1}{2}\tau_{ik}\dot{d}_{ji}\dot{d}_{jk}]\,\mathrm{d}V > 0 \tag{5.193}$$

for all $\dot{\tau}_{ij}$ *and* $\dot{d}_{ji}$ *which satisfy* (5.191) *and* (5.192).

When $X[\dot{\tau}_{ij}, \lambda_\gamma]$ *is also a homogeneous quadratic, an actual solution of* (5.152) *has unique* $\dot{\tau}_{ij}$ *and* $\dot{d}_{ji}$ *when* (5.193) *holds.*

Proof

The condition (5.190) holds for all μ_γ if $F_c[\dot{\tau}_{ij}, \mu_\gamma, \dot{d}_{ji}]$ has a positive minimum with respect to the μ_γ. This entails maximizing $Z[\dot{\tau}_{ij}, \mu_\gamma]$, which is strictly concave in the μ_γ by the strict concavity of its Legendre dual $X[\dot{\tau}_{ij}, \lambda_\gamma]$. Since

$$\lambda_\gamma = \frac{\partial Z}{\partial \mu_\gamma}, \qquad Z = \lambda_\gamma \mu_\gamma - X,$$

the maximum is stationary and achieved when all $\lambda_\gamma = 0$, with value $-X[\dot{\tau}_{ij}, 0]$. Hence (5.193) implies (5.190)

When $X[\dot{\tau}_{ij}, \lambda_\gamma]$ is quadratic this could have been anticipated alternatively by minimizing (5.175) with respect to $\lambda_{\gamma+} - \lambda_{\gamma-}$. The uniqueness of $\dot{\tau}_{ij}$ and $\dot{d}_{ij}$ then follows from Theorem 5.30. □

The strict concavity of $X[\dot{\tau}_{ij}, \lambda_\gamma]$ in the example (5.172) requires that the matrix with components $h_{\gamma\delta}$ be positive definite, whence so is that with components $h^{-1}_{\gamma\delta}$ in the $Z[\dot{\tau}_{ij}, \mu_\gamma]$ of (5.174), and (5.193) then contains

$$X[\dot{\tau}_{ij}, 0] = \tfrac{1}{2}M_{ijkl}\dot{\tau}_{ij}\dot{\tau}_{kl}, \tag{5.194}$$

whose coefficients are the elastic compliances only.

There seems to have been no serious attempt to exploit the exclusion functional $E_c[\dot{\tau}_{ij}, \dot{d}_{ij}]$, by comparison with the considerable literature associated with $E[\dot{u}_j]$ (reviewed by Hill, 1978, §3B). No doubt this is because the constraints (5.191) and (5.192) are harder to satisfy than merely $\dot{u}_j = 0$ on Σ_2, and because a bifurcation of $\dot{\tau}_{ij}$ and $\dot{d}_{ji}$ may be of less physical interest than a bifurcation of $\dot{u}_j$. One could introduce a homogeneous problem for the elastic solid, having $(5.152)_{5,6,7}$ replaced by $\lambda_\gamma = 0$ for all μ_γ. There will be a lower bound

$$L_\beta = -F_c[\dot{\tau}_{ij}, \mu_\gamma, \dot{d}_{ji}]$$

for this problem, by setting $c_j = 0$ in (5.171) and identifying it with (5.189). Maximization with respect to μ_γ is provided by

$$\mu_\gamma = v_{ij\gamma}\dot{\tau}_{ij}, \tag{5.195}$$

and it remains to find the stationary maximum of

$$L_\beta = -E_c[\dot{\tau}_{ij}, \dot{d}_{ji}],$$

subject to (5.191) and (5.192). Again this entails minimization of a homogeneous quadratic functional, now subject to homogeneous linear constraints.

(xii) Application of supplementary constraints

When a bound or an extremum principle has been established, it is always an option to impose more constraints than are needed in the original proof. In §2.7(iv) we called such extra constraints 'supplementary'. Strictly speaking they are unnecessary, and at face value they may appear to reduce flexibility in applying the bound. However, there may be compensating advantages in focussing on a particular method of optimizing the bound. In particular, supplementary constraints may allow a stationary extremum to replace a nonstationary one.

This effect is common in the elastic/plastic boundary value problem being studied in this section. In applying the upper bound L_α of Theorem 5.23 or 5.24, we can augment the kinematical conditions (5.152α) by the supplementary constraints $(5.152)_{5,7}$. In applying the lower bound $L_\beta - \overline{\lambda_\beta, (\partial L/\partial \lambda)|_\beta}$ of Theorems 5.23 or 5.28 we can augment the force-like conditions (5.152β) by the supplementary constraints $(5.152)_{6,7}$. Either set of supplementary constraints is effectively imposed when the internal variables λ_γ and μ_γ are eliminated from the outset, leaving the plastic constitutive equations written as multibranched relations between the external variables $\dot{\tau}_{ij}$ and $\dot{e}_{ij}$, as illustrated in (2.91)–(2.97). To see this, recall that the plastic flow conditions (2.85) are represented by the last trio in (5.152) or (5.159). The slack variables μ_γ can be eliminated immediately, e.g. by $\mu_\gamma = \partial X/\partial \lambda_\gamma$. Under suitable hypotheses the λ_γ may then be expressed in terms of the $\dot{\tau}_{ij}$ by using the trio, and the resulting relations are then substituted back into the right side of $(5.152)_2$. The result can be written

$$\dot{e}_{ij} = \frac{\partial V_c}{\partial \dot{\tau}_{ij}} \tag{5.196}$$

in terms of a multibranched (symmetrized) function $V_c[\dot{\tau}_{ij}]$ of the $\dot{\tau}_{ij}$ alone, as illustrated in (2.91) and (2.96).

For example, we see from (5.172) and (5.174) that if $m = 1$ (smooth yield surface) and $h \neq 0$ the composite potential (2.91) is

$$V_c[\dot{\tau}_{ij}] = \begin{cases} \frac{1}{2}M_{ijkl}\dot{\tau}_{ij}\dot{\tau}_{kl} + \dfrac{1}{2h}(v_{ij}\dot{\tau}_{ij})^2 & \text{if } \dfrac{v_{ij}\dot{\tau}_{ij}}{h} \geqslant 0, \\ \frac{1}{2}M_{ijkl}\dot{\tau}_{ij}\dot{\tau}_{kl} & \text{if } v_{ij}\dot{\tau}_{ij} < 0. \end{cases} \tag{5.197}$$

If $h > 0$ this potential is defined, single valued and continuous, with continuous first gradient, for all $\dot{\tau}_{ij}$, but its second derivatives are discontinuous across the hyperplane $v_{ij}\dot{\tau}_{ij} = 0$ in $\dot{\tau}_{ij}$-space. If $h < 0$ the potential is defined only in the half-space $v_{ij}\dot{\tau}_{ij} \leqslant 0$, where it is double valued and otherwise again piecewise C^2, like two half-parabolas joining at a cusp. If $m > 1$ we expect a composite potential with more than two branches, as indicated in (2.96), with domains between the m hyperplanes $v_{ij\gamma}\dot{\tau}_{ij} = 0$. Even then we can expect the composite $V_c[\dot{\tau}_{ij}]$ to be a piecewise C^2 function of the $\dot{\tau}_{ij}$, but again possibly multivalued.

The outcome of this prior elimination of the μ_γ and λ_γ is therefore a set of governing equations $(5.152)_{1,2,3,4}$ in which, on the right of $(5.152)_2$, the $\dot{\tau}_{ij}$-gradients of the single valued C^2 function $X[\dot{\tau}_{ij}, \lambda_\gamma]$ are replaced by those of the (possibly multivalued) piecewise C^2 function $V_c[\dot{\tau}_{ij}]$. In other words, $(5.152)_{2,5,6,7}$ are replaced by (5.196) (with (5.170)).

The subsequent developments, including the same version (5.156) of adjointness, lead to a new generating functional obtained by writing $V_c[\dot{\tau}_{ij}]$ in place of $X[\dot{\tau}_{ij}, \lambda_\gamma]$ in (5.158).

Theorem 5.32

The governing equations of the real elastic/plastic solid can be expressed as

$$\frac{\partial L}{\partial \dot{s}} = \frac{\partial L}{\partial \dot{t}} = 0, \quad (\alpha)$$
$$\frac{\partial L}{\partial \dot{u}} = \frac{\partial L}{\partial \dot{d}} = 0, \quad (\beta) \tag{5.198}$$

in terms of the gradients of the functional

$$L[\dot{s}, \dot{t}, \dot{u}, \dot{d}] = \int \left[\dot{s}_{ij}\frac{\partial \dot{u}_j}{\partial X_i} - \left(\dot{s}_{ij} - \dot{t}_{ij} - \dot{t}_{ik}\frac{\partial \dot{u}_j}{\partial X_k} \right) \dot{d}_{ji} + \tfrac{1}{2}\tau_{ik}\dot{d}_{ji}\dot{d}_{jk} - V_c[\dot{t}_{ij}] - b_j\dot{u}_j \right] \mathrm{d}V - \int a_j\dot{u}_j\,\mathrm{d}\Sigma_1 + \int n_i\dot{s}_{ij}(c_j - \dot{u}_j)\,\mathrm{d}\Sigma_2. \tag{5.199}$$

Proof
This parallels the proof of Theorem 5.20. Neither the multivaluedness of $V_c[\dot{t}_{ij}]$, nor the discontinuities permitted in its second derivatives, obviate computation of the first variation

$$\delta L = \left(\delta\dot{s}, \frac{\partial L}{\partial \dot{s}} \right) + \left(\delta\dot{t}, \frac{\partial L}{\partial \dot{t}} \right) + \left\langle \delta\dot{u}, \frac{\partial L}{\partial \dot{u}} \right\rangle + \left(\delta\dot{d}, \frac{\partial L}{\partial \dot{d}} \right). \tag{5.200}$$

□

Theorem 5.33 (Free stationary principle)
$L[\dot{s}, \dot{t}, \dot{u}, \dot{d}]$ is stationary with respect to arbitrary small variations of its arguments if and only if they are variations from an actual solution in the real elastic/plastic solid.

Proof
Sufficiency is clear, since (5.198) makes $\delta L = 0$ in (5.200). Necessity depends on the fundamental lemma of the calculus of variations. □

In place of (5.161) and (5.162) special significance will attach to

$$L - \left(\dot{s}, \frac{\partial L}{\partial \dot{s}} \right) - \left(\dot{t}, \frac{\partial L}{\partial \dot{t}} \right) = \int [V[\dot{e}_{ij}] + \tfrac{1}{2}\tau_{ik}\dot{d}_{ji}\dot{d}_{jk} - b_j\dot{u}_j]\,\mathrm{d}V - \int a_j\dot{u}_j\,\mathrm{d}\Sigma_1, \tag{5.201}$$

$$L - \left\langle \dot{u}, \frac{\partial L}{\partial \dot{u}} \right\rangle - \left(\dot{d}, \frac{\partial L}{\partial \dot{d}} \right) = \int [-V_c[\dot{t}_{ij}] - \tfrac{1}{2}\tau_{ik}\dot{d}_{ji}\dot{d}_{jk}]\,\mathrm{d}V + \int n_i\dot{s}_{ij}c_j\,\mathrm{d}\Sigma_2. \tag{5.202}$$

In (5.201) we have written

$$\dot{\tau}_{ij}\frac{\partial V_c}{\partial \dot{\tau}_{ij}} - V_c[\dot{\tau}_{ij}] = V[\dot{e}_{ij}] \tag{5.203}$$

by introducing the assumption, prompted by (2.95), that $V_c[\dot{\tau}_{ij}]$ has a Legendre dual function $V[\dot{e}_{ij}]$, which will also be piecewise C^2 and possibly multivalued.

For example, when $m = 1$ and $g \neq 0$, the example (2.93) is

$$V[\dot{e}_{ij}] = \begin{cases} \frac{1}{2}K_{ijkl}\dot{e}_{ij}\dot{e}_{kl} - \dfrac{1}{2g}(v_{ij}K_{ijkl}\dot{e}_{kl})^2 & \text{if} \quad \dfrac{v_{ij}K_{ijkl}\dot{e}_{kl}}{g} > 0, \\ \frac{1}{2}K_{ijkl}\dot{e}_{ij}\dot{e}_{kl} & \text{if} \quad v_{ij}K_{ijkl}\dot{e}_{kl} < 0. \end{cases} \tag{5.204}$$

This function is piecewise C^2, single valued when $g > 0$, but if $g < 0$ it is defined only over the half-space $v_{ij}K_{ijkl}\dot{e}_{kl} \leqslant 0$ and is double valued there – again like two half-parabolas meeting at a cusp. Since $g = h + K_{ijkl}v_{ij}v_{kl}$ and K_{ijkl} is positive definite in practice, it is possible to have $h < 0$ compatible with $g > 0$, and therefore to have double-valued $V_c[\dot{\tau}_{ij}]$ compatible with single valued $V[\dot{e}_{ij}]$.

We are continuing to regard $\dot{e}_{ij}$ as merely a shorthand for the function of $\partial\dot{u}_j/\partial X_i$ on the right of (5.170). We may put $u_k \equiv 0$ at will there, whenever we wish to regard the current configuration as the initial one.

Theorem 5.34 (Complementary stationary principles)

(a) The functional

$$J[\dot{u}_j] = \int\left[V[\dot{e}_{ij}] + \frac{1}{2}\tau_{ik}\frac{\partial\dot{u}_j}{\partial X_i}\frac{\partial\dot{u}_j}{\partial X_k} - b_j\dot{u}_j\right]\mathrm{d}V - \int a_j\dot{u}_j\,\mathrm{d}\Sigma_1 \tag{5.205}$$

is stationary with respect to arbitrary small variations of velocity which satisfy $\dot{u}_j = c_j$ *on* Σ_2 *if and only if they are variations from an actual solution.*

(b) The functional

$$-K[\dot{\tau}_{ij}, \dot{d}_{ji}] = \int[V_c[\dot{\tau}_{ij}] + \tfrac{1}{2}\tau_{ik}\,\dot{d}_{ji}\dot{d}_{jk}]\,\mathrm{d}V - \int n_i\left(\dot{\tau}_{ij} + \dot{\tau}_{ik}\frac{\partial u_j}{\partial X_k} + \tau_{ik}\dot{d}_{jk}\right)c_j\,\mathrm{d}\Sigma_2 \tag{5.206}$$

is stationary with respect to arbitrary small variations of $\dot{\tau}_{ij}$ *and* $\dot{d}_{ji}$ *which satisfy the continuing equilibrium equations and boundary conditions on* Σ_1, *i.e.* $(5.152)_{3,4}$, *if and only if they are variations from an actual solution.*

Proof

(a) This is a specialization of Theorem 5.33 subjected to the kinematical constraints (5.198α). That L can be written as the functional $J[\dot{u}_j]$ (cf. (5.169) and (5.87)) follows from (5.198α) inserted into (5.201). The constraint $\partial L/\partial \dot{t}=0$ is (5.196), which is deployed in inverted form (2.94), i.e.

$$\dot{t}_{ij}=\frac{\partial V}{\partial \dot{e}_{ij}}. \tag{5.207}$$

It is sufficient, but not necessary, to work within C^2 $\dot{u}_j$ fields.

(b) This is a specialization of Theorem 5.33 subjected to the force-like constraints (5.198β). That L can be written as the functional $K[\dot{t}_{ij},\dot{d}_{ji}]$ (cf. (5.171) and (5.88)) follows from (5.198β) inserted into (5.201). The constraint $\partial L/\partial \dot{d}=0$ is $(5.152)_4$, which has been used to eliminate $\dot{s}_{ij}$ from (5.206).

□

These complementary stationary principles, with piecewise C^2 constitutive potentials $V_c[\dot{t}_{ij}]$ and $V[\dot{e}_{ij}]$, are due to Hill (1959, 1962).

The saddle quantity (5.165), when redefined for the new $L[\dot{s},\dot{t},\dot{u},\dot{d}]$ of (5.199) in which there are no λ terms, turns out to be

$$S=\tfrac{1}{2}\int \tau_{ik}(\dot{d}_{ji+}-\dot{d}_{ji-})(\dot{d}_{jk+}-\dot{d}_{jk-})\,\mathrm{d}V + \int\left[V_c[\dot{t}_{ij-}]-V_c[\dot{t}_{ij+}]-(\dot{t}_{ij-}-\dot{t}_{ij+})\frac{\partial V_c}{\partial \dot{t}_{ij}}\bigg|_{+}\right]\mathrm{d}V \tag{5.208}$$

in place of (5.166).

Any single valued C^1 branch of $V_c[\dot{t}_{ij}]$ is strictly convex if the second integrand in (5.208) is positive for all distinct $\dot{t}_{ij+}$ and $\dot{t}_{ij-}$ in its (supposedly convex) domain. The same applies for any single valued piecewise C^2 function consisting of several such branches, in particular of (5.197) when $h>0$, since the second derivatives do not appear in (5.208). When $m>1$, even for quadratic single valued $V_c[\dot{t}_{ij}]$, it is not possible to be as explicit as (5.175) without enumeration of cases. Multivaluedness of $V_c[\dot{t}_{ij}]$ introduces ambiguity into the statement of convexity, e.g. for (5.197) with $h<0$, unless branches are discussed individually.

Evidently, then, a revised version of Theorem 5.22 applies in which condition (b) there is replaced by the requirement that $V_c[\dot{t}_{ij}]$ is a single valued piecewise C^2 (at least weakly) convex function.

Theorem 5.35 (Simultaneous upper and lower bounds)
When $V_c[\dot{t}_{ij}]$ is a single valued piecewise C^2 strictly convex function, and when $S\geqslant 0$ for every pair of distinct points in the domain of (5.199), *the bounds* (5.168) *are replaced by*

$$J[\dot{u}_j]\geqslant L_0\geqslant K[\dot{t}_{ij},\dot{d}_{ji}]. \tag{5.209}$$

Proof
The strict convexity of $V_c[\dot{\tau}_{ij}]$ permits the inversion (5.207) in terms of a strictly convex $V[\dot{e}_{ij}]$ satisfying (5.203). The proof of Theorem 5.23 can be repeated, leading this time to bounds $L_\alpha \geqslant L_0 \geqslant L_\beta$ in place of (5.168). Here L_α is the value of (5.201) for any solution of (5.198α), i.e. $L_\alpha = J[\dot{u}_j]$. Also L_β is the value of (5.202) for any solution of (5.198β), i.e. $L_\beta = K[\dot{\tau}_{ij}, \dot{d}_{ji}]$. Theorem 5.34 shows that the two bounds are stationary. □

The converse versions of (5.209) are also called dual extremum principles, in view of the Legendre transformation relating $V[\dot{e}_{ij}]$ and $V_c[\dot{\tau}_{ij}]$. The adjective 'dual' can also be justified in terms of a comparison between (5.161), (5.162) and Fig. 2.26.

The consequences of weaker hypotheses can be pursued as before, i.e. of supposing $S \geqslant 0$ or $S > 0$ over a smaller set of distinct point pairs. As an illustration, we consider pairs satisfying the kinematical constraints (5.198α), which consist of the original kinematical constraints (5.152α) (or (5.159α)) *supplemented* by $(5.152)_{5,7}$.

Theorem 5.36
When $V_c[\dot{\tau}_{ij}]$ is a single valued piecewise C^2 strictly convex function, and when $S > 0$ for every pair of distinct points which satisfy the kinematical conditions (5.198α),

(*a*) *the upper bound*

$$J[\dot{u}_j] \geqslant L_0 \tag{5.210}$$

holds, and the minimum is stationary: and

(*b*) *any actual solution of* (5.198) *has unique* $\dot{u}_j$.

Proof
Strict convexity of $V_c[\dot{\tau}_{ij}]$ implies that of its Legendre dual function $V[\dot{e}_{ij}]$, and we could rewrite (5.208) as

$$S = \frac{1}{2}\int \tau_{ik}(\dot{d}_{ji+} - \dot{d}_{ji-})(\dot{d}_{jk+} - \dot{d}_{jk-})\mathrm{d}V + \int\left[V[\dot{e}_{ij+}] - V[\dot{e}_{ij-}] - (\dot{e}_{ij+} - \dot{e}_{ij-})\frac{\partial V}{\partial \dot{e}_{ij}}\bigg|_{-}\right]\mathrm{d}V. \tag{5.211}$$

The hypothesis can be stated as requiring that $S > 0$ for every pair of distinct $\dot{u}_{j+}$ and $\dot{u}_{j-}$ fields satisfying the same boundary values c_j on Σ_2, and generating corresponding $\dot{d}_{ji}$ and $\dot{e}_{ij}$ (from $(5.152)_1$ and (5.170)). Proof of stationarity proceeds exactly as in the first part of Theorem 5.34, calling now only upon the weaker hypothesis of the present Theorem.

Proof of uniqueness follows that of Theorem 5.25, but is simpler to the extent that the λ terms are omitted from the current $L[\dot{s}, \dot{\tau}, \dot{u}, \dot{d}]$. Again we have strictly positive in the modified (5.180), but the left side would have to be zero (instead of nonpositive) if both points could be a solution. □

The remaining analysis approaches the exclusion functional $E[\dot{u}_j]$ via a Rayleigh quotient, and therefore in a rather different way from that via the minimization of $F[\dot{u}_j, \lambda_\gamma]$ with respect to the λ_γ, as we remarked after Theorem 5.27. The results in Theorem 5.36, and complementary results under equilibrium constraints, are due to Hill (1962). There have been many applications to bifurcation problems, as indicated in our references at the end of §§2.7(ii) and 5.10(x). Bushnell (1985) surveys practical results.

Exercises 5.6

1. Verify that the governing conditions (5.152) can be rewritten in the generalized Hamiltonian form

$$T^*\dot{u} = \frac{\partial H}{\partial \dot{s}}, \qquad 0 = \frac{\partial H}{\partial \dot{\tau}}, \qquad \lambda \geqslant 0, \quad (\alpha)$$

$$T\dot{s} = \frac{\partial H}{\partial \dot{u}}, \qquad 0 = \frac{\partial H}{\partial \dot{d}}, \qquad \frac{\partial H}{\partial \lambda} \leqslant 0, \quad (\beta)$$

$$\overline{\lambda, \frac{\partial H}{\partial \lambda}} = 0,$$

in terms of the adjoint operators (5.157) and the Hamiltonian

$$H[\dot{s}, \dot{\tau}, \dot{u}, \dot{d}, \lambda] = \int\left[\left(\dot{s}_{ij} - \dot{\tau}_{ij} - \dot{\tau}_{ik}\frac{\partial u_j}{\partial X_k}\right)\dot{d}_{ji} - \tfrac{1}{2}\tau_{ik}\dot{d}_{ji}\dot{d}_{jk} + X[\dot{\tau}_{ij}, \lambda_\gamma] + b_j\dot{u}_j\right]\mathrm{d}V - \int n_i\dot{s}_{ij}c_j\,\mathrm{d}\Sigma_2 + \int a_j\dot{u}_j\,\mathrm{d}\Sigma_1.$$

(See equation (119) of Sewell, 1973*b*).

2. By constructing suitable inner product spaces F and E (say) of column matrices from (respectively) vectors and symmetric second order tensors, verify that the (configuration-dependent) linear operators defined by

$$\begin{matrix} V \\ \\ \Sigma_1 \\ \Sigma_2 \end{matrix} \qquad T^*\dot{u} = \begin{bmatrix} \dfrac{1}{2}\left[\dfrac{\partial \dot{u}_k}{\partial X_j}(d_{ki} + \delta_{ki}) + \dfrac{\partial \dot{u}_k}{\partial X_i}(d_{kj} + \delta_{kj})\right] \\ 0 \\ -\tfrac{1}{2}[n_j\dot{u}_k(d_{ki} + \delta_{ki}) + n_i\dot{u}_k(d_{kj} + \delta_{kj})] \end{bmatrix},$$

$$\begin{matrix} V \\ \\ \Sigma_1 \\ \Sigma_2 \end{matrix} \qquad T\dot{\tau} = \begin{bmatrix} -\dfrac{\partial}{\partial X_i}[\dot{\tau}_{ij}(d_{kj} + \delta_{kj})] \\ n_i\dot{\tau}_{ij}(d_{kj} + \delta_{kj}) \\ 0 \end{bmatrix}$$

satisfy an adjointness property

$$(\dot{\tau}, T^*\dot{u}) = \langle \dot{u}, T\dot{\tau} \rangle$$

to be specified.

This specification and operator definitions are different from those in §5.10(vi), but still such that

$$T^*: F' \to E, \qquad T: E' \to F,$$

where F' and E' are subspaces constructed from sufficiently smooth (e.g. piecewise C^1) vectors and symmetric tensors. The presence of the given $d_{ki} = \partial u_k/\partial X_i$ describes the configuration dependence of T and T^*.

Verify that, if $\dot{s}_{ij}$ and $\dot{d}_{ji}$ are eliminated from the outset, equations $(5.152)_{1,2,3,4}$ reduce to

$$\begin{matrix} V \\ \\ \Sigma_1 \\ \Sigma_2 \end{matrix} \quad T^*\dot{u} = \begin{bmatrix} \dfrac{\partial X}{\partial \dot{\tau}_{ij}} \\ 0 \\ -\frac{1}{2}[n_j c_k(d_{ki} + \delta_{ki}) + n_i c_k(d_{kj} + \delta_{kj})] \end{bmatrix}, \quad (\alpha)$$

$$\begin{matrix} V \\ \\ \Sigma_1 \\ \\ \Sigma_2 \end{matrix} \quad T\dot{\tau} = \begin{bmatrix} \dfrac{\partial}{\partial X_i}\left(\tau_{ij}\dfrac{\partial \dot{u}_k}{\partial X_j}\right) + b_k \\ -n_i\tau_{ij}\dfrac{\partial \dot{u}_k}{\partial X_j} + a_k \\ 0 \end{bmatrix}. \quad (\beta)$$

Show that the functional

$$H[\dot{\tau}, \dot{u}, \lambda] = \int\left[X[\dot{\tau}_{ij}, \lambda_\gamma] - \frac{1}{2}\tau_{ik}\frac{\partial \dot{u}_j}{\partial X_i}\frac{\partial \dot{u}_j}{\partial X_k} + b_j\dot{u}_j\right]\mathrm{d}V + \int a_j\dot{u}_j\,\mathrm{d}\Sigma_1$$

$$- \int n_j c_k(d_{ki} + \delta_{ki})\dot{\tau}_{ij}\,\mathrm{d}\Sigma_2$$

has gradient $\partial H/\partial \dot{\tau}$ equal to the right side of equation (α) above, but the gradient $\partial H/\partial \dot{u}$ is equal to the right side of equation (β) plus an extra term arising from

$$-\int \delta\dot{u}_k n_i \tau_{ij}\frac{\partial \dot{u}_k}{\partial X_j}\,\mathrm{d}\Sigma_2.$$

Investigate circumstances when this is zero, so that the whole system of governing equations (5.152) can be rewritten

$$T^*\dot{u} = \frac{\partial H}{\partial \dot{\tau}}, \qquad \lambda \geqslant 0, \quad (\alpha)$$

$$T\dot{\tau} = \frac{\partial H}{\partial \dot{u}}, \qquad \frac{\partial H}{\partial \lambda} \leqslant 0, \quad (\beta)$$

$$\overline{\lambda, \frac{\partial H}{\partial \lambda}} = 0.$$

Finite element analyses are often based on discretizations of such equations (Sewell, 1973*b*, §7).

Show that the alternative approach in which λ is eliminated from the constitutive equations at the outset allows the multivalued piecewise C^2 $V_c[\dot{\tau}_{ij}]$ to be written in the above H in place of the single valued C^2 $X[\dot{\tau}_{ij}, \lambda_\gamma]$, leading to

$$T^*\dot{u} = \frac{\partial H}{\partial \dot{\tau}} \quad (\alpha), \qquad T\dot{\tau} = \frac{\partial H}{\partial \dot{u}} \quad (\beta)$$

as the complete set of governing equations in the circumstances indicated above, and with the configuration-dependent operators.

Construct generating saddle functionals and consequent upper and lower bounds for these various alternative formulations.

3. Prove that the gradients of $L[\dot{s}, \dot{\tau}, \dot{u}, \dot{d}, \lambda]$ which appear in Theorem 5.20 are

$$\begin{matrix} V \\ \Sigma_1 \\ \Sigma_2 \end{matrix} \quad \frac{\partial L}{\partial \dot{s}} = \begin{bmatrix} \dfrac{\partial \dot{u}_j}{\partial X_i} - \dot{d}_{ji} \\ 0 \\ n_i(c_j - \dot{u}_j) \end{bmatrix},$$

$$V \quad \frac{\partial L}{\partial \dot{\tau}} = \left[\frac{1}{2}\left(\dot{d}_{ij} + \dot{d}_{ji} + \dot{d}_{ki}\frac{\partial u_k}{\partial X_j} + \dot{d}_{kj}\frac{\partial u_k}{\partial X_i}\right) - \frac{\partial X}{\partial \dot{\tau}_{ij}}\right],$$

$$\begin{matrix} V \\ \Sigma_1 \\ \Sigma_2 \end{matrix} \quad \frac{\partial L}{\partial \dot{u}} = \begin{bmatrix} -\left(\dfrac{\partial \dot{s}_{ij}}{\partial X_i} + b_j\right) \\ n_i\dot{s}_{ij} - a_j \\ 0 \end{bmatrix},$$

$$V \quad \frac{\partial L}{\partial \dot{d}} = \left[\dot{s}_{ij} - \dot{\tau}_{ij} - \dot{\tau}_{ik}\frac{\partial u_j}{\partial X_k} - \tau_{ik}\dot{d}_{jk}\right],$$

$$V \quad \frac{\partial L}{\partial \lambda} = \left[-\frac{\partial X}{\partial \lambda_\gamma}\right].$$

0

References

(indicating pages where cited)

Akhiezer, N.I. (1962) *Calculus of Variations.* Blaisdell, New York.
248

Apostol, T.M. (1974) *Mathematical Analysis,* Second Edn. Addison-Wesley, Reading, Mass.
7, 10, 54, 202

Arnol'd, V.I. (1968) 'Singularities of smooth mappings'. *Russian Mathematical Surveys,* **23**, 1–43.
106

Arnol'd, V.I. (1974) *Proceedings of the International Congress of Mathematicians,* 19–39, Vancouver.
106, 130

Arnol'd, V.I., Gusein-Zade, S.M. and Varchenko, A.N. (1985) *Singularities of Differentiable Maps,* Volume I. Birkhäuser, Basel.
107

Arrowsmith, D.K. and Place, C.M. (1982) *Ordinary Differential Equations.* Chapman and Hall, London.
334

Arthurs, A.M. (1980) *Complementary Variational Principles,* Second Edn. Oxford University Press.
200, 255, 260, 292, 322

Atkins, A.G. and Mai, Y.-N. (1985) *Elastic and Plastic Fracture.* Ellis Horwood Ltd, Chichester.
123

Bailey, P.B., Champine, L.F. and Waltman, P.E. (1968) *Nonlinear Two-point Boundary Value Problems.* Academic Press, New York.
320

Baiocchi, C. (1972) 'Su un problema di frontiera libera connesso a questioni di idraulica'. *Ann. Nat. Pura. Appl.,* **92**, 107–27.
261

Barnsley, M.F. and Robinson, P.D. (1974) 'Bivariational bounds'. *Proc. Roy. Soc. Lond.,* **A338**, 527–33.
294

Barnsley, M.F. and Robinson, P.D. (1977) 'Bivariational bounds for nonlinear problems'. *J. Inst. Math. Applics.,* **20**, 485–504.
294, 319

Barrett, K.E. (1976) 'Minimax principle for magnetohydrodynamic channel flow'.

Z.A.M.P., **27**, 613–19
382

Barrett, K.E., Demunshi, G. and Shields, D.N. (1979) 'A minimax principle for the Navier-Stokes equations', pp. 401–8 of *Numerical Analysis of Singular Perturbation Problems*, edited by P.W. Hemker and J.J.H. Miller, Academic Press, London.
338

Bateman, H. (1929) 'Notes on a differential equation which occurs in the two-dimensional motion of a compressible fluid and the associated variational problems'. *Proc. Roy. Soc. Lond.*, **A125**, 598–618.
373

Bateman, H. (1930) 'Irrotational motion of a compressible inviscid fluid'. *Proc. Nat. Acad. Sci. U.S.A.*, **16**, 816–25.
373

Bateman, H. (1932) *Partial Differential Equations of Mathematical Physics*. Cambridge University Press.
376

Bazaraa, M.S. and Shetty, C.M. (1979) *Nonlinear Programming Theory and Algorithms*. Wiley, New York.
66, 72, 73

Benjamin, T.B. (1971) 'A unified theory of conjugate flows'. *Phill. Trans. Roy. Soc. Lond.*, **A269**, 587–647.
379

Benjamin, T.B. (1984) 'Impulse, flow force and variational principles'. *I.M.A.J. Appl. Math.*, **32**, 3–68.
376

Biot, M.A. (1965) *Mechanics of Incremental Deformations*. Wiley, New York.
411

Boot, J.C.G. (1964) *Quadratic Programming*. North-Holland, Amsterdam.
73

Britivec, S.J. (1973) *The Stability of Elastic Systems*. Pergamon Press, Oxford.
164, 175

Bushnell, D. (1985) *Computerized Buckling Analysis of Shells*. Martinus Nijhoff Publishers, Dordrecht.
164, 441

Callahan, J. (1974) 'Singularities and plane maps'. *Am. Math. Monthly*, **81**, 211–40.
159

Chadwick, P. (1976) *Continuum Mechanics*. George Allen & Unwin, Ltd, London.
364, 409

Charnes, A., Lemke, C.E. and Zienkiewiz, O.C. (1959) 'Virtual work, linear programming and plastic limit analysis'. *Proc. Roy. Soc. Lond.*, **A251**, 110–16.
406

Chynoweth, S., Porter, D. and Sewell, M.J. (1988) 'The parabolic umbilic and atmospheric fronts'. *Proc. Roy. Soc. Lond.* A**419**, 337–62.
157

Chynoweth, S. and Sewell, M.J. (1989) 'Dual variables in semigeostrophic theory'. *Proc. Roy. Soc. Lond.* A**424**, 155–86.
157

Chynoweth, S. and Sewell, M.J. (1990) 'Mesh duality and Legendre duality'. *Proc. Roy. Soc. Lond.* **A428**, 351–77.
157

Clebsch, A. (1859) 'Ueber die Integration der hydrodynamischen Gleichungen'. *J. reine und angew. Mathematik* **56**, 1–10.
376

Cohn, M.Z. and Maier, G. (1979) *Engineering Plasticity by Mathematical Programming.* Pergamon Press, Oxford.
406

Cole, R.J., Mika, J. and Pack, D.C. (1984) 'Complementary bounds for inner products associated with non-linear equations'. *Proc. Roy. Soc. Edin.*, **96A**, 135–42.
294

Collins, I.F. (1969) 'The upper bound theorem for rigid/plastic solids generalized to include Coulomb friction'. *J. Math. Phys. Solids*, **17**, 323–38.
407

Collins, I.F. (1982) 'Boundary value problem of plane strain plasticity'. pp. 135–84 of *Mechanics of Solids. The Rodney Hill 60th Anniversary Volume*, edited by H.G. Hopkins and M.J. Sewell, Pergamon Press, Oxford.
402

Collins, W.D. (1976) 'Upper and lower bounds for solutions of linear operator problems with unilateral constraints'. *Proc. Roy. Soc. Edin.*, **76A**, 95–105.
324

Collins, W.D. (1977) 'Dual extremum principles for the heat equation'. *Proc. Roy. Soc. Edin.*, **77A**, 273–92.
324, 334

Collins, W.D. (1979) 'Dual extremum principles and Hilbert space decompositions'. pp. 351–418 of *Duality and Complementarity in Mechanics of Solids*, Polish Academy of Sciences. Warsaw.
294

Collins, W.D. (1980) 'An extension of the method of the hypercircle to linear operator problems with unilateral constraints'. *Proc. Roy. Soc. Edin.*, **85A**, 173–93.
294, 302

Collins, W.D. (1982) 'Dual extremum principles and the method of the hypersphere for linear operator problems with convex constraints. *Department of Applied and Computational Mathematics Report, University of Sheffield.*
278, 294, 298, 302

Cottle, F., Gianessi, F. and Lions, J.L. (eds.) (1980) *Variational Inequalities and Complementarity Problems.* Wiley, New York.
271

Courant, R. and Hilbert, D. (1953) Methods of Mathematical Physics, Volume I, First English Edn. Interscience, New York.
233

Craggs, J.W. (1948) 'The breakdown of the hodograph transformation for irrotational compressible fluid flow in two dimensions'. *Proc. Camb. Phil. Soc.*, **44**, 360–79.
158

Davies, G.A.O. (1982) *Virtual Work in Structural Analysis.* Wiley, Chichester.
386

Diaz, J.B. (1960) 'Upper and lower bounds for quadratic integrals, and at a point, for

solutions of linear boundary value problems'. Reprinted from *Boundary Problems in Differential Equations*, edited by R. Langer, U.W. Press, Madison.
294, 315

Detyna, E. (1982) 'Variational principles and gauge theory: an application to continuous media'. *Mathematics Department Report, University of Reading.*
376

Duffin, R.J., Peterson, E.L. and Zener, C. (1967) *Geometric Programming–Theory and Application.* Wiley, New York.
61, 85, 88, 89

Durell, C.V. (1920) *Modern Geometry.* Macmillan and Co. Ltd, London.
108

Ekeland, I. and Temam, R. (1976) *Convex Analysis and Variational Problems.* North-Holland, Amsterdam.
272

Elliot, C.M. (1980) 'On a variational inequality formulation of an electrochemical machining moving boundary problem and its approximation by the finite element method'. *J. Inst. Maths. Applics.*, **25**, 121–131.
270

Elliott, C.M. and Ockendon, J.R. (1982) *Weak and Variational Methods for Moving Boundary Problems.* Pitman, London.
260, 271, 274

Ericksen, J.L. (1960) 'Tensor fields'. pp. 794–858 of *Handbuch der Physik*, III/1, edited by S. Flügge, Springer-Verlag, Berlin.
201, 376

Evans, D.V. and Morris, C.A.N. (1972) 'The effect of a fixed vertical barrier on obliquely incident surface waves in deep water'. *J. Inst. Maths. Applics.*, **9**, 198–204.
292

Faber, C. (1963) *Candela, The Shell Builder.* Architectural Press, London.
1

Farkas, J. (1902) 'Uber der Theorie der einfachen Ungleichungen'. *J. für die Reine und Angewandte Mathematik*, **124**, 1–27.
61

Fiszdon, W. (1962*a*) 'Application of variational methods to the solution of practical supersonic flow problems'. *Z.A.M.M.*, **42**, T134–44.
373

Fiszdon, W. (1962*b*) 'Known applications of variational methods to transonic flow calculations', pp. 362–9 of *Symposium Transonicum, Aachen*, edited by K. Oswattisch.
373

Fortin, M. and Glowinski, R. (eds.) (1962) *Méthodes de Lagrangien Augmenté. Application à la résolution de problèmes aux limites.* Dunod-Bordas, Paris.
168, 245

Fowler, D.H. (1972) 'The Riemann–Hugoniot catastrophe and van der Waals' equation'. *Towards a Theoretical Biology*, **4**, 1–7, edited by C.H. Waddington, Edinburgh University Press.
359

Friedman, A. (1971) *Advanced Calculus.* Holt, Rinehart and Winston.
208

Friedman, A. (1982) *Variational Principles and Free-Boundary Problems.* Wiley, New York.
271

Fujita, H. (1955) 'Contribution to the theory of upper and lower bounds in boundary value problems'. *J. Phys. Soc. Japan*, **10**, 1–8.
206, 277, 294, 295, 314

Gaydon, F.A. and Nuttall, H. (1959) 'Viscous flow through tubes of multiply connected cross sections'.*ASME J. Appl. Mech.*, 573–5.
397

Gregory, M.S. (1969) *An Introduction to Extremum Principles.* Butterworths, London.
393

Gurtin, M.E. (1964) 'Variational principles for linear initial value problems'. *Quart. App. Math.*, **22**, 252–6.
324

Gurtin, M.E. (1972) 'The linear theory of elasticity', pp. 1–295 of *Encyclopedia of Physics VIa/2* (*Mechanics of Solids II*), edited by C. Truesdell, Springer-Verlag, Berlin.
389, 391, 392

Gvozdev, A.A. (1960) 'The determination of the value of the collapse load for statically indeterminate systems undergoing plastic deformation'. *Int. J. Mech. Sci.*, **1**, 322–35.
405

Hammond, P. (1981) *Energy Methods in Electromagnetism.* Clarendon Press, Oxford.
380

Hargreaves, R. (1908) 'A pressure-integral as kinetic potential'. *Phil. Mag.*, 436–44.
373

Hashin, Z. and Shtrikman, S. (1962) 'On some variational principles in anisotropic and nonhomogeneous elasticity'. *J. Mech. Phys. Solids*, **10**, 335–42.
345, 347, 349, 352

Havner, K. (1982) 'The theory of finite plastic deformation of crystalline solids', pp. 265–302 of *Mechanics of Solids, The Rodney Hill 60th Anniversary Volume*, edited by H.G. Hopkins and M.J. Sewell, Pergamon Press, Oxford.
148

Hellinger, E. (1914) 'Die allgemeinen Ansätze der Mechanik der Kontinua'. *Eng. Math. Wis.*, **4**, 602–94.
417

Herrera, I. (1974) 'A general formulation of variational principles'. *Instituto de Ingeneria Report E10, Universidad Nacional Autonoma de Mexico.*
324, 339

Herrera, I. and Bielak, J. (1976) 'Dual variational principles for diffusion equations'. *Quart. Appl. Math.*, **34**, 85–102.
324, 341

Herrera, I. and Sewell, M.J. (1978) 'Dual extremum principles for non-negative unsymmetric operators'. *J. Inst. Maths. Applics.*, **21**, 95–115.
324, 341

Hill, R. (1948) 'A variational principle of maximum plastic work in classical plasticity'. *Q.J. Mech. App. Math.*, **1**, 18–28.
403, 405

Hill, R. (1950) *The Mathematical Theory of Plasticity.* Clarendon Press, Oxford.
399, 403, 405

Hill, R. (1951) 'On the state of stress in a plastic-rigid body at the yield point'. *Phil. Mag.*, **42**, 868–75.
405, 407

Hill, R. (1956*a*) 'New horizons in the mechanics of solids'. *J. Mech. Phys. Solids*, **5**, 66–74.
xvi, 146, 393, 397, 398

Hill, R. (1956*b*) 'The mechanics of quasi-static plastic deformation in metals, pp. 7–31 of *Surveys in Mechanics, The G.I. Taylor Volume*, edited by G.K. Batchelor and R.M. Davies, Cambridge University Press.
405

Hill, R. (1957) 'On uniqueness and stability in the theory of finite elastic strain'. *J. Mech. Phys. Solids*, **5**, 229–41.
411, 413

Hill, R. (1958) 'A general theory of uniqueness and stability in elastic/plastic solids'. *J. Mech. Phys. Solids*, **6**, 236–49.
153, 164, 431

Hill, R. (1959) 'Some basic principles in the mechanics of solids without a natural time'. *J. Mech. Phys. Solids*, **7**, 209–25.
439

Hill, R. (1962) 'Uniqueness criteria and extremum principles in self-adjoint problems of continuum mechanics'. *J. Mech. Phys. Solids*, **10**, 185–94.
151, 164, 439, 441

Hill, R. (1963) 'New derivations of some elastic extremum principles'. *Progress in Applied Mechanics*, pp. 99–106 (*The Prager Anniversary Volume*). Macmillan, New York.
345

Hill, R. (1966) 'Generalized constitutive relations for incremental deformation of metal crystals by multi-slip'. *J. Mech. Phys. Solids*, **14**, 95–102.
147, 154

Hill, R. (1967) 'The essential structure of constitutive laws for metal composites and polycrystals'. *J. Mech. Phys., Solids*, **15**, 79–95.
147

Hill, R. (1978) 'Aspects of invariance in solid mechanics'. *Adv. App. Mechanics*, **18**, 1–75,
148, 153, 241, 410, 411, 419, 431, 435

Hill, R. (1981) 'Invariance relations in thermoelasticity with generalized variables'. *Math. Proc. Camb. Phil. Soc.*, **90**, 373–84.
354

Hill, R. and Power, G. (1956) 'Extremum principles for slow viscous flow and the approximate calculation of drag'. *Quart. J. Mech. App. Math.*, **9**, 313–19.
397

Hill, R. and Sewell, M.J. (1960*a*) 'A general theory of inelastic column failure – I'. *J. Mech. Phys. Solids*, **8**, 105–111.
432

Hill, R. and Sewell, M.J. (1960*b*) 'A general theory of inelastic column failure – II'. *J. Mech. Phys. Solids*, **8**, 112–18.
432

Hill, R. and Sewell, M.J. (1962) 'A general theory of inelastic column failure – III'. *J. Mech. Phys. Solids*, **10**, 285–300.
432

Horne, M.R. (1971) *Plastic Theory of Structures.* Nelson.
405

Hunt, B. (editor) (1980) *Numerical Methods in Applied Fluid Dynamics.* Academic Press, London.
374

Hunt, J.C.R. and Shercliff, J.A. (1971) 'Magnetohydrodynamics at high Hartmann number'. *Ann. Rev. Fluid Mech.*, **3**, 37–62.
380

Hutchinson, J.W. (1974) 'Plastic buckling'. *Adv. App. Mechanics*, **14**, 67–144.
164, 419

Iri, M. (1969) *Network Flow, Transportation and Scheduling.* Academic Press, London.
62, 124, 175, 187

Isherwood, D.A. and Ogden, R.W. (1977) 'Towards the solution of finite plane-strain problems for compressible elastic solids'. *Int. J. Solids Structures*, **13**, 105–23.
412, 418

Jamal, N.G., Kamala, V., Prabhakara Rao, G.V. and Nigam, S.D. (1979) 'Complementary variational principles for Poisseuille flow'. *J. Math. Phys. Sci.*, **13**, 469–78.
397

Joedicke, J. (1963) *Shell Architecture.* Alec Tiranti, Ltd, London.
1

John, F. (1948) 'Extremum problems with inequalities as side conditions'. *Studies and Essays, Courant Anniversary Volume*, edited by K.O. Friedrichs, O.E. Neugebauer and J.J. Stoker. Wiley-Interscience, New York.
66

Johnson, M.W., Jr. (1960) 'Some variational theorems for non-Newtonian flow'. *Physics of Fluids*, **3**, 871–8.
397

Johnson, M.W., Jr. (1961) 'On variational principles for non-Newtonian fluids'. *Trans. Soc. Rheology*, **5**, 9–21, 1961.
397

Johnson, W. and Mellor, P.B. (1973) *Engineering Plasticity.* van Nostrand Reinhold Co., London.
402, 405, 407

Johnson, W., Sowerby, R. and Venter, R.D. (1982) *Plane Strain Slipline Fields for Metal Deformation Processes.* Pergamon Press, Oxford.
405, 406

Jones, D.S. (1979) *Methods in Electromagnetic Wave Propagation.* Clarendon Press, Oxford.
380

Kantorovich, L.V. and Akilov, G.P. (1964) *Functional Analysis in Normed Spaces.* Pergamon Press, Oxford.
209

Keller, J.B., Rubinfeld, L.A. and Molyneux, J.E. (1967) 'Extremum principles for slow viscous flows with applications to suspensions'. *J. Fluid Mech.*, **30**, 97–125.
396, 397

Kellogg, O.D. (1929) *Foundations of Potential Theory.* Springer-Verlag, Berlin.
314

Kinderlehrer, D. and Stampacchia, G. (1980) *An Introduction to Variational Inequalities*

and their Application. Academic Press, New York.
12, 271

Koiter, W.T. (1945) *On the Stability of Elastic Equilibrium.* Delft Thesis, H.J. Paris, Amsterdam. English translation available as NASA TT F 10–833, (1967).
164

Koiter, W.T. (1973) 'On the principle of stationary complementary energy in the nonlinear theory of elasticity', *SIAM J. Appl. Math.*, **25**, 424–34.
417

Koiter, W.T. (1976*a*) 'On the complementary energy theorem in nonlinear elasticity theory', pp. 207–32 of *Trends of Application of Pure Mathematics to Mechanics*, edited by G. Fichera, Pitman, London.
412, 417

Koiter, W.T. (1976*b*) 'Complementary energy, neutral equilibrium and buckling'. *Proc. Kon. Ned. Akad. von Wet.*, **B79**, 183–200.
417

Kuhn, and Tucker, A.W. (1951) 'Nonlinear programming', pp. 481–92 of *Proc. 2nd Berkeley Symposium on Mathematical Statistics and Probability*, edited by J. Neyman, University of California Press, Berkeley, California.
61, 64, 66, 70

Lamb, H. (1952) *Hydrodynamics*, Sixth Edn. Cambridge University Press.
396

Lanczos, C. (1949) *The Variational Principles of Mechanics.* Toronto University Press.
368

Lee, S.J. and Shield R.T. (1980*a*) 'Variational principles in finite elastostatics'. *Z.A.M.P.*, **31**, 437–53.
417

Lee, S.J. and Shield, R.T. (1980*b*) 'Applications of variational principles in finite elasticity'. *Z.A.M.P.*, **31**, 454–72.
417

Levinson, M. (1965) 'The complementary energy theorem in finite elasticity'. *J. Appl. Mech.*, **32**, 826–8.
417

Lions, J.L. (1971) *Optimal Control of Systems Governed by Partial Differential Equations.* Springer-Verlag, Berlin.
271

Liusternik, L.A. and Sobolev, V.J. (1961) *Elements of Functional Analysis.* Ungar.
209

Luenberger, D.G. (1969) *Optimization by Vector Space Methods.* John Wiley, New York.
90, 168, 209, 211

Lush, T.E. and Cherry, T.M. (1956) 'The variational method in hydrodynamics'. *Quart. J. Mech. Appl. Math.*, **9**, 6–21.
373

Mackie, A.G. (1969) 'A cavitation model in one-dimensional unsteady motion'. *J. Math. Anal. Applics.*, **27**, 390–404.
84

Mackie, A.G. (1977) 'An example from mechanics illustrating classical and modern mathematical techniques'. *S.I.A.M. Review*, **19**, 517–35.
82

Mackie, A.G. (1986) 'Complementary variational inequalities', to appear. *I.M.J. Appl. Maths.*
270

Maier, G. (1977) 'Future directions in engineering plasticity', pp. 631–48 of *Engineering Plasticity by Mathematical Programming*, edited by M.Z. Cohn and G. Maier, Pergamon Press, Oxford.
153

Maier, G. and Munro, J. (1982) 'Mathematical programming applications to engineering plastic analysis'. *App. Mech. Rev.*, **35**, 1631–43.
153

Maple, C.G. (1950) 'The Dirichlet problem: bounds at a point for the solution and its derivatives'. *Quart. App. Math.*, **8**, 213–28.
315

Markov, A.A. (1947) 'On variational principles in the theory of plasticity'. *P.M.M.*, **11**, 339–350.
405

Martin, J.B. (1964) 'A displacement bound technique for elastic continua subjected to a certain class of dynamic loading'. *J. Mech. Phys. Solids*, **12**, 165–75.
306

Martin, J.B. (1966) 'A note on the determination of an upper bound on displacement rates for steady creep problems'. *J. App. Mech.*, **33**, *Trans. ASME*, **88E**, 216–17.
305

Mei, C.C. and Black, J.L. (1969) 'Scattering of surface waves by rectangular obstacles in waters of finite depth'. *J. Fluid Mech.*, **38**, 499–571.
288

Miles, J.P. (1969) 'Bifurcation in rigid/plastic materials under spherically symmetric loading conditions'. *J. Mech. Phys. Solids*, **17**, 303–13.
406

Moreau, J.J. (1962) 'Décomposition orthogonale d'un espace hilbertien selon deux cônes mutuellement polaires'. *C.R. Acad. Sci. Paris*, **255**, 238–40.
294, 298

Moreau, J.J. (1966*a*) 'Principes extrémaux pour le problème de la naissance de la cavitation'. *J. de Mech.*, **5**, 439–70.
81, 83

Moreau, J.J. (1966*b*) 'Convexity and duality', pp. 145–9 of *Functional Analysis and Optimization*, edited by E.R. Caianiello, Academic Press, New York.
294, 298

Moreau, J.J. (1967) 'One-sided constraints in hydrodynamics'. *Nonlinear Programming*, Chapter 9, pp. 261–79, edited by J. Abadie, North-Holland Pub. Co., Amsterdam.
84, 236

Moreau, J.J. (1971) 'Weak and strong solutions of dual problems', pp. 181–214 of *Contributions to Nonlinear Functional Analysis*, edited by E. Zarantonello, Academic Press, New York.
294, 298

Morice, R. (1978) 'A variational principle and a finite element method for compressible flow with free boundaries'. *Arch. Mech. Stos.*, **30**, 517–30.
374

Morse, P.M. and Feshbach, H. (1953) *Methods of Theoretical Physics*, Volume I. McGraw Hill,
324

Neale, K.W. (1981) 'Phenomenological constitutive laws in finite plasticity'. *S.M. Archives*, **6**, 79–128.
413, 419

Needleman, A. and Tvergaard, V. (1982) 'Aspects of plastic postbuckling behaviour', pp. 458–98 of *Mechanics of Solids. The Rodney Hill 60th Anniversary Volume*, edited by H.G. Hopkins and M.J. Sewell, Pergamon Press, Oxford.
164

Nemat-Nasser, S. (1972*a*) 'On variational methods in finite and incremental elastic deformation problems with discontinuous fields'. *Quart. App. Math.*, **30**, 143–56.
387, 417

Nemat-Nasser, S. (1972*b*) 'General variational principles in nonlinear and linear elasticity with applications'. *Mechanics Today*, **1**, 214–61.
393, 417

Nemat-Nasser, S. (1977) 'A note on complementary energy and Reissner's principle in nonlinear elasticity'. *Iran. J. Sci. Tech.*, **6**, 95–101.
417

Neumann, and Witherspoon, (1971) 'Variational principles for fluid flow in porous media', *J. Eng. Mech. Div., A.S.C.E.*, **97**, 359–74.
344

Nixon, D. (editor) (1982) *Transonic Aerodynamics*. American Institute of Aeronautics and Astronautics, New York.
374

Noble, B. (1964) 'Complementary variational principles for boundary value problems I: basic principles, with an application to ordinary differential equations'. *Report 473, Mathematics Research Center, Univeristy of Wisconsin.*
167, 215, 224, 236

Noble, B. (1967) 'Complementary variational principles II: nonlinear networks'. *Report 643, Mathematics Research Center, University of Wisconsin.*
179

Noble, B. (1973) 'Variational finite-element methods for initial-value problems'. pp. 143–51 of *Proc. Finite Element Conference, Brunel University*, edited by J. Whiteman, Academic Press, London.
324

Noble, B. (1974) 'Pointwise bounds in nonlinear problems'. *Bull. Inst. Maths. Applics.*, **10**, 103–7.
305

Noble, B. (1981) 'Extremum principles for a class of equation that includes nonlinear heat conduction and Navier–Stokes'. Unpublished.
338

Noble, B. and Daniel, J.W. (1977) *Applied Linear Algebra*, Second Edition. Prentice-Hall Inc., Englewood Cliffs, New Jersey.
72

Noble, B. and Sewell, M.J. (1972) 'On dual extremum principles in applied mathematics'. *J. Inst. Maths. Applics.*, **9**, 123–93.
136, 137, 141, 145, 154, 191, 278

Noble, B. and Sewell, M.J. (eds.) (1978) 'Pointwise bounds in elasticity theory', by C. Weber. Unpublished.
315

Nuttall, H. (1965) 'The slow rotation of a circular cylinder in a viscous fluid'. *Int. J. Engng. Sci.*, **2**, 461–76.
397

Nuttall, H. (1966) 'The flow of a viscous incompressible fluid in an inclined uniform channel, with reference to the flow on a transporter belt'. *Int. J. Engng. Sci.*, **4**, 249–276.
397

Oden, J.T. and Kikuchi, N. (1980) 'Theory of variational inequalities with applications to problems of flow through porous media'. *Int. J. Eng. Sci.*, **18**, 1173–1284.
260

Oden, J.T. and Reddy, J.N. (1974) 'On dual-complementary variational principles in mathematical physics'. *Int. J. Engng. Sci.*, **12**, 1–29.
286

Oden, J.T. and Reddy, J.N. (1982) *Variational Methods in Theoretical Mechanics*, Second Edn. Springer-Verlag, Berlin.
393

Ogden, R.W. (1975) 'A note on variational theorems in nonlinear elastostatics'. *Math. Proc. Camb. Phil. Soc.*, **77**, 609–15.
417

Ogden, R.W. (1977) 'Inequalities associated with the inversion of elastic stress–deformation relations and their implications'. *Math. Proc. Camb. Phil. Soc.*, **81**, 313–24.
416

Ogden, R.W. (1978) 'Extremum principles in nonlinear elasticity and their application to composites – I'. *Int. J. Solids Structures*, **14**, 265–82.
412, 417

Ogden, R.W. (1982) 'Elastic deformations of rubberlike solids'. pp. 499–537 of *Mechanics of Solids, The Rodney Hill 60th Anniversary Volume*, edited by H.G. Hopkins and M.J. Sewell, Pergamon Press, Oxford.
412

Ogden, R.W. (1984) *Non-Linear Elastic Deformations*. Ellis Horwood Ltd, Chichester.
417

Ogden, R.W. and Isherwood, D.A. (1978) 'Solution of some finite plane-strain problems for compressible elastic solids'. *Q.J. Mech. App. Math.*, **31**, 219–49.
412

Oravas, G.A. and McLean, L. (1966) 'Historical development of energetical principles in elastomechanics'. *Applied Mechanics Reviews*, **19**, 647–58 and 919–33.
67, 95, 392

Palmer, A.C. (1967) 'A lower bound on displacement rates in steady creep'. *J. App. Mech.*, **34**, *Trans. ASME*, **89E**, 216–17.
306

Pearson, C.E. (1956) 'General theory of elastic stability'. *Quart. Appl. Math.*, **14**, 133–44.
418

Penfield, P., Spence, R. and Duinker, S. (1970) *Tellegen's Theorem and Electrical Networks*. M.I.T. Press, Cambridge, Mass.
177, 397

Pestel, E.C. and Leckie, F.A. (1963) *Matrix Methods in Elastomechanics*. McGraw-Hill.
175

Peyret, R. and Taylor, T.D. (1983) *Computational Methods for Fluid Flow*. Springer-Verlag, New York.
374

Pippard, A.B. (1985) *Response and Stability*. Cambridge University Press.
359

Ponter, A.R.S. (1975) 'General displacement and work bounds for dynamically loaded bodies'. *J. Mech. Phys. Solids*, **23**, 151–63.
315

Poston, T. and Stewart, I.N. (1976) *Taylor Expansions and Catastrophes*. Pitman, London.
13, 45

Poston, T. and Stewart, I.N. (1978) *Catastrophe Theory and its Applications*. Pitman, London.
107, 130, 164

Prager, W. (1954) 'Three dimensional plastic flow under uniform stress'. *Rev. Fac. Sci. Univ. Instanbul*, **19**, 23–6.
398

Prager, W. (1967) 'Variational principles of linear elastostatics for discontinuous displacements, strains and stresses', pp. 463–74 of *Recent Progress in Applied Mechanics. The Folke Odqvist Volume*. Wiley, New York.
387

Prager, W. and Hodge, P.G. (1951) *Theory of Perfectly Plastic Solids*. Wiley.
405

Rall, L.B. (1969) *Computational Solution of Nonlinear Operator Equations*. Krieger.
317

Rasmussen, R. (1974) 'Applications of variational methods in compressible flow calculations', pp. 1–35 of *Progress in Aerospace Science*, **15**, edited by D. Kuchemann, Pergamon Press, Oxford.
374

Reddy, B.D. (1981) 'Dual extremum principles in the dynamics of rigid perfectly-plastic bodies undergoing finite deformations'. *J. Mech. Phys. Solids*, **29**, 199–210.
84, 306, 408

Reissner, E. (1953) 'On a variational theorem for finite elastic deformations'. *J. Math. Phys.*, **32**, 129–35.
417

Reynolds, W.C. and McCormack, R.W. (eds.) (1981) *Seventh International Conference on Numerical Methods in Fluid Dynamics*. Springer-Verlag. Berlin.
374

Roach, G.F. (1970) *Green's Functions. Introductory Theory with Applications*. Van Nostrand Reinhold, London.
309

Robertson, S.A. (1977) 'Isometric folding of Riemannian manifolds'. *Proc. Roy. Soc. Edin.*, 275–84.
159

Robinson, P. and Barnsley, M.F. (1979) 'Pointwise bivariational bounds on solutions of Fredholm integral equations'. *SIAM J. Numer. Anal.*, **16**, 135–44.
315

Rockafellar, R.T. (1970) *Convex Analysis*. Princeton University Press.
62

Rockafellar, R.T. (1982) *Network Flows and Monotropic Optimization.* Wiley, New York.
175, 187

Sandhu, R.S. and Pister, K.S. (1971) 'Variational principles for boundary value and initial boundary value problems in continuum mechanics'. *Int. J. Solids and Structures*, **7**, 639–54.
324

Schatz, A. (1969) 'Free boundary problems of Stefan type with prescribed flux'. *J. Math. Anal. App.*, **28**, 569–80.
261

Serrin, J. (1959) 'Mathematical principles of classical fluid mechanics', pp. 125–263 of *Encyclopedia of Physics*, VIII/I, edited by S. Flügge, Springer-Verlag, Berlin.
373, 376

Sewell, M.J. (1963*a*) 'On reciprocal variational principles for perfect fluids'. *J. Math. Mech.*, **12**, 495–504.
362, 368, 373, 376

Sewell, M.J. (1963*b*) 'A general theory of elastic and inelastic plate failure – I'. *J. Mech. Phys. Solids*, **11**, 377–93.
406

Sewell, M.J. (1965) 'On the calculation of potential functions defined on curved boundaries'. *Proc. Roy. Soc. Lond.*, **A286**, 402–11.
418

Sewell, M.J. (1966) 'On the connexion between stability and the shape of the equilibrium surface'. *J. Mech. Phys. Solids*, **14**, 203–30.
107, 111

Sewell, M.J. (1967) 'On configuration-dependent loading'. *Arch. Rat. Mech. Anal.*, **23**, 327–51.
418

Sewell, M.J. (1968*a*) 'A general theory of equilibrium paths through critical points I'. *Proc. Roy. Soc. Lond.*, **A306**, 201–23.
71

Sewell, M.J. (1968*b*) 'A general theory of equilibrium paths through critical points II'. *Proc. Roy. Soc. Lond.*, **A306**, 225–38.
71

Sewell, M.J. (1969) 'On dual approximation principles and optimization in continuum mechanics. *Phil. Trans. Roy. Soc. Lond.*, **A265**, 319–51.
19, 81, 84, 398, 405, 408, 417

Sewell, M.J. (1972) 'A survey of plastic buckling', *Stability*, pp. 85–197, edited by H. Leipholz, University of Waterloo Press, Canada.
153, 419, 431

Sewell, M.J. (1973*a*) 'The governing equations and extremum principles of elasticity and plasticity generated from a single functional, I'. *J. Struct. Mech.*, **2**, 1–32.
168, 219, 385, 405, 406

Sewell, M.J. (1973*b*) 'The governing equations and extremum principles of elasticity and plasticity generated from a single functional, II'. *J. Struct. Mech.*, **2**, 135–58.
168, 219, 385, 441, 443

Sewell, M.J. (1974) 'On applications of saddle-shaped and convex generating functionals', pp. 219–45 of *Physical Structure in Systems Theory*, edited by J.J. van Dixhoorn and F.J. Evans, Academic Press.
148, 153

Sewell, M.J. (1977*a*) 'On Legendre transformations and elementary catastrophes'. *Math. Proc. Camb. Phil. Soc.*, **82**, 147–63.
115, 119

Sewell, M.J. (1977*b*) 'Degenerate duality, catastrophes and saddle functionals'. *Lectures to the Summer School on Duality and Complementarity in the Mechanics of Solids.* Polish Academy of Sciences, Warsaw, Published in *Mechanics Today*, **6**, 41–91, 1981.
359, 381

Sewell, M.J. (1978*a*) 'A pure catastrophe machine'. *Bull. Inst. Maths. Applics.*, **14**, 212–14.
111

Sewell, M.J. (1978*b*) 'On Legendre transformations and umbilic catastrophes'. *Math. Proc. Camb. Phil. Soc.*, **83**, 273–88.
127, 362

Sewell, M.J. (1978*c*) 'Some global equilibrium surfaces'. *Int. J. Mech. Engng. Educ.*, **6**, 163–74.
165

Sewell, M.J. (1980) 'Complementary energy and catastrophes'. pp. 163–8 of *Variational Methods in the Mechanics of Solids*, edited by S. Nemat-Nasser, Pergamon Press, Oxford.
123, 159

Sewell, M.J. (1981) 'Singularities and equilibrium surfaces', *Bull. Inst. Maths. Applics.*, **17**, 190–2.
159, 165

Sewell, M.J. (1982) 'Legendre transformations and extremum principles', pp. 563–605 of *Mechanics of Solids, The Rodney Hill 60th Anniversary Volume*, edited by H.G. Hopkins and M.J. Sewell, Pergamon Press, Oxford.
48, 95, 98, 113, 138, 142, 151, 168, 359, 371

Sewell, M.J. (1985) 'Properties of a streamline in gas flow'. *Physics in Technology*, **16**, 127–33.
364, 365, 379

Sewell, M.J. and Noble, B. (1978) 'General estimates for linear functionals in nonlinear problems'. *Proc. Roy. Soc. Lond.*, **A361**, 293–324.
294, 299, 300, 301, 304, 323

Sewell, M.J. and Porter, D. (1980) 'Constitutive surfaces in fluid mechanics'. *Math. Proc. Camb. Phil. Soc.*, **88**, 517–46.
363, 377, 379

Shanley, F.R. (1946) 'The column paradox'. *J. Aero. Sci.*, **13**, 678.
432

Shanley, F.R. (1947) 'Inelastic column theory'. *J. Aero. Sci.*, **14**, 261–7.
432

Simon, M.J. (1981) 'Wave-energy extraction by a submerged cylindrical resonant duct'. *J. Fluid Mech.*, **104**, 159–87.
292

Skalak, R. (1969) 'Extensions of extremum principles for slow viscous flows'. *Dept. Civil Engng. & Engng. Mech. Technical Report No. 4*, Columbia University, New York.
397

Smith, P. (1976) 'A note on dual extremum principles in magnetohydrodynamic pipe flow'. *J. Inst. Maths. Applics.*, **18**, 129–33.
382

Smith, P. (1978) 'Approximate operators and dual extremum principles for linear problems'. *J. Inst. Maths. Applics*, **22**, 457–65.
345

Smith, P. (1981) 'Extended dual extremum principles for nonlinear problems'. *Nonlinear Anal., Theory, Methods and Applics.*, **5**, 185–93.
345

Smith, P. (1985*a*) *Convexity Methods in Variational Calculus.* Research Studies Press, Letchworth.
345

Smith, P. (1985*b*) 'Extremum principles for a general class of saddle functionals'. *Acta App. Math.*, to appear.
345

Sokolnikoff, I.S. (1956) *Mathematical Theory of Elasticity.* McGraw-Hill, New York.
393

Strieder, W. and Aris, R. (1973) *Variational Methods Applied to Problems of Diffusion and Reaction.* Springer-Verlag, Berlin.
266

Synge, J.L. (1957) *The Hypercircle in Mathematical Physics*, Cambridge University Press.
78

Synge, J.L. (1960) 'Classical dynamics'. *Encyclopedia of Physics* III/1, pp. 1–225, edited by S. Flügge, Springer-Verlag, Berlin.
79, 84

Talbot, D.S.R. and Willis, J.R. (1985) 'Variational principles for inhomogeneous non-linear media'. *I.M.J. App. Math.*, **35**, 39–54.
266, 345

Temam, R. (1978) 'Mathematical problems in plasticity theory', pp. 357–73 of *Variational Inequalities and Complementarity Problems*, edited by F. Gianessi, R.W. Cottle and J.L. Lions, Wiley.
406

Temam, R. (1979) *Navier–Stokes Equations.* North-Holland, Amsterdam.
62

Thom, R. (1969) 'Topological models in biology'. *Topology*, **8**, 313–35.
106

Thom, R. (1975) *Structural Stability and Morphogenesis.* W.A. Benjamin, Inc., Reading, Mass.
106

Thompson, J.M.T. and Hunt, G.W. (1973) *A General Theory of Elastic Stability.* Wiley, London.
164

Thompson, J.M.T. and Hunt, G.W. (1984) *Elastic Instability Phenomena.* Wiley, London.
164

Thorndike, A.S., Cooley, C.R. and Nye, J.F. (1978) 'The structure and evolution of flow fields and other vector fields'. *J. Phys. A., Math. & General*, **11**, 1455–90.
158

Tonti, E. (1967) 'Variational principles in elastostatics'. *Meccanica*, **2**, 201–8.
393

Tonti, E. (1973) 'On the variational formulation for linear initial value problems'.

Annali di Matematica, **95**, 331–60.
324

Tonti, E. (1975) *On the Formal Structure of Physical Theories*. Instituto di Matematica del Politecnico di Milano.
286, 393

Troutman, J.L. (1983) *Variational Analysis with Elementary Convexity*. Springer-Verlag, New York.
12, 263, 264

Truesdell, C. (1956) 'Das ungelöste Hauptproblem der endlichen Elastizitätstheorie'. *Z. angew. Math. Mech.*, **36**, 97–103.
412

Truesdell, C. and Noll. W. (1965) 'The nonlinear field theories of mechanics', pp. 1–602 of *Encyclopedia of Physics* III/3, edited by S. Flügge, Springer-Verlag, Berlin.
412, 417, 418

Truesdell, C. and Toupin, R. (1960) 'The classical field theories', pp. 226–793 of *Encyclopedia of Physics* III/I edited by S. Flügge, Springer-Verlag, Berlin.
84, 353, 365, 386, 397, 411, 432

Vanderwalle, J.P. and Dewilde, P. (1975) 'On the minimal spectral factorisation of non-singular positive rational matrices'. *I.E.E.E. Trans. Information Theory*, **IT–21**, 612–18.
78

Walpole, L.J. (1974) 'New extremum principles for linear problems'. *J. Inst. Maths. Applics.*, **14**, 113–18.
345

Washizu, K. (1968) *Variational Methods in Elasticity and Plasticity*. Pergamon Press, New York.
393

Weatherburn, C.E. (1927) *Differential Geometry of Three Dimensions*, Volume I. Cambridge University Press.
6

Weber, C. (1931) 'Bestimmung des Steifigkeitswertes von Körpern durch zwei Näherungsverfahren'. *Z. angew. Math. Mech.*, **11**, 244–5.
315

Weber, C. (1938) 'Veranschaulichung und Anwendung der Minimalsätze der Elastizitätstheorie', *Z. angew. Math. Mech.*, **18**, 375–9.
315

Weber, C. (1941) 'Uber der Minimalsätze der Elastizitätstheorie'. *Z. angew. Math. Mech.*, **21**, 32–42.
315

Weber, C. (1942) 'Eingrenzung von Verschiebungen und Zerrungun mit Hilfe der Minimalsätze'. *Z. angew. Math. Mech.*, **22**, 130–6.
315

Whitaker, E.T. (1937) *A Treatise on the Analytical Dynamics of Particles and Rigid Bodies*, Fourth Edn. Cambridge University Press.
84, 253, 254

Whitney, H. (1955) 'On singularities of mappings of Euclidean spaces. I Mappings of the plane onto the plane'. *Ann. of Math.*, **62**, 374–410.
106, 158

Whittle, P. (1971) *Optimization under Constraints.* Wiley-Interscience, London.
90, 168

Willis, J.R. (1982) 'Elasticity theory of composites', pp. 653–86 of *Mechanics of Solids, The Rodney Hill 60th Anniversary Volume*, edited by H.G. Hopkins and M.J. Sewell, Pergamon Press, Oxford.
393

Willis, J.R. (1983) 'The overall elastic response of composite materials'. *J. Appl. Mech.*, **50**, 1202–9.
352

Willis, J.R. (1984) 'Variational estimates for the overall response of an inhomogeneous nonlinear dielectric'. *Proc. IMA Workshop on Homogenization and Composites*, Minneapolis, Springer-Verlag.
346

Woodcock, A.E.R. and Poston, T. (1974) *A Geometrical Study of the Elementary Catastrophes.* Springer-Verlag, Berlin.
116

Young, R.C. and Mote, C.D. (1976) 'Derivatives and their error bounds in second order problems by finite elements'. *Int. J. Num. Methods in Engng.*, **10**, 1065–75.
315

Zeeman, E.C. (1977) *Catastrophe Theory. Selected papers 1972–77.* Addison-Wesley, Reading, Mass.
106, 164

Subject index